New Concepts in Polymer Science

Cyclolinear Organosilicon Copolymers: Synthesis, Properties, Application

New Concepts in Polymer Science

Previous titles in this book series:

Modern Polymer Flame Retardancy
S.M. Lomakin and G.E. Zaikov

Polymeric Biomaterials. Part I. Polymer Implants
M.I. Shtilman

Biodegradation and Durability of Materials under the Effect of Microorganisms
S.A. Semenov, K.Z Gumargalieva and G.E. Zaikov

Fire Resistant and Thermally Stable Materials Derived from Chlorinated Polyethylene
A.A. Donskoi, M.A. Shashkina and G.E. Zaikov

Polymers and Polymeric Materials for Fiber and Gradient Optics
N. Lekishvili, L. Nadareishvili, G. Zaikov and L. Khananashvili

Polymers Derived from Isobutylene. Synthesis, Properties, Application
Yu.A. Sangalov, K.S. Minsker and G.E. Zaikov

Ecological Aspects of Polymer Flame Retardancy
S.M. Lomakin and G.E. Zaikov

Molecular Dynamics of Additives in Polymers
A.L. Kovarski

Structure and Properties of Conducting Polymer Composites
V.E. Gul'

Interaction of Polymers with Bioactive and Corrosive Media
A.L. Iordanskii, T.E. Rudakova and G.E. Zaikov

Immobilization on Polymers
M.I. Shtilman

Radiation Chemistry of Polymers
V.S. Ivanov

Polymeric Composites
R.B. Seymour

Reactive Oligomers
S.G. Entelis, V.V. Evreinov and A.I. Kuzaev

Diffusion of Electrolytes in Polymers
G.E. Zaikov, A.L. Iordanskii and V.S. Markin

Chemical Physics of Polymer Degradation and Stabilization
N.M Emanuel and A.L. Buchachenko

New Concepts in Polymer Science

Cyclolinear Organosilicon Copolymers:
Synthesis, Properties, Application

O.V. Mukbaniani and G.E. Zaikov

///VSP///

Utrecht • Boston - 2003

VSP
an imprint of Brill Academic Publishers
P.O. Box 346
3700 AH Zeist
The Netherlands

Tel: +31 30 692 5790
Fax: +31 30 693 2081
vsppub@compuserve.com
www.vsppub.com
www.brill.nl

First published in 2003

ISBN 90-6764-397-1

Printed in The Netherlands by Ridderprint bv, Ridderkerk.

Table of Contents

"If you perceive the Proper Way in the morning, you can quietly die at night with the sense of followed out duty."

Confucius, China
500 years A.C.

Preface

We are not sure that anybody sane will willingly agree to die, but everybody wants to know the "Proper Way" in "Science" and someone in "Religion" (and stay alive whenever possible). We also tried to do this on the example of Cyclolinear organosilicon copolymers.

The branch of science discussed in this book has particular applied objective: creation of new polymeric materials with improved properties (high heat and thermal stability, frost resistance, hydrophobicity, high dielectric indices, etc.).

As the famous storyteller Hoffmann said: "Dignity is in love, and yet more in suffering". We hope that the readers will understand the loving, with which we suffered out this monograph (we want to believe in this). However, it is known well that "he is lifeless who is faultless". Of course, some miscounts are also present here. We appreciate the readers for constructive notes and suggestions to the book, which will be taken into account in future.

To the point of view of the famous Russian writer Anton P. Chekhov "a man (here, a scientist – *O.M. and G.Z. note*) represents a fraction, where the numerator is what he really is, and denominator is what he thinks about himself". It would be likely to hope that fractions of the authors of the current monograph are greater than one.

Prof. Omar Mukbaniani, I. Javakhishvili Tbilisi State University, Tbilisi, Georgia
Prof. Gennady Zaikov, N.M. Emanuel Institute of Biochemical Physics, Moscow, Russia

Knowledge is force.

Frances Bacon (1561 – 1626), the Lord-Chancellor, philosopher-materials, England

Love to the Motherland is a good thing, but Love to the Truth is much better. Love to the Motherland makes Heroes, and Love to the Truth makes Sages. Love to the Motherland disengages people on the Earth, and Love to the truth unites the whole mankind.

Peter Chaadaev (19th century), the famous Russian poet and philosopher, Russia

Introduction

Development of new branches of sciences and technology applying high and super-low temperatures and high pressure and deep vacuum, broadening of electrification and motorization, development of machine-building industry, wireless and electronic engineering, creation of supersonic airplanes and artificial satellites of the Earth strictly demand materials with a set of specific operational properties. It is known that temperature on surfaces of machines at very high speeds rises to 300°C or higher. The only polymers capable of long resistance to such temperature are these, which chains are composed of thermally stable fragments. Among them of high interest are organosilicon copolymers with inorganic and organo-inorganic molecular chains.

Besides high thermal stability, organosilicon polymers possess high resistance to low temperatures, sunlight and humidity effects, atmospheric impacts, etc. Low variations of physicochemical indices in a broad temperature range are also typical of these polymers. That is why they are widely effectively applied to electric and radio engineering, coal, mechanical-rubber, aviation, metallurgic, textile and other industries. Besides industrial application, they are exclusively suitable for domestic and medical use, and their advantages in these branches can be hardly overestimated.

Hydrophobicity, which is useful for producing waterproof cloth, footwear and construction materials, can also be attributed to various materials with the help of organosilicon polymers. Application of organosilicon substances as antifoamers gives a possibility to eliminate foam, which can be hardly done in many branches of economy: drug manufacture, sugar production, wine industry and other branches of food industry. The modern medicine does also need them greatly: they help in elimination blood foaming during serious operations, which require temporary discharge of sufficient amount of blood from the organism. In this case, treatment of surgical instruments by liquid organosilicon oligomers eliminate occurrence of the smallest air bubbles in the blood and formation of air clots, which may be lethal for a patient.

Organosilicon oligomers are also widely used in production of hydraulic fluids and lubricants, which provide for machine operation in a broad temperature range from -140 to 300°C.

Poly(organosiloxanes) of both linear and cyclolinear structure, containing no classic mesogenic groups in macromolecules, are able to exist in a thermotropic mesomorphous state. Contrary to fully aromatic liquid-crystal polyesters, thermotropic mesomorphous poly(organosiloxane) possess isotropization temperature, which is much lower than the point of thermal degradation start. It is found that affinity of poly(organosiloxanes) to formation of a mesomorphous state and temperature borders of mesophase existence are defined by many factors, among which the most important ones are structure of both the backbone and side substituting groups and molecular mass of the polymer.

Year by year, the demand in organosilicon elastomers used in production of thermally resistant rubbers and organosilicon polymers for making non-metal composite materials is increased. At present, there are minor branches of industry and economy, which do not apply organosilicon polymers and materials derived from them.

This monograph presents a discussion of methods for synthesizing and studying properties of newly designed organosilicon polymers, namely, organosiloxane copolymers with cyclolinear structure of macromolecules.

Introduction of cyclic, specially aromatic and heterocyclic groups into the backbone of polymeric molecule induces sharp increase of its rigidity. Thermal stability of the chain is increased with its rigidity, because the rate of degradation processes, mainly induced by free-radical reactions, proceeding in vitrified and crystalline (solid) polymers is negligibly low. No wonder that cyclochain polymers became of great interest for creation of materials for specialized machinery.

The present book sums up the data on the methods of synthesizing primary linear and cyclic organosiloxanes of various structures, complicating structure and obtaining cyclolinear polymers. Molecular structure of cyclolinear poly(organosiloxanes) with different sizes of cycles, linked via oxygen atoms or other flexible bridges, depends on the number of flexible fragments both in the cycle and linear chain. A broad selection of cyclic structures in the chain (beads in a necklace), control of their size and linking type allow broad variations of properties of cyclolinear poly(organosiloxanes).

This book is designed for chemists working in the field of chemistry of organosilicons.

Chapter 1. Linear and cyclic organosiloxanes

1. Linear and cyclic organosiloxanes with assigned disposition of functional groups at silicon atoms

1.1. Polyfunctional linear organochlorosiloxanes and carbosilanes

Organosilicon monomers containing functional groups at silicon atoms are extremely inclined to various reactions of transformation, which produce ≡Si-O-Si≡ bonds [1]. At the present time, the most practically significant organosilicon polymers are industrially produced from organochlorosilanes of various functionalities [2]. To carry out directed synthesis of oligomers and block-copolymers with cyclic and polycyclic fragments in the macromolecular backbone, methods of linear di- and polyfunctional oligochlororganosiloxanes synthesis have been developed [3, 4], on the basis of which organocyclosiloxanes with the given disposition of chlorine atoms and other functional groups at atoms of silicon may be derived.

At the present time, linear di- and polyfunctional organosiloxanes are mostly synthesized by the following methods:

1. partial hydrolytic condensation of organochlorosilanes;
2. telomerization of organocyclosiloxanes with organochlorosilanes and silicon tetrachloride;
3. heterofunctional condensation of organochlorosilanes and organosiloxanes with dihydroxyorganosilanes, dihydroxyorganosiloxanes and alkoxyorganosilanes;
4. double replacement reactions;
5. hydrosilylation.

Application of partial hydrolytic condensation to synthesis of chlorine-containing alkyl- and arylsiloxanes is known since the old times. If insufficient amount of water affects on diorganochlorosilanes, the first stage of the reaction may be represented by the following scheme:

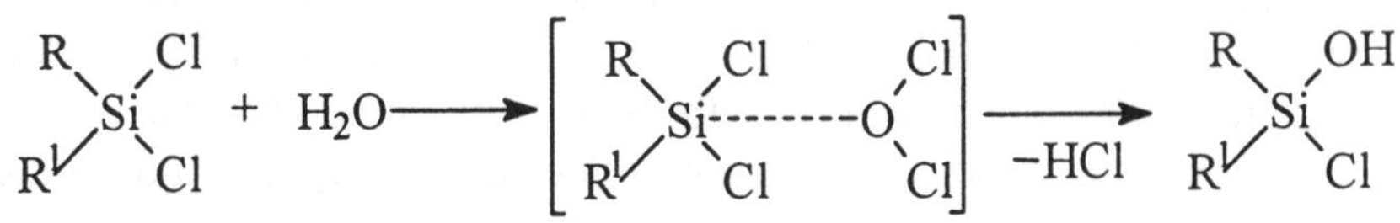

(1.1)

Future process of partial hydrolytic condensation proceeds stage by stage and leads to formation of organochlorosilanes of mainly linear structure:

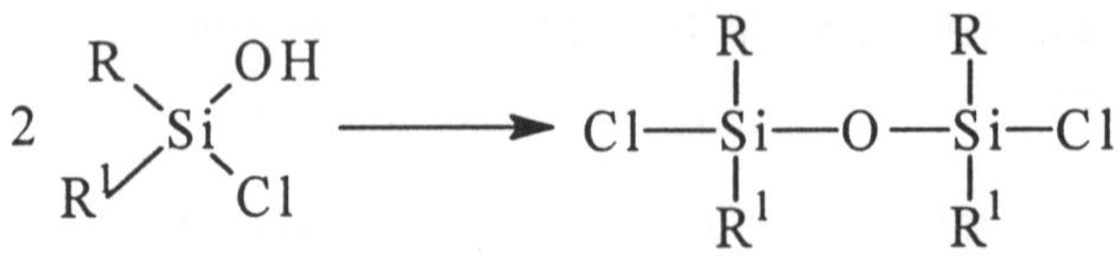

(1.2)

In the case if hydrolysis of dialkylchlorosilanes proceeds with water excess, complete replacement of chlorine atoms by hydroxyl groups and formation of dihydroxydiorganosilanes, which are easily condensed with water extraction, takes place already at the initial stage of the process [5]. The rate of hydrolysis and condensation of alkylchlorosilanes is so high that products of partial hydrolysis of dialkyldichlorosilanes which are organochlorosiloxanes may be obtained in the reaction with insufficient amount of water only [6]. To avoid further hydrolysis and condensation and preserve chlorine atoms at silicon, partial hydrolytic condensation is executed in highly diluted polar solvents. For example, for 1,3-dichlorotetramethyldisiloxane extraction, partial hydrolysis of dimethyldichlorosilane by water was executed in ether-dioxane solution [14]. In this case also, the yield of the target product did not exceed 28%. By partial hydrolytic condensation of dimethyldichlorosilane (diethyldichlorosilane) by insufficient amount of water in highly diluted solutions with further rectification, low amounts of the following compounds have been extracted [7]:

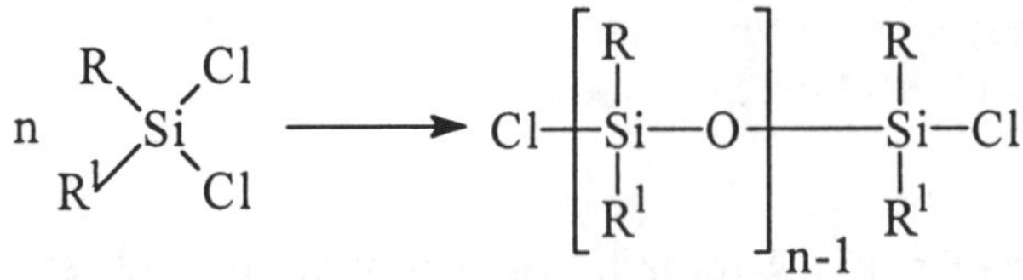

(1.3)

There are also data in the literature [8] indicating that 1,3-dichlorotetramethyldisiloxane with up to 30% yield is synthesized in the reaction of dimethyldichlorosilane with Fe_2O_3 under low boiling conditions of the reaction mixture during 8 hours. However, it has been found [9] that the mentioned reaction does not proceed with pure dimethyldichlorosilane and calcined (anhydrous) iron oxide. It needs iron oxide containing some amount of water (0.2%) or some amount of hydrogen chloride should be added at the beginning of the reaction. In this case, at low boiling of the reaction mixture during 5.5 hours, up to 30% of 1,3-dichlorotetramethyldisiloxane is formed. It is also known that α,ω-dichlorodimethylsiloxanes are formed by partial hydrolytic condensation of dimethyldichlorosilane in the presence of $FeCl_3$ [9] or crystalline *orhto*-phosphoric acid [10]. Similarly, telomerization of mixed organocyclosiloxanes in the presence of *orhto*-phosphoric acids proceeds [11]. This is the method of obtaining linear oligomers with a narrow molecular-mass distribution. Hydrolysis of α,ω-dichlorodimethylsiloxanes in a neutral medium gives corresponded α,ω-dihydroxydimethylsiloxanes of different polymerization degrees [12]:

$$Cl\left[(CH_3)_2SiO\right]_{n-1}Si(CH_3)_2Cl + H_2O \xrightarrow[-2Ac\bullet HCl]{2Ac} HO\left[(CH_3)_2SiO\right]_n H \quad (1.4)$$

$$n = 2 - 300$$

For the purpose of increasing selectivity of partial hydrolytic condensation of aralkyl(aryl)chlorosilanes, the vapor-phase method of polychlorosiloxanes production at high temperature in the absence of solvents has been suggested [13 - 16]. Partial hydrolytic condensation of organochlorosilanes in vapors gives an opportunity to release reaction products from the reaction mixture during the process and, consequently, control the composition increasing the process selectivity and yield of target products. In accordance with this method polychlorine-containing phenyl-, ethyl- and phenylhydride siloxanes were synthesized [16]. In all the cases, the initial stage of hydrolytic condensation proceeds in accordance with the scheme as follows:

$$RR'SiCl_2 + H_2O \rightarrow RR'SiCl(OH) + HCl \quad (1.5)$$

$$RR'SiCl(OH) + RR'SiCl_2 \longrightarrow Cl-\underset{R}{\overset{R}{\underset{|}{\overset{|}{Si}}}}-O-\underset{R}{\overset{R}{\underset{|}{\overset{|}{Si}}}}-Cl \quad (1.6)$$

where $R = C_6H_5$; $R' = Cl$; $R = C_2H_5$; $R = C_6H_5$, $R' = H$;

and gives 55 – 75% yield of 1,1,3,3-tetrachlorodiphenyl-, 1,1,3,3-tetrachlorodiethyl- or 1,3-dichloro-1,3-diphenyldisiloxanes. Besides the mentioned products with 10 – 26% yield, polychlorosiloxanes with higher condensation degree are also formed. This evidences that besides the main reaction, hydrolytic condensation is deeper. For example, in the case of phenyltrichlorosilane, further condensation proceeds by the schemes as follows:

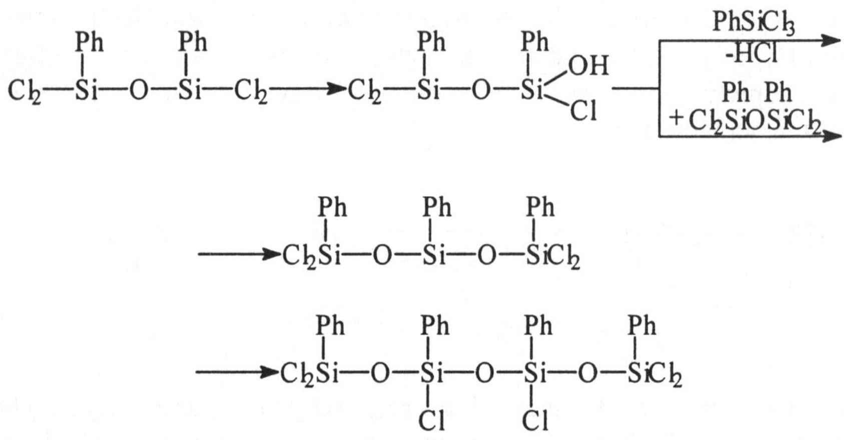

(1.7)

For the first time, the reaction of telomerization was used for obtaining bifunctional *α,ω*–dichlorodiorganosiloxanes [17]. The interaction between diorganocyclosiloxanes and diorganodichlorosilanes proceeded under pressure and at high temperature (380 - 400°C) by the scheme as follows:

$$[R_2SiO]_n + R_2SiCl_2 \xrightarrow{P,\,T} Cl\text{-}[\text{-}R_2SiO\text{-}]_n\text{-}SiR_2Cl \qquad (1.8)$$

where $n = 3 - 6$.

Further on, the mentioned method was studied in detail by K.A. Andrianov *et al.* [18 – 29].

It has been shown [18 – 20] that in telomerization under pressure and at high temperature (250 - 300°C) dialkyldichlorosilanes react with dialkylcyclosiloxanes, which gives *α,ω*–dichlorodialkylsiloxanes containing X_{n+1} atoms of silicon (where n is the number of Si atoms in the initial siloxane ring and $X = 1, 2, 3\ldots$).

The reaction mechanism may be represented as follows [18]: the first stage is that the ring scission proceeds with formation of α,ω–dichlorodialkylsiloxanes containing (n + 1) atoms of silicon in the chain:

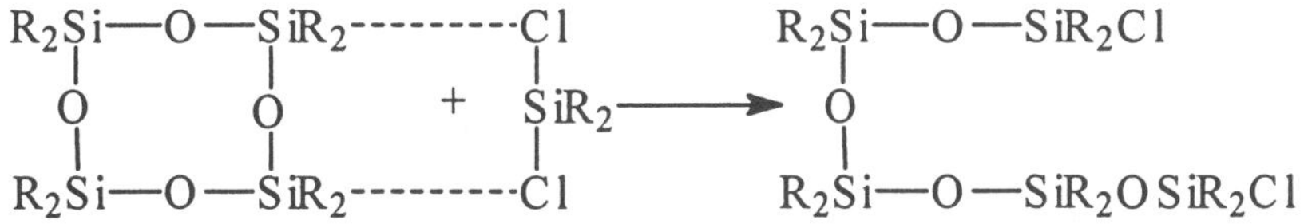

(1.8)

As dialkyl(aryl)cyclosiloxanes are telomerized by dimethyldichlorosilane in a steel autoclave or in the presence of $FeCl_3$ [21, 22], ring scission with formation of oligomers is also observed.

By analogy, dimethylcyclosiloxanes are telomerized with different organotrichlorosilanes [23, 24] and $SiCl_4$ [25] forming polyfunctional oligochlororganosiloxanes.

Among methods of synthesizing linear di- and polyfunctional organosiloxanes, of great interest is heterofunctional condensation (HFC) of dihydroxydiorganosilanes or dihydroxydiorganosiloxanes with chlorosilanes. In accordance with the ratio of initial components, in the presence or in the absence of liberated hydrogen chloride acceptor, the yield and structure of formed products may be various.

Recently, heterofunctional condensation reactions (HFC) are widely used for obtaining di- and polyfunctional organosiloxanes with the given disposition of functional groups at silicon atoms.

In accordance with the scheme as follows [30], in the presence of pyridine and chlorosilane excess, in diluted solution of anhydrous toluene in 0 – 5°C temperature range high yields of corresponded linear organochlorosiloxanes were synthesized by the HFC reaction of dihydroxydiphenylsilane with organotrichlorosinales:

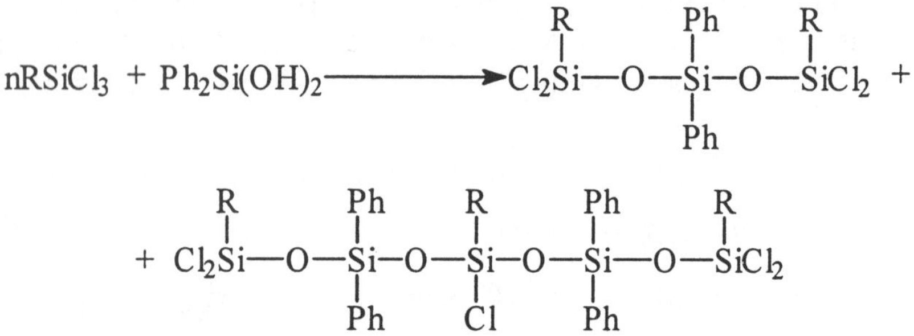

(1.10)

where R = Me, Ph.

The same investigators [31, 32] have also shown that as temperature of HFC reaction of organodi- and organotrichlorosilanes with dihydroxydiphenylsilane is decreased to -10°C, the yield of chlororganotrisiloxanes in the presence of amine acceptor is increased to 84%.

HFC reaction of dihydroxydiphenylsilane with phenyltrichlorosilane [33], silicon tetrachloride, trichlorosilane or vinyltrichlorosilane in the absence of HCl acceptor causes formation of various siloxane structures of both symmetrical and asymmetrical structure. As a result of interaction between liberated electrophilic hydrogen chloride with siloxane bond, the following reaction proceeds:

$$nPhSiCl_3 + Ph_2Si(OH)_2 \xrightarrow[-HCl]{} Ph_2SiCl_2 + Cl_2PhSiOSiPh_2Cl + Cl_2PhSiOSiPh_2SiPhCl_2 \qquad (1.11)$$

If acceptor of hydrogen chloride is used, secondary processes are reduced to the minimum and the yield of linear chlororganosiloxanes is increased [33].

HFC reaction of dihydroxydiphenylsilane with various organochlorosilanes was studied in the presence of co-reagents – organoaminosilanes or organocyclosilazanes [34]. It has been found that dihydroxydiphenylsilane interaction with excessive diorganodichlorosilanes in the presence of bis(diethylamino)organosilanes or cyclosilazanes at +5 - +15°C results in formation of appropriate 1,5-dichlorohexaorganotrisiloxanes with 60 - 84% yield in dioxane. The reaction proceeds in accordance with the scheme as follows:

$$2Ph_2Si(OH)_2 + 3R'R''SiCl_2 + R'R''Si(Net_2)_2 \rightarrow$$
$$\rightarrow 2ClR'R''SiOSiOSiR'R''Cl + 2Et_2NH{\cdot}HCl \qquad (1.12)$$

where R' = H, Me, Et; R" = R' or R" = Me, Et, Ph, Vin, CH_2, CH_2CF_3.

It has been found that the reaction proceeding is affected by the shielding action of framing groups at silicon atoms and the origin of co-reagents. The same investigators [34] have also shown in HFC of dihydroxydiphenylsilane with a mixture of diorganochlorosilanes and diorganodiaminosilanes that structure and yield of the products formed is strictly dependent on the ratio of initial components. This may result in formation of both linear and cyclic organosiloxanes. As concluded [34], the co-reagent containing ≡Si-N= groups plays the role of hydrogen chloride acceptor and reacts by ≡Si-Cl bond:

$$\equiv Si\text{-}N\text{-}Si\equiv + HCl \rightarrow \equiv Si\text{-}Cl + \text{-}NH\text{-}Si\equiv$$
$$\equiv Si\text{-}NH\text{-} + HCl \rightarrow \equiv Si\text{-}Cl + \text{-}NH_2$$
$$\text{-}NH_2 + HCl \rightarrow \text{-}NH_2{\cdot}HCl.$$

Further on, the process is resumed and terminated as one of the initial components of the mixture is exhausted.

The study of HFC reaction of 1,3-dihydroxytetraphenylsiloxane with silicon tetrachloride at equimolar ratio of the initial components has indicated [35] that besides organocyclosiloxanes, linear polyfunctional organosiloxanes are also produced in accordance with the scheme as follows:

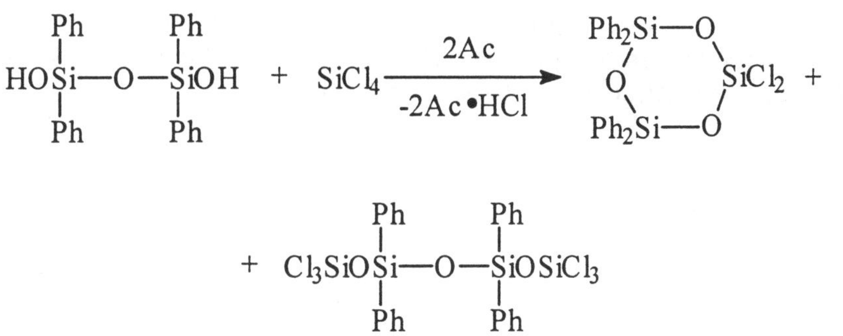

(1.13)

The yield of linear products is increased with the quantity of silicon tetrachloride. In a similar manner, equimolar HFC reaction of silicon tetrachloride with 1,3-dihydroxy-1,3-divinyl-1,3-diphenyldisiloxane has been studied [35].

The HFC reaction of 1,3-dihydroxytetraphenyldisiloxane with 1,5-dihydroxyhexaphenyltrisiloxane and organochlorosilanes at 1:3 ratio of initial components in anhydrous toluene solution at -5°C in the presence of pyridine under the above-mentioned conditions proceeds ambiguously and besides linear organochlorosiloxanes, the yield of which equals 70 – 87%, organocyclotrisiloxanes and organocyclotetrasiloxanes are also produced in low amounts (4 – 6%). As a consequence, the reaction proceeds in accordance with the general scheme as follows [36, 37]:

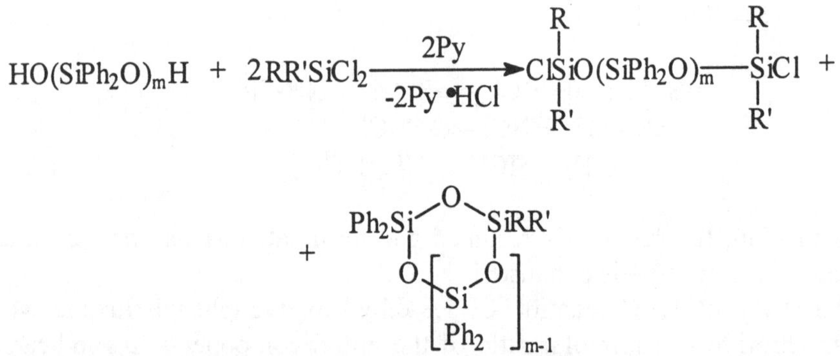

(1.14)

where $m = 2, 3$; R = Ph, Me, Vin, Cl; R' = Cl, H, Vin.

The HFC reaction between methyltrichlorosilane, dihydroxydimethylsilane and 1,3-dihydroxytetramethyldisiloxane at 1:1 molar ratio of initial components in the presence of acceptor (amine) mainly proceeds with formation of linear tetrachlororganosiloxanes [38]:

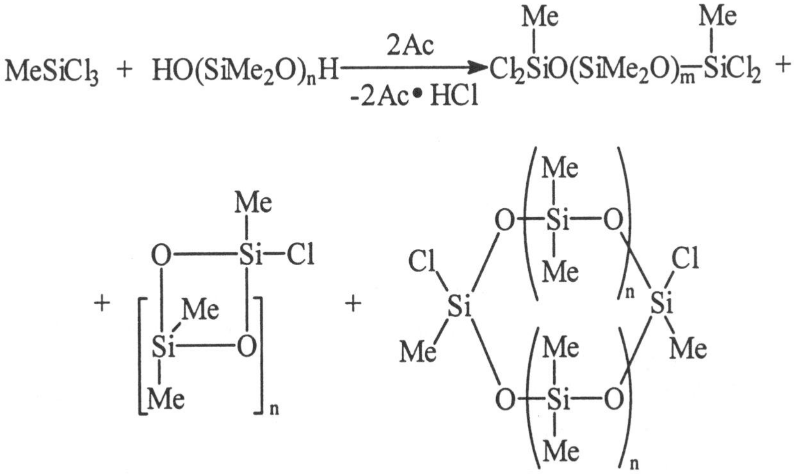

(1.15)

where $n = 1, 2, 3$.

The same investigators [39] have also obtained linear tetrachloroethylsiloxanes and cyclic ethylsiloxanes by the HFC reaction of ethyltrichlorosilane, dihydroxydiethylsilane with 1,3-dihydroxytetraethyldisiloxane in chlorosilane excess.

The HFC reaction of dihydroxydiorganosilanes disodium salts with organochlorosilanes causes formation of linear, cyclic and star-shaped siloxanes [40 – 43].

Investigation of the HFC reaction between 1,3-dihydroxy-1,3-dimethyl-1,3-diphenyldisiloxane and 1,5-disodiumoxyhexamethyltrisiloxane and silicon tetrachloride, methyl- and phenyltrichlorosilanes has displayed that with regard to the ratio of initial components both linear and cyclic compounds may be produced. The yield of linear reaction products is increased with excess of chlorosilanes in the primary reaction mixture, and the yield of cyclic compounds is decreased.

The HFC reaction between dialkyldichlorosilanes and dimethyldietoxychlorosilane has also been performed. It has been shown that in this case, the reaction proceeds according to the step-by-step mechanism as follows [45]:

$$R_2SiCl_2 + C_2H_5OSiMe_2Cl \rightarrow ClR_2SiOSiMe_2Cl + C_2H_5Cl \qquad (1.16)$$

$$ClR_2SiOSiMe_2Cl + C_2H_5OSiMe_2Cl \rightarrow$$

$$\rightarrow ClR_2SiOSiMe_2ClOSiMe_2Cl + C_2H_5Cl \quad (1.17)$$

where R = Me, C_2H_5.

Formation of rings in this reaction is absolutely excluded, and homopolycondensation of dimethyletoxychlorosilane proceeds at 80°C only. If HFC reaction proceeds without a catalyst, its rate is very low, and in the presence of a catalyst ($FeCl_3$, $AlCl_3$) the process is highly accelerated.

To synthesize polyfunctional linear carboxylic compounds, the reaction of hydride addition is also used.

Hydride addition of alkylchlorosilanes $R_nCl_{3-n}SiH$ to vinyl-containing chlorosilane $R_nCl_{3-n}SiH{=}CH_2$ in the presence of $Ph(PR^I_3)_3$, $Pd(PR^I_3)_3$, $CO_2(CO)_8$ and platinum catalysts has produced carbosiloxane compounds [37, 46 - 51]:

$$R_nCl_{3-n}SiCH{=}CH_2 + R_nCl_{3-n}SiH \xrightarrow{Cat} R_nCl_{3-n}SiCH_2CH_2SiCl_{3-n}R_n \quad (1.18)$$

where R = alkyl, $n = 0 - 3$.

More thorough studies of hydride addition reaction between vinyl-containing and hydride-containing silanes in the presence of the Spayer catalyst were performed [46]. It is suggested that siliconium ion is produced in the presence of catalysts which leads to formation of linear organosiloxanes.

It has been shown [47] that in the presence of nickel complexes the hydride addition reaction proceeds with formation of a mixture of α- and δ-adducts. Hydrosilylation of acetylene by silanes (general formula $HSiCl_nR_{3-n}$), in the presence of catalysts has produced silyl-substituted ethanes of the general formula $R_{3-n}Cl_nSiCH_2CH_2SiCl_nR_{3-n}$. Phosphine complexes and Ru^{+2}, Ru^{+3} or Ru^{+4} linked with silica were used as catalysts.

As is evident from the bibliography, HFC and hydrosilylation reactions are of precedence meaning for obtaining high yields of linear organosiloxanes and carbosilanes with functional groups of given disposition at silicon atoms, which are the initial products for synthesis of organocyclosiloxanes and organocyclocarbosiloxanes with functional groups.

Thus synthesis of linear organosiloxanes and carbosilanes with functional groups are mainly discussed in the present Chapter. Further on, these substances are used for producing organocyclosiloxanes and organocyclocarbosiloxanes with the given disposition of functional groups at silicon atoms and polymers derived from them.

1.2. Organocyclosiloxanes with given disposition of functional groups at atoms of silicon

Synthesis of organocyclosiloxanes with different disposition of functional groups at atoms of silicon is mainly reduced to two methods: hydrolytic condensation of chlorine-containing organosilanes and alkoxyorganosilanes and HFC reaction of silanes with siloxanes containing functional chlorine-, alkoxy-, amino and hydroxyl groups or to modification of previously prepared organocyclosiloxanes.

Hydrolytic condensation of dibutoxydichlorosilane in the presence of hydrogen chloride acceptor (tertiary amine) was used for synthesizing organocyclosiloxanes with butoxyl groups [52, 53]. Similarly, hydrolytic condensation of diethoxydichlorosilane proceeds in the presence of pyridine [54]. Partial hydrolytic condensation of tetraethoxysilane and hexaethoxydisiloxane [55, 56] induces formation of organocyclosiloxanes with ethoxy groups at silicon atoms.

Hydrolytic condensation of alkylalkoxydichlorosilanes in the presence of pyridine also produces organocyclosiloxanes containing, besides alkoxy groups, alkyl ones at silicon atoms [57, 58].

Joint reactions of $SiCl_4$ etherification and hydrolytic condensation with isopropyl alcohol produce organocyclosiloxanes with isopropyloxy groups at silicon atoms [59].

Joint hydrolytic condensation between dimethyldichlorosilane and methylalkoxydichlorosilanes in the presence of pyridine has produced organocyclosiloxanes with two alkoxy groups at different silicon atoms [60]:

$$2Me_2SiCl_2 + 2MeSi(OR)Cl_2 + 4\,H_2O + 8Py \rightarrow \qquad (1.19)$$

$$\rightarrow \underbrace{[Me_2SiO][MeSi(OR)O]_2}_{\text{cyclic}} + 8Py \bullet HCl$$

Hydrolytic co-condensation between diphenyldichlorosilane and phenyltrichlorosilane at 1:4 molar ratio of the initial components produces hydroxypentaphenylcyclotrisiloxane and 1,5-dihydroxyhexaphenylcyclotetrasiloxane [61]:

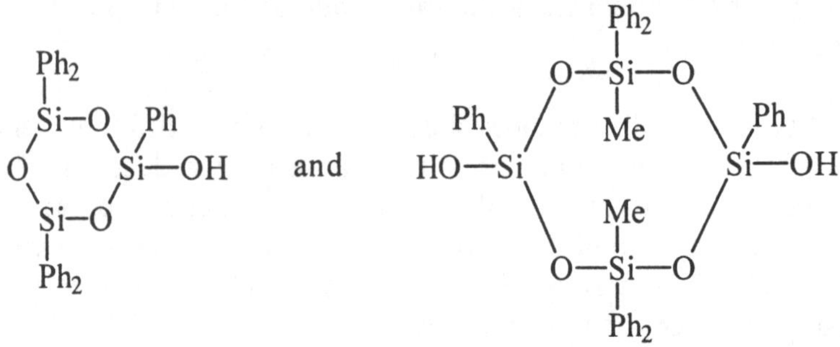

However, it should be noted that the yield of individual phenylcyclosiloxanes is negligible.

Joint hydrolysis of dimethyldichlorosilane and methylvinyldichlorosilane produced divinylhexamethylcyclotetrasiloxane [62]. Despite the application of efficient rectification towers as well as analytical chromatograph with preparative attachment, the authors have not managed to separate isomeric 1,3- and 1,5-divinylhexamethylcyclotetrasiloxanes, which may be produced by joint hydrolysis. The reaction proceeds in accordance with the general scheme as follows:

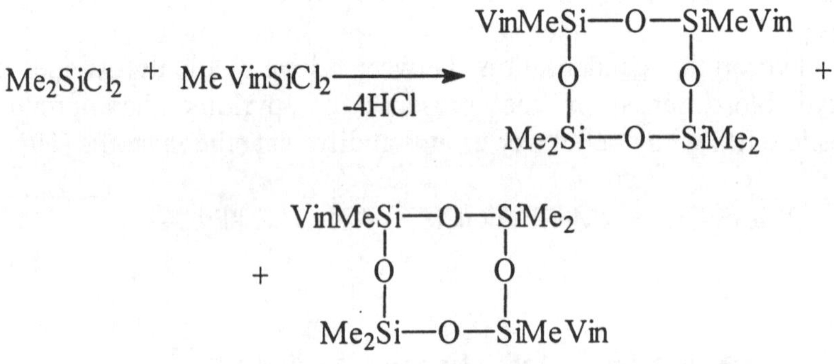

(1.20)

In accordance with the scheme as follows [63], 1-hydride-3-vinylhexamethylcyclotetrasiloxane was synthesized in co-hydrolysis reaction between 1,3-dichlorotetramethyldisiloxane and 1-chloro-3-chloro-3-vinyltetramethyldisiloxane:

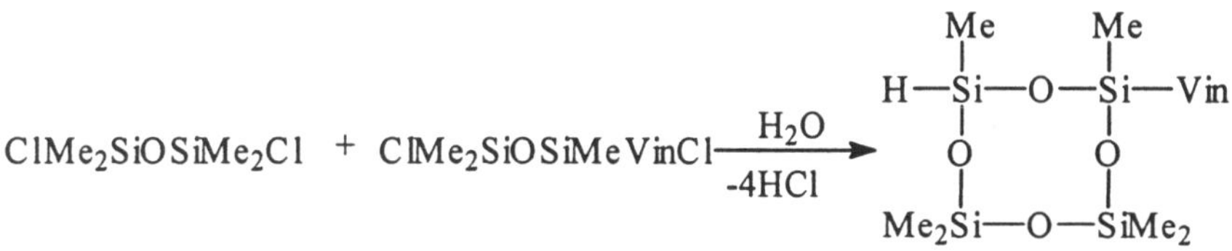

(1.21)

Hydrolytic condensation of phenyltrichlorosilane in aqueous solution of acetone, in acidic medium has synthesized cis-1,3,5,7-tetrahydroxy-1,3,5,7-tetraphenylcyclotetrasiloxane (tetrole) with high yield [64]. The reaction proceeds by the scheme shown below:

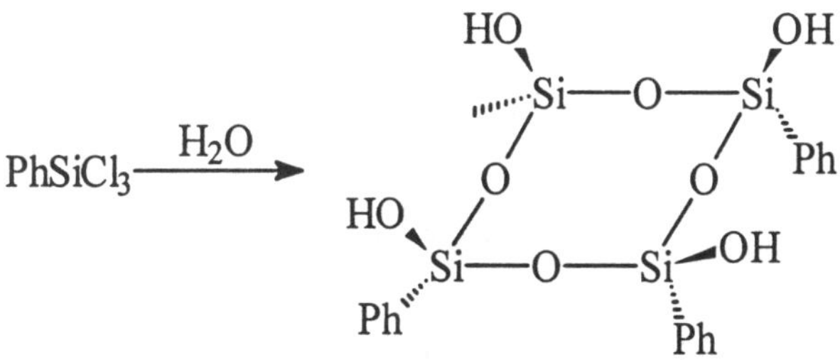

(1.22)

Later on, mutual configurative transformations of four stereo isomers of tetrole were detected. These stereo isomers are of the following structures:

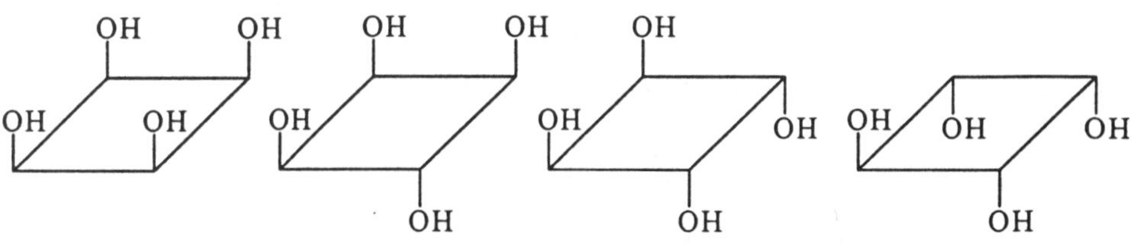

It has been shown [65] that isomerization of tetrasiloxane ring may proceed both with reversal of silicon atoms configuration and without it.

Tetrole-like compounds were also synthesized by hydrolytic condensation of (β-phenyl)phenethyltrichlorosilane, stilbene trichlorosilane and (α-phenyl)phenethyltrichlorosilane [66].

At 20 - 50°C in acetone, partial hydrolytic condensation of 1,1,3,3-tetrachloro-1,3-diphenyldisiloxane [67] proceeds with 20% yield of cyclotetra- and cyclohexasiloxane products, $[PhClSiO]_4$ and $[PhClSiO]_6$. Low yield of cyclic products is explained by formation of a significant quantity of linear products.

Besides 1,5-dimethyl-1,5-divinyl-3,3,7,7-tetraphenylcyclotetrasiloxane, the interaction between equimolar amounts of diphenylsilanediol and methylvinyldichlorosilane also synthesizes 1,3-dimethyl-1,3-divinyl-5,5-diphenylcyclotrisiloxane, 1-methyl-1-vinyl-3,3,5,5-tetraphenyltrisiloxane and high-molecular products in accordance with the scheme as follows [68]:

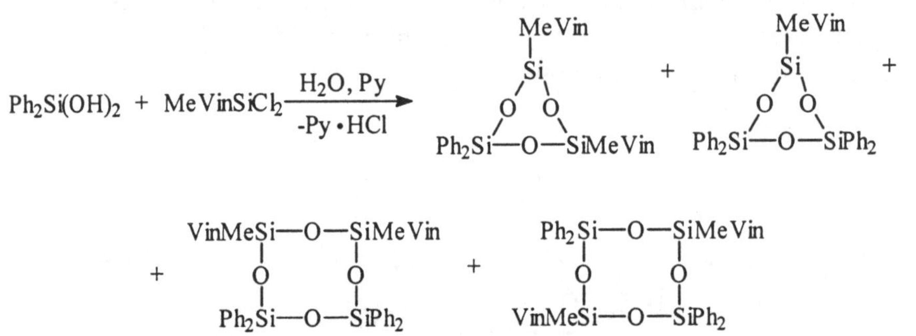

(1.23)

HFC reaction is more suitable for synthesizing organocyclosiloxanes with functional groups at silicon atoms. High yields of organocyclosiloxanes with a single functional group are produced in double replacement reaction of disodium salts of dimethylsiloxanes with ethylbutoxydichlorosilane [61], methylbutoxydichlorosilane [69] and methyltrichlorosilane [70].

The HFC reaction between 1,1,3,3-tetrachlorodiphenyldisiloxane and dihydroxydiorganosilanes in the presence of hydrogen chloride acceptor causes formation of 1,3-dichlorotetraorganocyclotrisiloxanes [71] in accordance with the scheme as follows:

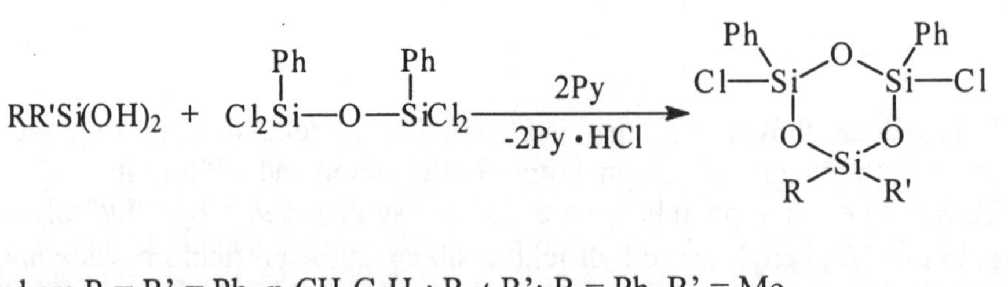

(1.24)

where R = R' = Ph, p-$CH_3C_6H_5$; R ≠ R'; R = Ph, R' = Me.

Hydrolysis of synthesized 1,3-dichlorotetraorganocyclotrisiloxanes has produced appropriate dihydroxy derivatives.

Condensation reactions of dihydroxydiphenylsilane or dihydroxydimethylsilane with methylvinyldichlorosilane and methyldichlorosilane [72, 73] in the presence of pyridine (1:1 molar ratio of the initial components) have given high yields of organocyclotetrasiloxanes with 1,5-disposition of functional groups at silicon atoms. The reaction proceeds as follows:

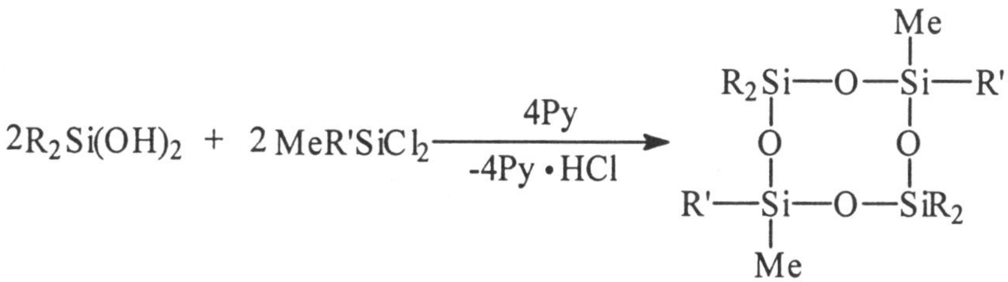

(1.25)

where R = Me, Ph, Et; R' = Vin, H.

Divinyl- and dihydride-containing organocyclosiloxanes with different disposition of functional groups at silicon atoms were synthesized by the HFC reaction of 1,3-dihydroxyorganodisiloxanes with dichlorodiorganosilanes and 1,3-dichlororganodisiloxanes in the presence of pyridine in accordance with the general scheme as follows [72, 74]:

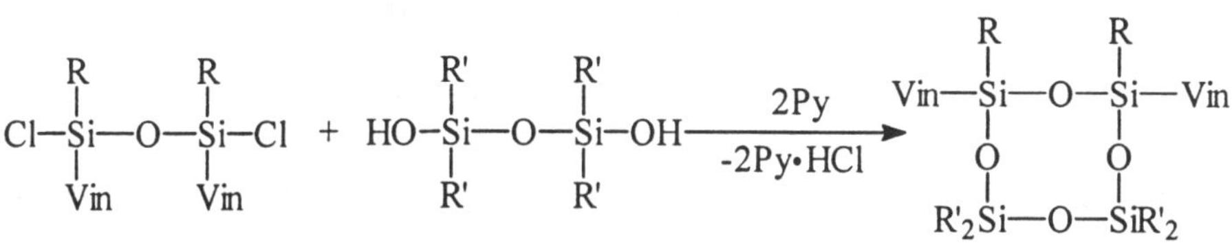

(1.26)

where R = R' = Me, Ph; R ≠ R'.

By analogy, heterofunctional condensation of dihydroxydiphenylsilane with methyltrichlorosilane [74, 75], silicon tetrachloride [76] and phenyltrichlorosilane [31, 77], as well as between diethyldihydroxysilane and ethylvinyldichlorosilane and ethyldichlorosilane, proceeds at equimolar (1:1) ratio of the initial components, in the presence of pyridine. The yield of target products with 1,5-disposition of chlorine atoms at silicon ones is varied from 41 to 55%. The reaction scheme is as follows:

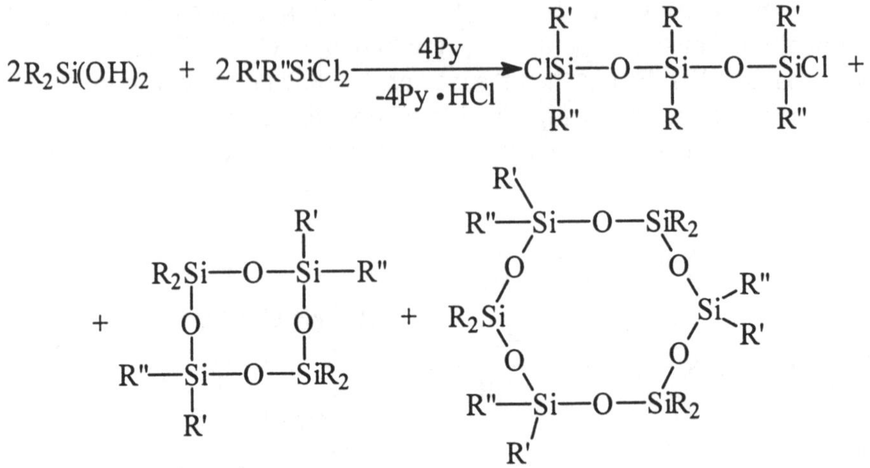

(1.27)

where R = Me, Ph, Et; R' = Me, Ph, Et, Cl,Vin; R" = Cl, H, Vin.

As shown in the above scheme, besides isomeric organocyclosiloxanes, linear tetrachlororganotrisiloxanes are also synthesized with the yield of 15 – 20%. For methyl-containing organocyclotetrasiloxanes, two singlet signals with almost equal peak ratio (53 – 47%) at 0.38 and 0.42 m.d. have been detected in ^{1}H NMR spectra. These ranges are typical of *cis-* and *trans-*forms of methyl groups. Shift for methyl protons to a weak field testifies about the electron-acceptor type of functional groups at the same silicon atom, where methyl groups are disposed.

To increase the yield of organocyclosiloxanes with 1,5-disposition of chlorine atoms at silicon ones, the two-stage synthesis method was suggested [32, 76, 77]. At the first stage, HFC of dihydroxydiphenylsilane with excess of organotrichlorosilane is carried out. At the second stage, HFC of obtained tetrachloroorganotrisiloxanes with dihydroxydiphenylsilane proceeds. The scheme of the reaction is as follows:

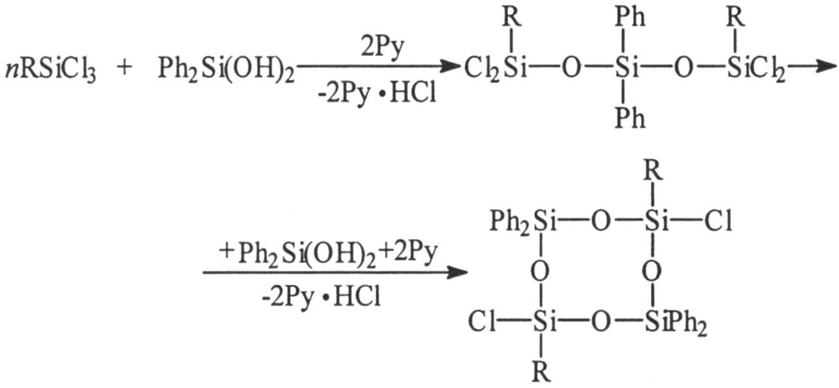

(1.28)

where R = Me, Ph, Cl.

It is shown [78] that depending on the ratio of initial reagents, diorganodiaminosilanes mixed with diorganodichlorosilanes interact with diphenylsilanediol giving cyclic or linear siloxanes. The yield of mixed cyclotetrasiloxanes is quantitative and almost independent of the type of framing at silicon atoms in amino- and chlorosilanes. Trisiloxane is synthesized with lower yield (~70%) due to a noticeable formation of secondary products, which are long-chain α,ω–dichlorosiloxanes.

Formation of long-chain siloxanes is much more typical of the condensation reaction proceeding with dihydroxydiphenylsilane excess. In this case, the yield of trisiloxane does not exceed 50%. The reaction proceeds by the schemes as follows:

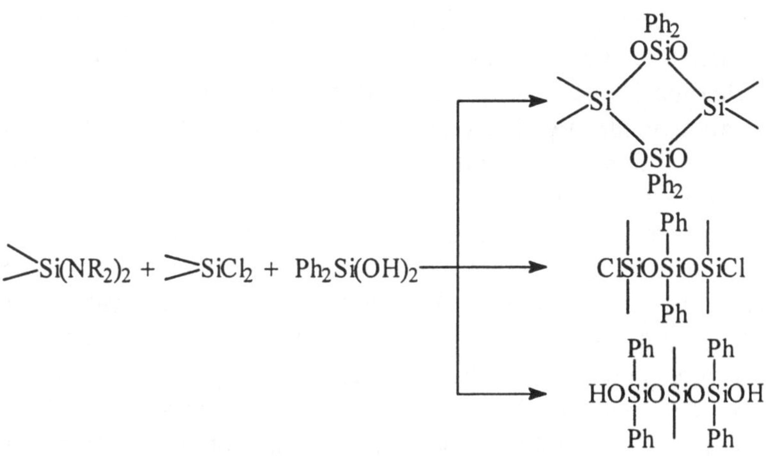

(1.29)

For obtaining high yields of synthesized methylcyclosiloxanes with given disposition of functional groups at silicon atoms, the two-stage synthesis was suggested [39, 79, 80] in accordance with the scheme as follows:

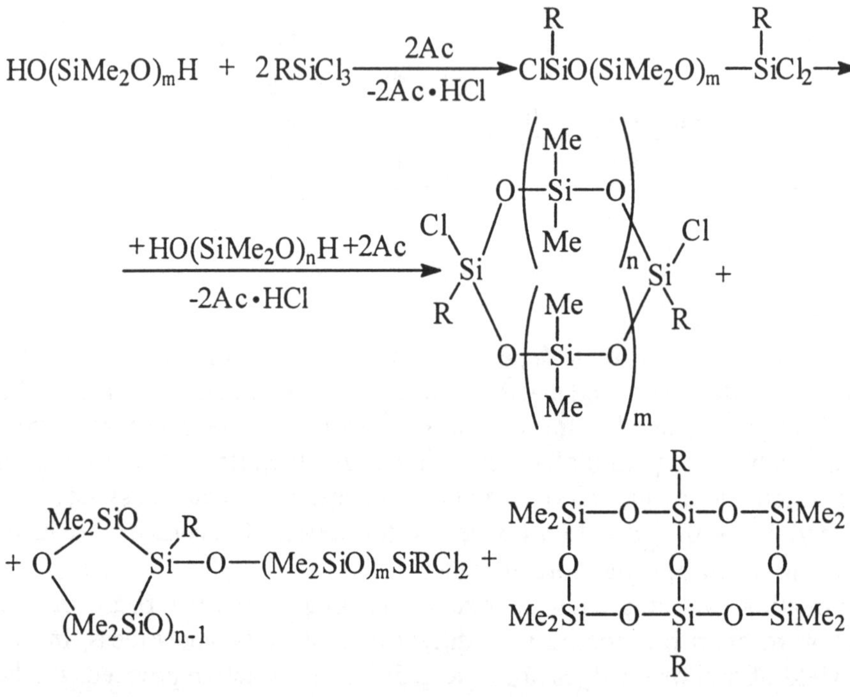

(1.30)

where $m = n = 1, 2, 3$; $m \neq n$; R = Me, Ph.

Two-stage reaction proceeding resulted in the yield of organocyclotetra(penta, hexa)siloxanes equal 28 – 51%. The ratio of dichlorocyclosiloxane isomers was monitored by ESR and mass-spectrometry methods. Basically, dichlororganocyclosiloxanes of 96.0 – 99.5% purity were produced. An attempt to separate dichlororganocyclotetra(penta, hexa)siloxanes to stereo-isomers was made. The majority of dichlororganocyclotetra(hexa)siloxanes are separated to *cis*- and *trans*-isomers. All attempts to separate stereo-isomers of dichloroctaorganocyclopentasiloxane and dichlorododecaorganocycloheptasiloxane to *cis*- and *trans*-isomers have failed. Hydrolysis and aminolysis of dichloromethylcyclo(tetra, penta, hexa, hepta, octa)siloxanes proceeding in the presence of hydrogen chloride acceptors have produced corresponded dixydroxy- and diaminomethylcyclosiloxanes. Inversion of isomers configuration during hydrolysis was not observed [79].

Separation methylcyclosiloxane dihydroxy-derivatives to *cis*- and *trans*-isomers was performed for 1,5-dihydroxyhexamethylcyclotetrasiloxane and 1,7-dihydroxydecamethylcyclohexasiloxane by recrystallization from heptane; under similar conditions, the rest hydroxy-derivatives cannot be separated. X-ray diffraction studies of the crystalline structure of 1,5-dihydroxydecamethylcyclohexasiloxane *cis*- and *trans*-isomers show that Si-O-Si bond angle in *cis*-isomers displays two values: 145.7 – 150.2° and 173.6 – 174.1°. The ring is tub-shaped with silicon atoms in the hydroxyl group distant from the middle atom plane by 0.96 - 0.98 Å. This allows OH-groups in *cis*-isomers forming intra- and intermolecular hydrogen bonds [81].

In the range of 3,200 – 3,800 cm^{-1} typical of valence oscillations of OH-groups, IR spectra of gihydroxymethylcyclo(tetra, hexa)siloxanes, mould with KBr and in CCl_4 solutions of various concentration display the absence of a broad absorption band at 3,300 – 3,500 cm^{-1}, which is present in spectra of all dihydroxymethylcyclosiloxanes mould with KBr, and the presence of a narrow band at 3,700 cm^{-1} typical of hydroxyl group with intermolecular hydrogen bonds, which confirms separation of pure *trans*-isomer of dihydroxymethylcyclosiloxanes. For *cis*-isomers, IR spectra of diluted solution possess two bands: a broad one in the range of 3,300 - 3,500 cm^{-1} typical of intramolecular hydrogen bond and a narrow one at 3,700 cm^{-1}.

Based on ^1H NMR spectra, chemical shifts and multiplicity of methyl protons of chloro(hydroxy)-substituted methylcyclotetra(penta, hexa)siloxanes were studied [82].

By similar two-stage HFC method, difunctional methyl(ethyl)cyclotetra(penta, hexa)siloxanes with defined disposition of chlorine, hydride or vinyl groups at silicon atoms were produced [83, 84].

HFC reaction of organodichlorosilane or organotrichlorosilane with 1,3-dihydroxytetraphenyldisiloxane under usual conditions (equimolar (1:1) ration of initial reagents, presence of hydrogen chloride acceptor – pyridine, diluted toluene solution, 0 - 5°C temperature range) mainly produces both linear organochlorosiloxanes and cyclic compounds. The presence of linear organochlorosiloxanes, which content in the reaction mixture did not exceed 6%, was confirmed by GLC analysis. The reaction proceeds as follows:

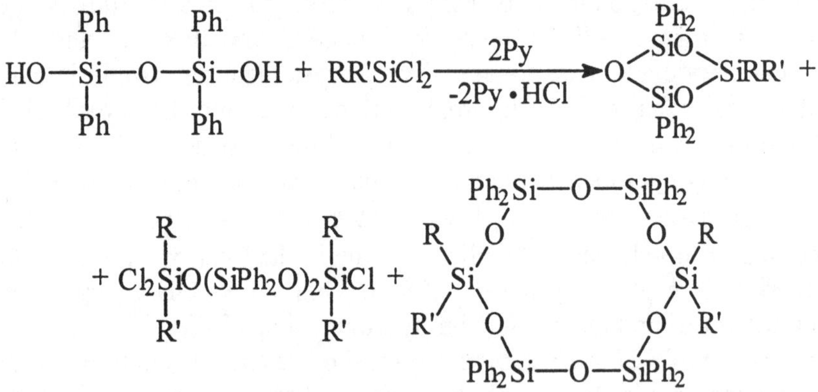

(1.31)

where R = Me, Ph, Cl; R' = Cl, Vin, H.

To the authors' point of view, similar to the case described in ref. [87], the reaction proceeds with formation of a transient formation:

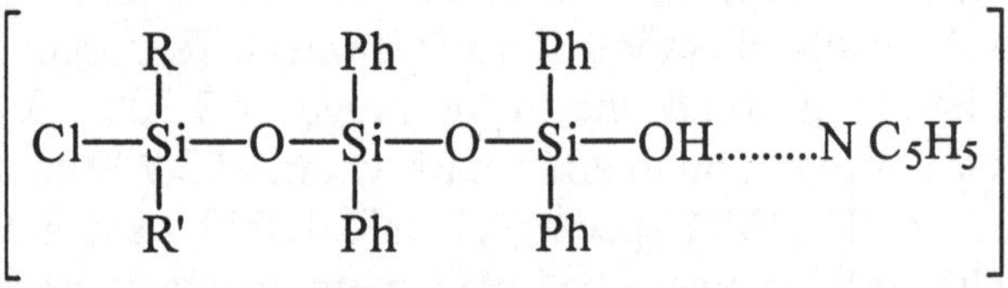

which then is subject to intramolecular condensation producing a hexatomic rings and intermolecular condensation with further cyclization of condensation products. The latter leads to formation of dodecatomic rings and higher boiling products. The study of ESR spectrum of 1,7-dichloro-1,7-dimethyloctapenylcyclohexasiloxane has indicated that HFC reaction of 1,3-dihydroxytetraphenyldisiloxane produces cyclohexasiloxane and its structural isomer:

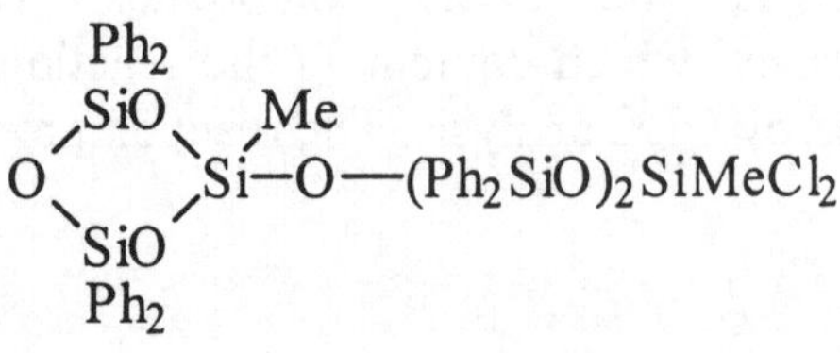

A singlet at 0.27 ppm is corresponded to methyl group of methyldichlorosiloxy-fragment; for methyl protons in silsesquioxane unit, a singlet at 0.46 ppm is also observed. Two singlets at 0.38 and 0.41 ppm, corresponded to two methyl groups of *cis*- and *trans*-forms, respectively, are observed in spectra. After recrystallization from toluene-heptane mixture, organocyclohexasiloxane product was purified from isomeric admixtures. The ratio of *cis*- and *trans*-isomers equaled 54 and 46%, respectively. Accordingly, after fractionation, HFC reaction products of organotrichlorosilanes and silicon tetrachloride with 1,3-dihydroxytetraphenyldisiloxane were recrystallized from the toluene-heptane mixture. HFC reaction of dichlorodiorganosiloxanes with 1,3-dihydroxytetraphenyldisiloxane did not produce isomeric products. At the single-stage synthesis, the yield of organocyclohexasiloxanes equaled 26 – 43%.

HFC reaction of tetrachloro- or dichlororganotetrasiloxanes with dihydroxyophenylsilane or 1,3-dihydroxytetraphenyldisiloxane under usual conditions (at equimolar ratio of initial components, in the presence of pyridine, in diluted anhydrous toluene solution) produced organocyclopenta- and organocyclohexasiloxanes with 1,5- and 1,7-disposition of chlorine atoms at silicon. The reaction scheme is as follows [88 – 91]:

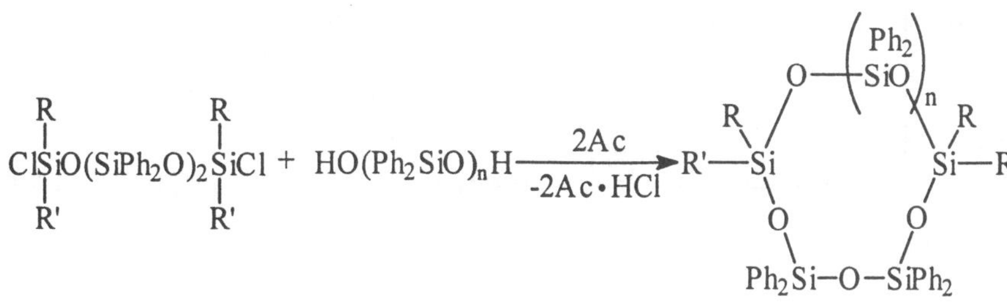

(1.32)

where $n = 1, 2$; R = Me, Ph; R' = Cl, H, Vin.

By HFC reaction of hexa(tetra, di)chlororganotetra(penta)siloxanes with dihydroxydiphenylsilane under usual conditions (in the presence of pyridine, at equimolar ratio of initial reagents, in diluted anhydrous toluene solution, in -5 - 0°C temperature range), high yields of difunctional organocyclopentasiloxanes and organocyclohexasiloxanes with 1,5-disposition of functional groups were obtained. Therefore, the yield of organocyclopenta(hexa)siloxanes is increased to 61 – 70%. General reaction scheme is as follows [89, 92, 93]:

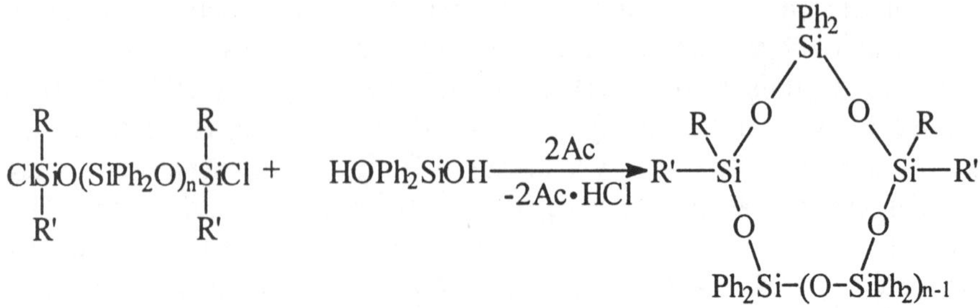

(1.33)

where $n = 2, 3$; R = Me, Ph, Cl; R' = Cl, H, Vin.

Hydrolysis of tetrachlororganocyclotetra(penta, hexa)siloxane proceeding in the presence of hydrogen chloride acceptor is described in the literature [32, 37, 94 – 97]. It is found that hydrolysis proceeding at room temperature produces polyfunctional oligomers, which are transformed into structured products. However, as the reaction temperature was decreased to 0 - 5°C, in the presence of sodium bicarbonate, tetrahydroxy-derivatives were synthesized by the following scheme:

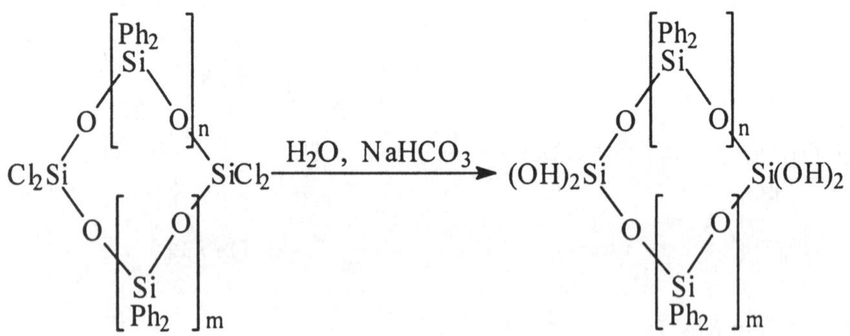

(1.34)

It is also found that 1,1,5,5-tetrahydroxytetraphenylcyclotetrasiloxane forms a stable structure with pyridine [95]:

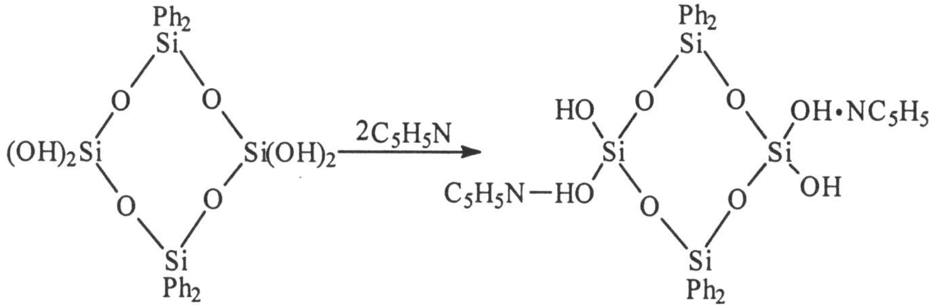

(1.35)

that was not observed for its isomer, cis-1,3,5,7-tetrahydroxytetraphenylcyclotetrasiloxane (tetrole).

It is assumed [98] that the structure formation is associated with increased acidity of hydrogen in hydroxyl groups, caused by interaction of *p*-electrons of oxygen atoms with silicon. Inertness of the second hydroxyl group to pyridine is explained by decrease of its acidity as a consequence of redistribution of electron density in the group:

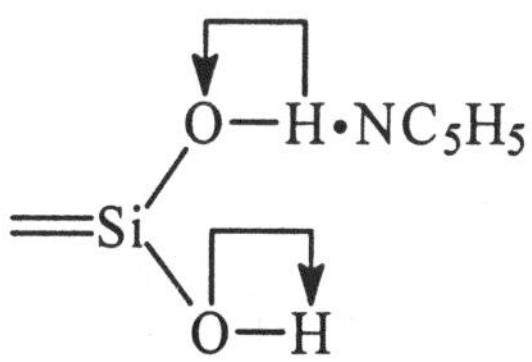

Dipyridine-1,1,5,5-tetrahydroxytetraphenylcyclotetrasiloxane structure enters HFC reaction with diorganodichlorosilanes or diorganoethoxychlorosilanes and forms corresponded tetrafunctional compounds by the following scheme:

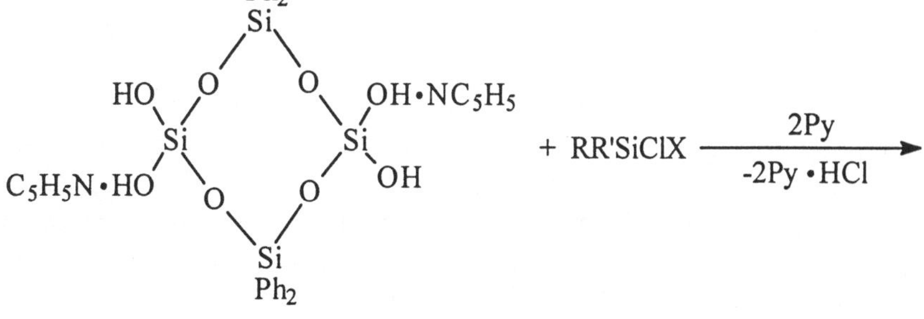

(1.36)

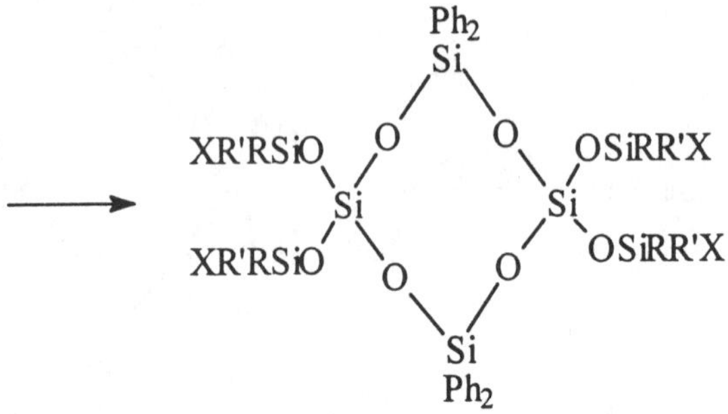

where R = Me, R' = Me, Ph; x = Cl, OEt.

The above-mentioned reaction produces quite pure products and, consequently, their production does not require repeated fractionation, which is necessary for the case of HFC of tetrahydroxytetraphenylcyclotetrasiloxane with diorganodichloro- or diorganoethoxychlorosilanes [99].

Hydrolysis of tetrachlorine-containing branched cyclic compounds with methyl and phenyl substituting agents at silicon atoms produced corresponded tetrahydroxy-derivatives [97]:

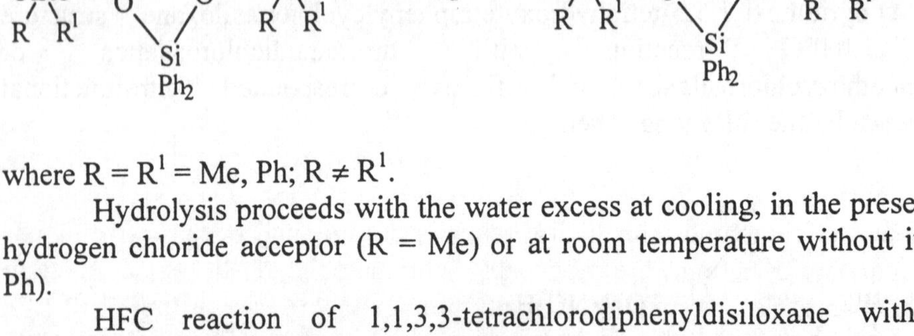

(1.37)

where $R = R^1$ = Me, Ph; $R \neq R^1$.

Hydrolysis proceeds with the water excess at cooling, in the presence of hydrogen chloride acceptor (R = Me) or at room temperature without it (R = Ph).

HFC reaction of 1,1,3,3-tetrachlorodiphenyldisiloxane with 1,3-dihydroxytetraphenyldisiloxane at 1:1 molar ratio of initial components in the presence of pyridine synthesizes organocyclotetrasiloxane with 1,3-disposition of chlorine atoms at silicon with ~50% yield, as well as a mixture of extended isomeric-structured rings. The reaction mainly proceeds in accordance with the scheme as follows [30, 75]:

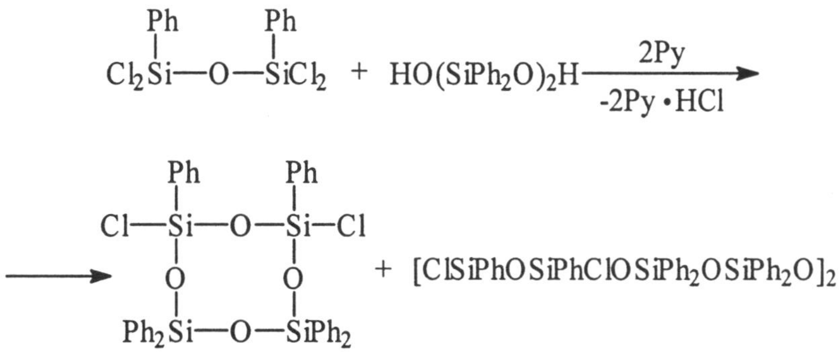

(1.38)

Hydrolysis of 1,3-dichlorohexaphenylcyclotetrasiloxane has synthesized an appropriate dihydroxy-derivative.

HFC reactions of 1,3-dihydroxytetraorganodi- and 1,5-dihydroxyhexaorganotrisiloxanes with silicon tetrachloride in the presence of amine or hydrogen chloride blowing off the reaction mixture have synthesized 1,1-dichlorotetraorganocyclotri- and 1,1-dichlorohexaorganocyclotetrasiloxanes in accordance with the scheme as follows [99, 100]:

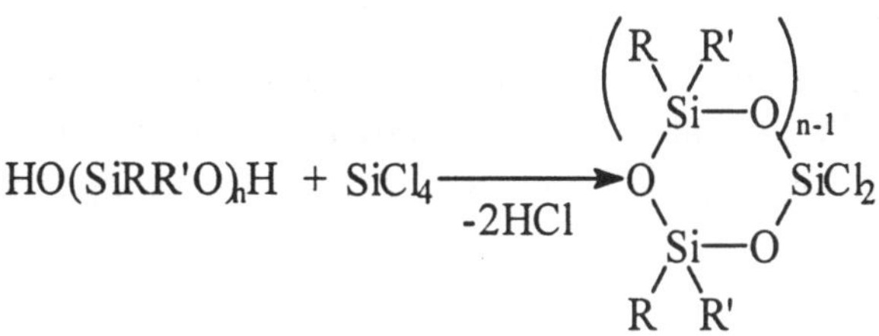

(1.39)

where $n = 1, 2$; R = R' = Me, Ph; R ≠ R'.

The yield of cyclotrisiloxanes is decreased with the size of framing groups at silicon atoms in the initial disiloxane. Application of an acceptor gives better results compared with hydrogen chloride blowing, because in this case the secondary reactions are noticeably eliminated and the yield of target cyclosiloxanes is increased. It has been shown that affinity of 1,3-dihydroxytetraorganodisiloxanes to homocondensation in HFC reactions increases with the decrease of the framing group size at silicon atoms, which results in reduction of the yield of hexatomic organochlorocyclosiloxanes. For example, if at R = R' = Ph the yield of 1,1-dichlorotetraphenylcyclotrisiloxane equals 62%, then at R = Me and R' = Ph the yield of the appropriate cycle does not exceed 10%.

Hydrolysis of synthesized dichlororganocyclosiloxanes produces appropriate dihydroxy-derivatives. However, it has been found that hydrolysis of dichlororganocyclosiloxanes is not limited by formation of dihydroxyorganocyclosiloxanes, because besides the main reaction, intermolecular condensation of the target product with formation of more complex cyclic structures does also proceed [101].

It has been shown [102, 103] that 1,1-dichlorotetraorganocyclotrisiloxanes easily enter reactions of chlorine substitution by alkylamino groups:

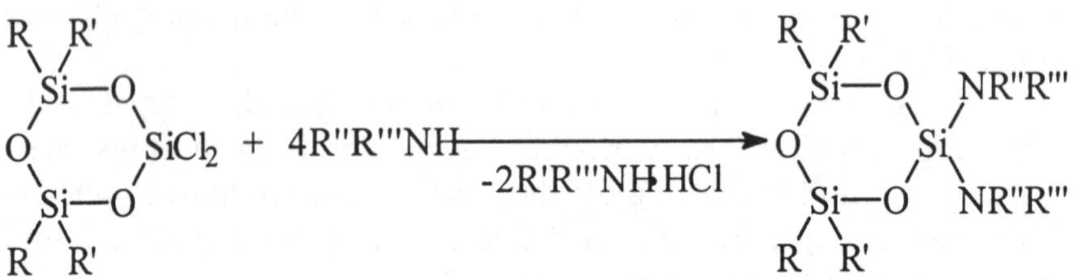

(1.40)

where R = Me, Et; R" = R"' = H, Me, Et; R" ≠ R"'.

The HFC reaction of 1,1-dichloorganocyclosiloxanes with 1,3-dihydroxytetraorganosiloxane at 1:1 ratio of initial components in the presence of pyridine has synthesized spiro cyclosiloxanes of the following structure [104, 105]:

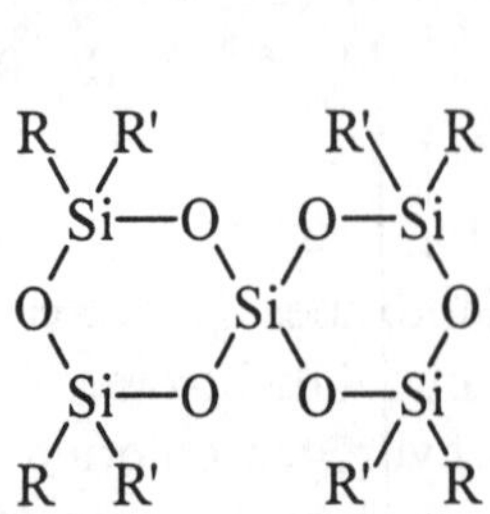

where R = R' = Me, Ph; R ≠ R'.

Further on, synthesized spiro cyclosiloxanes were used for synthesizing cyclolinear polymers.

Some physical and chemical parameters of synthesized organocyclotri(tetra, penta, hexa)siloxanes with different framing and functional groups are shown in Tables 1 – 6.

1.3. Organocyclocarbosiloxanes with functional groups at silicon atoms

For the purpose of synthesizing organocyclocarbosiloxanes with defined disposition of functional groups at silicon atoms, HFC reaction of 1,2-methylchlorosilyl-substituted ethane with dihydroxydiphenylsilane and 1,3-dihydroxytetraphenyldisiloxane in diluted anhydrous toluene solution in the presence of pyridine at 1:1 molar ratio of the initial components was studied [38, 106]. General reaction scheme is as follows:

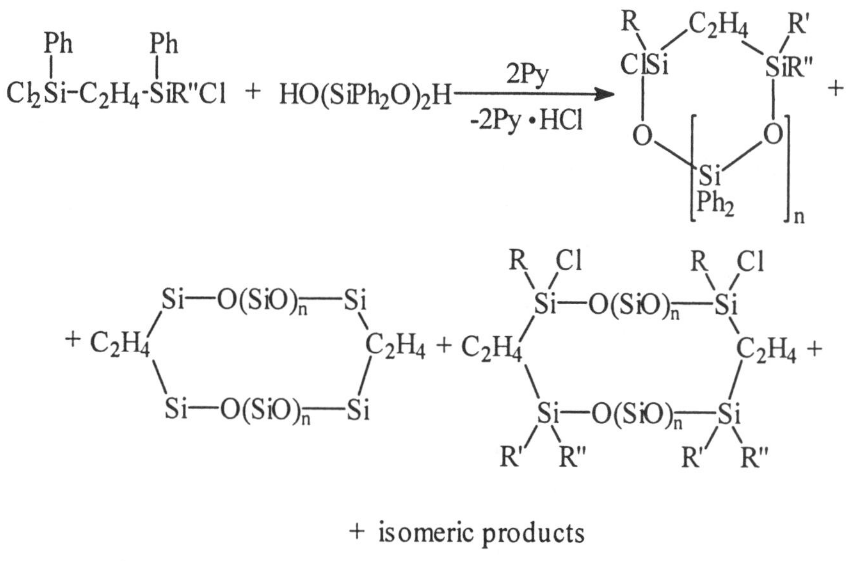

(1.41)

where $n = 1, 2$; R = R' = Me; R = Cl, R' = R" = Me; R = R' = R" = Cl.

The yield of organocyclocarbosiloxanes is seriously dependent on the reaction temperature. It is found that if the process proceeds at 20 - 25°C, total yield of cyclocarbosiloxanes does not exceed 70%, the yield of dichlorocarbosiloxanes being equal 55%. Reduction of the reaction temperature to -5 ÷ -10°C allows increase of total yield of cyclocarbosiloxanes up to 85% and dichlorocarbosiloxanes up to 68%. The yield of tetrachlorocyclocarbosiloxane equals 44%. Hydrolysis of di(tetra)chlorocyclocarbosiloxanes has synthesized corresponded di- and tetrahydroxy-derivatives in accordance with the scheme as follows:

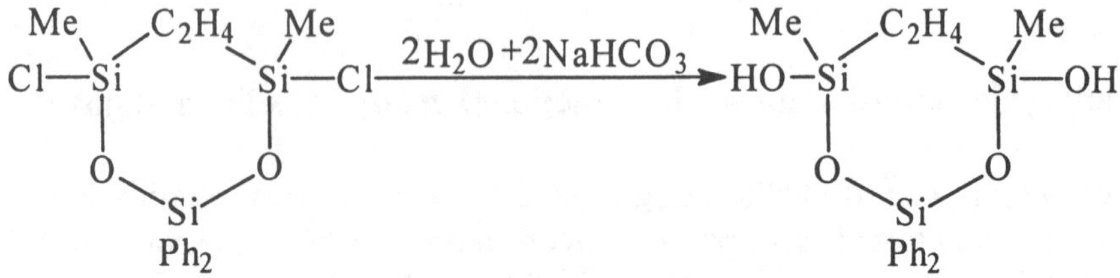

(1.42)

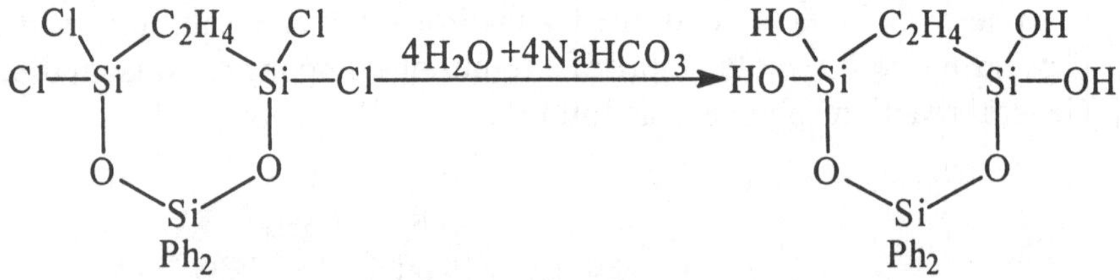

(1.43)

HFC reaction of 1,3-bis(dichloromethylsilyl)ethane and 1,3-bis(dichloromethylsilyl)propane with dihydroxydimethylsilane proceeded in diluted anhydrous ether solution, in the presence of pyridine at -5°C. However, the yields of produced methylcyclocarbosiloxanes are low, giving 10 – 22%. The reaction scheme is as follows [107]:

$$Cl_2MeSi\text{-}(CH_2)_n\text{-}SiMeCl_2 + HOMe_2SiOH \rightarrow$$

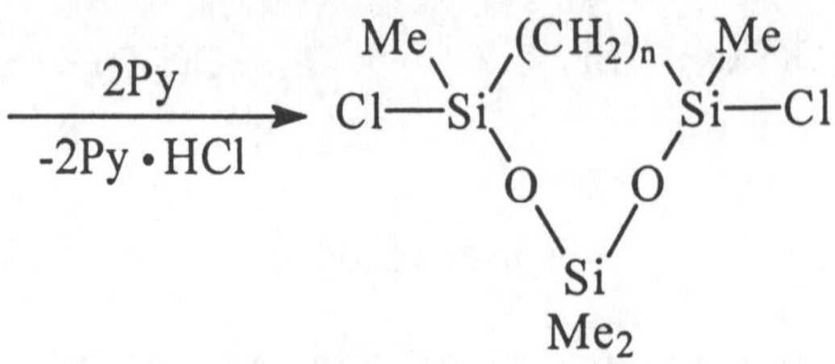

(1.44)

where $n = 2, 3$.

Extended dichloromethylcyclocarbosiloxane was synthesized by condensation of 1,1,7,7-tetrachloro-1,3,3,5,5,7-carbotetrasilocane with 1,3-dihydroxytetramethyldisiloxane under similar conditions in accordance with the scheme as follows [107]:

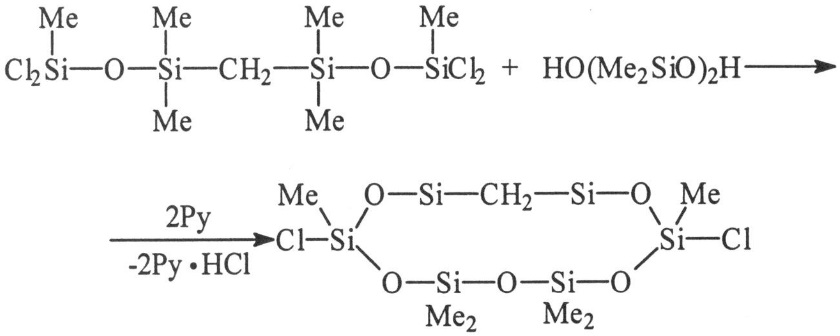

(1.45)

The yield of the above-mentioned dichloromethylcyclocarbosiloxane equals ~40%. Dihydroxy-derivatives were synthesized by hydrolysis of dichloromethylcyclocarbosiloxanes.

Physical and chemical parameters of di- and tetrafunctional organocyclosiloxanes and organocarbocyclosiloxanes are shown in Table 1.7.

Difunctional organosilicon heterocyclic compounds were successfully synthesized by pyrolytic ring formation of symmetrical disiloxanes with phenyl radicals. Pyrolysis in the gas phase in the temperature range of 680 - 690°C proceeds with benzene extraction, which leads to synthesis of representatives of a new class of heterocyclic compounds, which are organochlorine derivatives of 1,3-disila-2-oxaindane. This reaction proceeds by the scheme as follows [108, 109]:

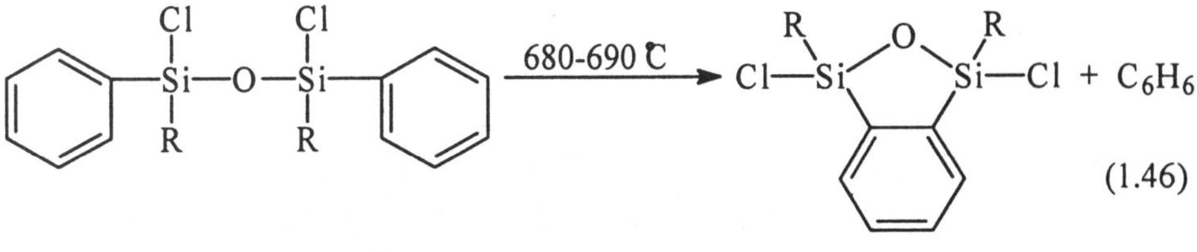

(1.46)

where R = Me, Ph.

Studying hydrolysis of disilaoxaindales, the authors of the work [110] have detected high reactivity of ≡Si-Cl bonds, which is explained by decrease of Si-O-Si valence angle in 1,3-disila-2-oxaindane. Lower reactivity of silaindane fragment with phenyl framing at silicon atoms is explained by both steric and inductive effects of the framing groups.

However, hydrolysis of 1,3-dichloro-1,3-disila-1,3-diphenyl-2-oxaindane under soft conditions in the temperature range of -5 ÷ -10°C in the presence of both sodium bicarbonate and aniline has synthesized not appropriate dihydroxy-derivatives, but oligomeric *α,ω*–dihydroxy-1,3-disila-1,3-1,3diphenyl-oxainadanes. The reaction proceeds in accordance with the scheme as follows [111]:

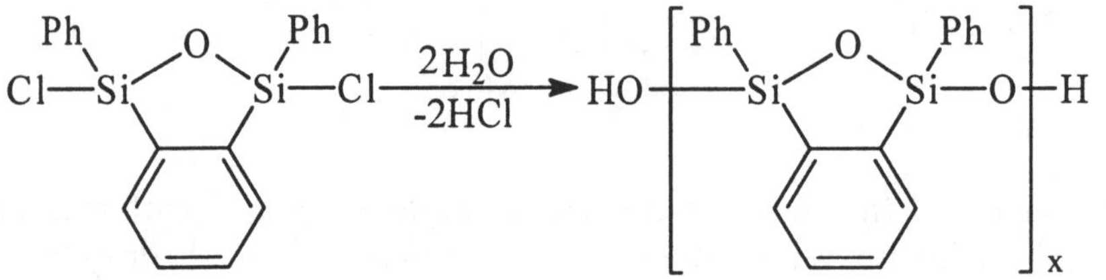

(1.47)

where $x = 4$.

Homocondensation of oligomeric dihydroxy-derivatives, boiled in toluene, has synthesized oligomers with higher transformation degree.

It is shown [112] that the gas-phase pyrolysis of *orhto*-dichlorosiloxydiphenyl easily synthesizes pentatomic silicon-containing heterocycle of 10,10-dichloro-10-sila-9-oxarophenanthrene with the yield of about 60%, which is caused by high reactivity of hydrogen *ortho*-atom. However, insignificant ring formation with simultaneous synthesis of dibenzofuran did also proceed. The general reaction scheme is as follows:

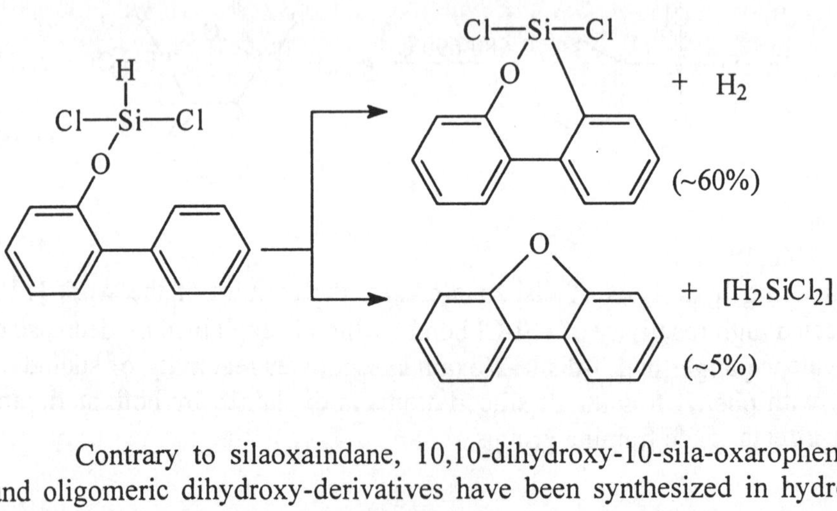

(1.48)

Contrary to silaoxaindane, 10,10-dihydroxy-10-sila-oxarophenanthrene and oligomeric dihydroxy-derivatives have been synthesized in hydrolysis of

10,10-dichloro-10-sila-9-oxarophenanthrene proceeded in the neutral medium in the presence of aniline. The process scheme is as follows [113]:

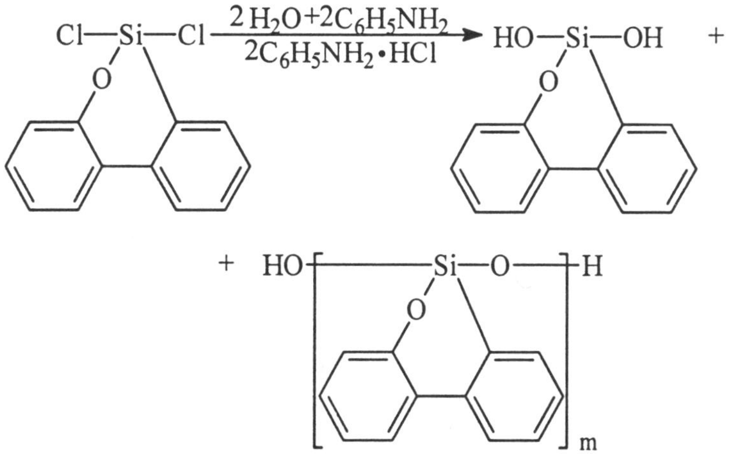

(1.49)

It is found that dichlorosilaoxarophenanthrene is not destroyed during boiling with aqueous or alcoholic alkaline, as well as with concentrated hydrochloric acid. At the same time, it is known well that usual Si-O-C bond is easily degraded by alkali and acids, and Si-C_{ar} bond is extremely sensitive to the action of acids [114]. Quantitatively without destruction of ≡Si-O- bond, dichlorosilaoxarophenanthrene is alkylated and arylated by Grignard reagents, restored by lithium aluminum hydride; when treated by hydrofluoric acid, it is transformed to difluorosilaoxarophenanthrene. Due to the authors' point of view, high chemical stability of silaoxarophenanthrene indicates a definite fragrance of this compound [115]. Some physical and chemical parameters of heterocyclic compounds with functional groups are shown in Table 1.8.

Table 1.1

Physical and chemical parameters of difunctional organophenylcyclotri(tetra, penta, hexa)siloxanes

No.	Compound	Yield, %	T_{boil}, °C (P, mmHg)	T_{melt}, °C	^{1}H (Me) chemical shift		Reference
					cis	*trans*	
1	2	3	4	5	6	7	8
1	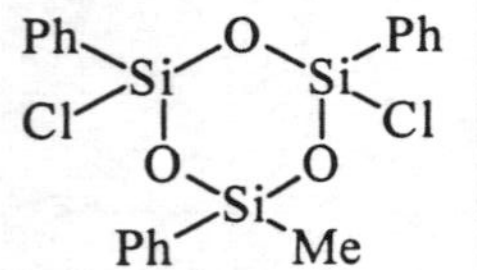	26	136 - 137 (0.1)	–	–	–	[71]
2	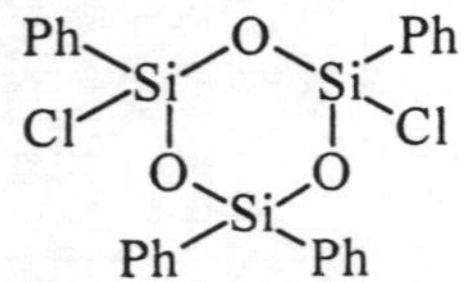	29	200 - 203 (0.001)	–	–	–	[71]
3	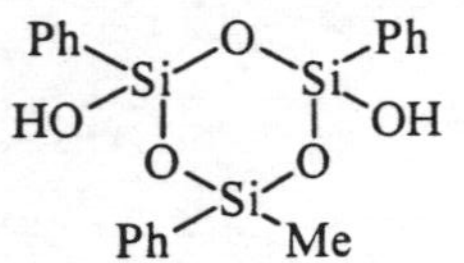	46	–	–	–	–	[71]
4	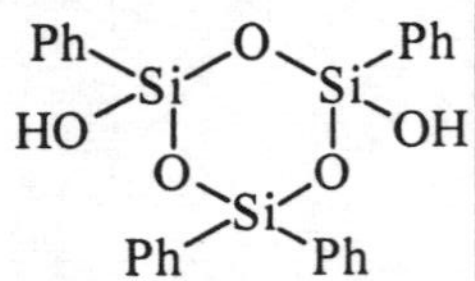	47	–	–	–	–	[71]

5	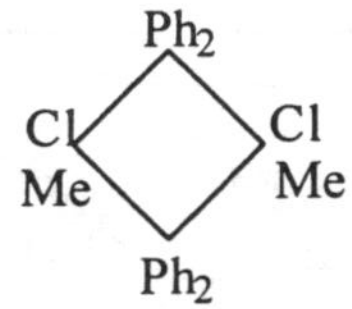	20 (66)*	247 - 250 (0.02)	–	0.38	0.42	[74, 75]
6	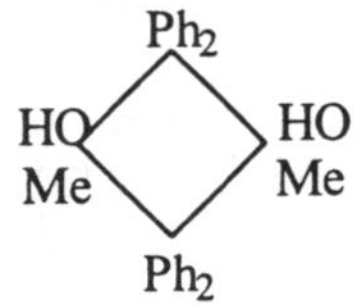	70	–	(*trans*) 141 - 142 (*cis*) 147 - 150	0.16	0.09	[75]
7	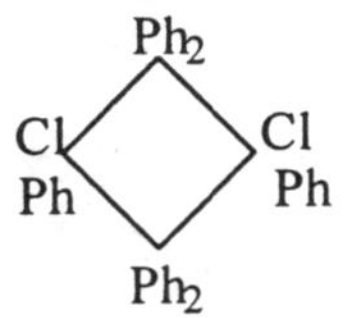	34.0 (65)	305 - 310 (0.02)	–	–	–	[30, 75]
8	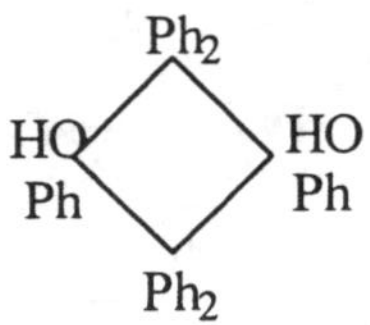	77	–	(mixture) 140 - 141	–	–	[75]
9	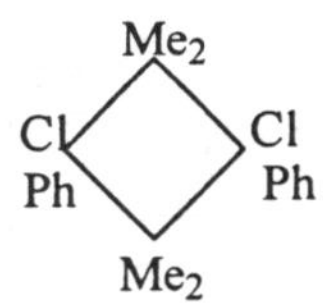	60	147 - 148 (0.5)	*trans* 116 - 117	0.42	0.29 0.16	[80]

10	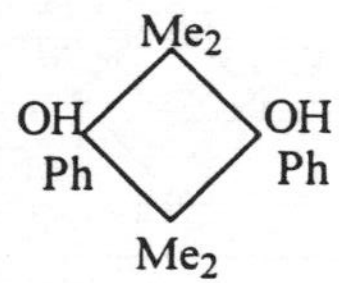	70	–	*trans* 126 - 127	–	0.12	[80]
11	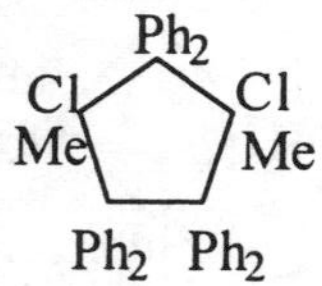	48 (79)	222 - 224 (760)	(mixture)	0.51	0.55	[86, 89]
12	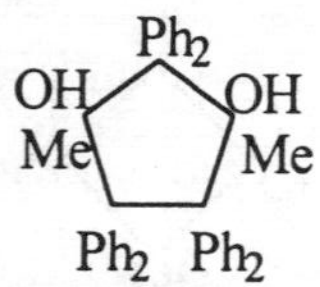	68	–	(mixture) 67 – 70	–	–	[86, 89]
13	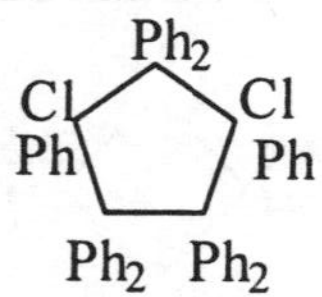	62	310 - 320 (0.04)	–	–	–	[86, 89]
14	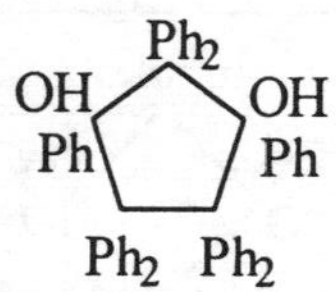	73	–	(mixture) 84 – 85	–	–	[86, 89]

15	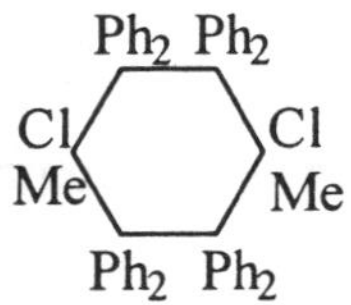	43 (61)	325 - 330 (0.01)	(mixture) 59 – 61	0.38	0.41	[85, 86, 88]
16	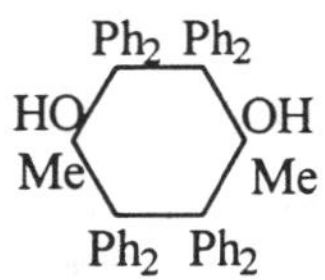	74	–	(mixture) 209 - 210	–	–	[85, 86, 88]
17	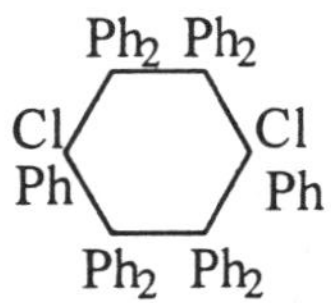	29 (65)	337 - 340 (0.01)	(mixture) 62 – 64	–	–	[85, 86, 88]
18	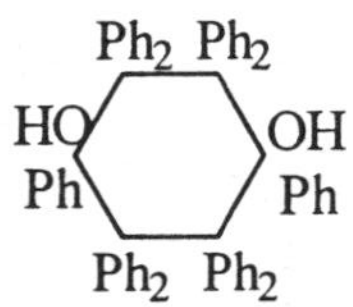	77	–	(mixture) 161 - 162	–	–	[85, 86, 88]
19	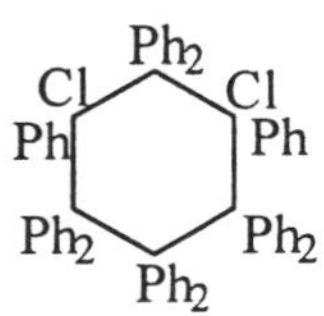	62 (65)	346 - 352 (0.01)	(mixture) 96 – 97	–	–	[36, 37, 86]

20	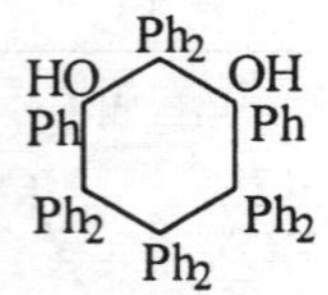	73	–	(mixture) 142 - 143	–	–	[36, 37, 86]
21	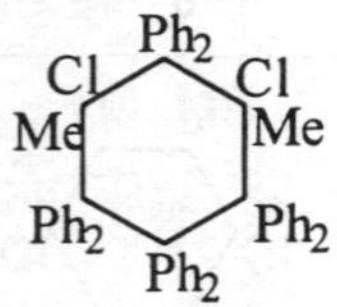	60 (65)	312 - 315 (0.01)	–	0.34	0.39	[36, 37, 86]
22	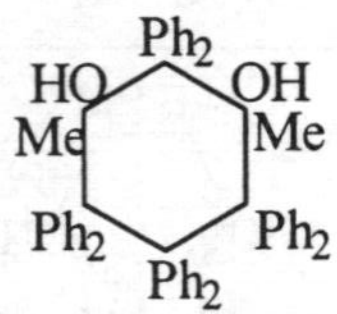	68	–	(mixture) 101 - 102	–	–	[36, 37, 86]

Note: * Shown in brackets are yields obtained for the two-stage method.

Hereinafter, cyclotetra-, cyclopenta- and cyclohexasiloxane fragments are denoted as follows:

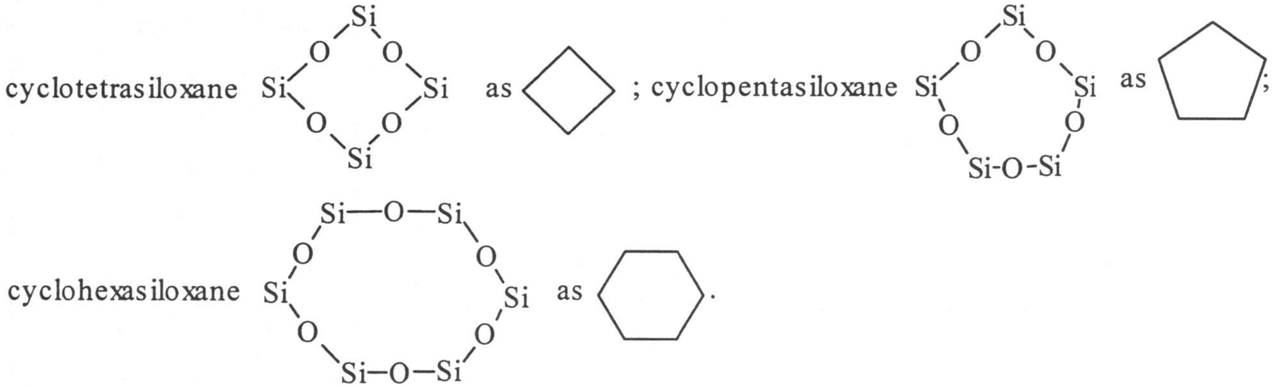

Table 1.2

Physical and chemical parameters of difunctional methylcyclosiloxanes

No.	Compound	Yield, %	T_{boil}, °C (*P*, mm Hg)	T_{melt}, °C	^{1}H (CH_3) chemical shift			Reference
					Number of Si atoms	*cis*	*trans*	
1	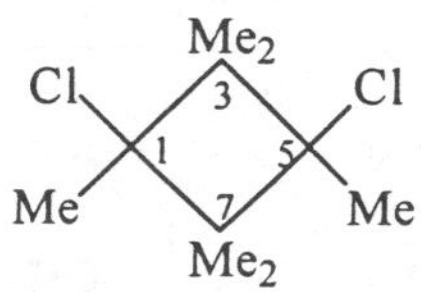	50	201 - 203 (760)	–	1, 5 3, 7[a]ь	0.461 0.170 0.296	0.449 0.230	[38]

2	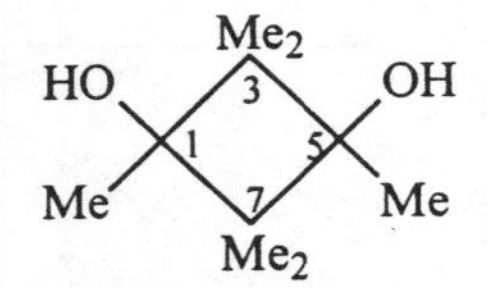	89	115 - 120 (4 - 5)	*trans* 119 – 121	1, 5 3, 7[a]$_{b}$	0.065 0.084 0.119	0.067 0.099	[38]
3	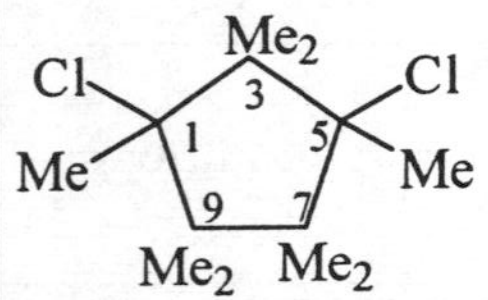	48	222 - 224 (760)	-	1, 5, 3[a] 3_b 7, 9[a] 7, 9_b	0.428 0.161 0.279 0.124 0.199	0.433 0.213 0.140 0,175	[38]
4	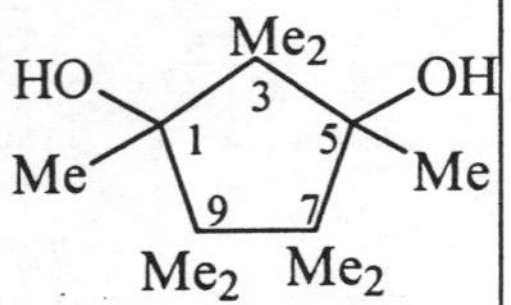	76	89 - 91 (3 - 4)	-	1, 5 3[a] 3_b 7, 9[a] 7, 9_b	0.065 0.091 0.127 0.082 0.110	0.072 0.100 0.100 0.084 0.108	[38]
5	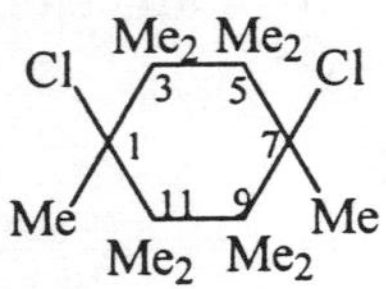	24	55 – 57 (1 - 2)	-6 ÷ -8	1, 7 3, 5[a] 9, 11_b	0.422 0.129 0.192	0.428 0.140 0.174	[38]
6	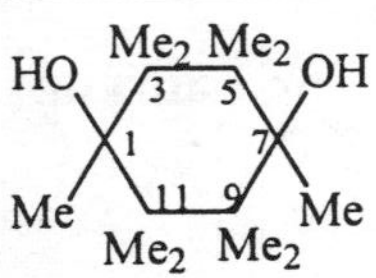	93	–	*trans* 96 – 99 *cis* 36 - 38	1, 7 3, 5, 9, 11[a] 3, 5, 9, 11_b	0.065 0.096 0.111	0.083 0.090 0.120	[38]

7	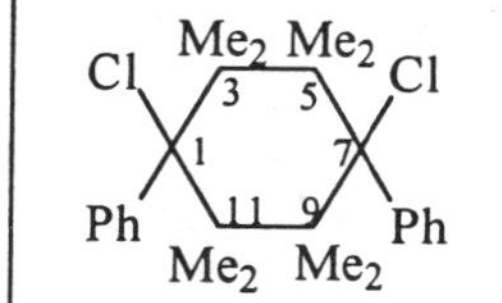	*cis* 13 *trans* 15	175 - 177 (1 – 1.5)	– *trans* 90 - 92	3, 5	0.267 0.069	0.192 0.134	[80]
8	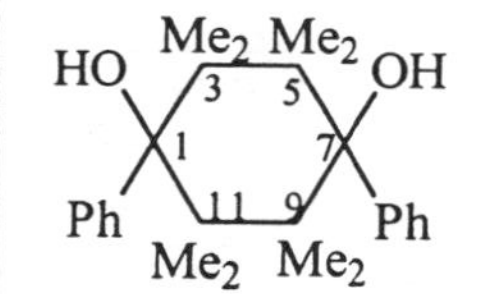	*cis* 89 *trans* 90	- -	*cis* 87 - 88 *trans* 72 - 73	3, 5	0.078 0.115	0.110 0.130	[80]
9	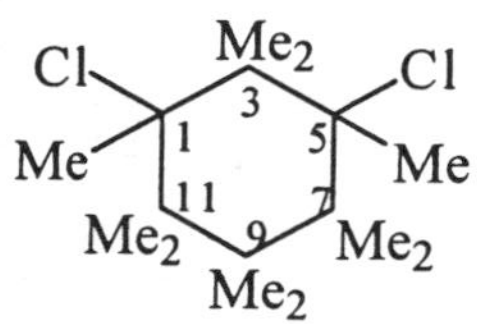	25	82 – 84 (1 - 2)	-	3a 3b 1, 5 9a 9b 7, 11a 7, 11b	0.170 0.274 0.426 0.085 0.113 0.123 0.181	 0.214 0.436 0.098 0.130 0.170	[38]
10	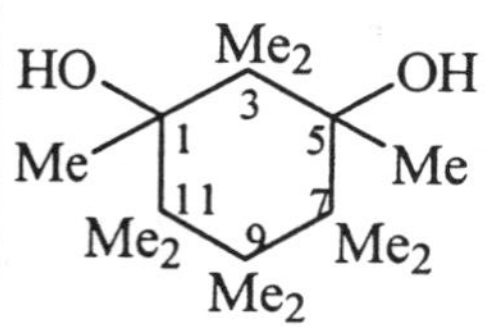	74	102 - 104 (0.03)	-	3a 3b 1, 5 7, 11a 7, 11b 9a 9b	0.106 0.136 0.074 0.089 0.125 below the reflex 0.095	 0.110 0.086 0.091 0.119 0.100	[38]

Note: a and b indicate detected signals of nonequivalent methyl groups at silicon atoms.

Table 1.3

Physical and chemical parameters of difunctional ethylcyclosiloxanes

No.	Compound	Yield, %	T_{boil}, °C (mm Hg) T_{melt}, °C	n_D^{20}	M^+-C_2H_5 molecular mass	^{29}Si NMR spectrum (δ, ppm) Si[1]	Si[2]	Si[3]	Reference
1	2	3	4	5	6	7	8	9	10
1	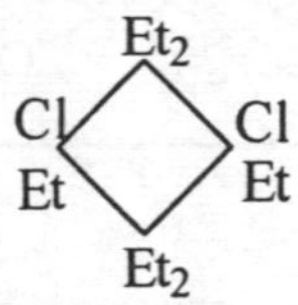	63	113 - 115 (1 – 1.5)	1.4355	391 / 421.6	*cis* -43.14 *trans* -42.87	*cis* -15 45 *trans* -15.03	- -	[39]
2	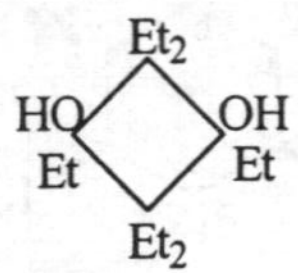	80	32	–	- / 384.7	*cis* 54.44 *trans* -55.53	*cis* -17.70 *trans* -18.60	- -	[39]
3	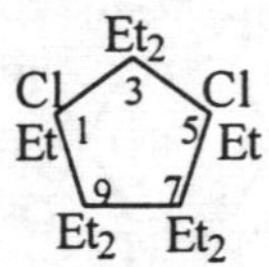	39	128 - 130 (1 - 2)	1.4385	493 / 523.8	*cis* -45.78 *trans* -45.68	*cis* -17.71 *trans* -17.51	*cis* -19.25 *trans* -19.33	[39]
4	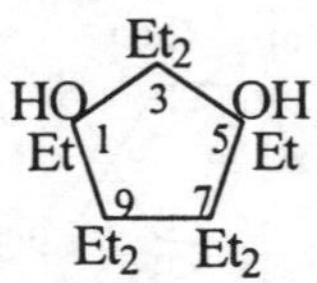	66	120 - 122 (0.02)	1.4400	- / 486.9	*cis* -56.46 *trans* -55.34	*cis* -18.42 *trans* -20.11	*cis* -21.08 *trans* -20.59	[39]

5	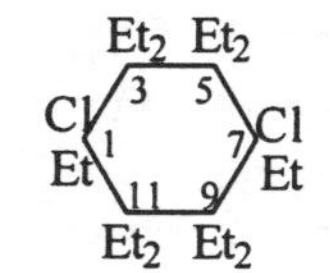	39	170 - 172 (1 - 2)	1.4425	595/626	*cis* -46.85 *trans* -46.90	*cis* -19.72 *trans* -19.81	- -	[39]
6	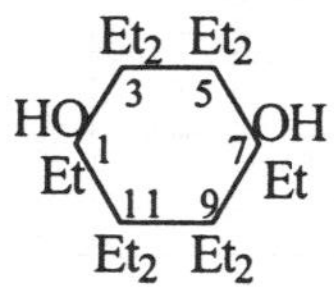	46	160 - 163 (0.02)	1.4420	-/589.1	-	-	-	[39]

Table 1.4

Physical and chemical parameters of vinyl-containing organocyclosiloxanes

No.	Structural formula	Yield, %	T_{boil}, °C P, (mm Hg) T_{melt}, °C	n_D^{20}	d_4^{20}	MR_D**	^{1}H, ppm (R) RXSi(O–)(O–)	References
1	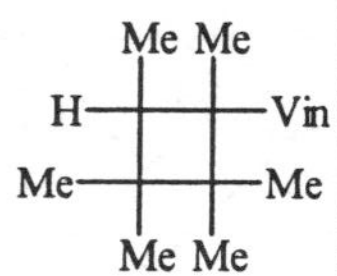	63	58 - 0 (8)	1.4017	0.9703	74.35/32.65	–	[63]
2	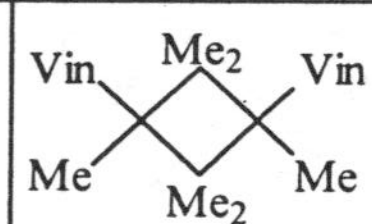	35	70 - 71 (11)	1.4166	0.9704	83.01/82.88	*cis* - 0.146 *trans* - 0.133 0.106 (Me_2)	[72]

3	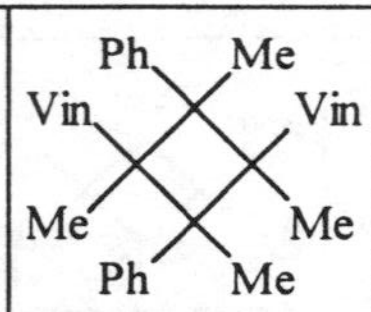	29	150 - 153 (1-2)	1.5000	1.0779	121.36 122.52	–	[72]
4	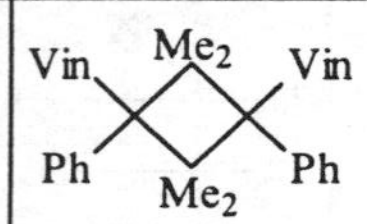	35	137 - 138 (2)	1.5024	1.0700	122.75 122.52	–	[72]
5	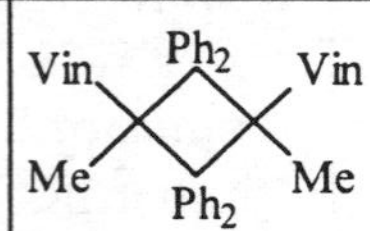	50	242 - 243 (1 - 2) 88-89	–	–	–	–	[72]
6	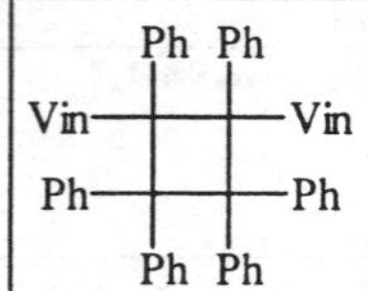	75	255 - 265 (2) 80 - 81	–	–	–	–	[73]
7	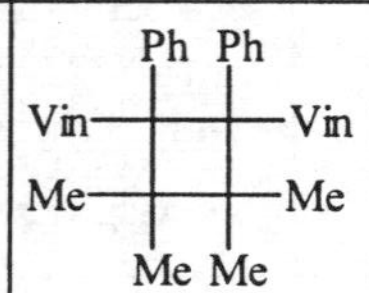	45	130 - 131 (2)	1.5017	1.0670	122.95 122.52	*cis* - 0.15	[73]
8	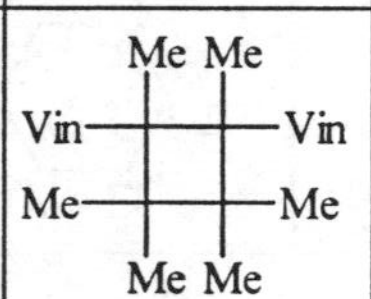	66	87 - 88 (10)	1.4171	0.9719	82.96 82.88	*cis* - 0.156 *trans* - 0.141 0.09*(Me_2)	[73]

9	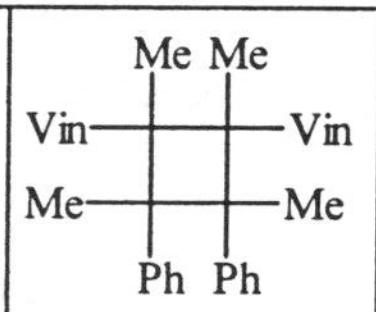	29	150 - 153 (1 - 2)	1.5000	1.0970	–	–	[72]
10	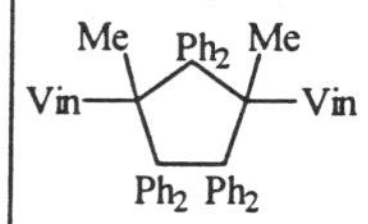	50	255 - 260 (0,02) 152 - 153	–	–	–	*cis* 0.20 *trans* 0.22	[32, 89, 91]
11	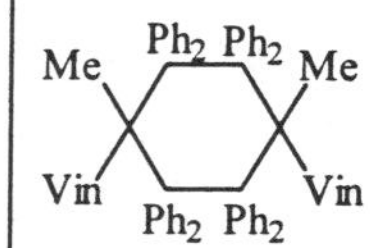	23 (62)	300-305 (0.02) 62-65	–	–	–	*cis* 0.20 *trans* 0.23	[89, 92]
12	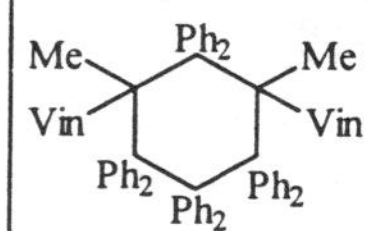	68	290 - 294 (0.02) 152 - 154	–	–	–	*cis* 0.17 *trans* 0.19	[37, 93]
13	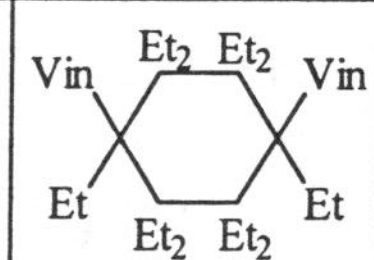	49	180 - 182 (1)	1.4443	–	–	0.45 *q* 0.85 *t*	[83]
14	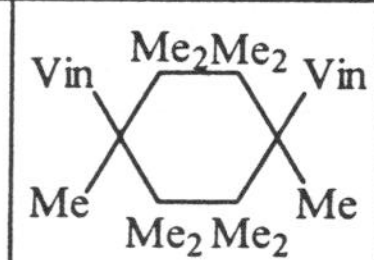	64	83 - 85 (1 - 2)	1.4128	–	–	*cis* 0.157 *trans* 0.150	[83]

15	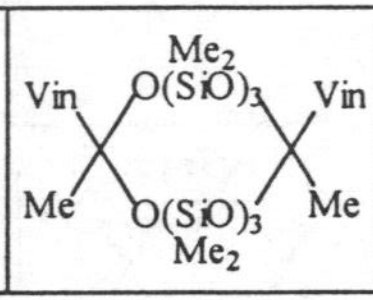	31	119 - 122 (2)	1.4139	–	–	*cis* 0.23 *trans* 0.25	[83]

Table 1.5

Physical and chemical parameters of hydride-containing organocyclosiloxanes

No.	Structural formula	Yield, %	T_{boil}, °C (*P*, mmHg) T_{melt}, °C	n_D^{20}	d_4^{20}	MR_D**	^{1}H, ppm	References
1	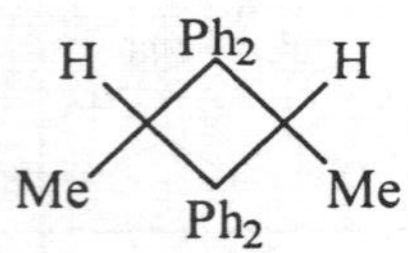	79	245 - 250 (3 - 4) 87 - 88	–	–	–	(mixture)	[72]
2	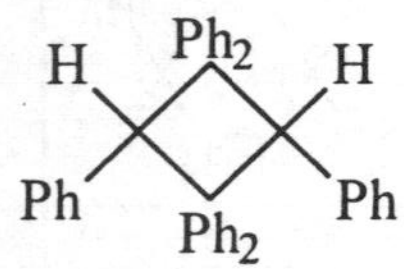	35	270 -280 (3) 60 - 61	–	–	–	(mixture)	[73]
3	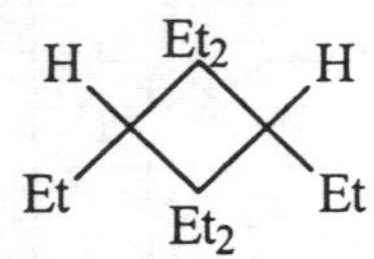	34	95 - 98 (2)	1.4230	–	–	(mixture) 0.62 *q* 1.07 *t*	[83]
4	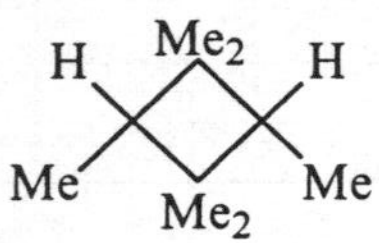	21	57 - 58 (18) -40	1.3930	0.9680	<u>65.88</u> 65.82	–	[12]

5	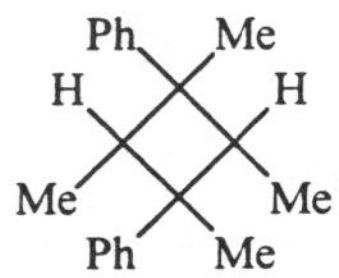	27	140 - 145	1.4914	1.0907	–	(mixture)	[72]
6	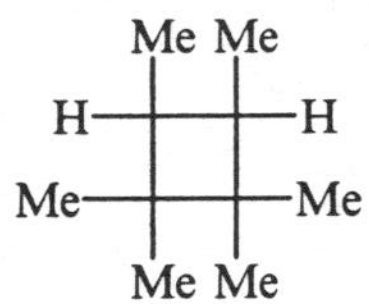	18	154 - 155	1.3927	0.9709	–	(mixture)	[116]
7	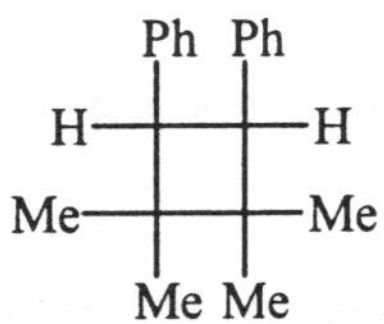	14	137 - 139 (1 - 2)	1.4876	1.0698	–	(mixture)	[117]
8	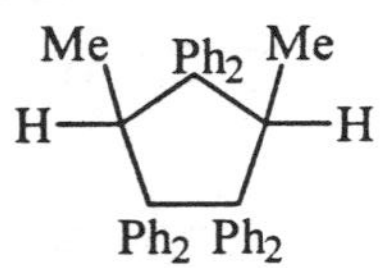	67	235 - 242 (0.02)	–	–	–	(mixture) *cis* – 0.25 *trans* – 0.28	[85,88]
9	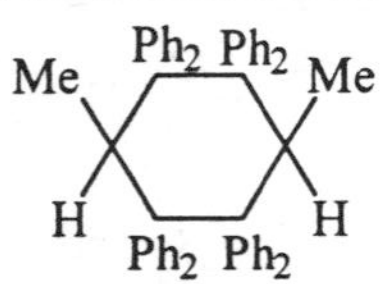	23 (67)	263 - 267 (0.01) 162 - 163	–	–	–	(mixture) *cis* – 0.26 *trans* – 0.30	[84, 88]

10	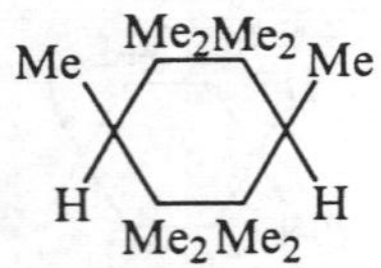	59	66 - 68 (1)	1.3971	–	–	(mixture) *cis* – 0.14 *trans* – 0.15	[82]
11	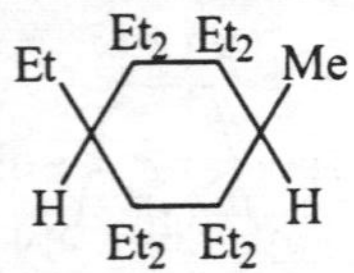	53	175 - 178 (1)	1.4341	–	–	(mixture) 0.6 *q* 1.0 *t*	[82]

Note: * Numerator shows calculated values and denominator – recorded ones.

Table 1.6

Physical and chemical parameters of tetrafunctional organophenylcyclosiloxanes

No.	Compound	Yield, %	T_{boil}, °C (P, mm Hg)	T_{melt}, °C	References
1	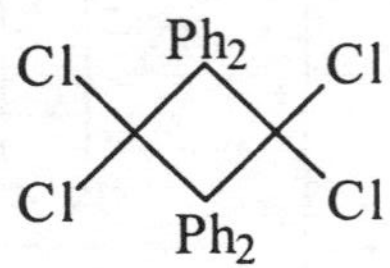	53	206 - 210 (0.002)	–	[32, 33, 94]
2	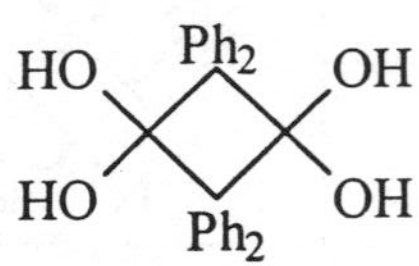	64	–	157 - 158	[32, 94]

3	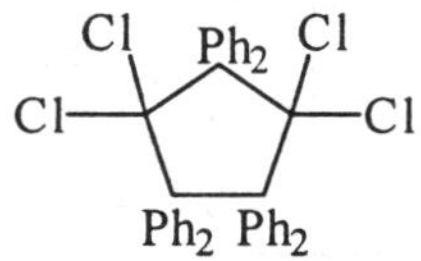	52 (70)	255 - 260 (0.02)	–	[32, 86]
4	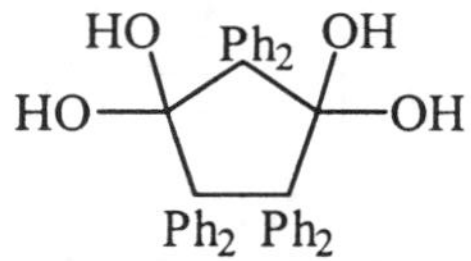	73	–	191 - 192	[32, 86]
5	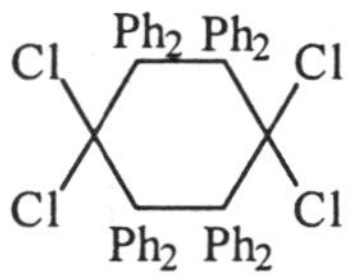	26	325 - 330 (0.06)	83 - 87	[85, 86, 90]
6	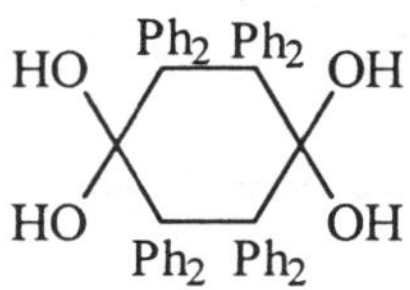	75	–	251 - 252	[86, 90]
7	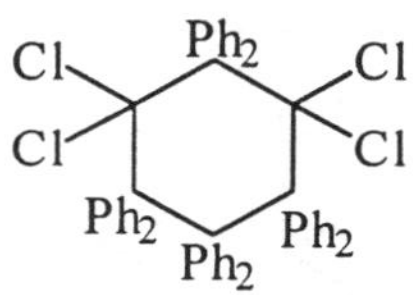	67	307 - 311 (0.01)	71 - 73	[37, 86]

8	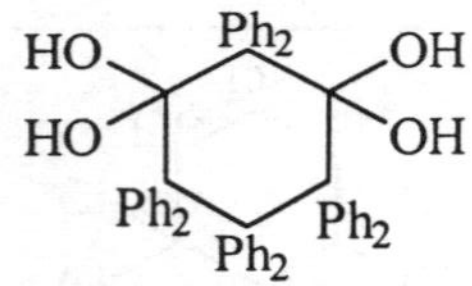	64	–	161 - 162	[37, 86]

Table 1.7

Physical and chemical parameters of difunctional organocyclocarbosiloxanes

No.	Compound	Yield, %	T_{boil}, °C (mm Hg)	T_{melt}, °C	^1H (CH_3) chemical shift *cis* *trans*	References
1	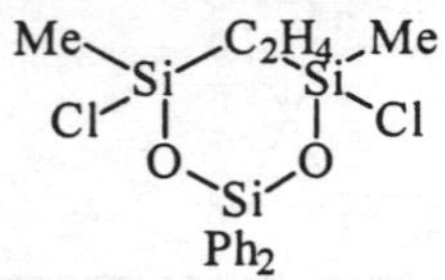	67.9	187 – 193 (0.01)	77 – 78	(mixture) *cis* - 0.47 *trans* - 0.52	[37, 83]
2	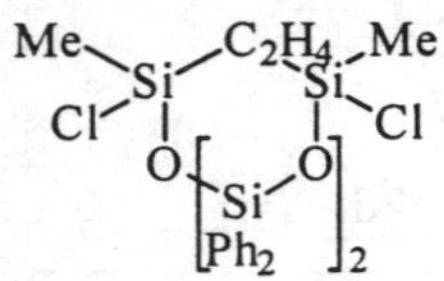	59.4	218 – 221 (0.02)	110 – 111	(mixture) *cis* - 0.41 *trans* - 0.49	[37, 83]
3	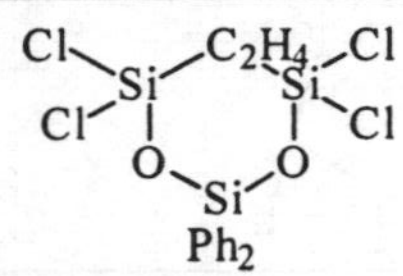	56	150 – 155 (0.01)	159 – 161	–	[83]

4	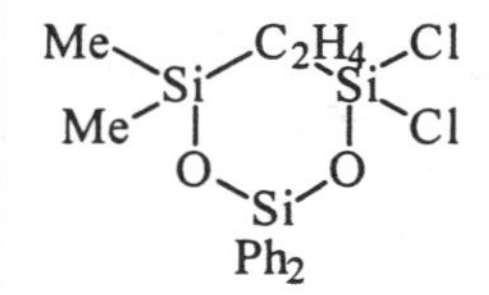	68	149 – 520 (0.01)	-	0.45 *s*	[37, 83]
5	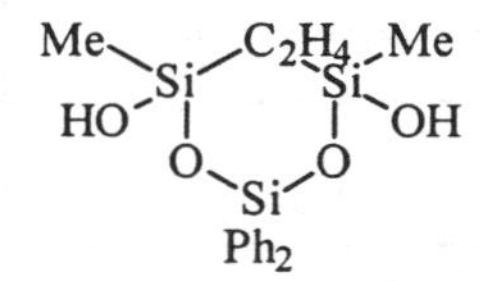	71	-	120 – 122		[37, 83]
6	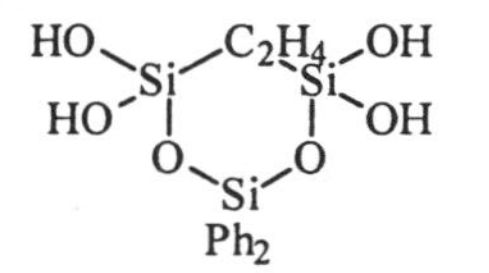	69	–	169 – 170	–	[37, 83]
7	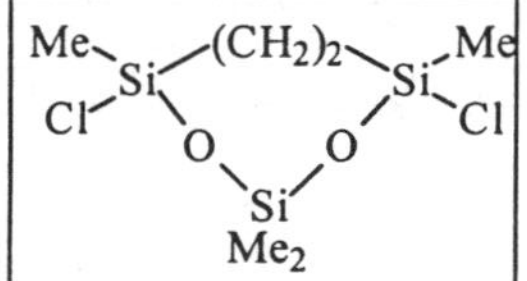	22	61 – 65 (5)	–	*cis* 0.88 *trans* 1.23	
8	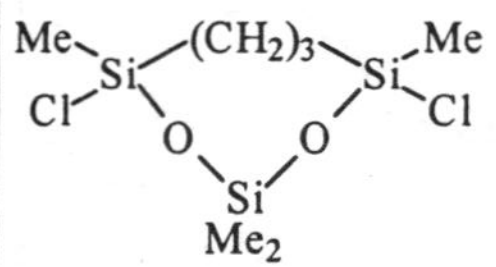	10	66 – 73 (4 - 5)	–	*cis* 0.38 *trans* 0.40	[106]
9	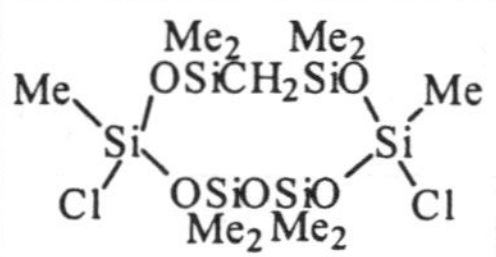	40	99 – 110	–	MeSiCl* 45.50	[106]

10	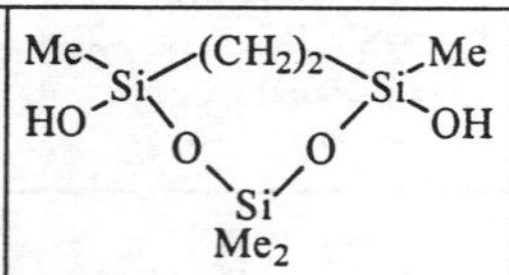	71	80 – 100 (4 - 5)		0.05	[106]
11	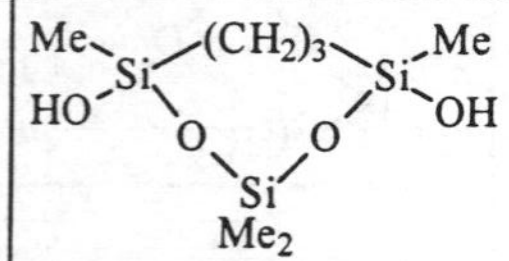	40	130 – 134		0.04	[106]
12	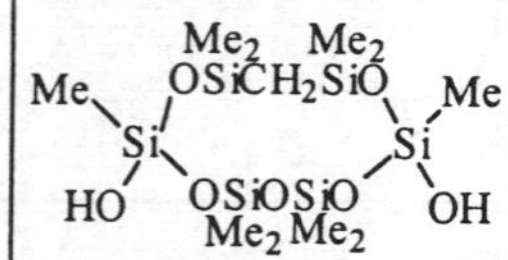	71	–	–	MeSi(OH)* 63.02	[106]

Note: * ^{29}Si spectrum.

Table 1.8

Physical and chemical parameters of heterocyclic organosiloxanes

No.	Compound	Yield, %	T_{boil}, °C (P, mm Hg)	d_4^{20}	n_D^{20}	References
1	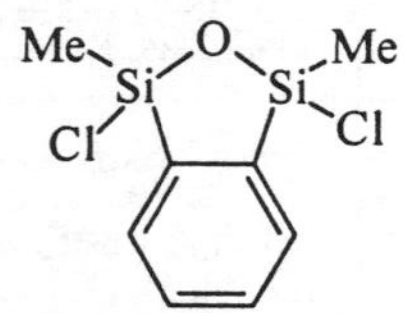	16	86 – 88 (4)	1.2220	1.5152	[108, 109]

2	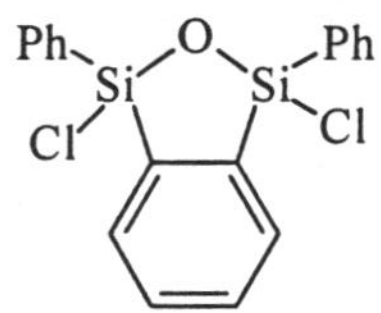	20	220 – 222 (3)	1.2521	1.6071	[108, 109]
3	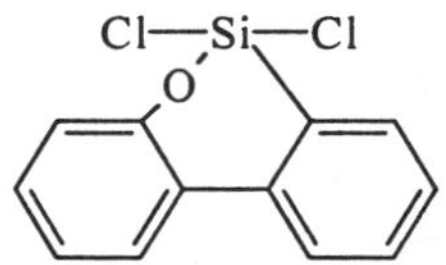	56	160 – 161 (1) 82 – 83	-	-	[112]
4	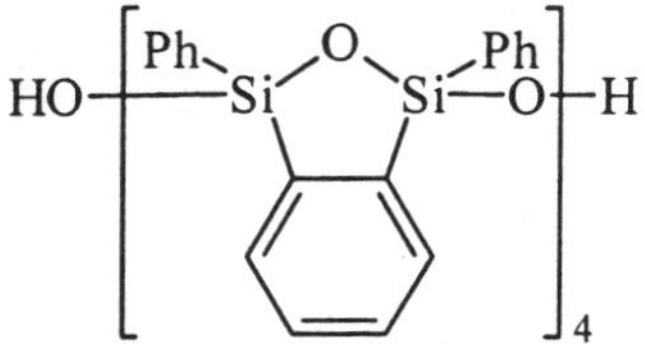	97	48 – 54	–	–	[110, 111]
6	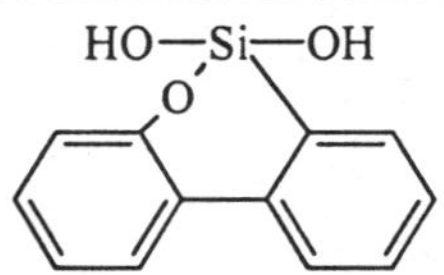	86	236 – 239	–	–	[113]
7	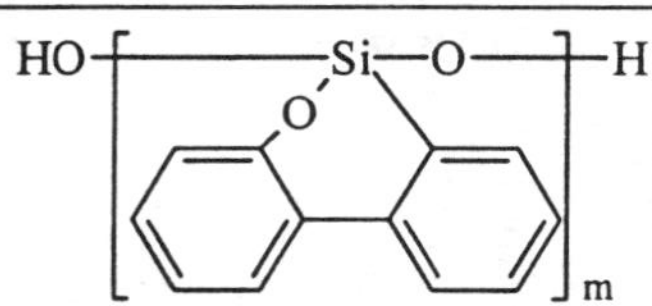	13	120 – 126	–	–	[113]

References

1. Andrianov K.A., *Organosilicon Compounds*, Moscow, Goskhimizdat, 1955. (Rus)
2. Andrianov K.A. and Khanashvili L.M., *Technology of Organic Monomers and Polymers*, Moscow, Khimia, 1998. (Rus)
3. Troenko V.I., *Candidate's Dissertation on Chemistry*, INEOS AS USSR, Moscow, 1971. (Rus)
4. Andrianov K.A., Emelianov V.N., and Troenko V.I., *Doklady AN SSSR*, 1969, vol. **189**(1), p. 91. (Rus)
5. Andrianov K.A. and Sobolevski M.V., *High-molecular Organosilicon Compounds*, Moscow, Oborongiz, 1949, p. 175. (Rus)
6. Andrianov K.A. and Pavlova S.A., *Doklady AN SSSR*, 1953, vol. **88**(5), p. 811. (Rus)
7. *Patent No. 1,444 (57)* (Japan), 1957; Baba H. and Wiskizakis C.A., *Alkylhalopolysiloxanes*, 1995, vol. **52**, p. 4678.
8. Frainnet E., Calas R., and Fritsch C., *Bull. Soc. Chem.*, 1960, p.1480.
9. Andrianov K.A., Khanashvili L.M., Khan En-tsze, and Tikhonov V.S., *Zh. Obshch. Khim.*, 1962, vol. **32**(12), p. 3951. (Rus)
10. *A.C. No. 696,034* (USSR), Karpov G.I., Sviridova N.G., Yuzhelevski Yu.A., Sergeeva E.P., Mileshkevich V.P., Lebedev E.P., and Sizkova V.A., 1977. (Rus)
11. Yuzhelevski Yu.A., Sviridova N.G., Karpov G.I., and Markasova A.V., *Zh. Obshch. Khim.*, 1981, vol. **51**(6), p. 1342. (Rus)
12. Sokolov N.N., *Zh. Obshch. Khim.*, 1959, vol. **29**(1), p. 258. (Rus)
13. Andrianov K.A., Emelianov V.N., and Troenko V.I., *Doklady AN SSSR*, 1969, vol. **189**(1), p. 91. (Rus)
14. Patnode W. and Wilcock D.F., *J. Amer. Chem. Soc.*, 1946, vol. **68**, pp. 358 - 363.
15. Elliot R.L. and Breed L.W., *Inorg. Chem.*, 1965, vol. **4**, p. 1455.
16. Andrianov K.A., Emelianov V.N., and Troenko V.I., *Doklady AN SSSR*, 1969, vol. **185**(2), p. 327. (Rus)
17. *Patent No. 2,421,653* (USA), 1947; Sommer R.O., *C.A.*, 1947, vol. **41**, p. 5544.
18. Andrianov K.A. and Severni V.V., *Doklady AN SSSR*, 1960, vol. **134**(6), p. 1347. (Rus)
19. Andrianov K.A., Severni V.V., and Zavin B.G., *Izv. AN SSSR, Ser. Khim.*, 1961, No. 7, p. 1456. (Rus)

20. Andrianov K.A. and Severni V.V., *Izv. AN SSSR, Ser. Khim.*, 1961, No. 10, p. 1788. (Rus)
21. Andrianov K.A. and Severni V.V., *Izv. AN SSSR, Ser. Khim.*, 1969, No. 7, p. 1237. (Rus)
22. Andrianov K.A. and Severni V.V., *Izv. AN SSSR, Ser. Khim.*, 1962, No. 12, p. 2133. (Rus)
23. Andrianov K.A., Severni V.V., and Izmailov B.A., *Izv. AN SSSR, Ser. Khim.*, 1963, No. 2, p. 282. (Rus)
24. Severni V.V. and Andrianov K.A., *J. Organometal. Chem.*, 1964, vol. **1**, p. 340.
25. Andrianov K.A. and Severni V.V., *Izv. AN SSSR, Ser. Khim.*, 1963, No. 1, p. 82. (Rus)
26. Andrianov K.A. and Severni V.V., *Izv. AN SSSR, Ser. Khim.*, 1964, No. 7, p. 1268. (Rus)
27. Andrianov K.A. and Severni V.V., *Doklady AN SSSR*, 1962, vol. **146**(3), p. 601. (Rus)
28. Andrianov K.A and Severni V.V., *J. Organometal. Chem.*, 1964, vol. **1**, p. 268.
29. Andrianov K.A. and Severni V.V., *Izv. AN SSSR, Ser. Khim.*, 1964, No. 7, p. 1271. (Rus)
30. Andrianov K.A., Nogaideli A.I., Slonimski G.L., Levin Yu.V., Makarova N.N., Kvachev Yu.P., and Mukbaniani O.V., *Vysokomol. Soedin.*, 1976, vol. **18B**, p. 359. (Rus)
31. Koyava N.A., Mukbaniani O.V., Khanashvili L.M., and Kutateladze G.Sh., *Soobshch. AN GSSR*, 1980, vol. **97**(3), p. 617. (Rus)
32. Koyava N.A., *Candidate's Dissertation on Chemistry*, I. Gzhavakhishvili Tbilisi State University, 1982, Tbilisi, Georgia. (Rus)
33. Andrianov K.A., Emelianov V.N., Muzafarov A.M., and Tot A., *Doklady AN SSSR*, 1976, vol. **226**, p. 827. (Rus)
34. Lebedev E.P., Lukashenko S.S., Baburina V.A., Saratov I.B., and Reikhsfeld V.O., *Zh. Obshch. Khim.*, 1978, vol. **48**(8), p. 1757. (Rus)
35. Andrianov K.A., Zachernyuk A.B., and Danielov S.A., *Khimia Geterotsikl. Soed.*, 1972, No. 7, p. 893. (Rus)
36. Achelashvili V.A., Mukbaniani O.V., Khanashvili L.M., and Kikolaldze V.S., *Zh. Obshch. Khim.*, 1986, vol. **56**(7), p. 1530. (Rus)
37. Achelashvili V.A., *Candidate's Dissertation on Chemistry*, I. Dzhavakhishvili Tbilisi State University, 1982, Tbilisi, Georgia. (Rus)
38. Makarova N.N., Petrova I.M., Godovski Yu.K., Lavrukhin B.D., and Zhdanov A.A., *Doklady AN SSSR*, 1983, vol. **269**(6), pp. 1368 – 1372. (Rus)

39. Makarova N.N., Lavrukhin B.D., Turkeltaub G.N., Kuzmin N.N., Petrova I.M., and Matukhina E.V., *Izv. AN SSSR, Ser. Khim.*, 1989, No. 6, p. 1351. (Rus)
40. Hyde J.F., *J. Amer. Chem. Soc.*, 1953, vol. **75**, p. 2167.
41. Sobolevski M.V., Chistyakova L.A., Nazarova D.V., and Kurilina, *Plast. Massy*, 1962, No. 10, p. 17. (Rus)
42. Suvorov A.L. and Spasski S.S., *Trudy Inst. Khimii AN SSSR*, Urals filial, 1966, No. 13, p. 25. (Rus)
43. *Patent No. 1,329,180* (UK), 1974.
44. Muzafarov A.M., *Candidate's Dissertation on Chemistry*, INEOS, Moscow, 1981. (Rus)
45. Sokolov N.N., In Coll.: *Methods of Polyorganosiloxanes Synthesis*, Moscow-Leningrad, VEI, 1959, 147 p. (Rus)
46. Sokolov N.N. and Andrianov K.A., *Izv. AN SSSR*, 1957, No. 7, p. 806. (Rus)
47. Vdovin V.M. and Petrov A.D., *Zh. Obshch. Khim.*, 1960, vol. **30**(3), p. 838. (Rus)
48. Petrashko A.I., Andrianov K.A., Korneeva G.P., Kuptsova Z.M., Moiseenko A.P., and Asnovich L.Z., *Vysokomol. Soedin.*, 1967, vol. **9A**(9), p. 2036. (Rus)
49. *Patent No. 124,589* (Poland), 1985.
50. Sheludyakov V.D., Zhun V.I., and Vlasenko S.D., *Zh. Obshch. Khim.*, 1985, vol. **55**(8), p. 1795. (Rus)
51. Sheludyakov V.D., Zhun V.I., Vlasenko S.D., Bochkarev V.N., and Slyusarenko T.F., *Zh. Obshch. Khim.*, 1985, vol. **55**(7), p. 1544. (Rus)
52. Iler R.K., *Ind. Eng. Chem.*, 1947, vol. **39**, p. 1384.
53. *Patent No. 2,547,944* (USA), 1951; Iler R.K.
54. Okawara R., Hotta S., and Watase T., *Nippon Kagaky Zasshi*, 1954, vol. **75**, p. 759.
55. Takatani T., *J. Chem. Soc. Japan, Pure Chem. Sect.*, 1953, vol. **74**, p. 890.
56. Takatani T., *J. Chem. Soc. Japan, Pure Chem. Sect.*, 1953, vol. **73**, p. 409.
57. Sprung M.M. and Guenter F.O., *J. Amer. Chem. Soc.*, 1955, vol. **77**, p. 4173.
58. Okawara R., Minami G., and Oku Z., *Bull. Chem. Soc. Japan*, 1958, vol. **31**, p. 22.
59. Okawara R., Hotta S., and Watase T., *J. Chem. Soc. Japan*, 1953, vol. **27**, p. 785.
60. Andrianov K.A. and Volkova L.M., *Vysokomol. Soedin.*, 1961, vol. **3A**(10), p. 1580. (Rus)

61. Andrianov K.A., Fromberg M.B., and Arsenieva E.N., *Zh. Prikl. Khim.*, 1970, vol. **43**(9), p. 2053. (Rus)
62. Zhdanov A.A., Andrianov K.A., and Malyikhin A.P., *Doklady AN SSSR*, 1973, vol. **211**(5), pp. 1104. (Rus)
63. Andrianov K.A., Sidorov V.I., Zaitseva M.G., and Khanashvili L.M., *Khimia Geterotsikl. Soedin.*, 1967, No. 1, p. 32. (Rus)
64. Brown J.F., *J. Amer. Chem. Soc.*, 1965, vol. **87**, p. 4317.
65. Klementiev I.Yu., Shklover V.E., Kulish M.A., Tikhonov V.S., and Volkova E.V., *Doklady AN SSSR*, 1981, vol. **259**(6), p. 1371. (Rus)
66. Andrianov K.A., Nogaideli A.I., Tskhovrebashvili V.S., Nakaidze L.I., and Zunkin D.Ya., *Doklady AN SSSR*, 1975, vol. **223**(1), p. 603. (Rus)
67. Andrianov K.A., Pavlova S.A., Tverdokhlebova I.I., Emelianov V.N., Larina T.A., and Rabkina A.Yu., *Vysokomol. Soedin.*, 1972, vol. **14**(10), p. 1580. (Rus)
68. Andrianov K.A., Blokhina O.G., and Volkova L.M., *Vysokomol. Soedin.*, 1976, vol. **46**(7), p. 1537. (Rus)
69. Andrianov K.A., Volkova L.M., and Tartakovskaya L.M., *Izv. AN SSSR, Ser. Khim.*, 1963, No. 1, p. 294. (Rus)
70. Andrianov K.A. and Volkova L.M., *Izv. AN SSSR, Ser. Khim.*, 1962, No. 1, p. 264. (Rus)
71. *Patent No. 3,340,289* (USA), 1967.
72. Zhdanov A.A., Andrianov K.A., Astapova T.A., and Vinogradova G.F., *Izv. AN SSSR, Ser. Khim.*, 1973, No. 11, p. 2628. (Rus)
73. Zhdanov A.A., Andrianov K.A., Malyikhin A.P., Dubchak I.L., and Strelkova T.V., *Izv. AN SSSR, Ser. Khim.*, 1976, No. 5, p. 1113. (Rus)
74. *Patent No. 1,246,945* (France), 1963; *C.A.*, 1964, vol. **60**, 12052.
75. Mukbaniani O.V., *Candidate's Dissertation on Chemistry*, I. Dzhavakhishvili Tbilisi State University, 1977, Tbilisi, Georgia. (Rus)
76. Andrianov K.A., Emelianov V.N., Muzafarov A.M., and Tot A., *Doklady AN SSSR*, 1975, vol. **222**(3), p. 603. (Rus)
77. Koyava N.A., Mukbaniani O.V., Khanashvili L.M., and Tsitsishvili V.G., *Zh. Obshch. Khim.*, 1980, vol. **50**(8), p. 1793. (Rus)
78. Lebedev E.P. and Fedorov A.D., *Zh. Obshch. Khim.*, 1978, vol. **48**, p. 931. (Rus)
79. Makarova N.N., Petrova I.M., Godovski Yu.K., Lavrukhin B.D., and Zhdanov A.A., *Doklady AN SSSR*, 1983, vol. **269**(6), p. 1368. (Rus)
80. Makarova N.N.Lavrukhin B.D., Timofeeva T.V., and Zelencheva V.N., *Izv. AN SSSR, Ser. Khim.*, 1985, No. 5, p. 1114. (Rus)

81. Furmanova H.G., Andrianov K.A., and Makarova N.N., *Zh. Strukt. Khim.*, 1987, vol. **28**, p. 113. (Rus)
82. Lavrukhin B.D., Makarova N.N., and Zhdanov A.A., *Izv. AN SSSR, Ser. Khim.*, 1986, No. 3, p. 559. (Rus)
83. Makarova N.N., Astapova T.V., and Lavrukhin B.D., *Izv. RAN, Ser. Khim.*, 1996, No. 4, p. 958. (Rus)
84. Makarova N.N., Petrova I.M., and Zhdanov A.A., *Doklady AN SSSR*, 1982, vol. **262**(3), p. 619. (Rus)
85. Meladze S.M., Mukbaniani O.V., and Khanashvili L.M., *Zh. Obshch. Khim.*, 1980, vol. **51**(7), p. 1624. (Rus)
86. Mukbaniani O.V., *Doctor's Dissertation on Chemistry*, I. Dzhavakhishvili Tbilisi State University, 1993, Tbilisi, Georgia. (Rus)
87. Yuzhelevski Yu.A., Kurlova T.V., Kogan B.G., and Suvorova M.V., *Zh. Obshch. Khim.*, 1972, vol. **42**(9), p. 2006. (Rus)
88. Koyava N.A., Mukbaniani O.V., and Khanashvili L.M., *Thes. Rep. V All-Union Conference on Chemistry and Application of Organosilicon Compounds*, Tbilisi, 1980, part 1, p. 20. (Rus)
89. Koyava N.A., Mukbaniani O.V., Khanashvili L.M., and Tsitsishvili V.G., *Zh. Obshch. Khim.*, 1981, vol. **51**(1), p. 130. (Rus)
90. Meladze S.M., *Candidate's Dissertation on Chemistry*, I. Dzhavakhishvili Tbilisi State University, 1980, Tbilisi, Georgia. (Rus)
91. Karchkhadze M.G., Mukbaniani N.O., Samsonia A.Sh., Tkeshelashvili R.Sh., Kvelashvili N.G., Chogovadze T.V., and Khananasvili L.M., *Bulletin of Georgian Academy of Sciences*, 1998, vol. **158**(1), p. 75. (Rus)
92. Karchkhadze M.G., Mukbaniani N.O., Khananasvili L.M., Meladze S.M., Kvelashvili N.G., and Doksopulo T.P., *Intern. J. Polym. Mater.*, 1998, vol. **41**, p. 89.
93. Mukbaniani N.O., Karchkhadze M.G., Samsonia A.Sh., Tkeshelashvili R.Sh., and Khananashvili L.M., *Bulletin of Georgian Academy of Science*, 1999, vol. **160**(1), p. 81. (Rus)
94. Andrianov K.A., Zhdanov A.A., Kurasheva N.A., and Khyinku E.S., *Doklady AN SSSR*, 1977, vol. **236**, p. 354. (Rus)
95. Andrianov K.A., Zhdanov A.A., Kurasheva N.A., and Khyinku E.S., *Doklady AN SSSR*, 1978, vol. **240**(3), p. 591. (Rus)
96. Andrianov K.A., Zhdanov A.A., Kurasheva N.A., and Khyinku E.S., *Thes. Rep. V Intern. Symp. Chemistry of Organosilicons*, FRG, Karlsrue, 1978, p. 72.

97. Kurasheva N.A. and Khyinku E.S., *Thes. Rep. V All-Union Conference on Chemistry and Application of Organosilicon Compounds*, Tbilisi, 1980, p. 155. (Rus)
98. Zhdanov A.A., Kurasheva N.A., and Khyinku E.S., *Izv. AN KazSSR*, 1981, No. 1, p. 38. (Rus)
99. Andrianov K.A. and Zachernyuk A.B., *Khim. Geterotsikl. Soed.*, 1972, No. 8, p. 1183. (Rus)
100. Zachernyuk A.B., Burlova E.A., and Zhdanov A.A., *Zh. Obshch. Khim.*, 1985, vol. **55**(6), p. 1368. (Rus)
101. Zachernyuk A.B., Zhdanov A.A., and Shkvoler V.E., *Zh. Obshch. Khim.*, 1987, vol. **57**(3), p. 562. (Rus)
102. Zachernyuk A.B., *Candidate's Dissertation on Chemistry*, INEOS, AS USSR, Moscow, 1974. (Rus)
103. Andrianov K.A., Zachernyuk A.B., and Danielov S.A., *Khim. Geterotsikl. Soed.*, 1972, No. 7, p. 893. (Rus)
104. Andrianov K.A. and Zachernyuk A.B., *Vysokomol. Soedin.*, 1974, vol. **16**(6), p. 1436. (Rus)
105. Burlova E.A., Zachernyuk A.B., Vinogradov G.V., and Yanovski Yu.G., *Thes. Rep. V All-Union Conference on Chemistry and Application of Organosilicon Compounds*, Tbilisi, 1980, part 2, p. 575. (Rus)
106. Inaridze I.A., *Candidate's Dissertation on Chemistry*, I. Dzhavakhishvili Tbilisi State University, 1994, Tbilisi, Georgia. (Rus)
107. Chizhova N.V., Makarova N.N., Godovski Yu.K., and Buzin A.I., *Vysokomol. Soedin.*, 2000, vol. **42A**(11), p. 1797. (Rus)
108. *A.C. No. 410,018* (USSR), 1974; Chernyishev E.A., Komalenkova N.G., Klochkova T.A., Kuzmina G.M., and Bochkarev V.N., *Otkr., izobr., prom., obr., tov. znaki*, 1974, No. 1. (Rus)
109. Chernyishev E.A., Komalenkova N.G., Klochkova T.A., Kuzmina G.M., and Bochkarev V.N., *Zh. Obshch. Khim.*, 1975, vol. **45**(10), p. 2299. (Rus)
110. Chernyishev E.A., Belkina T.V., Nikitin V.S., Komalenkova N.G., Shapalin A.S., and Zhinkin D.Ya., *Zh. Obshch. Khim.*, 1978, vol. **48**(3), p. 630. (Rus)
111. Mukbaniani O., Matsaberidze M., Karchkhadze M., and Khananashvili L., *Russian Polymer News*, 2002, vol. **7**(1), p. 7.
112. Chernyishev E.A., Krasnova T.L., and Shchepinov S.A., *Zh. Obshch. Khim.*, 1970, vol. **40**(8), p. 1735. (Rus)
113. Samsonia A.Sh., Mukbaniani O.V., and Karchkhadze M.G., *Georgian Engineering News*, 2000, No. 2, p. 142.
114. Dewar M.J.S. and Dietz R., *J. Chem. Soc.*, 1960, p. 1344.

115. Dewar M.J.S., Cubba V.P., and Pettit R., *J. Chem. Soc.*, 1958, p. 3073.
116. Okawara R., Takahashi U., and Sakiiyama M., *Bull. Chem. Soc. Japan*, 1957, vol. **30**, p. 608.
117. Andrianov K.A., Nogaideli A.I., Khananashvili L.M., and Nakaidze L.I., *Izv. AN SSSR, Ser. Khim.*, 1968, vol. ??, 828. (Rus)

Chapter 2. Organosilicon oligomers and copolymers of bead-shaped structure

Equilibrium rigidity of cyclolinear polyorganosiloxanes, macromolecules of which are composed of different sized rings linked by oxygen atoms or other flexible bond bridges, depends upon the number of flexible units both in the ring and linear chain. A broad selection of cyclic structures in the chain ("beads in the necklace"), regulation of their size and type of bonding allow significant variations in properties of cyclolinear poly(organosiloxanes).

Recently, contribution of dimethylsiloxane unit in the series of cyclic (D_3 - D_8) and linear (D_2 - D_{11}) oligodimethylsiloxanes with various end groups was studied [1 – 4]. It is indicated that absolute values of the activation energy of viscous flow of each cyclic compound are higher, compared with linear ones possessing the same number of dimethylsiloxane units in the molecule.

Later on, Andrianov *et al.* have studied contribution of cyclic groups in poly(dimethylsiloxanes) up to direct bonding of organocyclosiloxanes to one another [5]. Comparison of properties of compounds which are structural isomers have indicated that in the absence of dimethylsiloxane bridges evaporation temperatures and activation energies of viscous flow of bicyclic ones are higher than for their structural isomers possessing dimethylsiloxane units between cycles. The values of the activation energy obtained for bicyclic compounds with dimethylsiloxane bridges of different lengths indicate that higher cohesive energy and, consequently, stronger intermolecular interactions are typical of bis(organocyclosiloxy)oxides.

Discussed in the present Chapter are, mainly, the methods for obtaining bead-shaped oligomers and polymers by condensation techniques, but not by polymerization or exchange polymerization, because in the latter cases, primary siloxane procurements may not be preserved due to instability of ≡Si-O-Si≡ bond in cyclosiloxanes.

2.1. Bead-shaped organosiloxane oligomers and copolymers with organocyclotetra(penta, hexa)siloxane fragments in the backbone

The present Section discusses investigation results of the basic regularities and mechanisms of dihydroxyorganocyclosiloxanes' interaction with dihydroxy(dichloro, dihydride)organocyclosiloxanes, clears up synthetic abilities of homo- and heterofunctional condensation in the field of synthesizing cyclolinear oligomers, and considers surveys of thermal and other physical and chemical properties of synthesized oligomers.

Described in the literature is the method for obtaining cyclolinear oligomers by homocondensation of 1,3-dihydroxytetraphenylcyclotrisiloxane in the toluene solution [6]. Deep-level proceeding of condensation reaction is detected by elimination of ≡Si-H absorption band in IR-spectra. For obtaining organosiloxane oligomers of the bead-shaped structure, homocondensation of 1,3-dichlorotetraorganocyclotrisiloxane in the presence of sodium hydroxide in xylene solution has also been studied [6, 7]. The reaction proceeds in accordance with the scheme as follows:

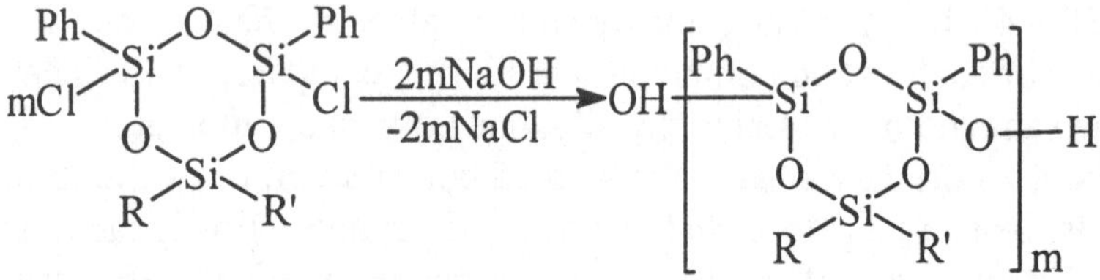

I

(2.1)

where R = Ph, p-$C_6H_4CH_3$; m = 3 - 50.

Indicated in the present work is that the reaction results in formation of difficultly soluble compounds, which apparently shows that the effect of sodium hydroxide on 1,3-dichlorotetraphenyl cyclotrisiloxane induces disclosure of organocyclotrisiloxane unit, because the catalyst used may not only induce homocondensation, but also acts as the anionic catalyst of organocyclosiloxanes' polymerization. As a result, composite materials derived from these oligomers were suggested as electrical insulating materials.

Homofunctional condensation of dihydroxyorganocyclosiloxanes was studied using catalysts of silanol condensation [8].

Despite high variety of catalysts accelerating silanes' condensation due to catalytic action, they may be divided into two groups, which are *equilibrating* and *non-equilibrating* catalysts [9 – 11]. To the first group the catalysts are corresponded, in the presence of which condensation proceeds with ≡Si-O-Si≡ bond cleavage, and the second group is composed of catalysts, which do not induce cleavage of the siloxane bond.

Studying synthesis of organosiloxane oligomers, the authors of the present monograph have surveyed reactions of homofunctional condensation of 1,5-dihydroxyorganocyclotetrasiloxanes [12, 13], 1,5-dihydroxyorganocyclopentasiloxanes [14] and 1,7-dihydroxyorganocyclohexasiloxanes [15] in different solvents in the presence of the catalyst (activated coal), and at boiling temperature of solvents used.

Homofunctional polycondensation of studied dihydroxy-containing organocyclotetra(penta, hexa)siloxanes proceeds in accordance with the scheme as follows:

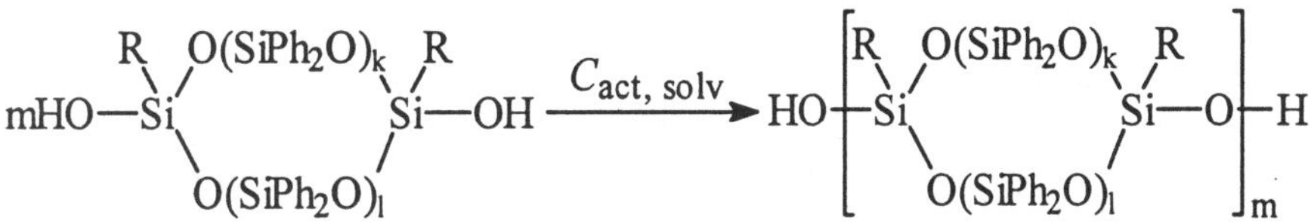

(2.2)

where R = Me, Ph; $l = k = 1, 2$; $l \neq k = 2$ (II); $m = 4 - 30$.

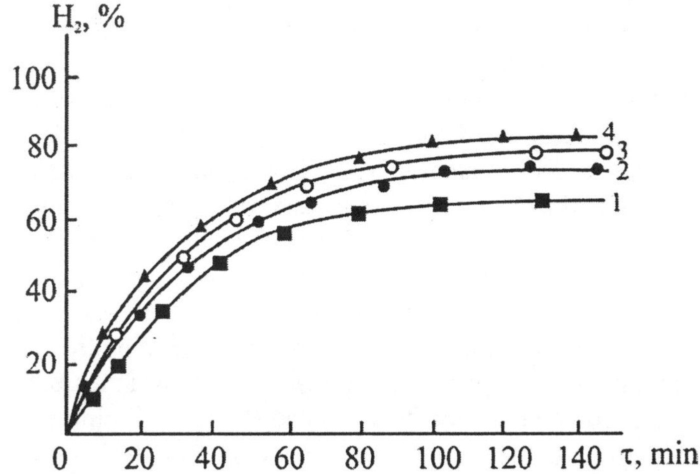

Figure 2.1. Hydrogen conversion in dehydrocondensation of 1,5-dihydride-1,5-dimethyltetraphenylcyclotetrasiloxane with 1,5-dihydroxy-1,5-dimethyltetra-phenylcyclotetrasiloxane: 1 – at 20°C; 2 – at 30°C; 4 – at 40°C. Curve 3 represents catalytic dehydrocondensation of 1,5-dihydride-1,5-dimethyltetraphenylcyclotetrasiloxane with 1,5-dihydroxyhexaphenylcyclotetra-silocane at 40°C (with H_2PtCl_6 as catalyst).

Synthesized copolymers of the bead-shaped structure are colored light-yellow transparent products, well-soluble in organic solvents. Application of solvents with different boiling temperatures to this reaction causes variation of the polymerization degree (*m*). The level of homofunctional polycondensation is increased with the boiling temperature of the solvent. The polycondensation level is also highly dependent on concentration of the catalyst applied to the reaction, which is activated coal. As the concentration of activated coal is increased from 7 to 15 wt.%, in the case of organocyclotetrasiloxanes, the polymerization degree is increased, approximately, by 10 units.

It should be noted that the increase of radical volume at silicon atom in organosil(sesqui)oxane unit of organocyclosiloxanes induces a decrease of the polymerization degree (*m*) of oligomers. As the volume of cyclic fragments is increased, i.e. at transition from organocyclotetrasiloxane to organocyclopenta- and -hexasiloxane fragments, the depth of homofunctional condensation is decreased.

To compare the depth of homo- and heterofunctional condensation of organocyclosiloxanes, catalytic dehydrocondensation of dihydroxyorganocyclotetra(penta, hexa)siloxanes with dihydride organocyclotetra(penta, hexa)-siloxanes was studied [6]. Recently, catalytic dehydrocondensation reactions were successfully applied to synthesis of linear organosiloxane copolymers [17 – 21]. The study of this reaction between tetramethylcyclotetrasiloxane and *cis*-1,3,5,7-tetrahydroxytetraphenylcyclotetrasiloxane in the presence of potassium or sodium methoxide is considered [22]. It indicates that insoluble polymer content is increased with the hydrogen conversion. In spite of contradictory data on the application of platinum-hydrochloric acid in dehydrocondensation reaction, present in the literature [23, 24], besides platinum-hydrochloric acid, catalytic quantity of anhydrous powder-like caustic potash as the catalyst was used in catalytic dehydrocondensation reactions of dihydroxyorganocyclosiloxanes with dihydride organocyclosiloxanes [16], which are to synthesize organosiloxane copolymers of the bead-shaped structure. Preliminary heating of initial organocyclosiloxanes during several hours at 40 - 50°C in the presence of anhydrous caustic potash (0.1 wt.% of total quantity of the initial components) and 0.1 M of platinum-hydrochloric acid solution in tetrahydrofurane ($\sim 5\times10^{-4}$ g per 1 g of the substance) has indicated that no polymerization of initial cycles proceeds; the more so, no changes for methyl protons in NMR spectra of organocyclosiloxanes happen. Apparently, in the case of use of anhydrous caustic potash and the above-mentioned conditions, no siloxane ring scission is observed, which is in agreement with the previously published data. Catalytic dehydrocondensation has been studied at different temperatures (20, 30 and

40°C) in the absolute toluene solution. It has been found that at initial stages of the reaction a short induction period (~1 – 2 min) is observed. Hydrogen conversion in the reaction with time has been studied. It has been found that in dehydrocondensation reaction, hydrogen conversion is increased from 66% (at 20°C) to 82% (at 40°C) with temperature.

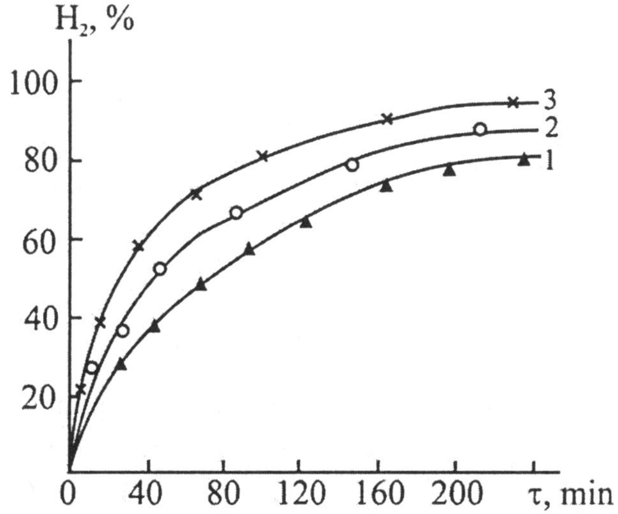

Figure 2.2. The hydrogen liberation rate in dehydrocondensation reaction of 1,5-dihydride-1,15-dimethyltetraphenylcyclotetrasiloxane with 1,5-dihydroxy-1,5-dimethyltetraphenylcyclotetrasiloxane, where 1 – at 40°C; 2 – at 30°C; 3 – at 20°C (with KOH as the catalyst).

To increase the level of catalytic dehydrocondensation at the final stage of the reaction, the reaction products were heated up to 80°C during 3 – 4 hours. General scheme of dehydrocondensation proceeding is as follows [16, 25]:

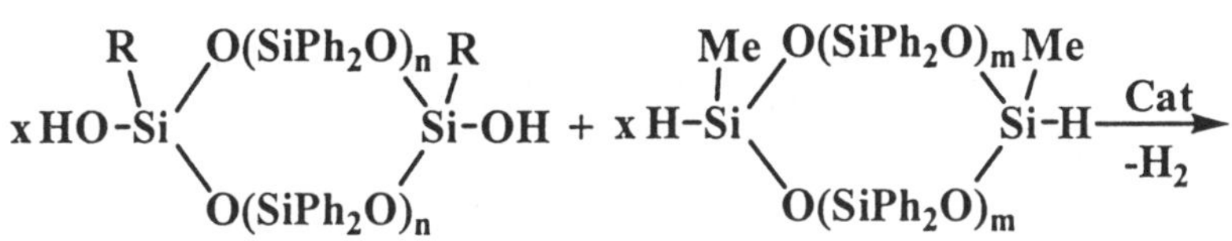

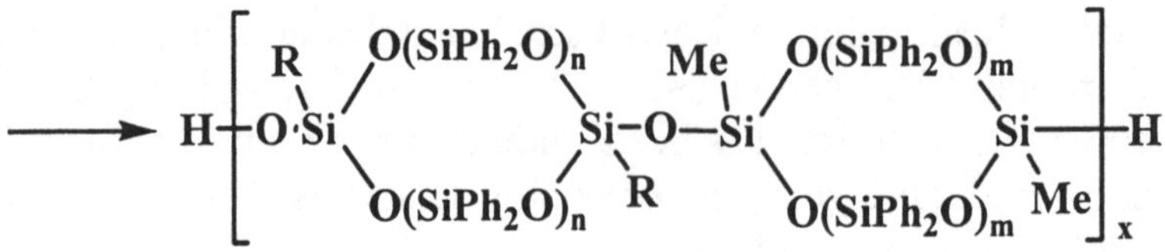

(2.3)

II, IV

where R = Me, Ph; $m = n = 1, 2$.

Resulting the above-mentioned reaction, solid transparent copolymers, well soluble in usual organic solvents with $\eta_{spec} = 0.06 - 0.13$, are obtained after reprecipitation. Some properties of obtained copolymers are shown in Table 2.1.

Since initial dihydride- and dihydroxyorganosiloxanes, used in dehydrocondensation, represent a mixture of *cis*- and *trans*-isomers, synthesized copolymers possess atactic structure.

As observed from the data, catalytic dehydrocondensation proceeds at a deeper level with formation of higher molecular products, than in the case of homofunctional products.

It is shown [16, 25] that besides the origin and quantity of the catalyst, temperature and origin of the solvent, etc., reactivity of ≡Si-H bond is highly affected by steric and inductive factors, induced by cyclic structures and framing groups disposed at silicon atoms. Figures 2.1 and 2.2 show hydrogen conversions for dehydrocondensation at different temperatures of catalysts. It has been found that catalytic dehydrocondensation reaction displays the second order. Dehydrocondensation reaction rate constants are determined, and catalytic dehydrocondensation activation energies are calculated: $E_{act} = 28.1 - 28.5$. As a consequence, for anhydrous caustic potash and platinum-hydrochloric acid application as the catalyst activation energies are almost the same.

It should be noted that in both cases formation of hindered compound of the following structure:

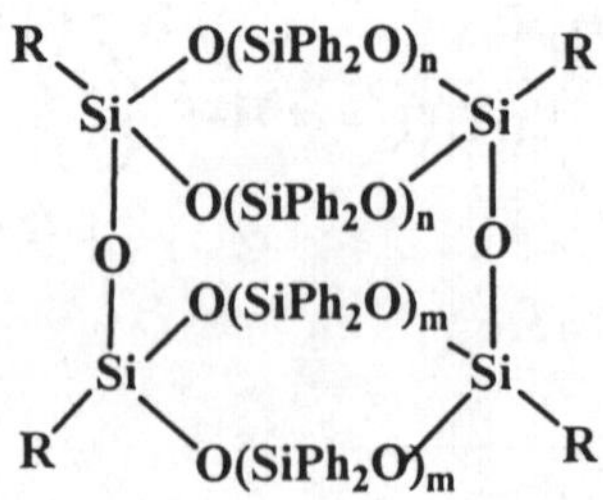

Table 2.1

Some physical and chemical parameters and the yield of bead-shaped structure polyorganosiloxanes, synthesized in catalytic dehydrocondensation

No.	Copolymer	Yield, %	η_{spec}**	*R*	*m*	*n*	T_{react}, °C	Catalyst	T_{soft}, °C	d_1, Å	$\bar{M}_{\omega}\times10^{-3}$ *
1	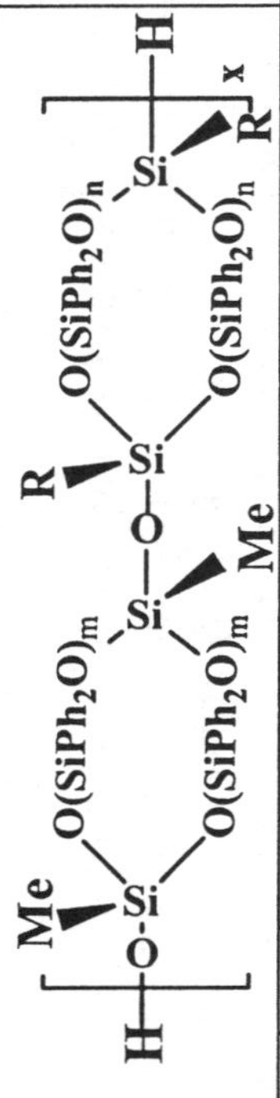	80	0.04	Me	1	1	20	H_2PtCl_6	-	-	-
2		83	0.06	Me	1	1	30	H_2PtCl_6	-	-	-
3		84	0.10	Me	1	1	40	H_2PtCl_6	75 - 85	10.20	55
4		91	0.07	Me	1	1	20	KOH	74 - 80	-	-
5		93	0.09	Me	1	1	30	KOH	-	-	-
6		94	0.13	Me	1	1	40	KOH	81 - 87	10.20	-
7		86	0.09	Ph	1	1	40	H_2PtCl_6	-	-	-
8		94	0.12	Ph	1	1	40	KOH	78 - 83	10.20	61
9		85	0.07	Me	2	1	40	H_2PtCl_6	62 - 69	-	-
10		93	0.10	Me	2	1	40	KOH	-	10.57	53
11		84	0.08	Ph	2	2	40	H_2PtCl_6	-	-	-
12		93	0.10	Ph	2	2	40	KOH	80 - 85	10.57	-
13		81	0.06	Ph	2	1	40	H_2PtCl_6	64 - 72	10.57	-
14		82	0.07	Ph	1	2	40	H_2PtCl_6	65 - 73	10.57	37

Notes: * Molecular masses were determined by the light scattering method.
** In toluene at 25°C.

by homo- and heterofunctional condensation reactions is eliminated [12 – 16, 25] due to steric hindrances and reaction proceeding in concentrated solutions. X-ray diffraction analysis indicates that the bead-shaped oligomers and copolymers are amorphous systems, and the interchain distance, d_1, is increased with the volume of cyclic fragments.

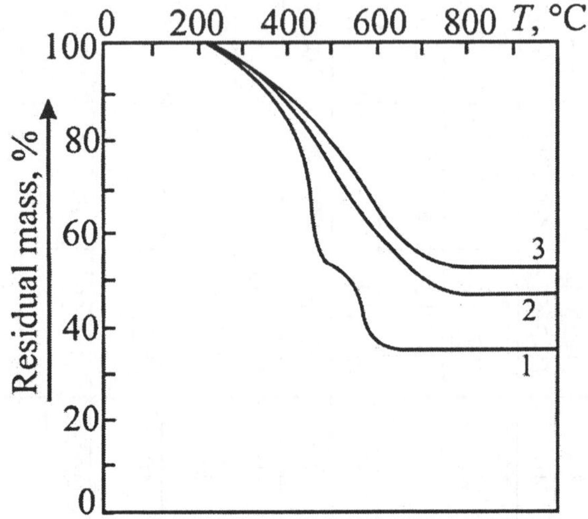

Figure 2.3. Thermogravimetric curves of copolymers with structure II; curve 1 – copolymer with R = Ph, n = 1; curve 2 – copolymer with R = Me, m = 1, n = 2; curve 3 – copolymer with R = Me, m = n = 1 (in air, heating rate 5 deg/min).

Thermogravimetric studies of copolymers have shown that at 300°C mass losses do not exceed 3 – 6%, but are regularly increased with temperature (Figure 2.3). The main degradation process proceeds in the temperature range of 450 - 700°C and final mass losses increase with the volume of the cyclic fragment, respectively. Copolymers with cyclotetrasiloxane fragments in the backbone are characterized by higher thermal oxidative stability, than copolymers with cyclohexasiloxane ones, which is explained by variation of silsesquioxane (T) and siloxane (D) fragments.

To synthesize cyclolinear bead-shaped polyorganosiloxanes, various HFC reactions in accordance with the following schemes are used [26]:

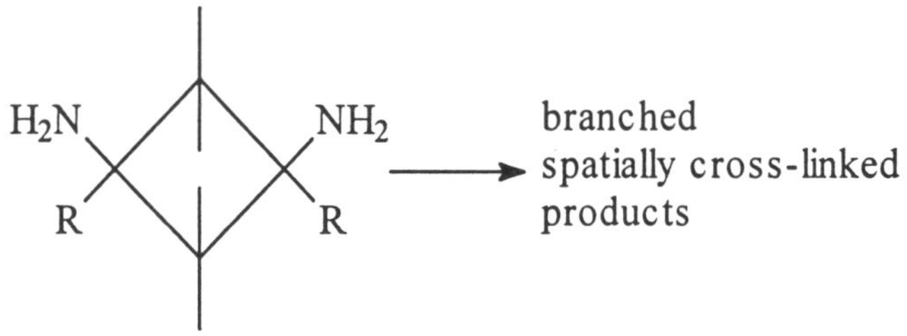

(2.4)

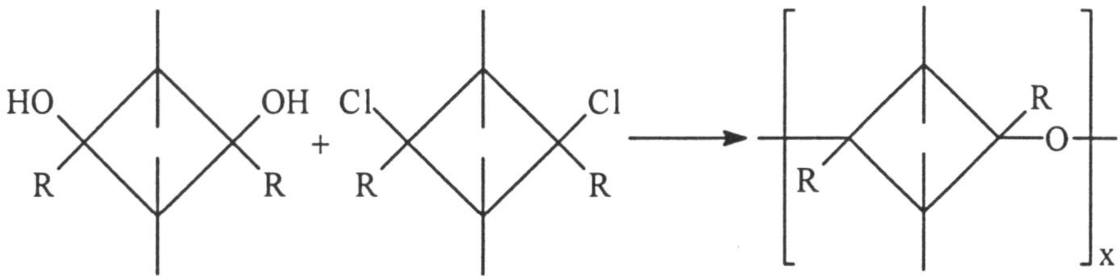

(2.5)

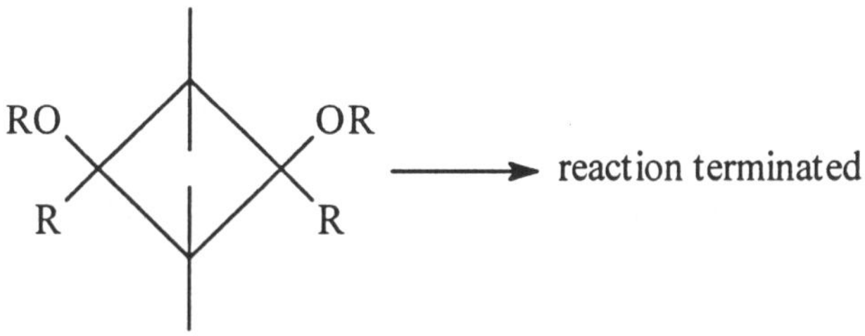

(2.6)

It has been shown by gel permeation chromatography (GPC) and NMR methods that cyclolinear structure of the backbone of synthesized polymers is obtained at the interaction of dichlororganocyclosiloxanea and dioxy-derivatives in the presence of HCl acceptors only. The conclusion about cyclolinear structure of synthesized polymers is based on the data of hydrodynamic studies, MMD values and results of equilibrium rigidity determination. Cyclolinear structure of the backbone is also proved by NMR-spectroscopy method on ^{29}Si nuclei.

HFC reaction between 1,5-dichlorohexaorganocyclotetrasiloxanes and appropriate dihydroxy-derivatives in the presence of hydrogen chloride acceptor has synthesized bead-shaped cyclolinear organosiloxane polymers [27 – 31] in accordance with the general scheme as follows:

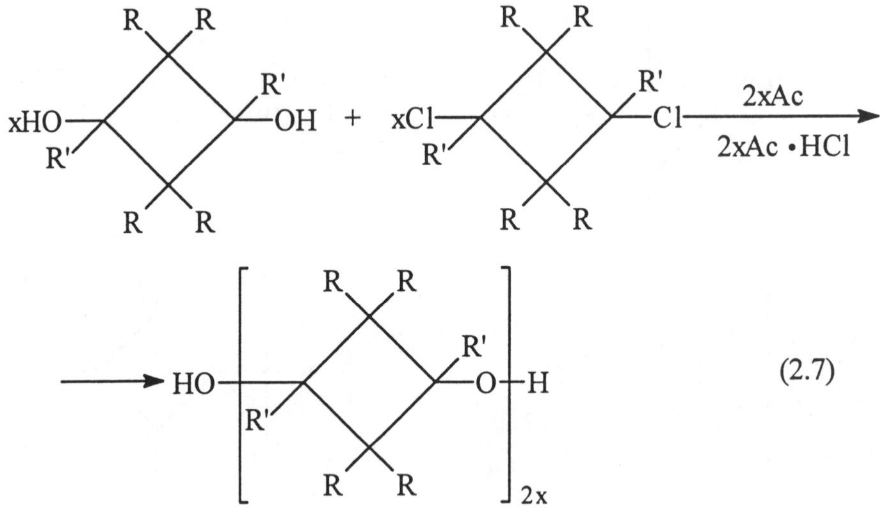

(2.7)

where R = R' = Me, Ph; R ≠ R'.

Table 2.2

Physical and chemical parameters of POCS-4 of the structural formula as follows:

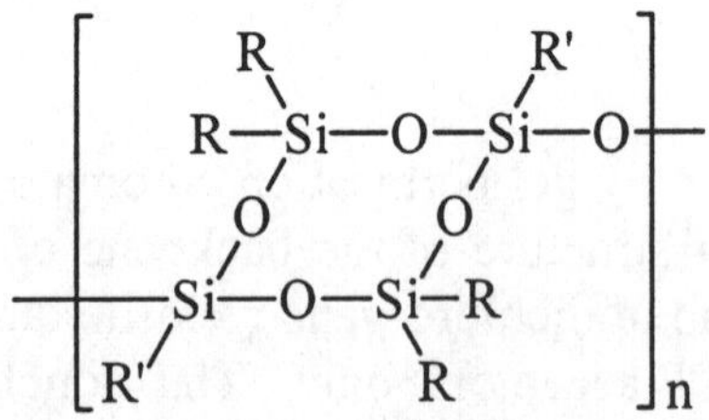

No.	R	R'	Isomeric composition, %		[η], dl/g	T_g, °C	T_{melt}, °C
			cis	*trans*			
1	Me	Ph	50	50	0.17	0	-
2	Me	Ph	0	100	0.1	+7	100*
3	Ph	Me	50	50	0.05	+43	-
4	Ph	Me	0	100	0.06	+46	142*
5	Ph	Ph	50	50	0.05	+220	-

Note: * Data of X-ray diffraction analysis prove the presence of crystalline phase.

Applied difunctional organocyclotetrasiloxanes, both pure *cis*- and *trans*-isomers and their mixtures with isomers in various ratios, were of interest as model compounds for synthesizing *stereo*-regular poly(organocyclotetrasiloxanes) (POCS-4): *trans*-tactic, *cis*-tactic and with strict regular alternation of *cis*- and *trans*-sequences.

For synthesizing POCS-4 (poly-1,5-dimethyltetraphenylcyclotetrasiloxane, poly-1,5-diphenyltetramethylcyclotetrasiloxane, poly(hexaphenylcyclotetrasiloxane) and poly(hexamethylcyclotetrasiloxane) – PMCS-4), *trans*-1,5-dichlorohexaorganocyclotetrasiloxane and appropriate *trans*-dihydroxy-derivatives, as well as several mixtures with different ratio of *cis*- and *trans*-isomers were used [27 – 29] in accordance with the scheme (2.5). Some physical and chemical parameters of polymethylphenylcyclotetrasiloxanes are shown in Table 2.2.

PMCS-4 with different ratios of *trans*- and *cis*-sequences in the backbone possess one and the same glass transition temperature, and the crystallinity degree is lower that for PDMS (Table 2.3). While polymeric chain is enriched by *trans*-units, crystallinity increases [27]. By methods of DSC, X-ray diffraction and thermooptical analysis, it have been detected that besides glass transition (T_g) and melting (T_{melt}), polymers synthesized from monomers containing up to 95 – 100% of *trans*-isomers display a phase transition in the temperature range of 70 - 90°C, which is simulated as a transition from mesomorphous to isotropic state (Figure 2.4).

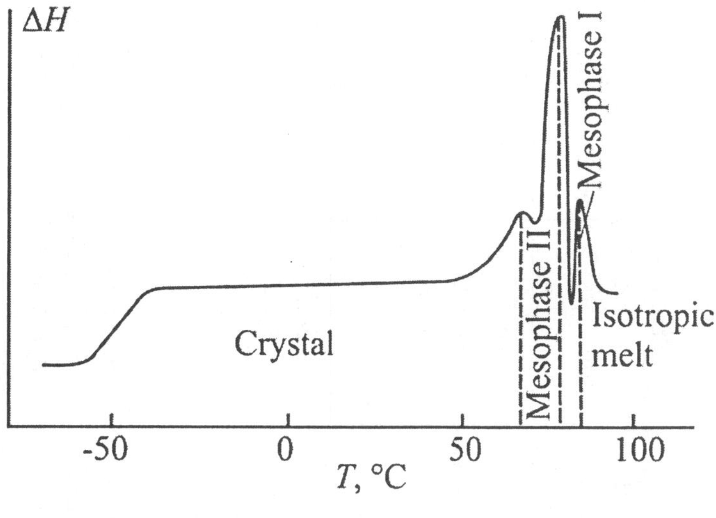

Figure 2.4. DSC curve for *trans*-microtactic PMCS-4

Table 2.3

Physical and chemical parameters of PMCS-4 and PMCS-b

Structural formula of polymer unit	*Trans*- and *cis*-units ratio *	Polymer yield, %	$\bar{M}_n \times 10^{-3}$	T_g, °C	Melting		Isotropization		[η] (25°C), dl/g
					T_{melt}, °C	ΔH, J/g	T_{iso}, °C	ΔH, J/g	
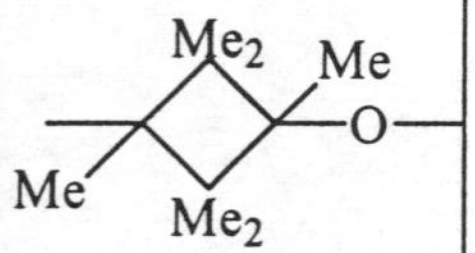 PMCS-4	70/30 87/13 100/0	43.0 20.0 56.0	37	-51 -51 -51	– – 77	– – 12.0	– – 82	– – 3.0	0.15 0.32 0.17
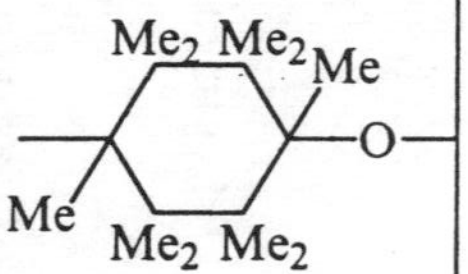 PMCS-6	75/25 85/15	74.0 71.0	25	-91 -91	** -53	5.5	320	10	0.10 0.15

Notes: * Data on the ratio of units in polymers are given on the basis of NMR-spectra of *trans*- and *cis*-isomers of initial dichloro(dihydroxy)cyclosiloxanes.

** Long-term exposure under optimal conditions induces crystallization, but the crystallinity degree is extremely low.

Detection of thermotropic mesophase presence in PMCS-4 has induced different opinion on poly(organocyclosiloxanes), because heretofore poly(dimethylsiloxane) had been considered one of the most flexible-chain polymers. More thorough study of the spatial structure of PMCS-4, performed by ^{29}Si NMR spectroscopy method, has shown that HFC reaction of 1,5-dichlorohexaorganocyclotetrasiloxanes with appropriate dihydroxy-derivatives proceeds (depending on selected conditions) with both preservation and inversion of isomers' configuration. This made the authors to clear up the effect of such factors as the reaction temperature, acceptor basicity and solvent polarity in the HFC reaction on formation of different configurations of PMCS-4 units by ^{29}Si NMR method.

Studied in ref. [26] is the effect of HFC reaction conditions on the configuration sequences in POCS-4. Since the mesomorphous state in PMCS-4 is formed in stereo-regular *trans*-tactic polymers only [27, 32] and spatial configuration of initial monomers is not always fully preserved in polymers, the effect of HFC conditions on transformation of ≡Si-Cl and ≡Si-OH centers in initial compounds has been studied.

^{29}Si NMR spectra of PMCS-4 contain Si reflexes from dimethylsiloxane (D) and methylsilsesquioxane (T) fragments, reflexes from D-fragments being divided into two components and reflexes from T-fragments into four components (Table 2.4). Relation of reflexes to *tt, tc, ct, cc* spatial configurations is performed based on a supposition that T-atoms' reflex splitting may be caused by spatial isomerism of backbone elements only. However, comparative analyses of ^{29}Si NMR spectra of PMCS-4, obtained for different conditions, have shown that configuration of initial rings is not fully preserved.

This is the reason why ^{29}Si spectra could not be related to spatial isomerism of the backbone units. Table 2.4 shows that, actually, two reflexes for D-units and three reflexes for T-fragments in PMCS-4 are observed, and their intensities depend on the synthesis conditions. Moreover, data shown in Table 2.4 also indicate proceeding of partial inversion of one of the initial compounds. In ^{29}Si NMR spectra of PMCS-4, respponses from T-atoms of silicon are the most informative ones. For these atoms, the following alternatives of spatial ringing are possible:

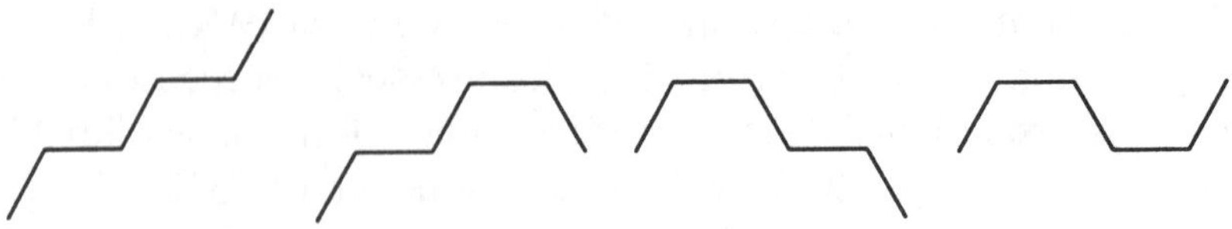

tt tc ct cc

The Table shows that at any variations of HFC reaction conditions, two reflexes of approximately equal intensity are always present in the T-area. These reflexes have constant disposition in spectra of polymers. Equality of *tc* and *ct* reflexes must be observed, because the number of atoms equals to the that of bonds between *cis*- and *trans*-units. Correlation in this pair of reflexes cannot be performed, because they posses equal intensities.

Because none of the examples shown in Table 2.4 displays preservation of *tt*-unit reflex intensities, which are close to 95%, a question about variation in configurations of initial rings in the HFC reaction is arisen. There are two centers capable of inversion in the initial molecules. In the case of full inversion of both centers, the molecule is reversed to the initial configuration. The latest example in Table 2.4 shows that if the *trans*-form of dichloro-derivative is preserved, *tt* and *tc* combinations only are possible, i.e. three reflexes from T-fragment should exist. Actually, four reflexes are observed and, consequently, *cc*-combination is formed that proves partial inversion of dichloride initial units. In the case of partial inversion of the initial diol, polymers 1 and 2 (Table 2.4) would display four reflexes, but three reflexes only for each polymer are observed.

The detected fact of cyclosiloxanes partial inversion at Cl atoms substitution at silicon was expected, as reported before [33, 34]. More detailed description of reflex correlation was carried out in ref. [35]. Symbols 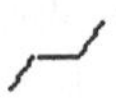 and  and mark projections of units and bonds to the plane perpendicular to the cycle plane.

Thus, ^{29}Si NMR spectra in PMCS-4 allow a conclusion about distribution of sequences from two units (*tt*, *tc*, (*ct*), and *cc* dyads) and the ratio of *trans*- and *cis*-units in the backbone. Correlation by reflexes from D-fragments can be performed in the presence of reflexes from T-fragments of *trans*- and *cis*-units, calculated by ratios. Table 2.4 shows that low-polar D reflexes in the whole series are more intensive than high-polar ones. That is why the former reflex is related to *trans*-units.

Table 2.4

Conditions of synthesis, chemical shifts and intensities of reflexes in ^{29}Si NMR spectra of PMCS-4

Polymer, No.	*Trans-/cis*-isomer ratio in initial monomers		HC1 acceptor	Relative intensity of δSi reflexes for dyad units and sequences					
	In chloride	In diol		D		T			
				-19.27	-19.31	*tt* -65.53	*tc* -65.47	*ct* -65.77	*cc* -65.68
1	0.95/0.05	0.95/0.05	$(C_2H_5)_3N$	0.80	0.20	0.78	0.11	0.20	0.01
2	0.95/0.05	0.95/0.05	C_5H_5N	0.62	0.38	0.61	0.18	0.19	0.02
3	0.95/0.05	0.95/0.05	C_5H_5N*	0.63	0.37	0.31	0.35	0.33	0.02
4	0.95/0.05	0.95/0.05	$(C_4H_9)_3N$	0.59	0.41	0.17	0.42	0.32	0.03
5	0.95/0.05	0.95/0.05	$(C_5H_{11})_3N$	0.54	0.46	0.15	0.41	0.40	0.04
6	0.70/0.30	0.70/0.30	C_5H_5N	0.64	0.36	0.40	0.22	0.25	0.13
7	0.95/0.05	0.50/0.50	C_5H_5N	0.60	0.40	0.32	0.25	0.27	0.16

Note: * The acceptor represents solvent, simultaneously; in all other reactions, benzene, toluene or diethyl ether is used as solvents.

Data on intensities of reflexes in T-fragment spectra of PMCS-4 indicate that HFC reaction of initial dichlorohexamethylcyclotetrasiloxane induces different reflexes of Cl atoms.

These data contradict to the previously published results on dioxy- and dialkoxy-derivatives, which indicate that the ring configuration does not change during the reaction [26, 28, 36], i.e. in spite of hydrolysis, in HFC reaction with appropriate dioxy-derivatives dichlorocyclotetrasiloxane is partly inverted.

Data in Table 2.5 show that the inversion degree varies in a broad range and for units in the case of $N(C_2H_5)_3$ acceptor equals 13%, and in the case of $N(C_5H_{11})_3$ is 88%.

Though inversion and estimation of stereospecificity of HFC reaction require more thorough studies, already existing data indicate that the reaction represent the multi-stage process.

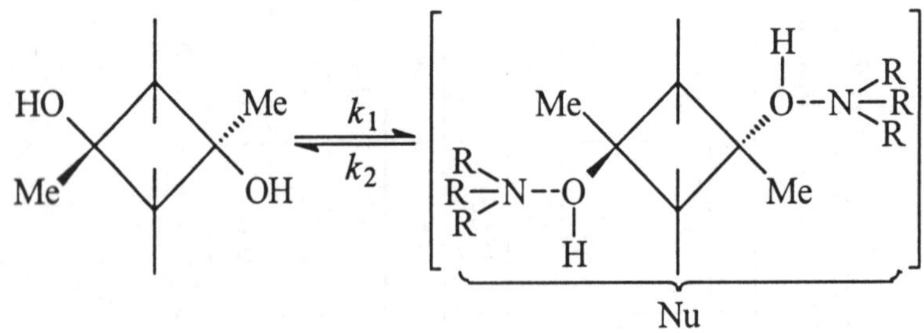

(2.8)

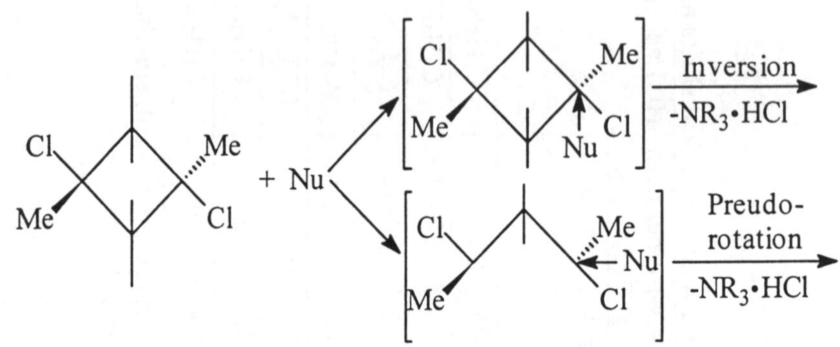

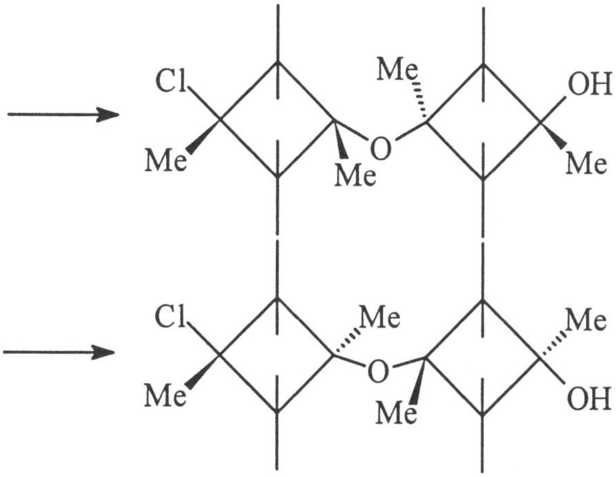

(2.9)

At the first stage, interaction between dihydroxycyclotetrasiloxane and amine forms a structure. The second stage (the structure interaction) proceeds under different conditions with regard to the structure stability. In cases, when structures with amines of similar basicity are formed, their stability is approximately the same. However, depending on transitional amine structure, the attack may proceed from the front via formation of a trigonal pyramid. The first path proceeds via inversion of ≡Si-Cl reaction center in the ring, and the second one via pseudo-rotation with the configuration preserved.

Table 2.5

Dependencies of the fracture of *cis*-units on the quantity of units, formed from dichlorohexamethylcyclotetrasiloxane, and the degree of inversion of dichloride units with regard to HCl acceptor used

Polymer, No.	Acceptor	Fracture of *cis*-units	Degree of inversion of units
1	$(C_2H_5)_3N$	0.17	0.13
2	C_5H_5N	0.36	0.34
3	C_5H_5N	0.67	0.69
4	$(C_4H_9)_3N$	0.81	0.84
5	$(C_5H_{11})_3N$	0.84	0.88

As different configurations of the backbone units are formed from the same *trans*-isomers of initial dichloro- and dihydroxy-derivatives, the PMCS-4 chain configuration is generally regulated by chosen HCl acceptor.

Moreover, the inversion degree of every Si-Cl center in dichlorocyclotetrasiloxane is different; in the cases of polymers 3 – 5 (Table 2.5), it exceeds 50%, but at equal degrees of inversion of all centers its maximum is 50%, because further increase of it for every ≡Si-Cl center converts the cyclosiloxane unit back to the initial configuration. Hence, it should be taken into account that the difference in stereochemistry of Cl atoms' substitution at both centers is stipulated by different conditions: the first Cl atom is substituted in the structure of dichlorohexamethylcyclotetrasiloxane free molecule, and the second one at the position of the end group of oligomers, i.e. when steric hindrances and equilibrium rigidity of the backbone vary due to increased length of the oligomers.

Studies of HFC reaction of 1,5-dichloro-1,5-diphenyl-3,3,7,7-tetramethylcyclotetrasiloxane with the appropriate dioxy-derivative in the presence of HCl acceptor by ^{29}Si NMR spectroscopy method have indicates that reflexes from phenylsilsesquioxane T-fragment fall within the area of -79 ppm and are split into four components; reflexes from D-fragment are observed at -17 ppm and split into two components.

If the same selection of HCl acceptors and solvents are used in HFC reaction, lower affinity of units from dichlorocyclosiloxane to inversion is observed (Table 2.6). While $(C_2H_5)_3$ and C_5H_5N acceptors are used, the inversion degree of ≡Si-Cl centers in dichloromethylphenylcyclotetrasiloxane equals 4 – 6%, and for $N(C_4H_9)_3$ acceptor - 30%. Reduction of inversion of ≡Si-Cl centers in the ring, all other conditions being absolutely the same, may be associated with the effect of bulky phenyl substituting agent only, which renders difficulties to inversion of the reactive centers. For spatially hindered molecules of bi- and polycyclic structures, it is observed that substitution of Cl atom at Si atom proceeds with preservation of the configuration [37]. Data shown in Table 2.6 indicate that synthesized PMPCS-4 may display different tacticity, and stereospecificity of the reaction reaches 95%.

The effect of tacticity of the polymeric backbone on properties of POCT-4 has been searched for [26]. Figure 2.5 shows curves for PMCS-4 with different backbone structure, obtained by the differential scanning calorimetry (DSC) method (curves 1 – 4). For atactic polymer the only transition, corresponded to T_g, is observed on curve 1, above which, in accordance with the data of X-ray diffraction analysis, the polymer is amorphous (two amorphous haloes at $2\theta = 8 - 11°$ and 20 - 35°). As the polymeric backbone is enriched with *tt*-sequences (curves 2 – 4), additional phase transitions are observed. For different cooling rates and annealing temperatures, it has been observed that depending on PMCS-4 enrichment by *tt* units the number of endothermal peaks and the temperature range between endoeffects from T_{melt} to T_{inv} vary.

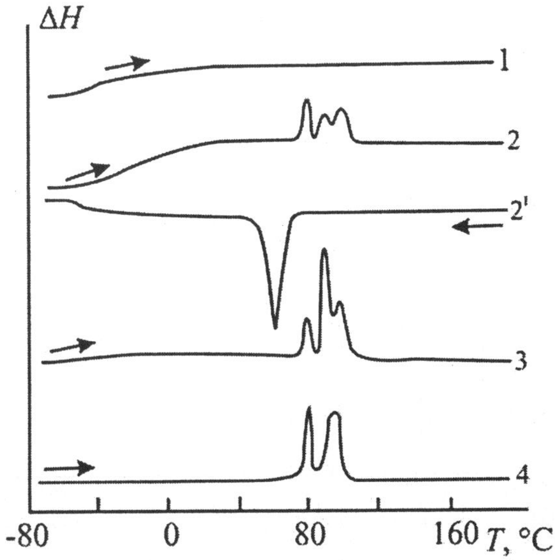

Figure 2.5. DSC curves for PMCS-4 (Table 2.3) of various stereoregular structure: 1 – atactic; 2, 2' – *trans*-tactic polymer 2 (2- heating, 2' – cooling); 3 – heating of polymer 3; 4—heating of polymer 4

Table 2.7 shows correlations between *tt* and *cc* units in polymers and temperature and heats of phase transitions. If the sample has transited to the amorphous phase, independently of the heating rate of it, crystallization at 66°C with ΔH = -11.5 J/g is observed; other exothermal effects were absent, except for glass transition at T_g = -50°C. Though polymer 2 (Table 2.7) is more low-molecular, X-ray diffraction analysis displays its higher crystallinity degree (80%) (Figure 2.6a). At the ratio of *tt*- and *cc*-units equal to polymer 2, diffraction pattern of polymer 4, which possesses higher molecular mass, displays lower resolution (Figure 2.6).

Table 2.6

Phase transitions in POCS-4 containing different amounts of *tt*-sequences

No.	POCS-4	*tt*-sequence fracture in POCS-4	[η] (25°C), dl/g	T_g, °C	Melting			Mesomorphous transition			Isotropization		
					T_{melt}, °C		ΔH, J/g	$T_{MII\text{-}MI}$, °C		ΔH, J/g	T_{inv}, °C		ΔH, J/g
					DSC	RSA		DSC	RSA		DSC	RSA	
1	PMCS	0.46	0.13	-51	-	-	-	-	-	-	-	-	-
2	"	0.87	0.07	-	55-70	60-70	5.9	80-90	85-97	-	95-100	100-110	1.5
3	"	0.83	0.17	-5	67-77	-	12.0	-	-	-	85-95	-	3.0
4	"	0.87	0.23	-52	80-87	-	5.3	-	-	-	92-96	100-106	7.3
5	PMPCS	0.96	0.39	+5	180-205	190-210	24.8	-	-	-	-	-	-
6	PPCS	0.80	0.09	+58	-	-	-	170-180	180-190	1.0	>260	>270	-

Table 2.7

Conditions of synthesis, chemical shifts and intensity of reflexes in ^{29}Si NMR spectra of PMPCS-4

Polymer, No.	The ratio of *trans-/cis*-isomers in initial monomers		HCl acceptor	Relative intensity of δ_{Si} reflexes for dyad units and sequences				
				D		T		
	In chloride	In diol		-17.49 (*trans-*)	-17.27 (*cis-*)	*tt* -79.65	*tc* -79.50	*ct* -79.46
1	100/0	100/0	$(C_3H_5)N$	0.95	0.05	0.90	0.05	0.05
2	100/0	100/0	$(C_4H_5)N$	0.72	0.28	0.44	0.28	0.28
3	100/0	100/0	C_5H_5N	0.95	0.05	0.90	0.05	0.05
4	0/100	100/0	$(C_2H_5)_3N$	0.53	0.47	0.06	0.47	0.47

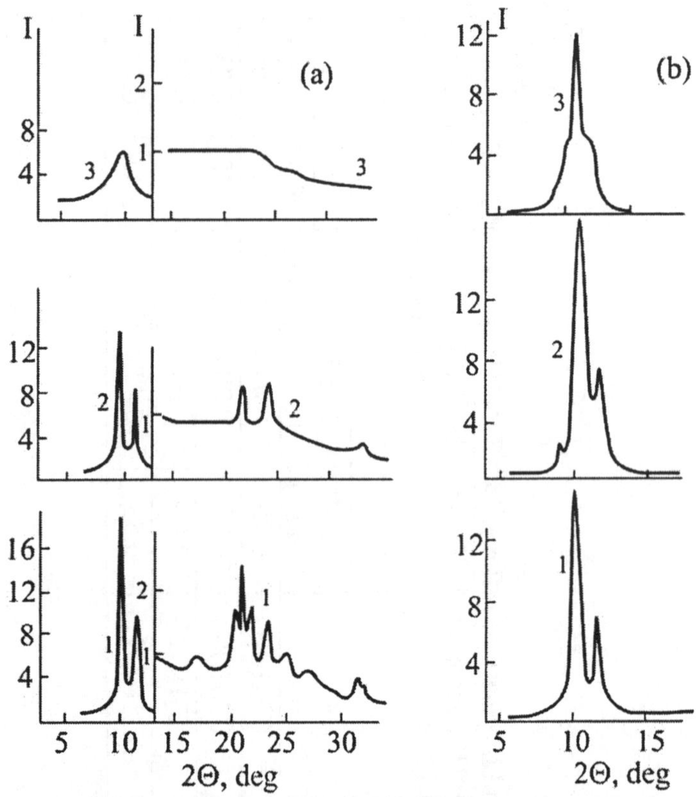

Figure 2.6. Diffraction patterns for PMCS-4 (Table 5): a – polymer 2 at 12°C (1) and 74°C (2, 3); b - polymer 4 at 20°C (1), 86°C (2) and 96°C (3) (*I*, dimensionless units)

Heating of *trans*-tactic PMCS-4 at 60 - 70°C induces gradual elimination of the greater part of reflexes, four of which only remain on the diffraction pattern already at 74°C (Figure 2.6). The character of $2d$ temperature dependence and correlation between angular positions of these reflexes allow a supposition that they relate to regulation in the basic plane of the rectangular cell with dimensions a = 17.27 Å and b = 8.46 Å. The first transition is not accompanied by significant variations in the basic plane package, but is associated with the loss of the farther translation order. Temperature increase above 74°C leads to elimination of low-intensive reflexes in angles (2θ = 14 – 30°). Fusing of reflexes at 2θ= 10 – 11° into a single one is observed on diffraction patterns in the temperature range of two following endothermal transitions (Figure 2.5, curve 2). This may testify about packing variation in the basic plane. This effect is screened by isotropization (Figure 2.6). Final isotropization of polymer 2 proceeds at 110°C.

Thus in spite of atactic PMCS-4, transition of *trans*-tactic PMCS-4 from crystalline to amorphous state is preceded by polymorphous transition of the mesophase II - mesophase I type.

To obtain fuller picture on the cyclolinear polyorganosiloxane tacticity effect on formation of crystalline and mesomorphous states, PMPCS-4 and PPCS-4 were synthesized and their properties were studied.

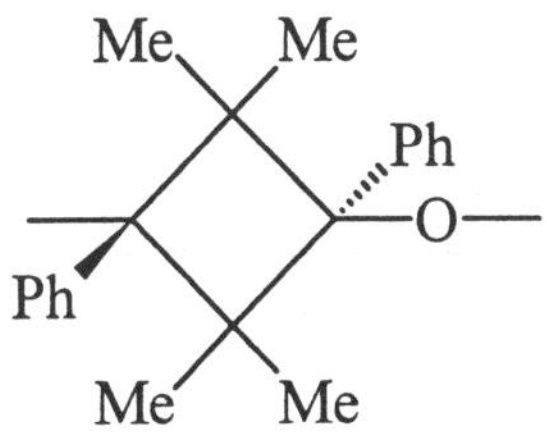

PMPCS-4

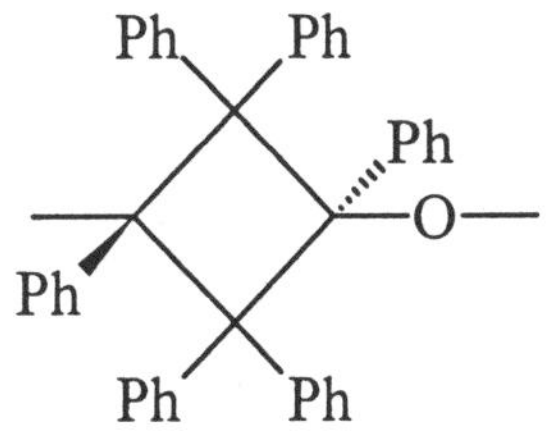

PPCS-4

Figure 2.7 shows DSC curves for *trans*-PMPCS-4 and *trans*-PPCS-4. In spite of the above-considered *trans*-PMCS-4, *trans*-PMPCS-4 displays lower crystallinity degree (~60%) at room temperature, though its saturation by *tt*-sequences is higher. As temperature increases, a single endothermal transition at 197 - 202°C only is observed in it. In accordance with RSA data, this transition is associated with polymer melting. Under different modes of annealing, cooling and repeated heating of PMPCS-4 sample, cold crystallization at 36°C and crystal II – crystal I transition at 119°C were observed. At 56 – 58°C, the DSC curve for PPCS-4 enriched with *tt*-sequences (Figure 2.7, curve 1) displays the area of glass transition, and at 170 - 180°C – low endothermal effect with ΔH = 1.0 J/g. In accordance with RSA data, T_g of PPCS-4 is about 60°C. In the range of 60 -160°C, the diffraction pattern possesses two narrow reflexes above the first amorphous halo: 2θ = 8.02° and 8.39°. The second amorphous halo is localized at 2θ = 15° - 30°. At 160 -180°C, reflex 8.39° is decreased and reflex 8.02° increased. Above 220°C, the third amorphous halo is observed (2θ = 14.20°, $d \approx$ 6.23 Å); further increase of temperature does not change the diffraction image. Comparing data on RSA and DSC, the authors concluded that below 250°C PPCS-4 exists in the mesomorphous state. In accordance with the data of thermooptical analysis, isotropization of PPCS-4 is observed in the temperature range of 320 - 340°C, simultaneously with degradation of the polymer.

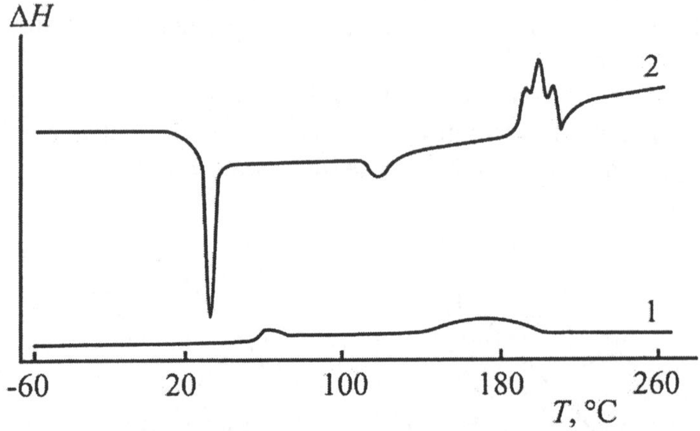

Figure 2.7. DSC curves for PPCS-4 (1) and PMPCS-4 (2)

More detailed description of PPCS-4 polymesomorphism and calculations of the cell parameters at chain packing in the basic plane are given in ref. [38]. Negligible variations proceeding in the basic plane at the mesophase II – mesophase I transition correlate with low heats of it. Limitation of mobility of irregularly distributed phenyl groups in cyclic unit of the polymer must induce occurrence of sufficient anisotropy of intermolecular forces and, consequently, deformation of the lattice.

Thus considered data on three POCS-4 possessing different spatial structures do not allow an unambiguous conclusion that enrichment of the polymer by *tt*-sequence units always causes occurrence of a mesophase, but hence, in the majority of cases the crystallinity degree is increased. Apparently, the data obtained should indicate that higher contribution to stabilization of the mesophases is provided by strong intermolecular interactions; therefore, the side framing is the determining factor for realization of polymesophase transitions in cyclolinear POCS-4.

The authors have also studied the effect of organic framing on properties of POCS-4 polymers. It is common knowledge that introduction of side groups of various origin causes changes in properties of linear poly(diorganosiloxanes) [39 – 42]. The above-mentioned data on PMCS-4 and PMPCS-4 show similar dependence of the phase states of polymers on the origin of organic substituting agents in POCS-4 polymers. That is why for better understanding of the influence of intermolecular interactions on POCS-4 properties, poly(hexaalkyl-cyclotetrasiloxanes) (PACS-4) were synthesized [43].

There are two reasons for studying properties of PACS-4 on the example of atactic polymers. The first one is that synthesis of pure *trans*- and *cis*-isomers of dichloro- and dihydroxy-derivatives of hexaethyl(*n*-propyl)cyclotetra-siloxanes is impossible. The use of dichloro- and dihydroxy-derivatives with different enrichment level in the HFC reaction by scheme (5) meant introduction of an additional factor, which is different ratios of *tt*- and *cc*-sequences which, as shown above, play an independent role. The second reason is that atactic polymers, synthesized in accordance with scheme (5), possess closer molecular masss and molecular mass distribution values. Table 2.8 shows main parameters of PACS-4. Studies of PACTS properties by RSA method have shown that atactic PACS-4 with ethyl (PECS-4) and *n*-propyl substituting agents (PPCS-4) are not crystallized, but a single narrow reflex at $2\theta = 10 - 11°$ and amorphous halo at $2\theta = 20°$ are observed on the diffraction pattern in a broad temperature range. If methyl substituting agent is changed by ethyl or propyl one, the narrow reflex is shifted towards low angles. The character of PACS-4 distribution image in the temperature range from T_g to 230°C for ethyl substituting agent and to 270°C for *n*-propyl one does not change. RSA data and clarification temperatures, T_{cl}, indicate that above T_g, atactic PACS-4 (except for methyl one) exists in the mesophase and transits to isotropic melt either at 220 - 230°C (as for PECS-4) or at polymer degradation temperature (as for PPCS-4). Since for these polymers such factors as MM, MMD and tacticity of polymeric chain are close, one may conclude that occurrence and formation of the mesophase is stipulated by the increase of intra- and intermolecular interactions between alkyl substituting agents in cyclolinear polymeric chain.

Table 2.8

Physical and chemical parameters and interchain distances of atactic POCS-4

No.	R = R' in POCS-4	T_g, °C	T_{cl}, °C	d_1, Å
1	CH_3	-51	-	8.4
2	C_2H_5	-110	280	8.8
3	n-C_3H_7	-55	300	10.4
4	C_6H_5	+58	> 300	11.2

Comparing phase diagrams of PACS-4 and linear polydialkylsiloxanes, one may easily observe general tendency of T_g and T_{inv} variation in organosiloxane polymers, for PACS-4 all temperature transitions being shifted towards higher temperatures [43, 44].

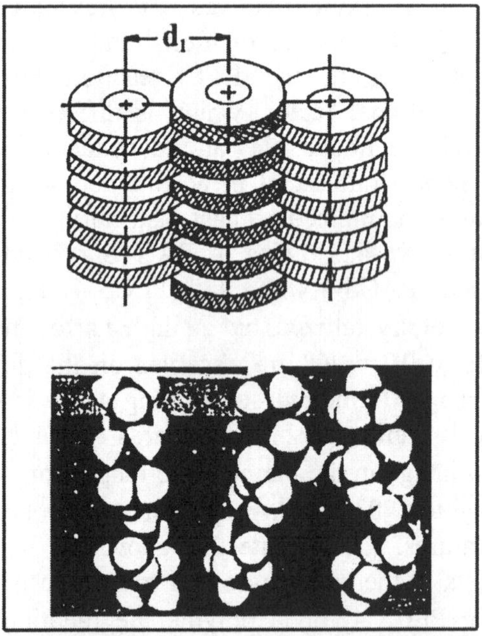

Figure 2.8. Hypothetical model of POCS-4

For polymeric chains from *tt*-sequences, analysis of PACS-4 structure, carried out on Stewart – Briegleb models, indicates a tendency of plane zigzag formation by the siloxane backbone, formed from intercyclic oxygen atoms [43]. Evidently, PACS-4 may be approximated by cylinders, diameters of which increase with the volume of alkyl substituting agent. Interchain distance, d_1, indicated by X-ray diffraction data, varies from 8.4 (PMCS-4) to 10.4 Å (PECS-4). Figure 2.8 shows a hypothetical model of POCS-4.

Basing on the present model, one may suppose existence of a package representing somewhat peculiar packs of polymeric disks. However, for PACS-4, interchain distance, detected by the RSA method, is by 3.0 Å shorter than for the model suggested. In this connection, two variants are possible: the first one suggests occurrence of side groups in neighboring chains by 1.5 Å [43]; the second one drives rings to some angle to the main axis of the macromolecule. Based on atom – atomic approximation of the idealized structure model, one may decide which alternative is more real for PMCS-4 at the atomic level. The first part of such modeling is calculation of conformations of isolated polymeric unit and chain fragments [45]. The second part of modeling is packing of polymeric chains in the crystal with low energy of the lattice. Computer modeling was carried out using POLYM software. Calculations performed indicate that the second alternative is more real, when rings along the backbone are disposed at some angles repeated in next nearest units:

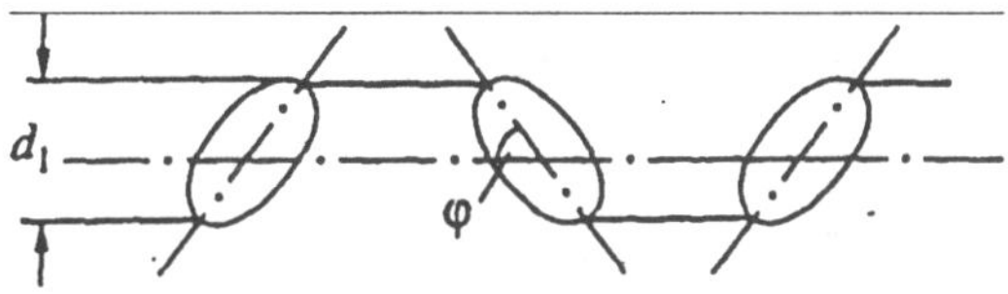

The type of property variation at transition from PMCS-4 to PPCS-4 at gradual increase of the number of phenyl substituting agents in the rings for *trans*-tactic polymers is shown in Table 2.6. Figure 2.10 shows diagrams of the phase state of PMPCS-4 and PPCS-4. Contrary to PACS-4, for PMPCS-4 clear dependence of temperature range of the mesophase existence on the number of phenyl substituting agents is absent. Apparently, irregular distribution of bulky substituting agents in cyclic fragment in every particular case introduces its own specific features. Among all the above-considered POCS-4, one polymer only possesses the mesophase obtaining it from the monomer - octaphenylcyclotetrasiloxane [38].

Summing up the data on POCS-4 phase transitions, unambiguous answer cannot be given, if side substituting agents stabilize the mesomorphous state of the polymer or not. Melting temperature of the polymer increases with the number of phenyl groups in the unit, and the range of existence of one- and two-dimensional regular structures is converged, because the initial degradation temperature is about 300°C.

The effect of POCS-4 molecular mass on the phase states is detected. Among works devoted to the study of MM effect on isotropization temperature variation, T_{iso}, in organic LC-polymers, the ones by Blumstein *et al.* should be outlined [46, 47]. These works show that T_{iso} increases with MM, for organic LC-polymers with mesogenic groups in the backbone in a quite narrow range of molecular masses.

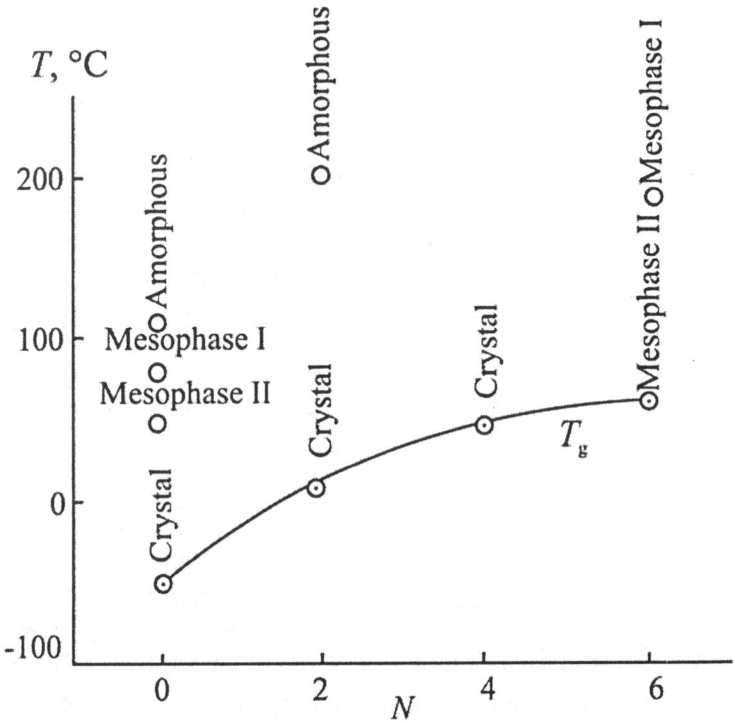

Figure 2.9. Influence of the number of phenyl groups, *N*, in POCS-4 unit on phase transitions

Study of the molecular mass effect on the mesophase – isotropic melt phase transition, carried out for polydiethyl- (PDES) and polydipropylsiloxanes (PDPS), has shown that border values of T_{iso} are reached for PDES at the polymerization degree above 3,000 [42] and for PDPS at 600 [41]. For PMCS-4, the effect of molecular mass was considered for three samples. Data from Table 2.7 indicate that the mesophase - isotropic melt temperature transition occurs and reaches border values at much lower polymerization degrees. Data from works [29, 43] on T_{iso} variation for PACS-4 prove that border values by T_{iso} for these polymers are also reached at much lower molecular masses.

For the purpose of studying the effect of cyclic unit on polymeric chain, the HFC reaction of 1,5-dichlororganocyclopentasiloxanes with 1,5-dihydroxyorganocyclopentasiloxanes in the presence of hydrogen chloride acceptor and synthesized polyorganocyclosiloxanes (POCS-5) were studied in accordance with the scheme as follows [27, 36, 48, 49]:

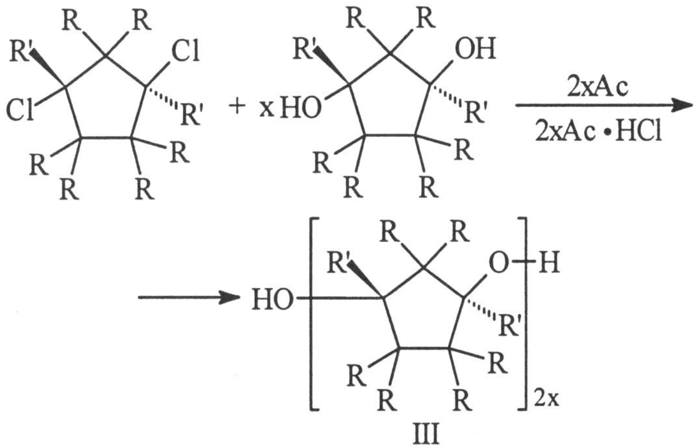

(2.10)

where R = R' = Me [27], Et [48], *n*-Pr [48]; R ≠ R'; R = Me; R' = Ph [49].

Forasmuch as neither of initial compounds was separated into spatial isomers (though the affinity to enrichment of a mixture of *trans*-isomers was observed), synthesized polymers possessed atactic structure.

Table 2.9 shows physical and chemical parameters of synthesized POCS-5. Data of RSA and DSC indicate variations of T_g of POCS-5 analogous to PACS-4 with regard to the length of alkyl substituting agents. However, the difference in T_g of PMCS-4 and PMCS-5 should be outlined, which equals approximately 20°C. Hence, T_g of PECS-4 and PECS-5 are equal (Tables 2.7 and 2.9).

Table 2.9

Physical and chemical parameters of POCS-5

Polymer, No.	R in POCS-5	Yield, %	$[\eta]$ (25°C)	$\bar{M}_n \times 10^{-3}$	T_g, °C
1	Me	75	0.27	30.12	-72
2	Et*	57	0.11	-	-111
3	*n*-Pr	60	0.09	-	-80
4	Ph**	59	0.11	-	-35

Notes: * T_{iso} = 40 - 50°C.

** POCS-5 with methyl groups in organosilsesquioxane fragments partly substituted by phenyl ones.

All POCS-5, except for PECS-5, are amorphous above T_g; according to RSA data, PECS-5 is mesomorphous above T_g, which is proved by three narrow reflexes at $2\theta = 8.98°$, 9.57° and 9.93°. This diffraction is preserved up to 40 - 50°C. At this temperature on the DSC curve, termination of heat absorption is observed. For PECS-5, variations on the diffraction pattern at 20°C correlate well with the beginning of heat absorption on the DSC curve (Figure 2.10). At the same temperatures (5 - 20°C) texture disappearance is observed on a polarization microscope.

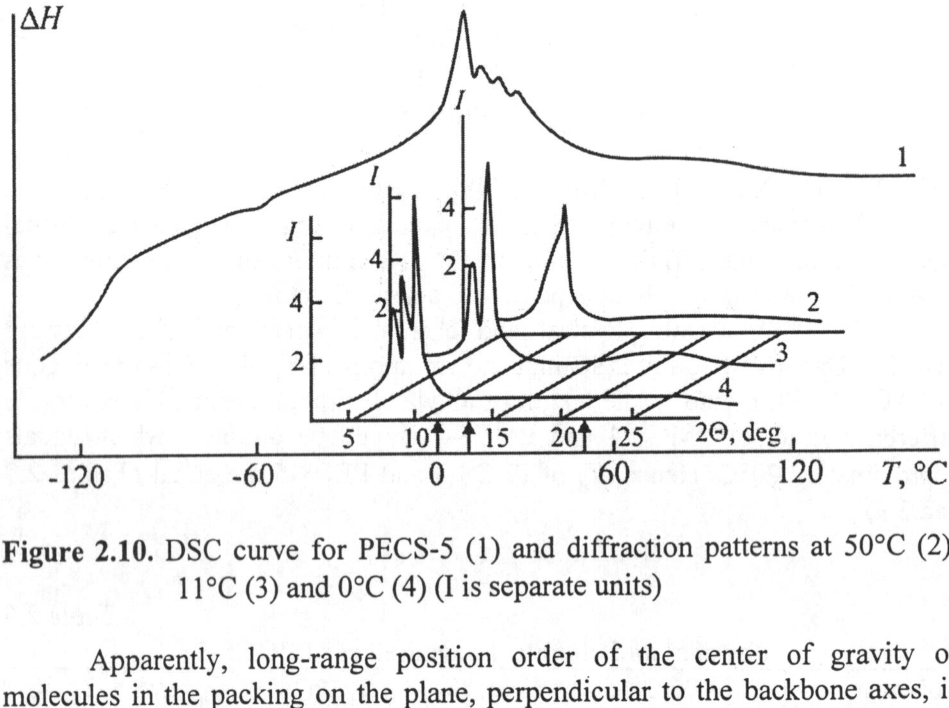

Figure 2.10. DSC curve for PECS-5 (1) and diffraction patterns at 50°C (2), 11°C (3) and 0°C (4) (I is separate units)

Apparently, long-range position order of the center of gravity of molecules in the packing on the plane, perpendicular to the backbone axes, is realized for PECS-5. As not many reflexes are observed above and beneath the transition point on diffraction patterns of PECS-5, determination of the type of packing in the basic plane is difficult.

For low-temperature phase of PECS-5, rectangular cell is the most probable in the basic plane; it is also probable 10°C above the transition point. Thus for PECS-5, which unit is represented by asymmetrical rings, formation of highly ordered crystalline as well as one- and two-dimensional ordered structures is hindered. However, due to intensification of intra- and intermolecular interactions of organic groups, conditions for overlapping between macromolecules may occur. Finally, introduction of polar groups may induce a mesogenic fragment to become more symmetrical.

More representatives of bead-shaped cyclolinear polymers are POCS-6, backbones of which are composed of cyclohexasiloxanes. They differ from POCS-5 by inclusion of one more diorganosiloxane fragment to the ring and the ratio between diorganosiloxane fragments to diorganosilsesquioxane ones reaching the value of two. In spite of difunctional octaorganocyclopentasiloxanes, in the majority of cases, dichloro- and dihydroxy-derivatives of decaorganocyclohexasiloxanes are easily divided into *cis-* and *trans*-isomers, which gives an opportunity to obtain polymers with various sequences in the configuration. HFC reaction of 1,7-dichlorodecaorganocyclohexasiloxanes with 1,7-dihydroxydecaorganocyclohexasiloxanes mainly proceeds by the scheme as follows:

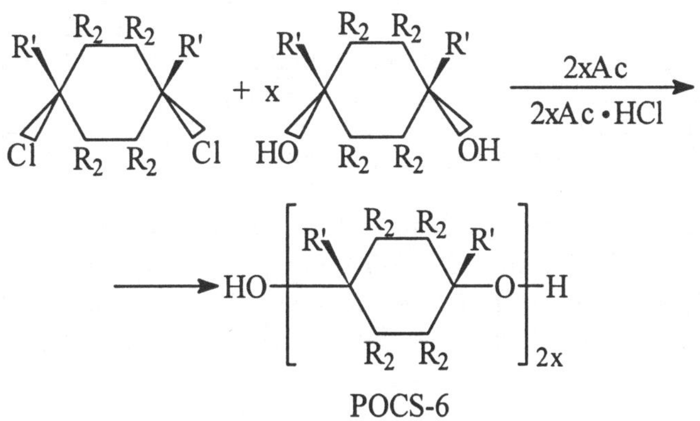

(2.11)

where R = R' = Me [27], Et [36], Ph [50]; R ≠ R'; R = Me; R' = Ph [51]; R = Ph, R' = Me [50, 51].

Synthesized in this HFC reaction were copolymers, completely soluble in usual organic solvents with $[\eta] = 0.10 - 0.24$. The authors of the works have studied the influence of HFC reaction conditions on configuration sequences of synthesized polymer. ^{29}Si NMR spectra of pure *trans*-isomers and mixtures with different ratios of *cis-* and *trans*-isomers have indicated that spatial configuration of initial monomers is not fully preserved, and in some cases, inversion of ≡Si-Cl centers exceeds 50% [35]. Factors affecting variations in configuration of initial difunctional organocyclohexasiloxanes during HFC reaction were also studied by ^{29}Si NMR analysis. By analogy, relation of reflexes in ^{29}Si NMR spectra to spatial *tt*, *ct*, *tc* and *cc* configurations is based on the fact that fission of silsesquioxane atom may be stipulated by spatial isomerism of the backbone components only.

Data from Table 2.10 on PMCS-6 prove that HFC reaction in the presence of HCl acceptor (pyridine) proceeds with partial inversion of dichlorodecamethylcyclohexasiloxane, because three reflexes are detected in spectra. In the case of partial inversion of dioxy-derivatives, polymers 1 and 2 would contain all possible combinations of units, i.e. four reflexes of T-fragments.

Table 2.10

Conditions of synthesis, chemical shifts and intensity of reflexes in ^{29}Si NMR spectra for PMCS-6

Polymer, No.	*Trans-/cis*-isomer ratio in initial monomers		Relative intensity of δ_{Si} reflexes for dyad units and sequences				
	In dichloride	In diol	D	T			
			-21.95	-67.60	-67.63	-67.55	-67.58
1	0.52/0.48	1.00/0	-	0.28	0.35	0.37	-
2	0.52/0.48	0/1.00	-	-	0.23	0.21	0.56
3	0.52/0.48	0.52/0.48	-	0.23	0.23	0.24	0.30

A series of experiments studying the effect of acceptor and solvent origins in the HFC reaction during PMCS-6 synthesis give results analogous to the ones, previously executed for PMCS-4. However, one more additional factor affecting on the inversion degree of the initial dichlorodecamethylcyclo-hexasiloxane was observed: configuration of preceding unit in the backbone (configuration of dihydroxy-derivative). As *trans*-isomer of the initial monomer diol (Table 2.10) is used, the ratio of *cis*- and *trans*-units in the polymer becomes equal 0.72:0.28 from the initial one in dichloro-derivatives, which equals 0.52:0.48 [35].

Pure *cis*-dihydroxy-derivative was one of the initial products for synthesizing polymer 2, and units formed from dichloro-derivativepreserved the configuration ratio, present in the initial monomer. Because minimal inversion equals 12 – 13%, it is quite probable that ≡Si-Cl center is inverted not in the initial molecule of cyclic dichloride, but in oligomer composed of *tt*-sequences. This molecule consists of ~5 units (as shown below, equilibrium rigidity of *trans*-tactic PMPCS-6 is higher than that of atactic polymers). That is why for somewhat decrease of the backbone rigidity, one ≡Si-Cl center is inverted in the HFC reaction starting from a definite length of oligomers at consecutive act of addition of initial diol or interaction of two oligomers (of any length). There are no grounds for considering inversion proceeding at the interaction of two initial organocyclohexasiloxanes only, as it has been suggested for PMCS-4. But simultaneously, there are no data disproving the supposition that the inversion proceeds at the stage of oligomerization. The latest results of synthesis of oligomers with strictly defined polymerization degree (three, five, seven repeated units) prove the case, when ≡Si-Cl center is inverted as the end group in oligomers [52].

The effect of chain tacticity on the phase states of POCS-6 has been studied. On the example of two POCS-6 polymers, the influence of configuration sequences on thermal transitions has been investigated [45]. For this purpose, *cis*- and *trans*-isomers of 1,7-dihydroxydecamethylcyclohexasiloxane, 1,7-dichloro-1,7-diphenyl-3,3,5,5,9,9,11,11-octamethylcyclohexasiloxane and its pure *cis*- and *trans*-dihydroxy-derivatives were used. Polymers enriched with *tc*- and *ct*-sequences were synthesized by interaction of mixture of dichlorodecamethylcyclohexasiloxane *cis*- and *trans*-isomers with *trans*-diol (Table 2.11, polymer 1), *cc*-sequences being completely absent.

Polymer 3 (Table 2.10), in which *tt*-sequences were absent, was synthesized from pure *cis*-dihydroxydecamethylcyclohexasiloxane. Structures of polymers 1 and 3 are the following:

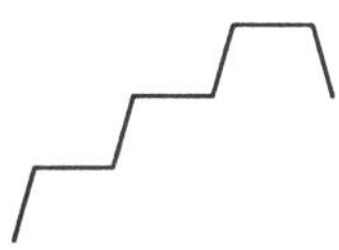

tt-tc (polymer 1)

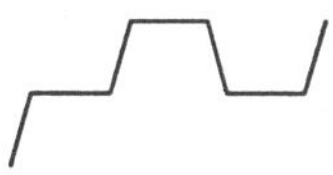

tc-cc (polymer 3)

The image of sequences in the cyclolinear chain represents projections of cyclic units and bonds connecting them to the plane perpendicular to the ring plane.

The structure of polymer 1 is named *trans*-microtactic one, and structure 3 is *cis*-microtactic one. Atactic polymer 3 contained all types of *tt*, *tc*, *ct* and *cc* configuration sequences in approximately equal fractures.

Four cyclolinear PBPCS-6 polymers of different structures were synthesized from pure *cis*- and *trans*-isomers of dichloro- and dihydroxy-derivatives of diphenyloctamethylcyclohexasiloxanes: the first one is enriched with *tt*-sequences in the absence of *cc*-combinations (*trans*-tactic polymer); the second one is enriched with *cc*-sequences in the absence of *tt* ones (*cis*-tactic polymer); the third one is mostly enriched with *tc*-sequences from *cis*-dichloro-

Table 2.11

Some physical and chemical parameters of PMCS-6

Polymer, No.	*Trans-/cis*-isomer ratio in initial monomers		Yield, %	[η] at 25°C, dl/g	$\bar{M}\times 10^{-3}$	T_g, °C	Mesophase-isotropic melt transition			T_{cl}, °C
	In dichloride	In diol					T_{iso}, °C	ΔH_{iso}, kJ/mol	ΔS, J/mol·deg	
1	0.70/0.30	1.00/0	71	0.15	46.0	-91*	320	4.3	7.2	295-305
2	0.55/0.45	1.00/0	79	0.24	-	-91	310	5.4	-	270-290
3	0.40/0.60	0.03/0.97	76	0.15	-	-91	250	2.9	5.5	250-260
4	0.50/0.50	0.50/0.50	80	0.17	24.1	-91	294	4.6	8.1	220-240

derivative and trans-diol (due to partial inversion of dichloride, *tt*-sequences are present; that is why spatial structure of the third polymer is similar to the first one); the fourth one represents atactic polymer. In the first two cases, *tc*-sequences are present in the NMR spectrum due to partial inversion of initial dichloro-derivative. All polymers with different tacticity types displayed high yields, except for the polymer of *cis*-tactic structure, the yield of which is much lower due to intramolecular cyclization reaction, which induces formation of tricyclosiloxane (in the case of methyl groups at silicon atoms).

The phase state of PMCS-6 polymers of different tacticities was identified by the methods of DSC, RSA and thermooptical analyses. Figure 2.11 shows DSC curves for PMCS-6 polymers of various tacticities. Thermal characteristics of all transitions in PMCS-6 are shown in Table 2.10. Therefore, T_g is defined by chemical structure of the polymer unit and is independent of its spatial structure. The melting point is observed only for *trans*-microtactic PMCS-6 (polymer 1), the melting heat and, consequently, degree of crystallinity of PMCS-6 (polymer 1) being low. Above T_g, atactic and *cis*-microtactic PMCS-6 polymers exist in the mesomorphous state up to 300°C; *trans*-microtactic PMCS-6 polymer is the only one transiting into the crystalline phase above T_{melt}. Such interpretation of phase states is proved by RSA data.

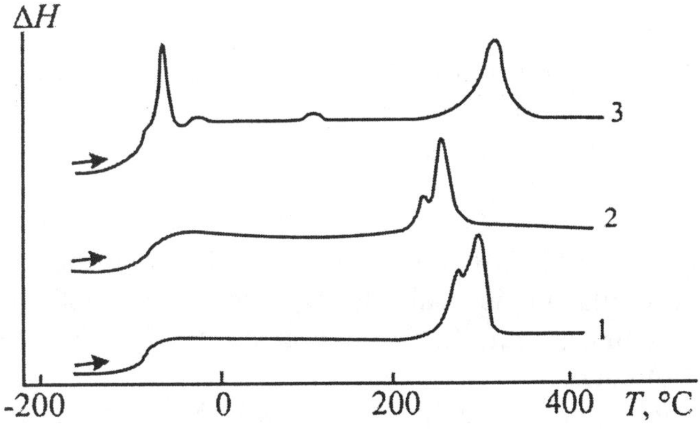

Figure 2.11. DSC curves for PMCS-6 of different structures: 1 – atactic polymer; 2 – *cis*-microtactic polymer; 3 – *trans*-microtactic polymer

PMCS-6 diffraction patterns in the temperature range from T_g to 250 - 300°C displays a single narrow reflex (1/2Δ = 20′) of high intensity at 2θ =

10.5°, corresponded to the interchain distance. Optical observations on a polarization microscope indicate that the polymers possess the birefringence property [32].

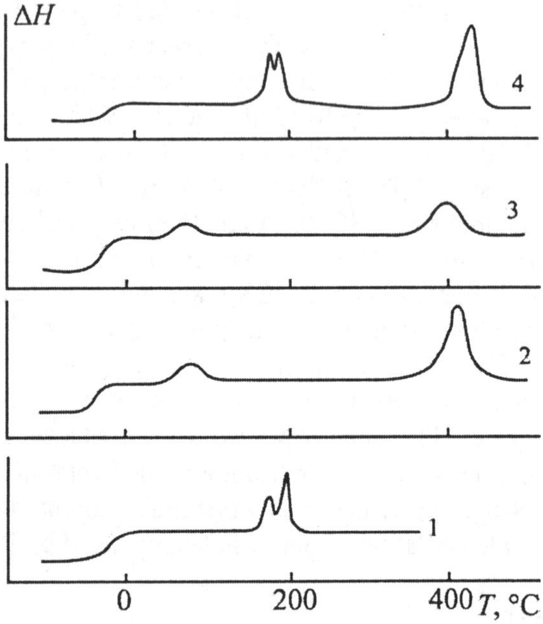

Figure 2.12. DSC curves for PMPCS-6 of different structures: 1 – *cis*-microtactic polymer; 2 – polymer with regular alternation of *trans*- and *cis*-sequences; 3 – atactic polymer; 4 – *trans*-microtactic polymer

In spite of PMCS-6, DSC curves of PMPCS-6 independently of the tacticity type of polymeric chain display additional endothermal transitions, induced by melting (Figure 2.12). Proceeding of PMPCS-6 crystallization is proved by RSA data.

The ability to crystallization of PMPCS-6 samples depends on the cooling rate, but the mesophase is formed in these polymers even at 300 deg/min cooling rate, and mesomorphous glass is formed below T_{melt}. The mesophase – isotropic melt phase transition was detected for *cis*-tactic PMPCS-6 polymer only; the rest polymers remain in the mesophase in the temperature range from T_{melt} to 300°C.

Hence, it was concluded [27, 36, 51] that independently of dominance of one of the sequence types (*tt*, *cc* or *ct*) or they all exist simultaneously (atactic polymer), all POCS-6 polymers may exist in the mesomorphous state in a wide temperature range. Actually, the specific ability to form the mesophase in poly(organocyclohexasiloxanes) should be corresponded exclusively to occurrence of regular one- and two-dimensional formations, induced by additional specific interactions of the cyclohexasiloxane unit, because other factors (organic substituting agents, chain tacticity, MM and MMD) for all synthesized cyclolinear polymers with different cyclic units are almost the same.

These investigators [27, 36, 51] have also studied the effect of organic framing at silicon atoms on properties of POCS-6 polymers. The influence of organic framing on temperature and heat of thermal transitions in POCS-6 polymers is studied on the example of *trans*-tactic and *trans*-microtactic poly(methylphenylcyclohexasiloxanes) with different content of phenyl substituting agents in the units of homopolymers [53]:

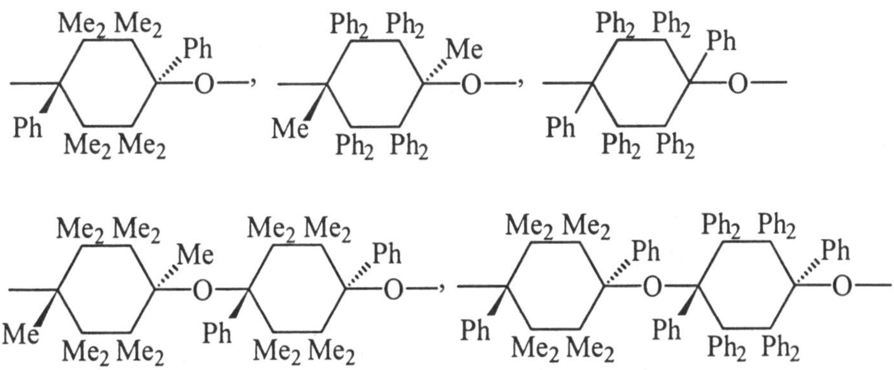

Dependencies of T_g, T_{melt} and T_{iso} on content of phenyl groups in PMPCS-6 polymers and copolymers are shown in Figure 2.13. The phase diagram of PMPCS-6 indicates that sequential substitution of methyl groups by phenyl ones causes a monotonous increase of T_g.

Differences between T_g of PMCS-6 and linear PDMS polymer are low, and some increase of T_g reflects restrictions of the local mobility caused by the presence of cyclic sequences in the chain. T_g values of homopolymers and PMPCS-6 copolymers fall within the range, limited by glass transition temperatures of PDMS and PDPS polymers. All PMPCS-6 polymers possess the ability to crystallization. Values of T_{melt} of PMCS-6 polymer falls within the range typical of PDMS melting. Introduction of phenyl substituting agents increases T_{melt} of PMPCS-6, which is the most abrupt at introduction of initial ones.

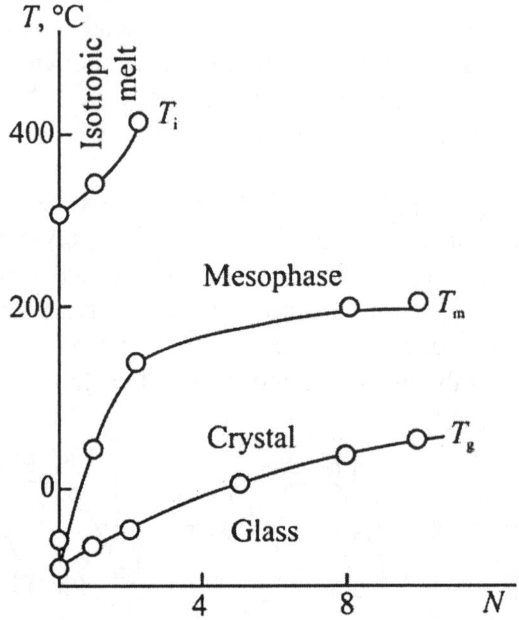

Figure 2.13. Dependencies of T_g, T_{melt} and T_{iso} on the quantity of phenyl groups, *N*, in POCS-6 polymers

Crystalline and mesomorphous phases in PMPCS-6 are identified with the help of RSA data, which prove the fact of their partial crystallization. Above T_{melt}, all PMPCS-6 polymers and copolymers display the mesomorphous phase. All diffraction patterns on PMPCS-6 display narrow, extremely intensive reflex at $2\theta = 8 - 10°$, which is typical of all siloxane polymers forming the mesophase. Values of T_{melt} and T_{iso}, determined from DSC and RSA data, correlate well with one another. Figure 2.14 shows the dependence of melting and isotropization heats for PMPCS-6 on the quantity of phenyl groups in the polymer unit. The heat of PMPCS-6 melting is sharply increased with the quantity of phenyl groups in the ring, approaching the melting heat of linear PDPS [54]. In spite of melting heat, dependence of the heat of PMPCS-6 isotropization on the content of phenyl groups in the unit is abruptly decreased after introduction of initial phenyl substituting agents.

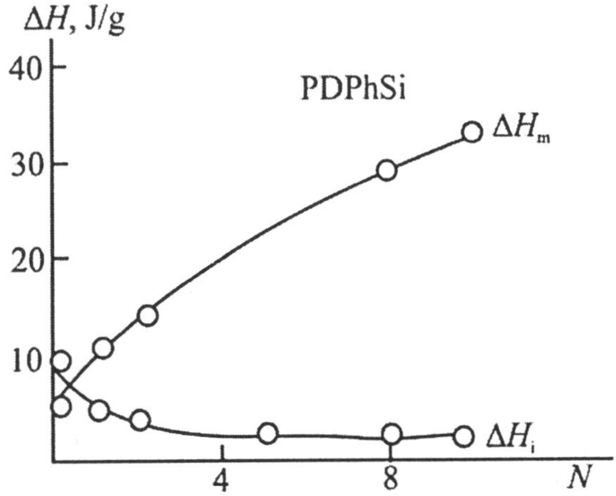

Figure 2.14. Dependence of melting and isotropization heats on the quantity of phenyl groups, *N*, in units of PMPCS-6 polymers and copolymers

Anomalously low values of ΔH_{melt} and ΔH_{iso} of PMPCS-6 copolymers were detected. This anomaly is explained by the fact that for many PMPCS-6, the upper temperature border of the transition from the mesophase to isotropic melt often falls within the temperature range of thermal decomposition. The data shown indicate that variation of the side framing in cyclolinear chain of organosiloxane polymers with a cyclohexasiloxane unit does not eliminate the mesophase, but just varies the temperature range of its existing and, apparently, may cause an effect on the mesophase degree.

The effect of molecular mass on occurrence of the mesomorphous phase in POCS-6 polymers is detected. Results displayed for POCS-6 polymers, in which configuration sequences of units alternate irregularly along the backbone, indicate that the ability to crystallization in them is disturbed. For the mesomorphous phase, strict regularity of units is not the necessary condition of its occurrence. In connection with the difference between studied POCS-6 polymers by the type of tacticity and molecular mass, the authors have separated effects induced by these two factors and detected more accurate, which type of configuration sequences creates conditions for mesophase stabilization.

To study dependencies of POCS-6 translucence temperatures on molecular mass (MM), PMPCS-6 and PMCS-6 were fractioned, and molecular masses and isotropization temperatures of the fractions were determined [43, 55, 56]. Data in Table 2.12 show that atactic PMCS-6 possesses the mesophase in all fractions. The isotropization temperature is decreased slowly first and then abruptly with the polymerization degree.

Such type of T_{iso} dependence on MM of cyclolinear polymers is quite different from the analogous dependence for PDES [42] and PDPS [41], in which (as mentioned above) the ability to form the mesophase is observed at polymerization degrees as follows: $P > 200$ for PDPS and $P > 500$ for PDES. For fractioned PMCS-6 samples, isotropization heat and entropy are much higher and the temperature range is much narrower, which may be induced by the solvent applied and conditions of sample precipitation during fractionation.

A sharp dependence of T_{iso} on polymerization degree, mesophase occurrence in low-molecular fractions, high values of T_{iso} and a broad range of mesomorphous phase existence are the indirect factors proving quite high equilibrium rigidity of atactic PMCS-6 polymer.

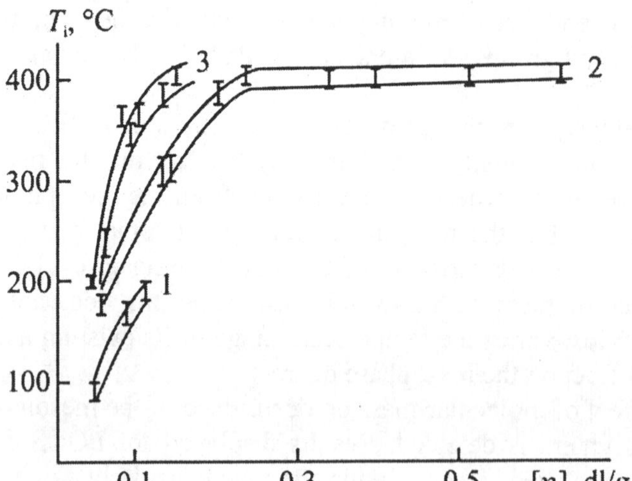

Figure 2.15. Dependence of T_{iso} on $[\eta]$ for PMPS-6 polymers: 1 – *cis*-microtactic polymer; 2 – atactic polymer; 3 – *trans*-microtactic polymer

Figure 2.15 shows dependencies of T_{iso} on inherent viscosity for PMPCS-6 of different chain tacticities. Clearly the highest T_{iso} values are inherent to cyclolinear polymers with *tt*- and *cc*-sequences, i.e. *trans*-tactic polymers. Hence, for the polymers with regular *tc* alternation, isotropization temperature is increased with molecular mass. The lowest T_{iso} values are displayed by *cis*-tactic PMPCS-6 polymers. For polymer fractions with regular

Table 2.12

Description of PMCS-6 fractions

Fraction No.	$[\eta]$, dl/g (25°C)	$\bar{M}_{\omega}$	Polymerization degree	T_g, °C	T_g, °C	ΔH_{iso}, kJ/mol	ΔS_{iso}, kJ/mol·deg
1	0.07	8,250	19	-88	202	4.40	9.70
2	0.09	8,750	20	-88	257	4.50	8.54
3	0.11	14,400	34	-88	277	4.60	8.40
4	0.13	16,000	37	-89	286	3.76	6.84
5	0.17	29,000	67	-88	294	5.65	10.0
6, 7	0.18	34,000	79	-89	295	5.50	9.80
8	0.24	44,000	102	-90	310	5.65	9.80
9	0.28	76,000	176	-89	313	5.56	9.66
Non-fractioned sample	0.14	24,000	61	-90	279	4.40	8.05

alternation of *trans-cis*-sequences, T_{iso} falls within the range between T_{iso} of *trans*-tactic and atactic polymers.

The packing structure of POCS-6 chains in the mesomorphous phase is of special interest. Consider the structure and conformation of decaorganocyclohexasiloxane unit on the Stewart-Brigleb models and compare with the temperature dependence of d_1 reflex.

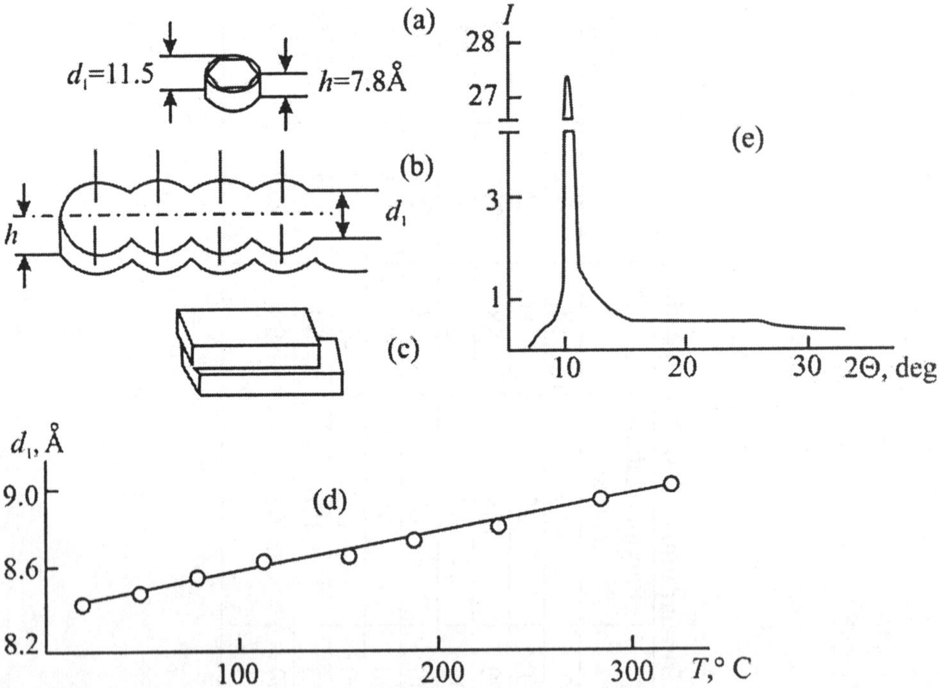

Figure 2.16. A hypothetical model of POCS-6 polymer unit (a), POCS-6 backbone fragment (b), and POCS-6 chain packing (c). Temperature dependence of interchain distance, d_1, for mesomorphous component of PMCS-6 polymer (d) and typical diffraction pattern of POCS-6 in the mesomorphous phase (e) (*I*, dimensionless units)

Figure 2.16a shows a disk-like structure of the unit with d_1 = 11.5 Å at the cycle layer thickness d_1 = 7.8 Å. Disks combines in a chain form plates looking like the so-called sanidic mesophase [57] in organic LC-polymers. The height of POCS-6 macromolecule layer is close to that determined for PMCS-6 polymer (d_1 = 8.4 Å). The presence of long-range order in the packing with the lower coherence border at 400 Å indicates that packing types shown in Figures 2.16b and 2.16c are exclusively possible at the current intermolecular distance (though similar distances are realized in the ring, too, however, so high long-range order cannot exist in them; moreover, temperature dependencies of inter- and intramolecular distances are different). The model suggested indicates presence of a single maximum corresponded to the long-range order in one direction, perpendicular to the plane of the ring, which is the polymer backbone unit.

Cyclolinear structure of synthesized POCS-4, POCS-5 and POCS-6 polymers was proved by hydrodynamic study of properties of their solutions [58]. The expected results of these investigations should indicate the predominant influence of the equilibrium flexibility of macromolecules and other specific properties of polymeric chains in block on occurrence of the thermotropic mesophase in cyclolinear organosiloxanes.

The polymers were fractioned into 8 – 14 components by coacervate extraction from the benzene – methanol system. For fractions and non-fractioned polymers, characteristic viscosities, $[\eta]$, were measured on a viscosimeter with suspended leveler at 25°C. Molecular mass was measured by the light scattering method on Fica Co. photogoniodiffuser at non-polarized light wavelength of 546 nm.

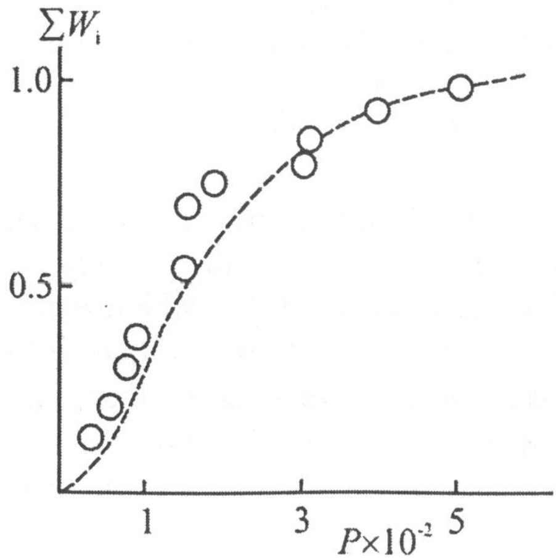

Figure 2.17. Theoretical (dashed) and experimental (dotted) dependencies of MMD for PMCS-5 polymers. *P* is the polycondensation degree

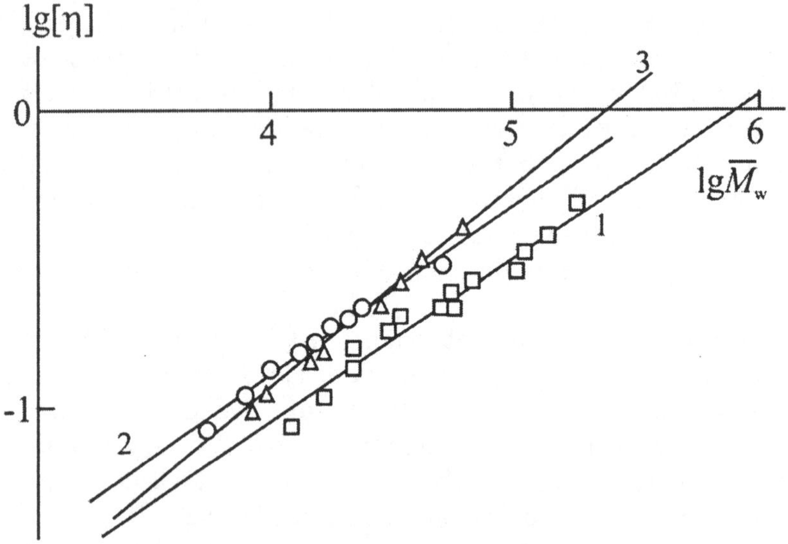

Figure 2.18. Dependence of lg[η] on $\lg \overline{M}_\omega$ for: 1 – PMCS-5 polymer; 2 – PMCS-4 polymer; 3 – PMCS-6 polymer

The device was calibrated for benzene. The refraction index increment, d*n*/d*c*, was measured on Pulphricha type refractometer equipped with a cuvette for differential measurements. For PMCS-4 polymer d*n*/d*c* = -0.079, for PMCS-5 polymer d*n*/d*c* = -0.080, and for PMCS-6 polymer d*n*/d*c* = -0.082. All experimental results were treated by the method of least squares. Figure 2.17 shows the MMD curve with polydispersion index two for PMCS-5 polymer, which is very close to the most probable one.

Because that was the first example of studying conformations of macromolecules of this type in diluted solutions, authors of the work [58] paid mucg attention to selection of an equation, which would adequately describe hydrodynamic behavior of polymeric chains. Figure 2.18 shows dependencies of [η] on molecular mass (MM), represented in double logarithmic coordinates. Parameters of the Mark-Kuhn-Hauvink equation for toluene medium at 25°C were determined from the slope and disposition of the straight lines.

Table 2.12

Conformation parameters of poly(organosiloxanes)

Polymer	Unit structural formula	a*	Kuhn segment, A	References
PDMS	~~~Me_2SiO~~~	0.5 - 0.8	10	[59]
PMCS-4	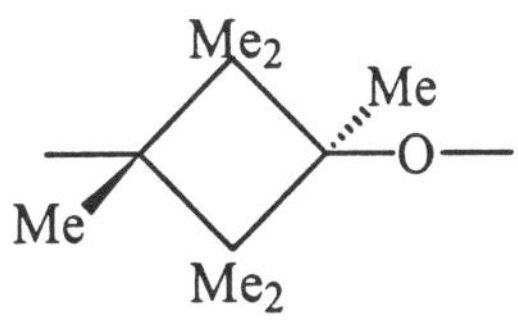	0.71	29	[58]
PMCS-5	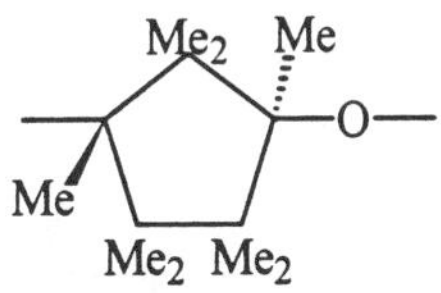	0.64	23	[58]
PMCS-6	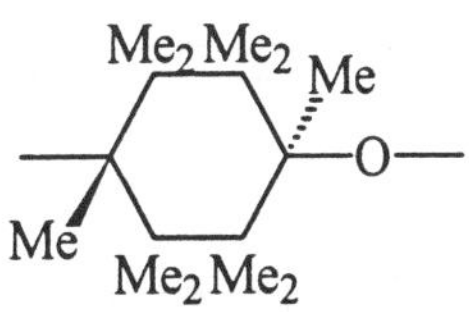	0.78	38	[58]
PMPCS-6 (atactic)	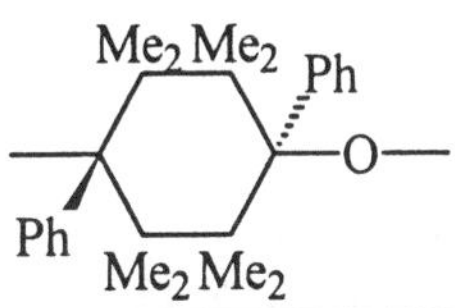	0.98	80	[26]
PMPCS-6 (*trans*-microtactic)	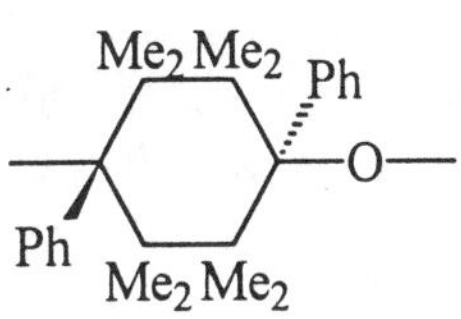	1.16	115	[26]
Poly(organo silsesqui-oxanes)	~~~$RSiO_{1.5}$~~~	0.5 - 1.8	50 - 130	[61 - 63]

Note: * In the equation $[\eta] = KM^a$.

Deviation of parameter *a* from 0.5 in this equation may be caused by volumetric effects in good solvents or partial hydrodynamic permeability of the coil, promoted by skeletal rigidity of macromolecules. Selection of a model for quantitative estimation of rigidity by these properties is usually complicated by the absence of reliable criterion, especially in the cases, when the value of parameter *a* falls within the range of 0.5 – 0.8. That is why to describe behavior of macromolecules in toluene, the authors [58] have used the model of perturbed Gaussian non-permeable coil and the model of worm-like backbone.

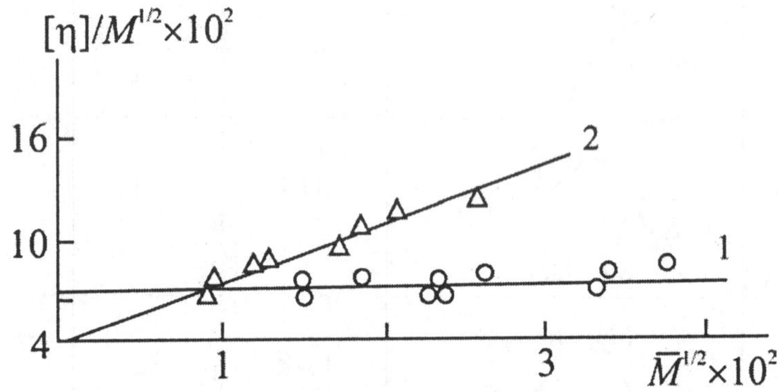

Figure 2.19. Extrapolation by the Fixman-Stokmayer method: 1 - PMCS-6 polymer; 2 - PMCS-5 polymer

Equilibrium rigidity of polymeric backbone was quantitatively estimated by the Fixman – Stockmayer technique based on experimentally determined hydrodynamic parameters within the frames of the Gaussian model of impermeable coil. This technique considers volumetric effects and neglects permeation of the coils [59].

Figure 2.19 shows experimental dependencies of $[\eta]/M^{1/2}$ on $M^{1/2}$ for PMCS-5 and PMCS-6 samples, parameters *a* of which are mostly different. As a consequence, conformation parameters and the value of the Kuhn segment, *A*, values of which are shown in Table 2.12, are determined.

Found values of conformation parameters of these polymers indicate that equilibrium rigidity of macromolecules is increased with the size and symmetry of siloxane unit. Hydrodynamic behavior of PMCS-4 and PMCS-6 give no reason for corresponding these polymers to classic flexible chain ones [60]. Introduction of phenyl substituting agents to organocyclohexasiloxane unit increases rigidity of macromolecules (the Kuhn segment 80 A). As for *trans-*

microtactic PMPCS-6, further on, the macromolecule rigidity is increased, which is comparable with rigidity of ladder poly(organosilsesquioxanes) [60 – 63].

Simultaneously, thermomechanical investigations were performed [28], obtained under continuous impact of compressing stress (100 g/cm^3) on the sample in accordance with the technique [64]. Glass transition temperatures, shown in the Tables, are determined from the primary deviation of thermomechanical curve run in the area of positive deformations. Depending on the spatial structure of the cycle, synthesized atactic and syndiotactic polymers are characterized by almost equal glass transition temperatures, T_g.

Thermogravimetric studies of synthesized polymers indicate different types of copolymers' degradation with due regard to substituting agents at silicon atom. For example, PMCS-4 methylsiloxane copolymer with regular alternation of dimethylsiloxane and methylsilsesquioxane units displays S-shaped curve of the mass loss, which reaches 98 – 99% at 650°C, and differential thermal curve displays a single maximum. Determination of volatile degradation product composition under isothermal conditions at 500 - 700°C has indicated that polymethylbicyclosiloxanes and polymethyltricyclosiloxanes of T_2D_n and T_4D_n composition (n = 2, 3, 4) are main products of the process. However, besides polymethylcyclosiloxanes, D_3 and D_4 ones were detected, D_4 content being below 3 – 4%. The data shown indicate degradation proceeding by the depolymerization mechanism. For copolymers containing diphenylsiloxane and phenylsiloxane units, the mass loss reaches 28 – 30% at 900°C, and DTA curves display three maximums at 490, 530 and 565°C. In the case of PMCS polymers, the composition of volatile degradation products displays benzene, methane and hexamethylcyclotrisiloxane. At simultaneous presence of methyl and phenyl substituting agents, the composition of volatile products indicates more complicated proceeding of degradation. However, if formation of benzene and methane can be explained by radical-type Si-C bond break, D_3 formation cannot be explained in the frames of previously suggested depolymerizational mechanism due to formation of transient four-center complex [65, 66].

As shown in the literature, recently, a new type of polymers was synthesized which, besides three usual (solid, glassy and liquid) phases, display liquid-crystal and mesomorphous phases [67, 68]. Two types of thermotropic liquid-crystal polymers were of special interest, i.e. polymers with mesogenic groups in the backbone and polymers containing mesogenic groups in comparatively short side chains. On the other hand, some polymers are known, capable of forming thermotropic mesophase without mesogenic groups.

Polydiethyl(propyl)siloxanes [69, 70] and polyphosphazenes [72] are the most well-known polymers of these group. In this relation, the above-mentioned *stereo*-regular cyclolinear methyl- and methylphenylsiloxane copolymers with different-size rings and framing groups in the macromolecular backbone are of special interest. Similar to poly(diethylsiloxanes) and poly(phosphazenes), mesophases in cyclolinear organopolysiloxanes are formed in the absence of mesogenic groups, which makes representatives of this class quite interesting.

2.2. Bead-shaped methylphenylsiloxane copolymers with ethylene bridges between cycles

Synthesis of carborganosiloxane copolymers and polymers is based on hydride polyaddition of organohydrosiloxanes to organoalkenyl silanes [72]. Recently, this reaction is of great interest in the field of obtaining complex monomers [73], as well as in the field of study of addition mechanisms on various Pt, Pd, Co and metal carbonyl catalysts [74 – 77] and specificity of actions of one or other catalytic systems [78]. Besides carborganosiloxane oligomers and linear polymers, other compounds were also synthesized by this method [79 – 81].

Discussed in the present Chapter are synthesis and investigation results on properties of carbosiloxane copolymers containing, besides organocyclotetrasiloxane fragments, organocyclopenta- and organocyclohexasiloxane fragments. The initial substances, from which carbosiloxane copolymers with cyclosiloxane fragments in the backbone are synthesized, divinyl- and dihydride-containing organocyclosiloxanes were considered in Chapter 1.

The first cyclolinear organosiloxane polymer with ethylene bridges between rings was synthesized by hydrosilylation of 1-hydro-3-vinylhexamethylcyclotetrasiloxane. Hydrosilylation performed in CCl_4 medium at 75°C in the presence of the Spire catalyst ($H_2PtCl_6{\cdot}6H_2O$) has synthesized a polymer, viscous-flow at room temperature, in accordance with the following scheme [82]:

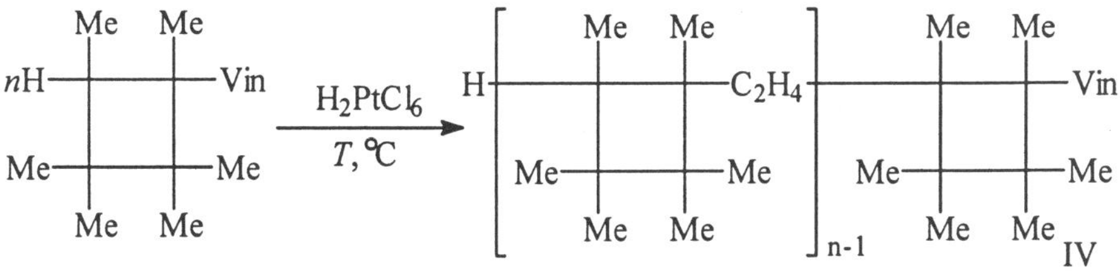

(2.12)

It is found that hydrosilylation proceeds without breaking cyclotetrasiloxane ring and a low-molecular polymer with molecular mass M = 2,260 is synthesized in the reaction, well-soluble in benzene and other organic solvents. Besides the peak displayed by ≡Si-Me groups, the NMR spectrum possesses one non-splitted peak typical of ≡Si-CH_2- groups with chemical shift (in relation to ≡Si-Me groups) equal δ = 0.35 ppm

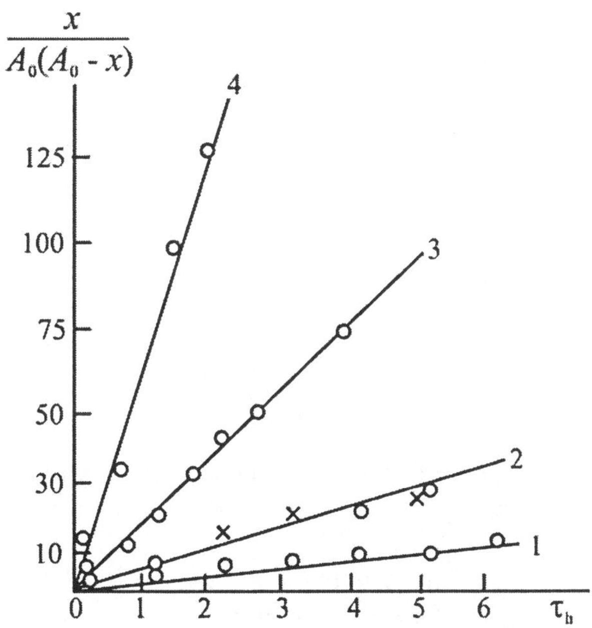

Figure 2.20. Dependence of concentration decrease with time at temperature, T (°C): 1 - 50°C, 2 - 65°C ($C_{cat} \approx 1.4 \cdot 10^{-4}$ mol/l); 3 - 50°C, 4 - 75°C ($C_{cat} \approx 4.2 \cdot 10^{-4}$ mol/l)

Kinetics of polyaddition reaction was studied by measuring concentration of active ≡Si-H groups at different temperatures and catalyst concentrations. The measurements were executed by the spectroscopic method - by calculation of integral intensity of the absorption band by IR-spectroscopy method. Figure 2.20 shows time dependence of variation of the initial cyclotetrasiloxane concentration at different temperatures; the catalyst concentration varies from $1.4 \cdot 10^{-4}$ to $4.2 \cdot 10^{-4}$ mol/l.

Figure 2.20 shows that polyaddition reaction is described by equation of the second order reaction. In accordance with the technique, described in ref. [83], rate constants of the reaction were calculated. The results of calculations are shown in Table 2.13. Figures 2.21 and 2.22 show dependencies of the rate constant of polyaddition on reverse temperature and catalyst concentration.

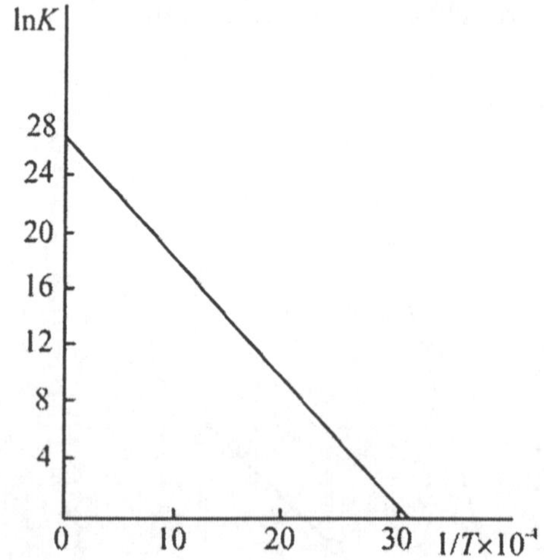

Figure 2.21. Dependence of the rate constant of polyaddition on reverse temperature of the reaction

From Figure 2.21, the activation energy (E = 17 kcal/mol) and the Arrhenius factor ($A = 2.9 \cdot 10^{8}$ l/mol·s) for catalyst concentration ($C_{cat} = 1.4 \cdot 10^{-4}$ mol/l) were calculated by the graphical method. Figure 2.22 shows dependence of the reaction rate constant on the catalyst concentration.

The dependence shown is linear and indicates that polyaddition reaction proceeds via the stage of an intermediate compound formation of the monomer with the catalyst and is the first order reaction by the catalyst concentration.

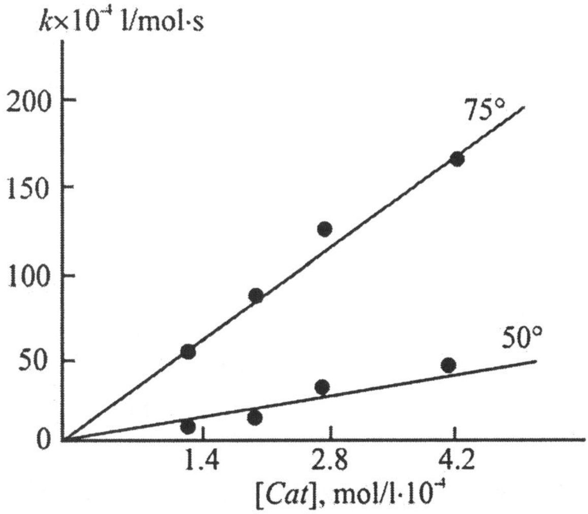

Figure 2.22. Dependence of the rate constant of polyaddition on the catalyst concentration at 50 and 75°C

Table 2.13

Rate constants of 1-hydro-3-divinyl-hexamethylcyclotetrasiloxane polyaddition

No.	Catalyst concentration, $\times10^{-4}$ mol/l	$k\cdot10^{-3}$			
		50°C	65°C	75°C	85°C
1	1.4	0.675	1.72	5.7	7.8
2	2.1	1.76	-	9.45	-
3	2.8	3.59	-	13.75	-
4	4.2	5.38	-	13.87	-

Synthesized cyclolinear polymers consist of tetrasiloxane rings as elementary units, bound by ethylene bridges. Recently, synthesis of cyclolinear polymers was described in the literature; the investigators have also studied hydrosilylation of dihydride-containing organocyclotetrasiloxanes with divinyl-containing organocyclotetrasiloxanes [85].

The above-described (Chapter 1) organocyclotetrasiloxanes with different disposition of functional (vinyl and hydride) groups were the initial monomers for synthesizing current polymers.

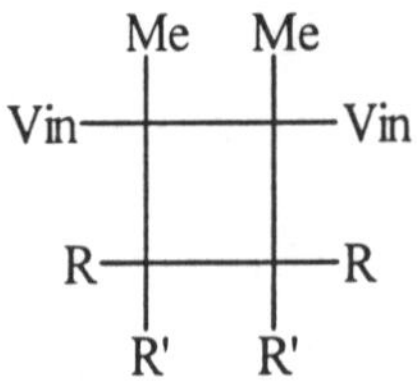

R=R'=Me (1a)
R=Me; (1b)
R'=Ph (1c)
R=R'=Ph

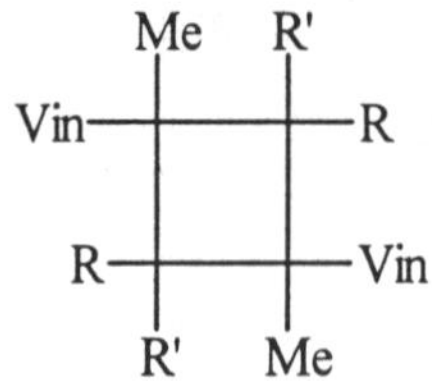

R=R'=Me (2a)
R=Me; (2b)
R'=Ph (2c)
R=R'=Ph

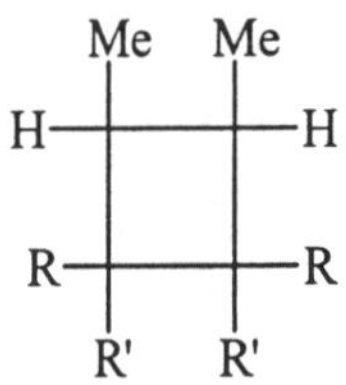

R=R'=Me (3a)
R=Me; (3b)
R'=Ph (3c)
R=R'=Ph

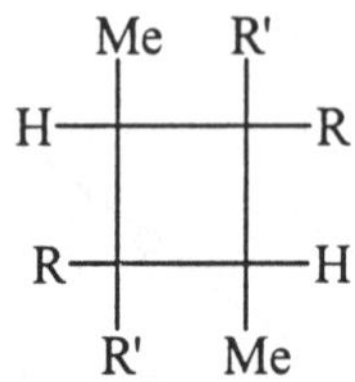

R=R'=Me (4a)
R=Me; (4b)
R'=Ph (4c)
R=R'=Ph

The polymers were synthesized by hydride polyaddition with 0.01*N* platinum-hydrochloric acid-tetrahydrofurane solution as the catalyst. The catalyst concentration equaled $5 \cdot 10^{-6}$ g per 1 mole of vinyl component in the temperature range of 60 - 150°C. It has been found that the reaction proceeding at high temperatures displays formation of branched structures. The effect of initial monomers' structure on the formation rate of polymers, as well as on physical and chemical properties and structural features of synthesized polymers are traced. The reaction proceeds by the general scheme as follows [85]:

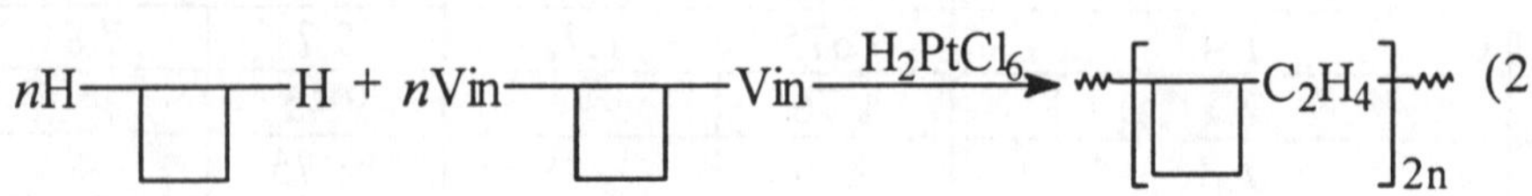

(2.13)

1 3 (Polymers D, F, G)

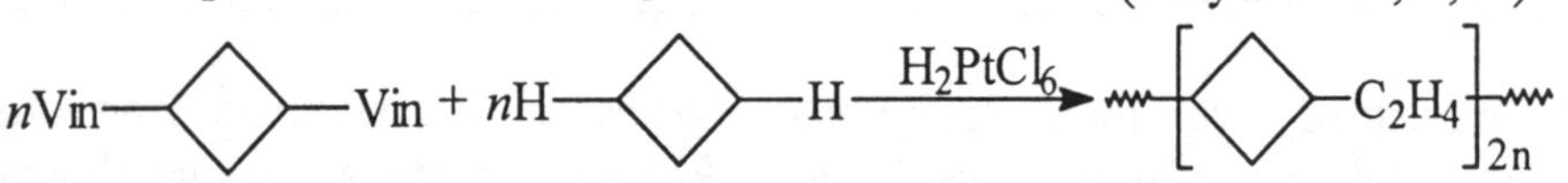

(2.14)

2 4 (Polymers A, B, C)
V

The reaction proceeding was traced by viscosity increase of 1% polymer solution in toluene. It is found that at the same framing and under equal conditions rings possessing functional groups in 1,5-position enter the reaction of polyaddition at higher rate and form polymers with higher molecular masses. However, it has been observed that the rings with functional groups in 1,5-

position are not only higher reactive, but also subject to side reactions. For example, soluble polymer A is formed at 60°C only, polymer B – at 60 - 70°C, and already at 120°C branching proceeds. For polymer C, the loss of solubility is observed at temperature above 80°C. Polymers D, F and G can be easily synthesized at 50 - 80°C, and the loss of solubility is not observed in this case.

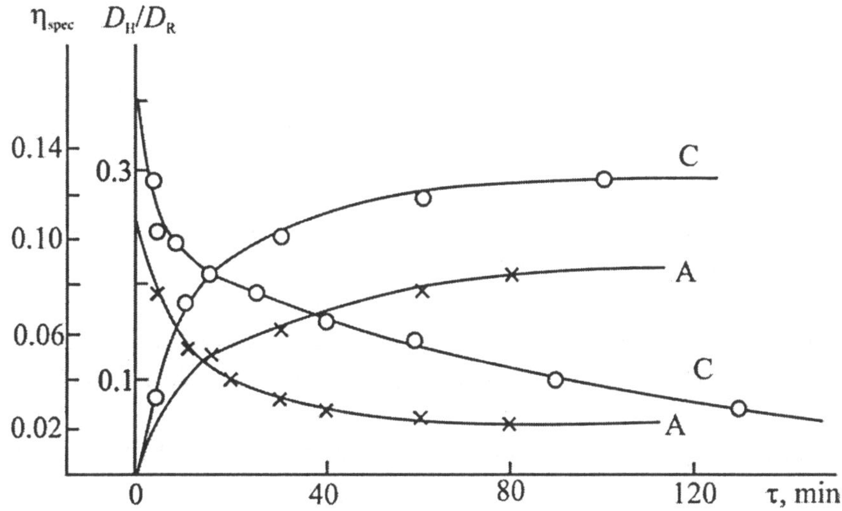

Figure 2.23. Dependence of optical density and specific viscosity variation of ≡Si-H bond on time for polymers A and C

Figure 2.23 shows the dependence of variation of ≡Si-H bond optical density on time, wherefrom it is obtained that up to 50% of ≡Si-H groups are consumed during initial 5 – 10 minutes of the reaction proceeding, which induces sharp increase of viscosity. Further on, smoother decrease of optical density of the valence oscillation band of ≡Si-H bonds proceeds. When over 70% of ≡Si-H groups are converted, the error of optical density variation is high. That is why ≡Si-H group concentration in equilibrated and reprecipitated polymers was determined by the method of comparing ≡Si-H group optical density in the reference solution and in solution of the polymer. Table 2.14 shows calculation results. For polymers with high viscosity and molecular masses, quantitative determination of ≡Si-H groups is difficult, because this requires application of concentrated solutions. Concentration of residual ≡Si-H groups in polymer C in the case of high viscosity cannot be determined due to poor solubility of this polymer.

Molecular masses of polymers measured by light scattering and IR-spectroscopy methods coincide well, which confirms the cyclolinear structure of the polymers.

As follows from the data shown, at the same framing and under equal conditions the rings possessing functional groups in 1,5-position enter polyaddition reaction at high rate and form polymers with high molecular mass. It is shown that the rings with functional groups in 1,5-positions (at the same framing and under equal conditions) are not only reactive, but also subject to side reactions. For example, soluble polymer A is synthesized at 60 and 100°C, and already at 120°C branching processes are observed.

There are data in the literature indicating that as heated with silicones [86] and silanes [87], catalysts from the platinum group are capable of detaching methyl or phenyl group at the catalyst concentration from 0.01 to 1%. Spire *et al.* [88] has observed detachment of methyl group at hydride addition of bis(trimethylsiloxy)methylsilane to hexane-2 at 50°C and $H_2PlCl_6{\cdot}6H_20$ concentration equal $5{\times}10^{-5}$ mole per vinyl group. It is shown [85] that as heated with polymer 4a under the reaction conditions, platinum-hydrochloric acid does not detach methyl group, and the ring does not enter regrouping or disproportionation reaction. Possible side reactions were estimated by GLC and ^{1}H NMR-spectroscopy methods.

In polyaddition of 2a to 4a (polymer A) proceeding at 100°C and catalyst concentration equal $5{\times}10^{-4}$ mole per vinyl group no methane was liberated, but above 120°C hydrogen liberation was observed. IR-spectra of insoluble polymers displayed the band at 3,600 cm^{-1}, typical of associated hydroxyl ≡Si-OH group. This proves the possibility of hydrogen detachment at polymer synthesis in accordance with the scheme as follows:

$$\equiv Si\text{-}H + H_2PtCl_6{\cdot}6H_2O \rightarrow \equiv Si\text{-}OH + H_2. \qquad (2.15)$$

The authors also suppose that branching is associated with scission of the cyclosiloxane ring.

The study of thermal oxidative degradation of synthesized polymers at heating with the rate equal 5 deg/min indicates that the temperature of decomposition beginning of polymer depends on the origin of framing groups. Polymers A and B, which contain methyl groups only, begin degrading at 250°C independently of molecular mass within its range from $6{\times}10^4$ to $2{\times}10^5$. Substitution of a part of methyl groups by phenyl ones induces the increase of

Table 2.14

Some physical and chemical parameters of cyclolinear polymers with ethylene bridges bonding rings

No.	Initial monomers	Reaction temperature, °C	η^{*}_{spec}	$[\eta]\cdot 10$, m^3/kg	$M\cdot 10^{-3}$		Polymer
					Light scattering	By ≡Si-H groups, %	
1	2	3	4	5	6	7	8
1	2a + 4a (R = R' = Me)	80 60	0.06 0.12	0.08 0.14	67 100	53 -	A
2	1a + 3a (R = R' = Me)	60 150	0.12 0.26	0.12 0.22	200 616	142 -	D
3	1b + 3b (R = Me, R' = Ph)	100	0.09	0.07	53	48	F
4	2b + 4b (R = Me, R' = Ph)	60 100	0.12 0.16	0.12 -	154 -	108 -	B
4a	2c + 4c (R = R' = Ph)	100 150 180	0.12 0.22 0.25	0.13 - -	51 172 250	40 - -	C
5	1c + 3c (R = R' = Ph)	100	0.04	0.04	7	6.5	G

Note: * At 25°C in toluene.

the initiation temperature of decomposition up to 350 - 370°C, which is associated with increased resistance of phenyl groups to oxidation, as well as their inhibiting effect of methyl groups on oxidation [89].

Besides 1,5-divinyl(dihydride)-1,5-dimethyltetraphenylcyclotetra-siloxane, the authors [90] have used 1,5-divinyl(dihydride)-1,5-dimethylhexaphenylcyclopentasiloxane and 1,7-divinyl(dihydride)-1,7-dimethyloctaphenylcyclohexasiloxane as initial compounds in hydride polyaddition reaction, by which cyclolinear organosiloxane copolymers with ethylene bridges between rings containing organocyclotetrasiloxane, organocyclopentasiloxane and organocyclohexa-siloxane fragments are synthesized. Data on these compounds are shown in Tables 1.4 and 1.5 (Chapter 1).

As the catalyst for polyaddition, 0.01 M solution of platinum-hydrochloric acid in tetrahydrofurane (5 - 9×10^{-5} g per 1 g of initial substance) was used. Polyaddition proceeds in argon at equimolar ratio of initial substances (1:1) in the absence of solvent and in the temperature range of 75 - 115°C. It is found that the above-mentioned conditions do not induce scission of the siloxane ring. As a consequence, hydride polyaddition under selected conditions proceeds in accordance with the scheme as follows [90, 91]:

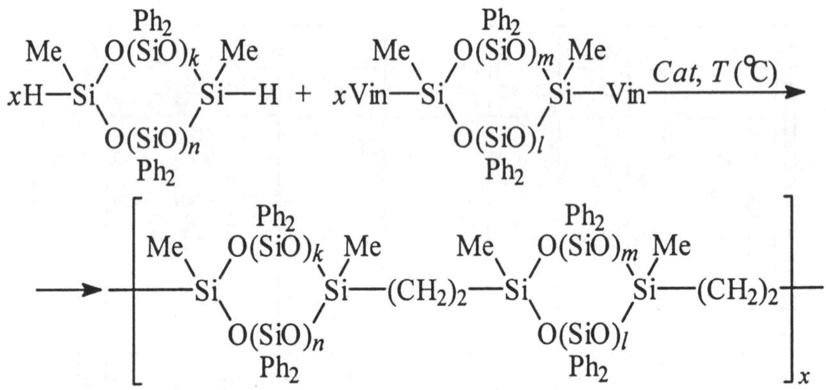

(2.16)

where $k = n = m = 1, l = 2$ (VI); $k = n = 1, m = l = 2$ (VII); $k = m = 1, n = l = 2$ (VIII); $k = 1, n = m = l = 2$ (IX); $k = n = m = l = 2$ (X).

Copolymers with $\eta_{spec} = 0.13 - 0.18$ representing slightly yellow glassy-like transparent products, which are soluble in usual organic solvents, were synthesized in the reaction. The reaction proceeding was traced by the increase of 1% toluene solution viscosity (Figure 2.24). It has been found that viscosity growth rate is increased with temperature increasing from 95 to 115°C, and increase of the cyclic fragment volume induces deceleration of the viscosity increase.

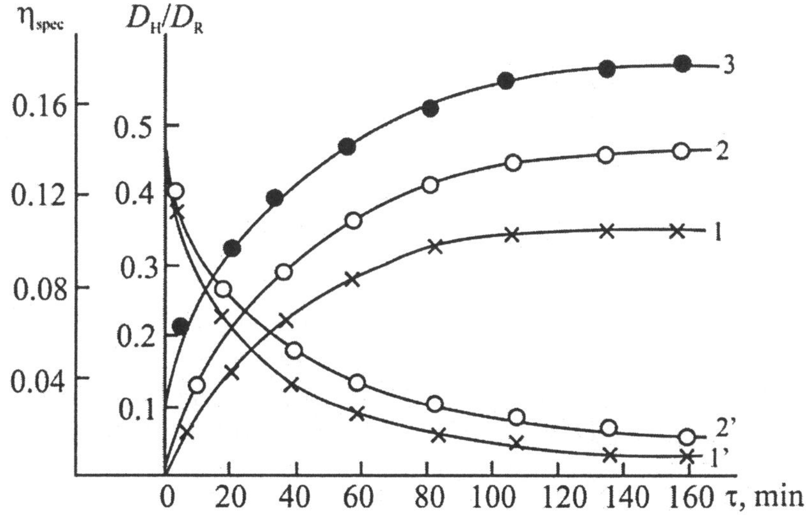

Figure 2.24. Variation of optical density of ≡Si-H bond and specific viscosity with time by reaction (17): 1 and 1' – copolymer 4 (Table 2.15) at 75°C; 2 and 2' – for copolymer 1 at 95°C; 3 – for copolymer 1 (Table 2.15) at 115°C

Figure 2.25 shows time dependence of ≡Si-H bond concentration variation in dihydrorganocyclopentasiloxane polyaddition to divinylorganocyclopentasiloxane at different temperatures. Similar to the synthesis of organocyclotetrosiloxanes, the level of polyaddition is increased with temperature of the reaction.

The current reaction has displayed the order two. The following rate constants of polyaddition reaction were detected: $k_{75°C} = 1.44\times10^{-2}$; $k_{95°C} = 2.33\times10^{-2}$ and $k_{115°C} = 3.20\times10^{-2}$. The activation energy was also calculated ($E \approx 24.7$ kJ/mol), almost the same as per reaction (2.12).

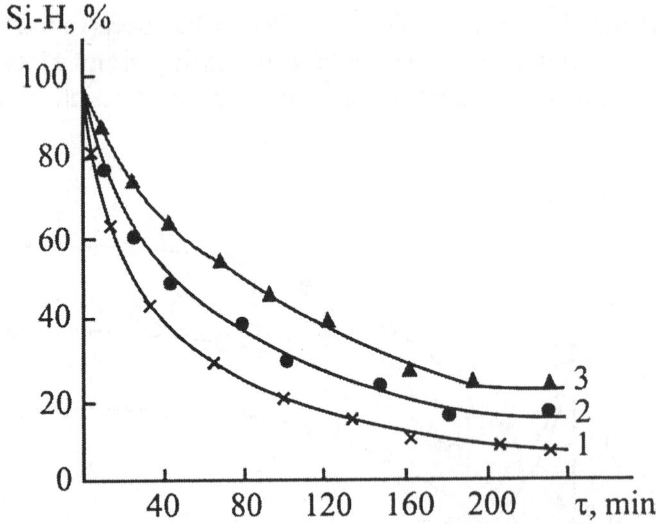

Figure 2.25. Variation of ≡Si-H bond concentration with time during polyaddition of dihydrorganocyclopentasiloxane to divinylorganocyclopentasiloxane: 1 – at 115°C; 2 – at 95°C; 3 – at 75°C.

^{1}H NMR spectral studies display that polyaddition mainly proceeds by the Farmer rule with formation of dimethylene bridges between cyclic fragments. Copolymer 1 (Table 2.15) displays a complex multiplet in the range of 6.9 - 7.6 ppm typical of protons of phenyl groups, and the reflex centered at 0.28 ppm belongs to protons of methyl group. The reflex with chemical shift at 0.32 ppm is corresponded to protons of methylene group ($-CH_2-CH_2-$). Moreover, the spectrum displays reflexes from protons of non-reacted vinyl groups (a complex multiplet in the area of 5.6 – 6.1 ppm) and ≡Si-H groups (at 4.4 ppm). The spectrum also displays a doublet reflex with chemical shift centered at 1.05 ppm, which may be related to methyl protons in $=CH-CH_3$ group with the quantity below 8%. Table 2.15 shows some properties of carbosiloxane copolymers with cyclic fragments in the backbone, represented by the following formula:

Table 2.15

Some physical and chemical parameters and yields of copolymers containing cyclotetra(penta, hexa)siloxane fragments in the backbone

No.	k	n	m	l	Yield, %	Reaction temperature, °C	η_{spec}	T_{soft}, °C	d_1, Å	$\overline{M}_n \times 10^{-3}$
1	1	1	1	2	94	95	0.15	48-50	9.40	97
	1	1	1	2	95	115	0.18	51-54	-	-
2	1	1	2	2	91	115	0.16	43-46	9.50	-
3	1	2	1	2	90	75	0.11	-	9.50	-
	1	2	1	2		95	0.14	-	-	-
	1	2	1	2		115	0.18	45-48	-	87
4	1	2	2	2	92	95	0.10	42-44	-	-
5	2	2	2	2	90	115	0.11	39-41	9.72	51

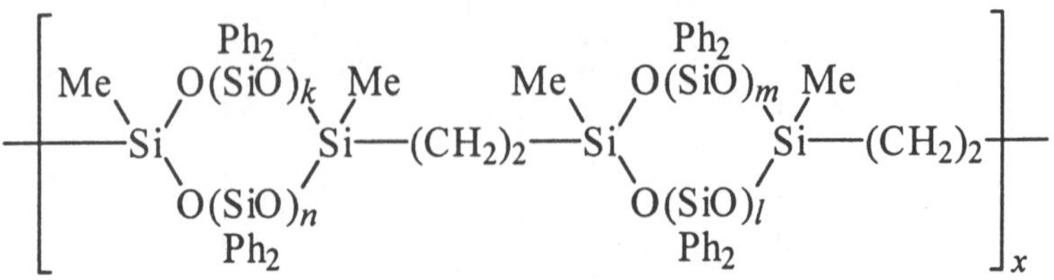

The copolymers were thermographically studied with the help of Seteram Co. thermoweighing machine B-60 in argon (at the heating rate of 5 deg/min) with simultaneous selection and analysis of gaseous degradation products, and the effect of introduction of bulky cyclic fragments into the backbone on their thermal stability was traced.

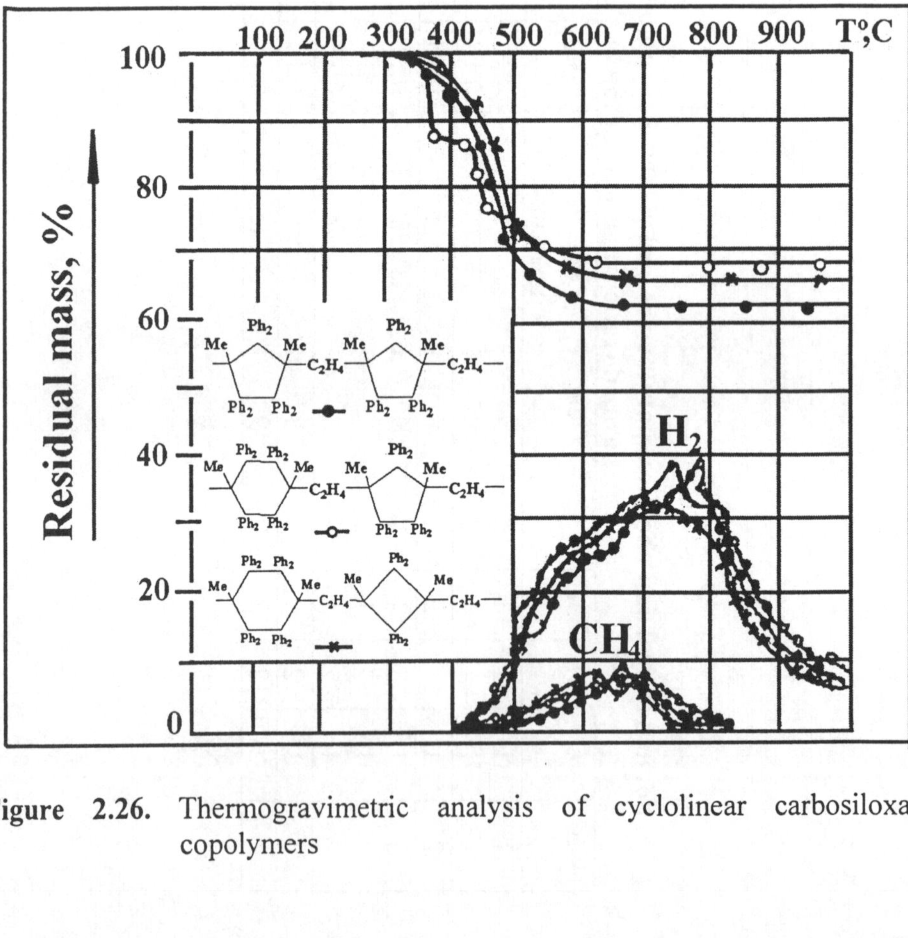

Figure 2.26. Thermogravimetric analysis of cyclolinear carbosiloxane copolymers

Figure 2.26 shows the investigation results indicating that initial mass losses are observed at 280 - 350°C depending on the volume of cyclic fragments in the chain. In the temperature range of 400 - 450°C hydrogen and methane are liberated, which is caused by ≡Si-C and C-H bond break and leads to cross-linking by methyl and phenyl groups [92, 93]. Similar effect is also observed at thermal degradation of both carbosiloxane [94] and organosiloxane copolymers [95]. Degradation proceeds by the radical mechanism with formation of oligomeric products [96, 97].

This gives an opportunity to suggest [90, 91] that besides the backbone break, radical cross-linking reactions by methyl and phenyl groups proceed at thermal pyrolysis of carbosiloxane copolymers, which is also accompanied by liberation of hydrogen, methane and other low-molecular organic compounds of benzene and other types. Above 600°C, breakthrough of the mass loss curves happens. Final mass losses fall within the range of 34 – 38%.

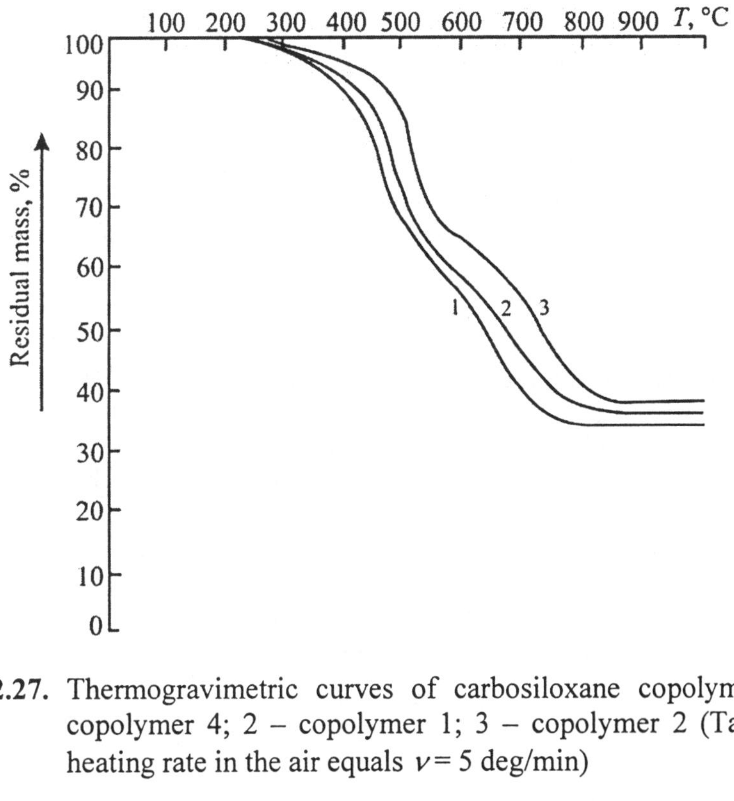

Figure 2.27. Thermogravimetric curves of carbosiloxane copolymers: 1 – copolymer 4; 2 – copolymer 1; 3 – copolymer 2 (Table 2.15, heating rate in the air equals $v = 5$ deg/min)

Figure 2.27 shows thermogravimetric curves for carbosiloxane copolymers. The curves shown indicate that thermal oxidative stability of synthesized copolymers is below that of carbosiloxane ones containing cyclotetrasiloxane fragments only [85], i.e. thermal oxidative stability of copolymers is decreased with the increase of the cyclic fragment volume. Moreover, thermal oxidative stability of carbosiloxane copolymers is below that of pure siloxane analogs. Already at 300°C 5% mass losses of copolymers with ethylene bridges are observed, whereas oligomers with oxygen bridges display losses of this level at 400°C.

Thermomechanical studies have indicated that the glass transition temperature of synthesized copolymers is decreased as the volume of cyclosiloxane ring in the chain is increased. The results of X-ray structural analysis indicate that the copolymers represent amorphous systems, and increase of cyclic fragment volume leads to an insignificant increase of the interchain distance.

2.3. Bead-shaped methyl(ethyl)siloxane polymers with ethylene bridges between rings

To detect the effect of the size of cyclosiloxanes, cross-linking bridges between them and delimitation of their roles in ability of polymers to self-organize as mesomorphous structures, investigators [98] have studied polyaddition of dihydroalkylcyclotetra(hexa)siloxanes to divinylorganocyclotetra(hexa, octa)siloxanes in the presence of different complex platinum catalysts (dicyclopentadienyl-platinum dichloride, the Carsted catalyst) and their reduced forms by the scheme as follows:

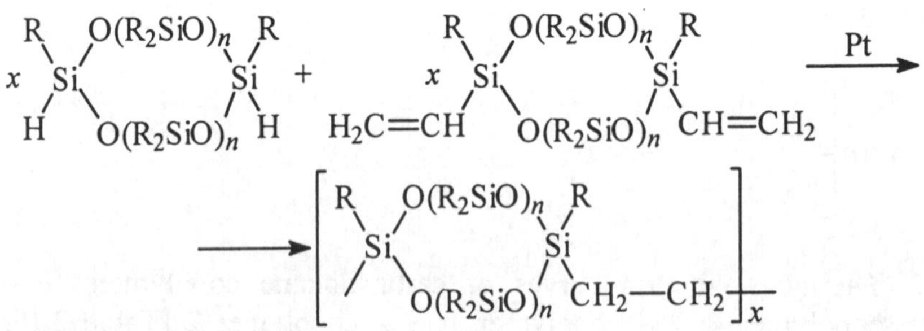

(2.17)

where $n = 1 - 3$, R = Me, Et.

Polyaddition reactions were performed in a retort with mixers at stoichiometric ratio of the components. The catalyst concentration equaled 5×10^{-5} g of Pt per 1 mole of vinyl-containing component. Products of polyaddition reaction were dissolved in benzene and precipitated by triple excess of methanol. Precipitated polymer was separated and dried in vacuum to a constant mass. The yield of the polymer equaled 70 – 80%. All synthesized polymers are completely soluble in usual solvents.

The effect of platinum colloid forms used in polyaddition reaction on molecular structure of synthesized cyclolinear polyalkylcarbosiloxane polymers has been studied. In accordance with the opinion [98], application of platinum-hydrochloric acid [85] to polyaddition of dihydromethylphenylcyclotetrasiloxanes to divinylmethylphenylsiloxanes induces side reactions, which disturbs the regularity. This is proved by discrepancy between molecular masses and viscosity parameters. The possibility of side reactions proceeding in the case of polyaddition is indicated in the work [99].

Taking into account the data obtained [85, 99, 100] for the synthesis of cyclolinear alkylcarbosiloxanes, special attention was paid to determination of optimal reaction conditions of polyaddition [98], in particular, selection of a catalyst providing for formation of polymers of strict cyclolinear structure.

Dicyclopentadienyl-platinum dichloride (CPDP) and the Carstedt catalyst representing a complex of platinum with divinyltetramethyldisiloxane were selected as catalysts for polyaddition.

Polymers with increased $\overline{M}_\omega/\overline{M}_n$, i.e. the ones with seriously distirbed chain linearity, are formed in the interaction of dihydroalkylcyclotetra(hexa)siloxane compounds with corresponded divinylcyclotetra-(hexa)siloxanes in the presence of CPDP or the Carstedt catalyst. In some cases, formation of cross-linked polymers is observed. Table 2.16 shows parameters of synthesized cyclolinear carbosiloxane polymers with regard to conditions of polyaddition proceeding, which indicate that polymers No. 1, 4 and 7 with broad MMD are formed both at 25 and 130°C. Primarily, during the synthesis of polymers No. 1, 4 and 7, $\overline{M}_\omega$ and $\overline{M}_n$ values increased, and then $\overline{M}_\omega$ increased sharply and $\overline{M}_n$ remained almost unchanged. This proves proceeding of macromolecular chain branching.

Formerly, Levis [101] has found that hydrosilylation proceeds with high efficiency on platinum complexes, reduced to the colloid form. Preliminary results show that polyaddition of 1,5-dihydrohexaethylcyclotetrasiloxane to 1,5-divinylhexaethylcyclotetrasiloxane on reduced CPDP in the form of colloid platinum decreases polydispersion degree of carbosiloxane polymer from 8.21

Table 2.16

Some parameters of cyclolinear poly(alkylcarbosiloxanes) with regard to synthesis conditions

Polymer, No.	R	*n*	Reaction conditions			$[\eta]$ toluene 25°C dl/g	$\bar{M}_\omega \times 10^{-3}$	$\bar{M}_\omega / \bar{M}_n$	$\bar{M}_z / \bar{M}_\omega$
			Catalyst*	Reaction temperature, °C	Time, hour				
1	2	3	4	5	6	7	8	9	10
1	Me	1	NRF	70	1	0.31	46.0	-	-
2	Me	1	YC	70	1	0.21	-	-	-
3	Me	1	BC	70	20	0.12	-	-	-
4**	Et	1	NRF	70	10	0.08	-		
5	Et	1	YC	70	20	0.15	49.5	4.34	2.81
6	Et	1	BC	70	47	0.08	26.0	3.10	2.4
7	Me	2	NRF	25	168	0.80	322.1	10.1	6.27
8	Me	2	YC	25	144	1.30	462.2	7.10	5.2-
9	Me	2	BC	70	50	0.13	29.0	2.08	1.72
10	Et		NRF	130	9	0.14	50.9	13.7	6.51
11	Et		YC	130	10	0.24	67.3	3.30	2.2
12	Me		NRF	70	10	0.12	46.9	-	-
13	Me		YC	70	10	0.14	34.0	-	-
14	Me		BC	70	10	0.06	21.0	-	-

Notes: * NRF is the non-reduced form of dicyclopentadienyl-platinum dichloride (CPDP); YC is the yellow colloid of CPDP; BC is the black colloid of CPDP.

** After 10 hours polymer transforms into the cross-linked form.

*** Polymers represent cyclolinear carbosiloxane copolymers with regular alternation of decamethylcyclohexasiloxane and tetradecamethylcyclooctasiloxane fragments.

to 3.10. Taking into account encouraging results of the work [100], a sequence of cyclolinear carbosiloxane polymers were synthesized [98] in the presence of three different forms of platinum catalysts, introduced into the reaction mixture after preliminary formation.

Table 2.16 shows that the forms of platinum compounds display different catalytic activities. Usually, yellow colloid form is more active compared with the black one, but lower polydispersion degrees are typical of polymers, synthesized in the presence of the latter one. As a consequence, the use of black colloid for obtaining polymers of regular structure is optimal, though due to low catalytic activity of it either high temperature of the reaction or high reactivity of initial reagents is required, which, for example, takes place for polymer 9 (Table 2.16).

For synthesized carbosiloxane polymers, MMD curves were determined by the gas permeation chromatography (GPC) method. Table 2.16 shows $\bar{M}_\omega$, $\bar{M}_\omega / \bar{M}_n$ and $\bar{M}_z / \bar{M}_\omega$ values for polymers with different sizes of cyclic fragment, from cyclotetrasiloxane to cyclooctasiloxane. As an example, Figure 2.28a shows curves of molecular-mass distribution for copolymers with decamethylcyclohexasiloxane fragments, differed by the catalyst form used in the synthesis. Clearly polymers No. 7 and 8 are characterized by broader MMD curves, than No. 9. The latter displays symmetrically shaped MMD curve, though its molecular mass is lower, than MM of polymers No. 7 and 8. MMD curve 1 for polymer No. 7, synthesized on non-reduced catalyst, and curve 2 of polymer No. 8, synthesized on reduced catalyst of yellow colloid form, are shifted towards higher MM. They also possess small peak squares that indicates the presence of some molecules with high molecular mass in the polymer composition.

Because MM values of polymers No. 7 and 8 are by the order of magnitude higher than for polymer No. 9, apparently, the former ones contain branched fractions of the cyclolinear structure. Figure 2.28b shows molecular-mass distribution curves for copolymers with ethyl substituting agents in cyclohexasiloxane fragments. All MMD curves display monomodal distribution with no regard to the kind of platinum catalyst (non-reduced or reduced), used in polyaddition. Simultaneously, polymers with ethyl substituting agents, synthesized on reduced yellow colloid, possess narrower MMD curve (Figure 2.28b, curve 1), than cyclolinear carbosiloxane polymers with methyl substituting agents, for example, polymer No. 8 (Figure 2.28a, curve 2), synthesized at 25°C on yellow colloid.

For polymer No. 11, similar technique was used for synthesizing a series of polymers on reduced yellow colloid catalyst. Molecular masses detected fall within the same range and $\bar{M}_\omega / \bar{M}_n$ in the range from 2.2 to 4.7. To put it differently, for both polymers with methyl and ethyl substituting agents in the presence of reduced yellow colloid, some tendency to increasing polydispersion degree is observed.

Data on MMD (Table 2.16) indicate that cyclolinear polymers with the most homogeneous structure and composition are formed on the reduced catalyst. The more so, every particular case displays its own optimal type of the reduced form, i.e. catalytic system for synthesizing cyclolinear polymers should be selected with regard to activity of dihydrorganocyclosiloxane in polyaddition reaction. It should be noted that as yellow colloid is applied as the catalyst, the reaction temperature has no effect on the shape of MMD curves for ethyl-substituted polymers with tetra- and hexasiloxane cyclic fragments.

Cyclolinear structure of synthesized polymers is proved by the results of NMR-spectroscopy studies. For all polymers independently of MMD, ^{29}Si NMR spectra display two groups of reflexes in the area of -20.0 – 23.0 ppm, corresponded to atoms of silicon in R_2SiO and $RSiCH_2CH_2$ groups (Table 2.16). Reflexes typical of silicon atoms in the branching centers are absent. The quartet of reflexes from $RSiCH_2CH_2$, observed in ^{29}Si NMR spectra of polymers No. 8 and 9 (Table 2.16) confirms atactic structure of synthesized products.

Analysis of NMR spectra of polyaddition products indicates that the reactions proceed by the Farmer rule forming CH_2-CH_2 bridges between rings in more than 95% of cases.

By the methods of differential scanning calorimetry (DSC), X-ray structural analysis (RSA) and optical microscopy, parameters of thermal transitions in synthesized cyclolinear polymers with different structures of units, MM and polydispersion degree were determined. Table 2.17 shows that T_g of all other polymers are higher than T_g of cyclolinear siloxane polymers, but this difference is decreased with increase of the quantity of diorganosiloxane groups in cyclosiloxanes. Regularities observed for T_g shift with variations of the cyclosiloxane size and length of side substituting agents of carbosiloxane polymers are similar to these, previously detected for pure cyclolinear polysiloxanes at analogous variations of the chemical structure [49].

Independently of molecular mass and polydispersion degree, all cyclolinear carbosiloxane polymers with cyclic tetrasiloxane fragments (Table 2.16, polymers No. 1 – 6) and the ones with decamethylcyclohexasiloxane fragments (Table 2.16, polymers No. 7 – 9) do not display mesomorphous properties. RSA data (Figure 2.29) indicate that polymers No. 2 – 9 are amorphous, because their diffraction patterns are similar by type to these of amorphous polyorganosiloxanes. They contain two amorphous halos:

symmetrical intensive one at $2\theta_1$ ($\Delta_{1/2}$ = 1.8°) and diffuse low one at $2\theta_1$. Distribution of amorphous scattering intensity, in particular, angular $2\theta_1$ position, depends on the sizes of cyclic fragment and side substituting agent (Figure 2.29).

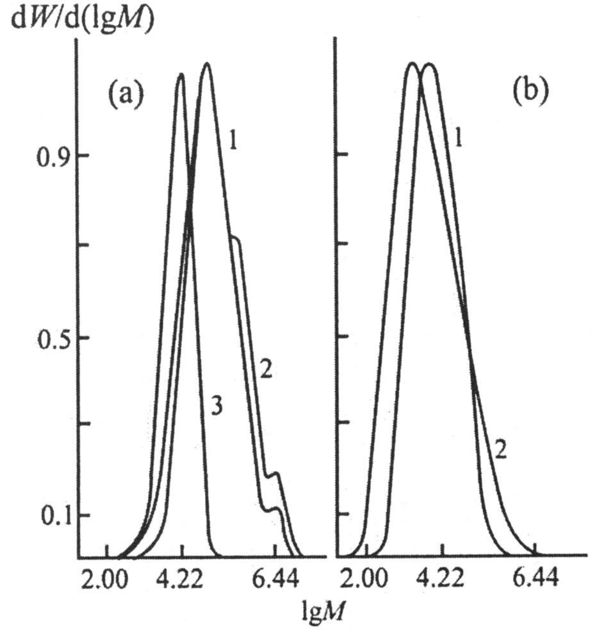

Figure 2.28. Molecular-mass distribution (MMD) curves: a – polymers No. 7 (1), 8 (2) and 9 (3); b – polymers No. 10 (1) and 11 (2) (Table 2.16)

Below 73°C, polymer No. 1 is crystalline (Figure 2.29); according to RSA data, the crystallinity degree of this polymer is below 30%. Above T_{melt} = 73°C, polymer No. 1 transits into isotropic melt.

Because compounds representing the mixture of *cis*- and *trans*-isomers were used in polyaddition reactions, carbosiloxane polymers should possess the atactic structure. Such conclusion is also proved by the above-mentioned NMR data. In this connection, the absence of ability to crystallize in the majority carbosiloxane polymers is quite explainable. However, the absence of mesomorphous properties in polymer with decamethylcyclohexasiloxane fragments is the unexpected result. The case is that at the polymerization degree $P \geq 5$, atactic non-crystallizing cyclolinear polyorganosiloxanes with decamethylcyclohexasiloxane fragments are mesomorphous, temperature range of mesophase existence increasing with molecular mass (MM) [56]. It should be noted also that the average intermolecular distance $d_l \approx 8.3$ Å for polymers No. 7 – 9 (calculated from $2\theta_{al}$) is practically coincident with the similar value $d_l \approx 8.6$ Å for amorphous cyclolinear poly(organosiloxanes) with decamethylcyclohexasiloxane fragments [102]. The latter are characterized by the polymerization degree, *P*, below 5.

Comparison of the above-shown data on cyclolinear carbosiloxanes with previously published results on the phase structure of cyclolinear organosiloxane homo- and copolymers [103] indicates that introduction of different groups between cyclic siloxanes (flexible junctions) causes a significant decrease of isotropization temperature.

Table 2.17

Calorimetric and structural parameters of cyclolinear carbosiloxane copolymers

Polymer*, No.	T_g, °C	T_{melt}, °C	2θ, deg	Chemical shifts δ ^{29}Si NMR, ppm	
				R_2SiO	$RSiCH_2CH_2$
2**	-30	42	12.0	-18.99	-19.14
6	-74	-	13.0	-20.85	-20.22
9	-69	-	10.6	-21.76 21.77	-21.59 -21.61 -21.66 -21.69
11	-104	220***	9.80	-22.95	-23.62 -23.69
13	-75	-	9.37	-	-

Notes: * Polymer numeration in accordance with Table 2.16.
** Melting heat Q_{melt} = 167.6 J/mol.
*** Isotropization temperature.

At some critical distance between cyclic rings in considered carbosiloxane copolymers, as well as in cyclolinear siloxane copolymers [48, 104], the ability of molecules to form the mesomorphous phase is suppressed: above T_{melt}, crystallizing copolymers transit into the melt, and non-crystallizing ones are amorphous in the whole temperature range.

For example, copolymers with $–R_2SiO-$ junction preserve mesomorphous properties, whereas copolymers with $-O(CH_3)_2SiCH_2CH_2Si(CH_3)_2O-$, $-(R_2SiO)_2-$ and $-CH_2-CH_2-$ junctions lose these properties.

There is another fact of special interest, which may affect the mesomorphism of copolymers: lower flexibility of $-CH_2CH_2-$ groups compared with oxygen.

By increasing the elementary unit to tetradecamethylcyclooctasiloxane, an attempt was made [98] to change the ratio of distances between rings and junctions. Carbosiloxane copolymers No. 12 – 14 with regular alternation of decamethylcyclohexasiloxane and tetradecamethylcyclooctasiloxane fragments in the chain were synthesized by polyaddition on different non-reduced and reduced colloid forms of the CPDP catalyst (the attempt to synthesize dihydrotetradecamethylcyclooctasiloxane has failed [105]).

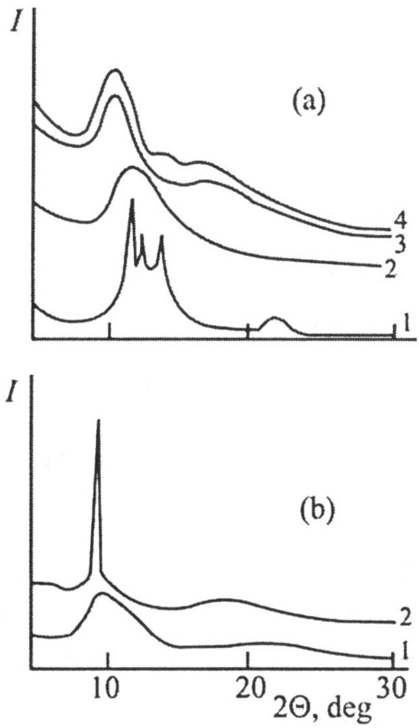

Figure 2.29. Diffraction patterns: a – polymer No. 3 at 20°C (1) and 73°C (2), polymer No. 9 at 20°C (3), copolymer No. 12 at 20°C (4); b – polymers No. 6 (1) and 11 at 20°C (2)

Figure 2.29a shows the diffraction pattern of polymer No. 12 (curve 4) indicating that in this case, polymers are amorphous, too.

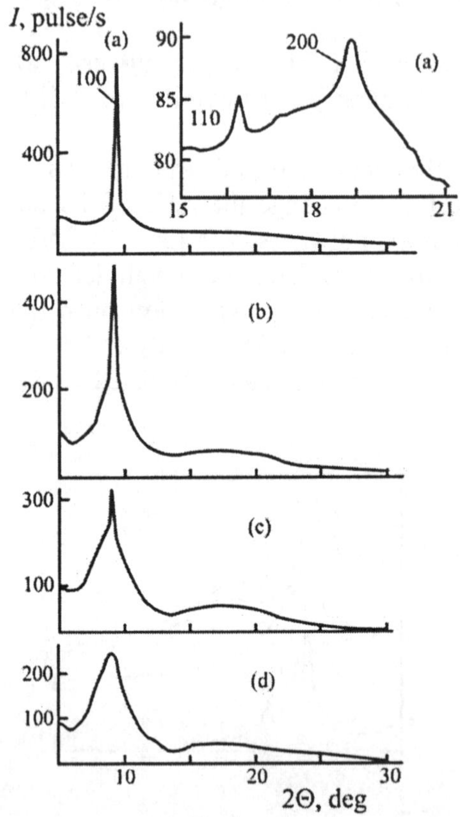

Figure 2.30. Diffraction patterns for polymer No. 11 at 10 (a), 180 (b), 240 (c) and 260°C (d)

Mesomorphous properties of CLPCS with decaethylcyclohexasiloxane fragments are detected (Table 2.16, polymers No. 10 and 11). Diffraction patterns for polymer No. 11, obtained at temperatures below T_{iso} are typical of mesomorphous systems (Figure 2.30). They display a single narrow reflex at $2\theta_m$ ($2\theta_m = 9.43°$; $d_m = 9.37$ Å at 20°C). In the same range of angles amorphous halo I ($\Delta_{1/2} = 2.2°$) with the maximum at $2\Theta_I$ is located. Very low amorphous halo II is observed at $2\theta_{II} = 16 - 25°$. Below 10°C, in addition to intensive reflex at $2\theta_m$ two narrow low reflexes at 2θ and $2\theta'$ are observed in the range of the second amorphous halo (Figure 2.30a); in this case, $\sin^2\theta_m : \sin^2\theta : \sin^2\theta' =$ 1:3:4. The latter result indicates that the mesophase is characterized by 2D-packing with hexagonal symmetry. Isotropization of polymers No. 10 and 11 proceeds in a broad temperature range; hence, gradual decrease of relative intensity of the reflex at $2\theta_m$ with simultaneous increase of the fracture of amorphous scattering in the area of $2\theta_l$ is observed (Figure 2.30c).

Figure 2.30 shows temperature dependencies of the interplane distance d_{100} for polymer No. 11 (line 1) and mesomorphous cyclolinear polymers with hexagonal 2D-structure of the column type in the mesophase. This polymer possesses atactic backbone. This fact as well as high gradient of $d_m(T)$ dependence (Figure 2.31) testifies about intermolecular type of interaction between polymer chains in the mesophase.

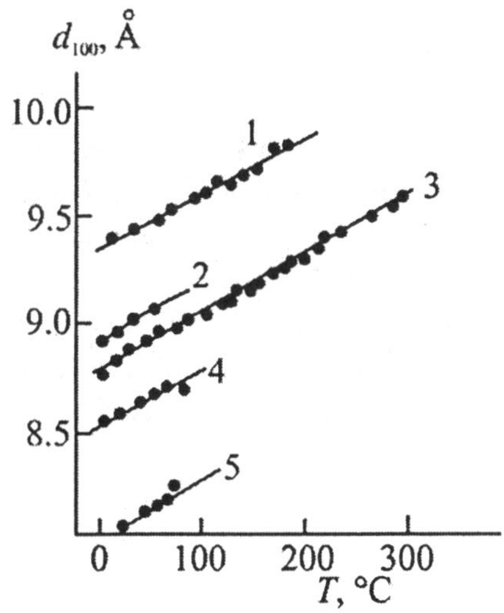

Figure 2.31. Temperature dependencies of the interplane distance, d_{100}: 1 – polymer No. 11, 2 - 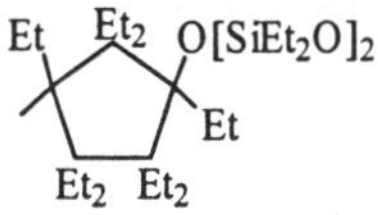, 3 - 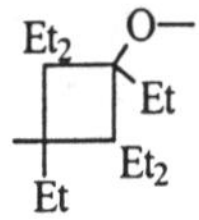, 4- 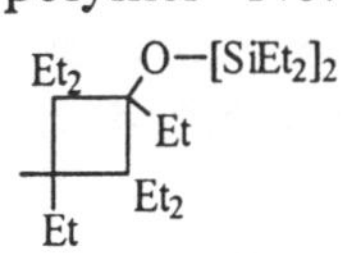, 5 – PDMS

Figure 2.31 shows that the data on the mesophase of polymers No. 10 and 11 correlate well with the results, previously obtained for cyclolinear poly(organosiloxanes) with the column-type structural organization of the mesophase [103]. The latter clearly displays the tendency to increase of d_m at introduction and increase of the size of the cyclic part, i.e. by some characters the mesophase 10 – 11 may be related to the column-shaped 2D-structure, where isolated cylindrical macromolecule is the structure-forming unit. This is also proved by decrease of isotropization temperature range, ΔT_{iso}, with molecular mass of the polymer.

For example, RSA data display the following values of ΔT_{iso}: polymer No. 10 – 122 - 206°C, polymer No. 11 – 200 - 265°C.

Numerical estimation of the average intramolecular distances, l^*, between closest cyclic fragments in the mesophase, performed in the work [103], displayed $l^* = 6.15$ Å at T_g. These values of l^* indicate the presence of intramolecular correlation of cyclic fragments. Apparently, side radicals of cyclolinear carbosiloxane copolymers tend to form an organic blanket for collapsed polymeric chain. Analogous tendencies are typical of mesomorphous linear and cyclolinear polyorganosiloxanes of the column-type 2D-packing, where mesomorphous ordering processes are also associated with self-consistent ordering at intra- and intermolecular levels.

Thus, substitution of oxygen bridges between cyclic fragments by -CH_2CH_2- groups leads to the loss of the ring ability to intra- and intermolecular correlations and, consequently, to the loss molecule ability to self-organization and formation of mesomorphous structures. Introduction of ethyl substituting agents at hexasiloxane rings form additional intra- and intermolecular interactions, due to which the ability of macromolecules to self-organization occurs.

2.4. Oligomers and copolymers with organocyclocarbosiloxane fragments in the chain

To synthesize oligomeric organocyclocarbosiloxanes, the reaction of dihydroxyorganocyclocarbosiloxane homofunctional condensation in 60% toluene solution was studied [106] on the Dyne–Stark device both in the presence (7%) and in the absence of activated coal. As a consequence, dihydroxy-containing organocyclocarbosiloxane oligomers were synthesized in accordance with the scheme as follows [106, 107]:

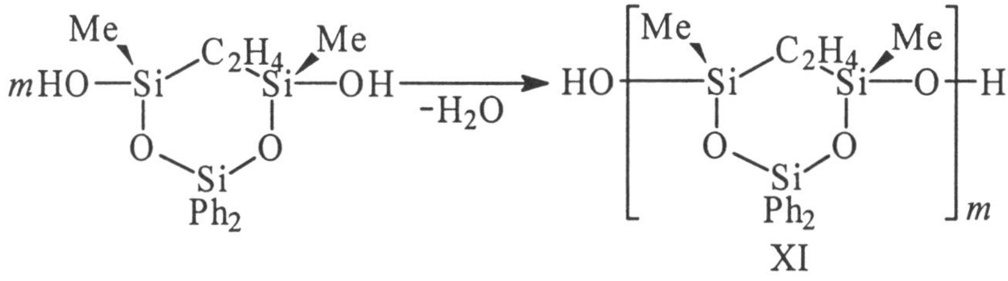

(2.18)

where $m = 8, 11$.

Organocyclocarbosiloxane oligomers obtained represent solid products, well soluble in various organic solvents. It is found that "non-equilibrating" catalyst which is activated coal, used in homofunctional condensation, does not induce scission of the organocyclocarbosiloxane ring.

X-ray diffraction analysis indicates that oligomers are amorphous systems with the interchain distance equal $d_1 \approx 8.64$ Å. Thermogravimetric studies show that by thermal oxidative stability oligomers are behind polyorganocyclotetrasiloxanes only.

For the purpose of synthesizing organocyclocarbosiloxane oligomers and polymers with different ratio of SiO/SiC groups in the complex repeated unit and studying the effect on their ability to self-organization in bulk and in ultrathin films, HFC reaction of dihydroxymethylcyclosiloxalkanes with dichloromethylcyclo-siloxalkanes proceeding in the presence of pyridine acceptor in accordance with the scheme as follows was studied [108]:

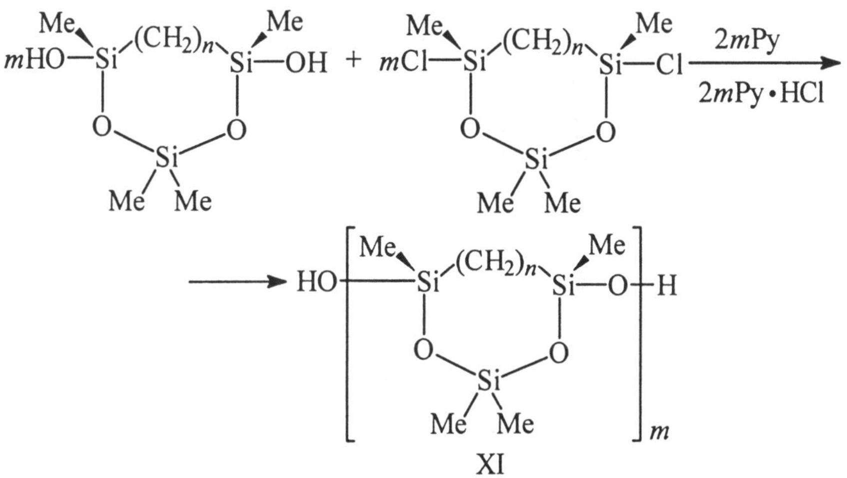

(2.19)

where $n = 2, 3$;

Table 2.18

Physical and chemical and heat physical parameters of Langmuir thin films

No.	Reaction product	n	Yield, %	$\bar{M}_n$	$\bar{M}_\omega$	$\bar{M}_\omega / \bar{M}_n$	T_g, °C	T_{iso}, °C	Langmuir thin film parameters	
									Number of layers	πÅ, mN/m
1	(2.20)	3	65	12,900	127,300	8.10	-60	-	Bilayer	6.0
2	(2.20)	2	24	3,270	5,660	1.73	-60	-	Monolayer	8.0
3	(2.21)	-	35	4,400	6,000	1.38	-89	68-85	Bilayer	9.0

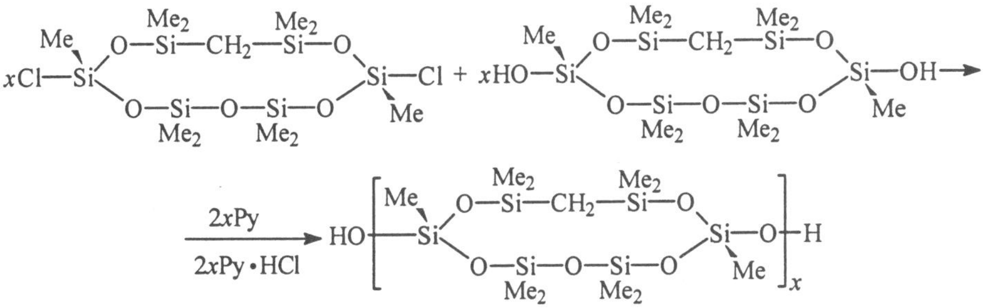

(2.20)

XII

Composition and structure of synthesized oligomers is proved on the basis of the ultimate analysis and by IR, ^{1}H and ^{29}Si NMR spectral data. Spectral data (oligomers No. 1 – 3, Table 2.18) indicate that the presence of *cis-trans*, *trans-trans* and *trans-cis* combinations of rings in the polymer chain is similar to polydecaorganocyclohexasiloxanes [26]; in the case of *trans*-dihydroxyorgano-cyclohexasiloxane, *cis-cis*-combination in the polymer chain is absent. They display atactic structure of cyclolinear chain, enriched with *trans-trans*-sequences.

Table 2.18 shows physical, chemical and heat physical parameters of compounds and some characteristics of ultrathin films. Clearly in the case of compound No. 1, high yield of the fraction possessing η_{cl} = 0.27 dl/g is observed. Due to GPC data, this fraction possesses polydispersion degree equal 8.1, and MMD curve is of a low bimodal type in the area of high molecular masses. Oligomer No. 2 displays narrow MMD ($\bar{M}_\omega / \bar{M}_n$ = 1.7). RSA data indicate that both products are amorphous, and DSC data display equality of their T_g values. Apparently, the absence of self-organization processes in the studied samples 1 and 2 is provided by asymmetrical structure of the elementary unit and atactic structure of oligomeric chain, because *trans*-isomer of dihydroxy-derivative was the only one used in heterofunctional polycondensation.

As to the authors' point of view, substitution of a single oxygen atom in organocyclohexasiloxanes by methyl group between atoms of silicon in oligomers No. 3 (Table 2.18) must induce properties inherent to previously described poly(dimethylcyclohexasiloxane) [26]. In spite of the fact that the only product obtained is an oligomer with M = 6,000 and narrow MMD ($\bar{M}_\omega / \bar{M}_n$ = 1.38), thermooptical analysis performed at 68 - 85°C have displayed a phase transition from mesomorphous to isotropic phase. At the initial heating (20 deg/min rate), DSC data display a jump increase of temperature in the area of -90°C, corresponded to the glass transition temperature. Cooling down displays no reflexes and repeated heating does not change the DSC curve shape.

Figure 2.32 shows curves displaying dependencies of the area per one monomeric unit on surface pressure of the layer (π-Å-isotherms) for polymers No. 1, 2 and 3 (Table 2.18). Figure 2.32a shows that at substitution of a single ≡Si-O-Si≡ group by $-(CH_2)_3-$ one in the ring, the polymer remains able to form the proper Langmuir monolayer on the water surface.

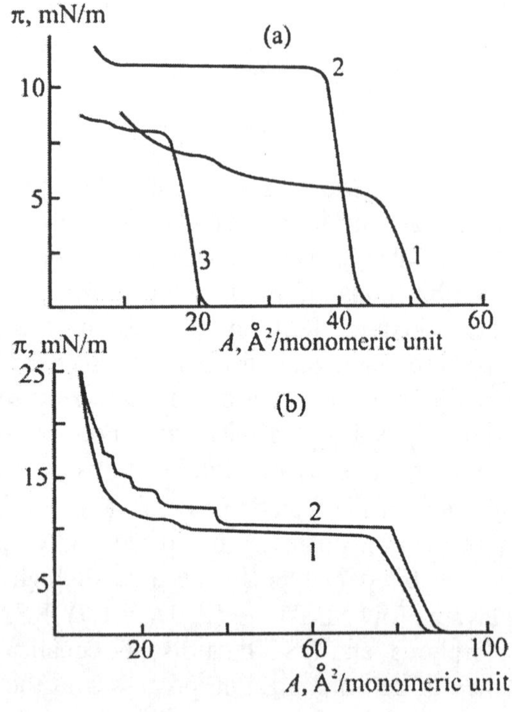

Figure 2.32. Dependence of surface pressure, π, on the layer area, Å^2, per one monomeric unit: a) curve 1 - oligomer No. 1 (Table 2.18), 2 – polycyclotetrasiloxane [109], 3 – PDMS [110]; b) curve 1 - oligomer No. 3, 2 – polydimethylcyclohexasiloxane [111]

Areas per monomeric unit in the dense monolayer of these polymers are comparable and equal 50 Å^2. The monolayer collapse pressure is decreased from 11 to 6 mN/m. This is associated with the decrease of relative content of oxygen atoms, which give the main part of the polymer interaction with water surface. As per PDMS (Figure 2.32a, curve 3), the second grade of surface pressure on the π-Å isotherm of the sample No. 1 is observed. Presumably, the presence of it cannot be explained by transition to the spiral conformation, as suggested for PMDS [110].

Figure 2.32b shows that the oligomer No. 3 (Table 2.18) and its closest analog, poly(dimethylcyclohexasiloxane), [111] possess equal areas per monomeric unit and the monolayer collapse pressure, but in contrast with the latter polymer, compound No. 3 forms Langmuir-Blodgett bilayer only, probably, due to lower molecular mass of the sample No. 3.

2.5. Silarylene cyclosiloxane and silarylene carborganocyclosiloxane oligomers and copolymers

Introduction of phenylene unit into the polyorganosiloxane chain causes a significant variation of the chain configuration and flexibility, which affects physical and chemical parameters of the polymer. Moreover, different combinations of linear ≡Si-O-Si≡ or silarylene fragments with cyclic ones may change conformational flexibility of the chain [112]. As indicated [113, 114], the origin (chemical and geometrical) of substituting agent at the atom of silicon may induce increased chain rigidity, if substituting agents between siloxane chains are composed of –CH_2-CH_2- and ≡Si-C_6H_4-Si≡ fragments [115, 116].

For the purpose of synthesizing dihydroxy-containing silarylene cyclosiloxane oligomers, HFC reaction of 1,4-bis(chlorodimethylsilyl)benzene, synthesized by the known technique, with 1,5-dihydroxyhexaphenylcyclotetrasiloxane and 1,7-dihydroxydecaphenylcyclohexasiloxane was studied [91, 117 – 119] at the molar ratio of reagents as follows: 1:0.9, 1:0.95 and 1:1, respectively. The reaction proceeded in 60 - 70% solution of anhydrous toluene in the presence of pyridine, the acceptor of hydrochloric acid, at room temperature with consequent boiling of the solution up to the boiling point of the solvent used.

Generally, the reaction proceeds by the scheme as follows:

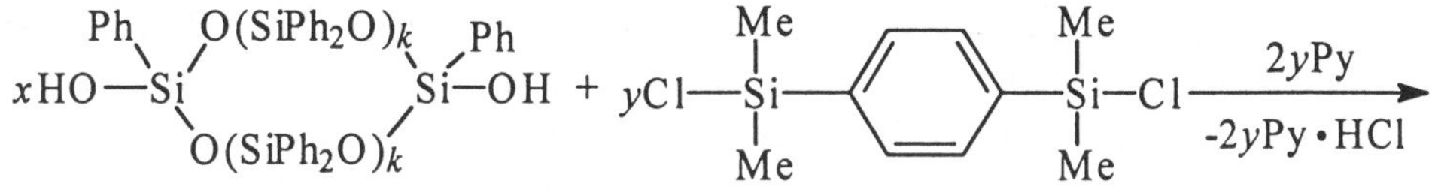

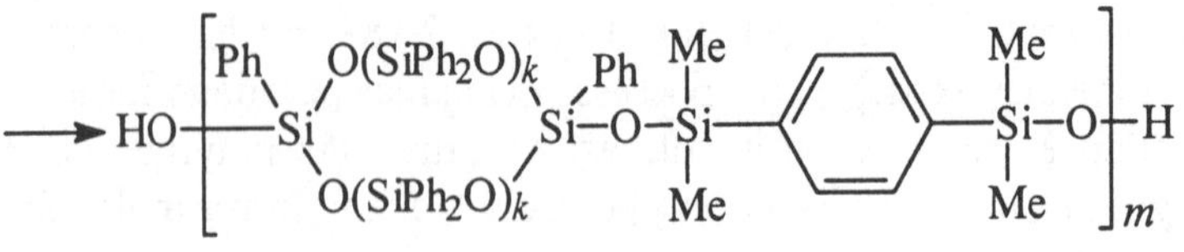

(2.21)

XIII, XIV

where $k = 1, 2$; $m = 4 - 16$; $k = 1$ (XIII), 2 (XIV).

The HFC depth can be varied by changing the ratio of initial components. Equimolar ratio of the initial components gives maximal depth of HFC reaction. After partial over-precipitation by methanol from toluene solution, synthesized oligomers become yellow or light-brown transparent products, well-soluble in various organic solvents. Some parameters of the oligomers are shown in Table 2.19.

Silarylenecyclosiloxanes of structure XIV with higher molecular mass were synthesized by catalytic dehydrocondensation of 1,7-dihydride-1,7-dimethyloctaphenylcyclohexasiloxane and 1,4-bis(hydroxydimethylsilyl)-benzene [118, 119].

For the purpose of synthesizing silarylenecyclohexasiloxane copolymers, catalytic dehydrocondensation of 1,7-dihydride-1,7-dimethylocta-phenylcyclohexasiloxane with 1,4-bis(hydroxydimethylsilyl)benzene, at different temperatures (30, 40 and 50°C) in the solution of anhydrous toluene (C = 0.4686 mol/l) in the presence of powder-like caustic potash has been investigated. Recently, it has been shown that during catalytic dehydrocondensation of linear α,ω-dihydride-diorganosiloxanes with α,ω-dihydroxydiorganosiloxane in the presence of caustic potash, depending on the length of siloxane fragments, both individual organocyclosiloxanes and linear copolymers are synthesized [120]. Besides, it was shown that during catalytic dehydrocondensation of dihydride-organocyclosiloxanes with dihydroxyorgano-cyclosiloxanes in the presence of nucleophilic anhydrous potassium hydroxide, the break of cyclosiloxane rings does not take place [121]. Figure 2.33 shows the time dependence of ≡Si-H group concentration decrease. As is observed, with the temperature rise in the reaction of catalytic dehydrocondensation, the hydrogen conversion increases from 85% (at 30°C) to 96% (at 50°C). Dehydrocondensation reaction proceeds according to the following scheme [118,119]:

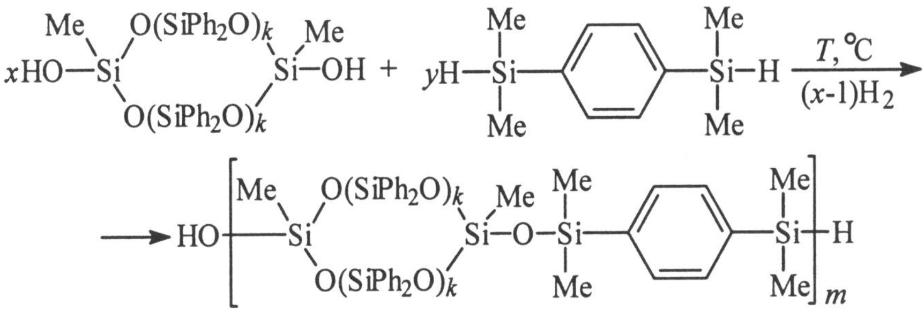

(2.22)

where $T = 30, 40, 50°C$.

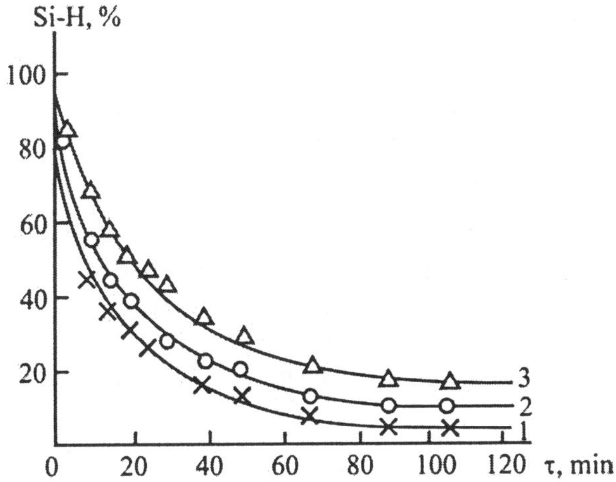

Figure 2.33. Hydrogen conversion in dehydrocondensation of 1,7-dihydride-1,7-dimethyloctaphenylcyclohexasiloxane with 1,4-bis(hydroxydimethylsilyl)benzene: 1 - 50°C; 2 – 40°C; 3 – 30°C.

During catalytic dehydrocondensation of 1,7-dihydrideorganocyclohexasiloxane with 1,4-bis(hydroxydimethylsilyl)benzene in the presence of potassium hydroxide, the reaction order, rate constants and activation energy were determined. Catalytic dehydrocondensation is the second order reaction. The reaction rate constants were calculated at different temperatures: $k_{30°C} = 0.9823\times10^{-1}$, $k_{40°C} = 1.5550\times10^{-1}$, $k_{50°C} = 2.1435\times10^{-1}$. Some physical and chemical parameters of low-molecular copolymers are shown in Table 2.19.

Table 2.19

Yield and physical and chemical parameters of α,ω–dihydroxysilarylenecyclo(tetra)hexasiloxanes

No.	Oligomer structural unit	K	Structure	Yield, %	m	η_{spec}	T_g, °C	d_1, Å
1	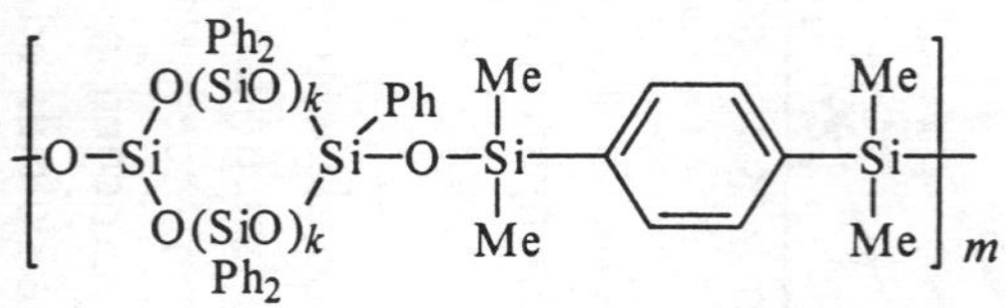	1	XIII	76	4	0.04	55-61	-
2		1	XIII	81	6	0.04	61-65	9.98
3		1	XIII	85	16	0.06	63-68	-
4		2	XIV	80	4	0.04	61-65	10.04
6		2	XIV	86	12	0.05	67-71	10.08
7*		2	XIV	93	13	0.08	61-65	10.04
8*		2	XIV	93	-	0.09	64-66	-
9*		2	XIV	95	16	0.09	67-71	10.04

Note: * The oligomers were synthesized by dehydrocondensation reaction.

It is shown that the temperature coefficient of this dehydrocondensation reaction equals $\gamma \approx 1.5$. From the dependence of the reaction rate constants logarithm on reverse temperature, the activation energy of the reaction was calculated, which equals $E_a = 32.55$ kJ/mol. This value of the activation energy is close to that of catalytic dehydrocondensation of dihydridorganocyclosiloxane with dihydroxyorganocyclosiloxane [111]. For copolymers No. 7* and 9* (Table 2.19), quantitative values of $\bar{M}_n$, $\bar{M}_\omega$, $\bar{M}_z$ and $\bar{M}_\omega / \bar{M}_n$ were determined by gel permeation chromatography methods, which equal $\bar{M}_n = 1.05 - 1.62 \times 10^4$ and $\bar{M}_\omega \approx 1.69 - 1.98 \times 10^4$; polydispersion degrees, D, of copolymer No. 7* and 9* (Table 2.19) equal ~1.46 and 1.22, respectively. In the ^{1}H NMR spectra of copolymer No. 9* in the range of $\delta = 0 - 0.3$ ppm, one can observe various resonance strips for various ≡Si-Me groups. The correlation integral intensity of ≡Si-Ph and ≡Si-Me groups is suitable for elementary fragments.

Arylenecyclosiloxane oligomers were synthesized in HFC reaction of 1,4-bis(dichlororganosilyl)benzene with dihydroxydiphenylsilane and dihydroxydiphenylsiloxanes in 50 – 60% anhydrous toluene solution at 1:2 ratio of the initial components both in the presence and in the absence of pyridine, which is hydrogen chloride acceptor. The reaction proceeded at room temperature; at the final stage the reaction mixture was heated up to 100°C. The reaction scheme is as follows [122]:

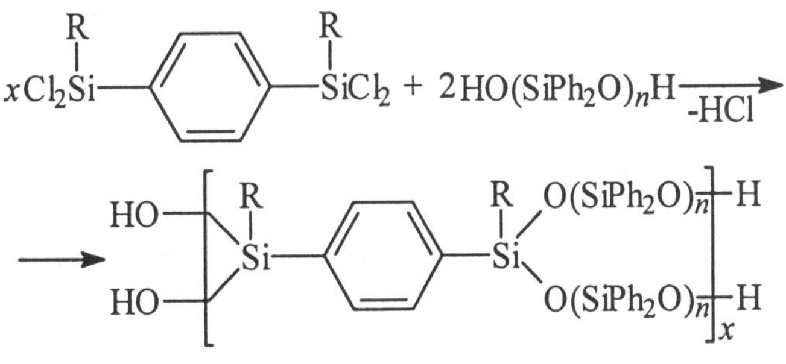

(2.23)

where R = Me, Ph; $n = 1 - 3$.

After over-precipitation from toluene solution with methyl alcohol, the oligomers synthesized represent solid products with $\eta_{spec} = 0.05 - 1.0$. It is shown that at short length of linear diphenylsiloxane unit, $n \leq 2$, oligomers completely soluble in various organic solvents are obtained. The yield of soluble part of oligomers is decreased with the length increase, n.

Synthesized silarylenecyclosiloxane oligomers and copolymers represent amorphous systems, and the interchain distance in oligomers equals $d_1 = 9.98$ - 10.08 Å, approximately. The results of thermogravimetric analysis indicate that 5% of the mass is lost at 430 - 450°C; the main degradation process proceeds in the range of 500 - 700°C, and in the area of 800°C the mass loss curves become evanescent. Further on, dihydroxyarylenecyclosiloxanes were used for synthesizing cyclolinear organosilicon block-copolymers.

To clear up action of the origin of organic radicals in organocyclotetrasiloxane fragments on polyaddition process, as well as to determine contribution of ethylene and phenylene units containing organocyclotetrasiloxane fragments of 1,5- and 1,3-position into cyclolinear copolymers, investigators have studied 1,4-bis(dimethylsilyl)benzene hydride polyaddition to 1,5-divinyl-1,5-dimethyltetraorganocyclotetrasiloxane and 1,3-divinyl-1,3-dimethyltetraorganocyclotetrasiloxane in the presence of the catalyst (0.1 M solution of platinum-hydrochloric acid in tetrahydrofuran) [123].

In both cases, the reaction proceeds in accordance with the scheme as follows:

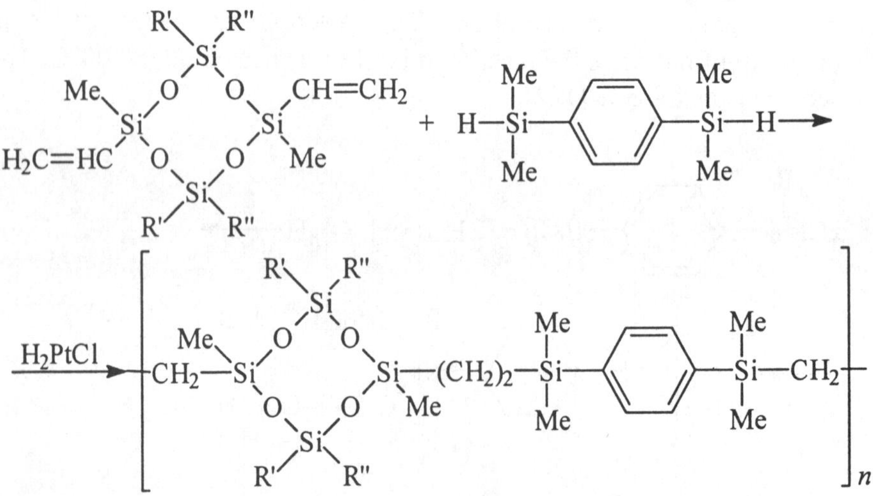

(2.24)

XV

where R' = R" = Me; R' = Me, R" = Ph; R' = R" =Ph.

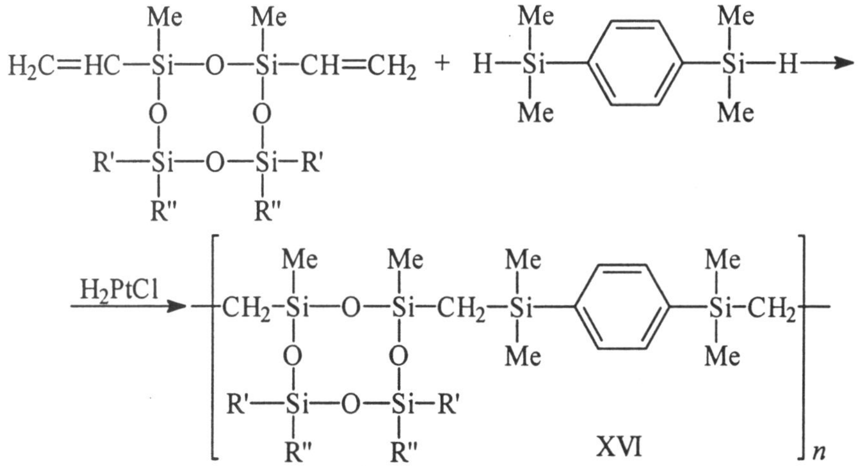

(2.25)

where R' = R" = Me; R' = Me, R" = Ph; R' = R" = Ph.

Initial divinylcyclosiloxanes, used in polyaddition reactions, represent a mixture of *cis*- and *trans*-isomers (at 46:54 ratio). Because these isomers could not be separated by rectification or fractional over-precipitation [123], synthesized silarylenecarboorganocyclosiloxanes are considered to be atactic copolymers.

Hydrosilylation conditions depend on reactivity of the initial monomers. As 1,4-bis(dimethylsilyl)benzene is quite active hydrosilylating agent, already at 60 - 70°C it interacts with divinylorganocyclotetrasiloxanes containing even four phenyl groups at atoms of silicon; in the case of methyl substituting agents only the reaction proceeds at 50°C. All synthesized copolymers are transparent colorless compounds which, in the unique case of methyl substituting agents at atoms of silicon in initial cyclosiloxanes, represent highly viscous-flow polymers; in the case of diphenyl substituting agents, they are transparent solids forming friable soluble films. Table 2.20 shows physical and chemical parameters of silarylenecarborganocyclosiloxanes. Note that the highest values of $[\eta]$ = 0.42 and 0.48 dl/g are displayed by polymers containing rings with methyl and phenyl groups at single atom of silicon, which are copolymers No. 1 and 5 (Table 2.20).

In its turn, location of vinyl groups in 1,3- or 1,5-position in OVCS induces clearly determinable regularity. In all considered cases, PFOS viscosity is higher, when two vinyl groups are located in 1,5-position of tetrameric ring. This is apparently associated with disposition of vinyl groups more suitable for

Table 2.20

Physical and chemical parameters of silarylenecarborganocyclosiloxanes

No.	Copolymer structure	R′	R″	[η], dl/g	T_{degr}* of 5% mass loss, °C	Coke-like residue (800°C), %	T_g, °C
1	2	3	4	5	6	7	8
1	XV	Me	Me	0.36	450 (250)	50	-11
2	XV	Me	Ph	0.48	470 (450)	48	+20
3	XV	Ph	Ph	0.40	450 (350)	45	+27
4	XVI	Me	Me	0.25	325 (300)	57	-5
5	XVI	Me	Ph	0.42	420 (400)	48	+16
	XVI	Ph	Ph	0.26	400 (375)	40	+50

Note: * TGA data on copolymers treated by heptamethylvinylcyclotetrasiloxane. Shown in brackets are T_{degr} for copolymers untreated by heptamethylvinylcyclotetrasiloxane.

hydrosilylation and formation of more symmetrical structure in copolymers No. 1 – 3 (structure XV).

Table 2.21

Chemical shifts of ^{29}Si structural fragments in silarylenecarborganocyclosiloxanes*

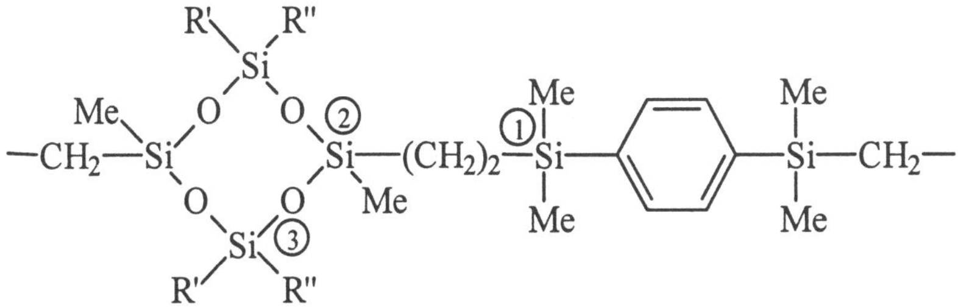

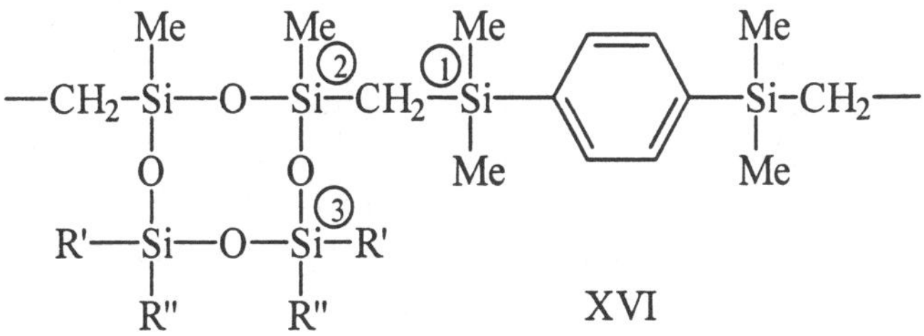

XVI

No.	Copolymer structure	R′	R″	Chemical shifts δ (ppm) for Si atoms**		
				1	2	3
1	XV	Me	Me	-1.16	-19.16	-18.98
2	XV	Me	Ph	-1.23	-18.05	-32.32
3	XV	Ph	Ph	-1.26	-15.98	-45.15
4	XVI	Me	Me	-1.22	-19.30	-19.05
5	XVI	Me	Ph	-1.20	-18.32	-31.48
6	XVI	Ph	Ph	-1.15	-17.32	-43.92

Note: * NMR spectra are registered on WP-200 device, solvent - benzene.

Reflexes from silicon atoms of initial monomers differ from these of silarylenecarborganocyclosiloxanes and one may suggest that hydrosilylation proceeds with formation of ethylene and silarylenephenylene bridges, which is proved by corresponded Si^1 reflex in the range of -1.15 … -1.26 ppm From the data shown in Table 2.21, one may determine several features associated with the shift of reflexes from atoms of silicon due to the composition and the structure of cyclosiloxane fragment. For example, if the reflex induced by atom of silicon in the group (shown below) in polymers containing diphenylsiloxane rings falls in the area of -15.97 and -17.32 ppm, then for polymers containing methyl substituting agents only at atoms of silicon in the ring, this reflex is shifted to -19.16 and -19.30 ppm

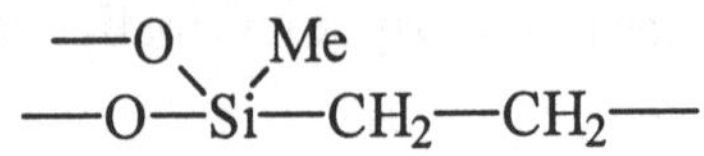

In the case of two -CH_2CH_2- groups' location at neighboring atoms of silicon, the reflex is shifted towards more negative values comparing with the case of their location at opposite atoms of silicon in the ring. For O_2SiPh_2 and $O_2SiMePh$ fragments depending on disposition of vinyl groups at atoms of silicon in initial carbocyclosiloxanes, corresponded shift of reflexes from silicon is observed.

In ^1H NMR spectra of silarylenecarborganocyclosiloxanes in the area of 5.8 and 4.5 ppm, reflexes from protons belonged to ≡Si-CH=CH_2 and ≡Si-H groups of initial monomers are absent. The reflex at 1.0 ppm, corresponded to protons of –CH_2-CH_2- group, formed in hydrosilylation reaction. Generally, location of reflexes and the ratio of their intensities in ^{29}Si and ^1H NMR spectra prove the supposed structure of silarylenecarborganocyclosiloxane.

In the case of equimolar ratio of the components, under selected conditions hydrosilylation is almost completed: by NMR and IR-spectroscopy methods no residual end groups were detected. High-molecular product synthesized possesses MM within the range from 1.0×10^5 to 5.0×10^5. At the same time, taking into account high volatility of 1,4-bis(dimethylsil)benzene compared with divinylcyclotetrasiloxane, the quantity of introduced dihydride-derivative was increased by 0.1% of the theoretical one, because ≡Si–H groups are present at the ends of the chain. To eliminate the effect of end groups on physical and chemical parameters, on thermal stability of silarylenecarborganocyclosiloxane, in particular, copolymer solution in benzene is treated by the excess of vinylheptamethylcyclotetrasiloxane before over-precipitation in accordance with the scheme as follows:

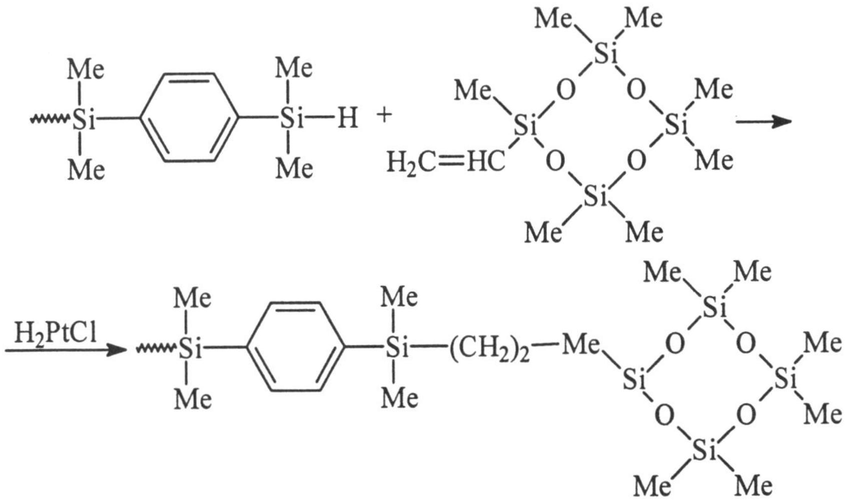

(2.26)

Variation of the ratio of reagents for hydrosilylation gives an opportunity to obtain both high-molecular products and oligomers with defined chain length, which is very important for synthesis of block-copolymer and polymer networks.

After polymer treatment by heptamethylvinylcyclotetrasiloxane, heat resistance of silarylenecarborganocyclosiloxanes is slightly increased. The reason is that residual end ≡Si-H groups promoting high-temperature polymer degradation are substituted by higher heat resistant heptamethylcyclotetrasiloxane fragments, which causes an increase of thermal stability of the polymer.

Estimation of the glass transition temperature of silarylenecarborganocyclosiloxanes and its variation with regard to the origin of substituting agents at atoms of silicon in cyclic rings and introduction of disilphenylene unit into the polymer backbone has indicated that at quite high molecular mass (up to 5.0×10^5) no high elasticity plateau on thermomechanical curves of the samples is present. This is probably associated with higher rigidity of silarylenecarborganocyclosiloxanes compared with linear poly(organosiloxanes).

For the purpose of carbosiloxane network synthesis, at the first stage in accordance with scheme (2.25), silarylenecarborganocyclosiloxanes with 1,4-bis(dimethylsilyl) benzene excess and narrow MMD were obtained [124]. Hence, synthesized oligomers of the structure XV are transparent, colorless and viscous products with *MM* = 800 - 2,500. At the second stage, polyaddition of reaction (2.23) product with tetramethyltetravinylcyclotetrasiloxane is performed in accordance with the scheme as follows:

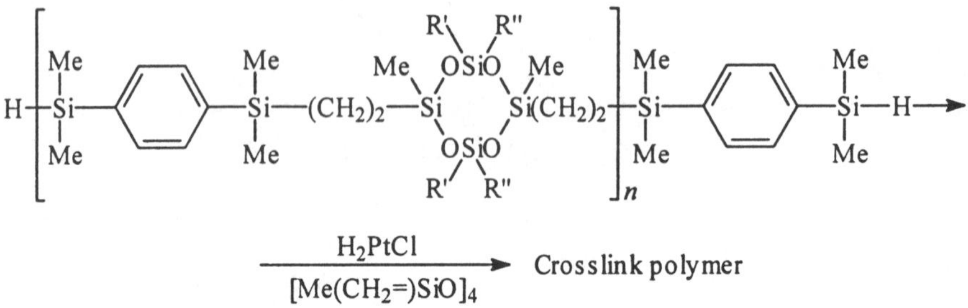

(2.27)

where R' = R" = Me, Ph; R' ≠ R".

Polymer networks depending on the structure of initial monomers and oligomers are characterized by different framing at atom of silicon and different regulated distance between network points. For the purpose of studying some physical and chemical properties, formation of a network structure directly during film formation from the solution is the unique method of obtaining film samples of polymer networks.

To obtain films of studied polymers, 50% solution of the oligomers in tetrhydrofurane (THF) is added by specified quantity of tetramethyl-tetravinylcyclotetrasiloxane and mixed at 20°C until a homogeneous mixture is obtained. The mixture is poured out on a cellophane support 60 mm in diameter and exposed during 1 hour at 20°C; thereafter, it is places to a thermostat, in which temperature was increased gradually: 60°C during 1 hour, 80°C during 2 hours and 120°C during 10 hours. This mode of heating induced formation of transparent colorless films with 85 – 95% of gel fraction in them. Further exposure at 120°C did not cause mass increase of the films that testified about complete elimination of the solvent.

All synthesized polymers are amorphous. At transition from methylphenyl to diphenyl substituting agents at atoms of silicon in the ring, d_1 value varies from 8.5 to 8.8 Å. For polymers with dimethyl substituting agents at atoms of silicon in the ring, unique determination of d_1 value is impossible due to overlapping of intensive diffraction maximums in the area of 5 and 7 Å. However, one may consider that d_1 value of this polymer is the lowest in the sequence of studied objects.

Thermomechanical studies of obtained polymer networks have detected quite non-trivial fact of T_g independence of the type of chemical framing and the size of linear fragments between chain branching rings. For all studied polymers, T_g equals ~30°C. Qualitative explanation may be represented by a random compensation of influence of different framing groups and the length of interpoint fragments on their mobility and packing. However, this requires future investigations.

Table 2.22 shows values of strength at break, σ_b, deformation at break, ε_b, and elasticity coefficient, *E*, of studied polymer networks. The data indicate that substitution of methyl groups by phenyl ones and variation of the distance between chain branching centers is accompanied by a significant change of deformation and strength properties of studied polymers. Note that substitution of both methyl groups by phenyl ones is accompanied by the increase of the elasticity parameter from ~3 to ~600 MPa and the strength from ~1 to ~20 MPa. Besides the possibility of broad variation of mechanical properties, in some cases high mechanical properties may be realized, for example, for a polymer with R' = R" = C_6H_5 groups at atoms of silicon (*n* = 2), which displays high strength (~20 MPa) and elasticity parameter (~600 MPa) combined with quite high deformability (~30%).

Table 2.22

Deformation and strength parameters of films based on silarylenecarborganocyclosiloxanes of the structure XV:

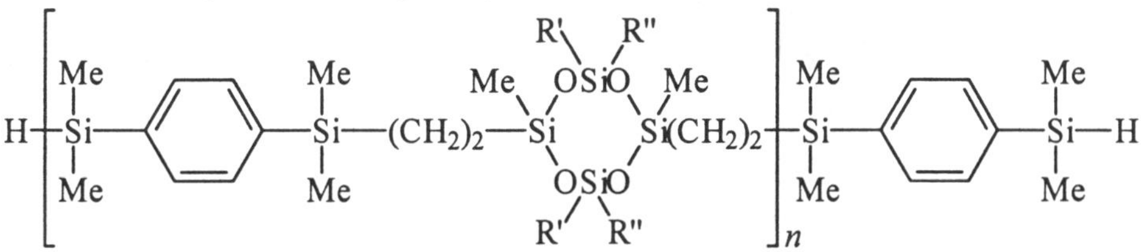

No.	R^I	R^{II}	*n*	σ_b, MPa	ε_b, %	*E*, MPa
1	Me	Me	1	1.1	36	3.1
2	Me	Me	2	0.6	52	1.5
3	Me	Ph	1	10.8	158	43.0
4	Me	Ph	2	0.9	164	6.8
5	Me	Ph	3	0.8	209	0.6
6	Me	Ph	4	0.6	234	0.3
7	Ph	Ph	1	21.3	9	610
8	Ph	Ph	2	19.3	30	570
9	Ph	Ph	3	9.5	50	350

For the purpose of estimating conformational parameters of silarylenecarborganocyclosiloxane of the structure XV, two approaches were used [125]: computerized mathematical simulation using the Monte-Carlo method and experimental estimation of the flexibility parameters in solution under natural conditions. In the first case, mathematical simulation has determined the skeletal flexibility of the molecule in the absence of substituting agents at atoms forming the backbone. In the second case, all fragments of the chain, possible interactions with the solvent and temperature effect have been taken into account in the flexibility estimation.

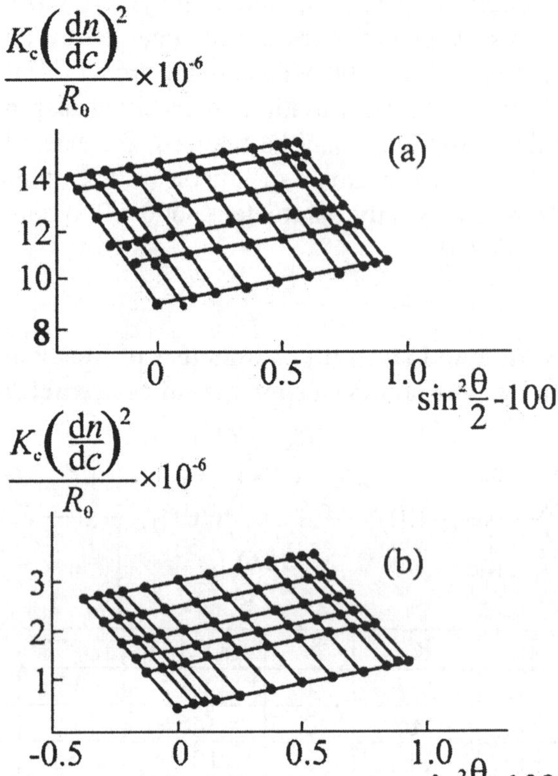

Figure 2.34. Zimm diagrams of fractions 5 (a) and 9 (b) of silarylenecarborganocyclosiloxane copolymer 2 of structure XV (Table 2.20)

For the purpose of experimental estimation of the flexibility parameters in solution, silarylenecarborganocyclosiloxane copolymer 2 (Table 2.20) with structure XV was fractioned from the benzene (solvent) – ethanol (coagulant) system into 14 fractions. At 25°C in toluene, characteristic viscosity values, $[\eta]$, of these fractions fall within the range of 5.1 – 0.07 dl/g in toluene. For nine fractions, sufficient quantity of which was obtained, by the light scattering method, values of $\bar{M}_\omega$ at the angle of 90° in chloroform ($dn/dc \approx 0.95$) on Fika photogoniodiffusometer were measured. For four most high-molecular fractions (Table 2.22), $\bar{M}_\omega$ and $\langle R^2\rangle^{1/2}$ values at different angles were measured (Figure 2.34).

Polydispersity coefficient of the copolymer was estimated by the GPC method (on KhZh-1302 gel chromatograph). For silarylenecarborgano-cyclosiloxane copolymer 2, GPC data display $\bar{M}_\omega / \bar{M}_n = 1.34$. Table 2.23 shows values of $\bar{M}_\omega$ and $\langle R^2\rangle^{1/2}$ for silarylenecarborganocyclosiloxane fraction, measured by the light scattering method. As for fractions 4 – 9 $\bar{M}_\omega$ values are high, they are measured in non-polarized and polarized light. As these values are close, Table 2.23 shows the average one for these fractions.

Table 2.23

Values of $\bar{M}_\omega$, scattering asymmetry and the second virial coefficient, A_2, for silarylenecarborganocyclosiloxane fraction

No.	$\bar{M}_\omega \times 10^{-2}$	$\langle R^2\rangle_z^{1/2}$	Scattering asymmetry	$A_2 \times 10^{-5}$, $cm^2 \cdot mol/g^2$
1	206	-	1.03	0.940
2	252	-	1.06	0.281
3	400	-	1.03	0.221
4	883	-	1.09	0.264
5	1,072	298.6	1.05	0.236
6	1,381	405.7	1.07	0.202
7	1,866	-	1.06	0.140
8	644.8	644.8	1.20	0.135
9	12,500	1,221.4	1.20	0.093

Small differences in $\bar{M}_\omega$ values, measured in polarized and non-polarized light, as well as low scattering asymmetry values close for different fractions testify about low polydispersity of the fractions and coil-like conformation of macromolecules, which is of importance for experimental estimation of flexibility parameters by data on $[\eta]$ and molecular mass of the fractions [126]. Low polydispersity of fractions and the absence of branched molecules in them are also proved by linear type of the Zimm network (Figure 2.34).

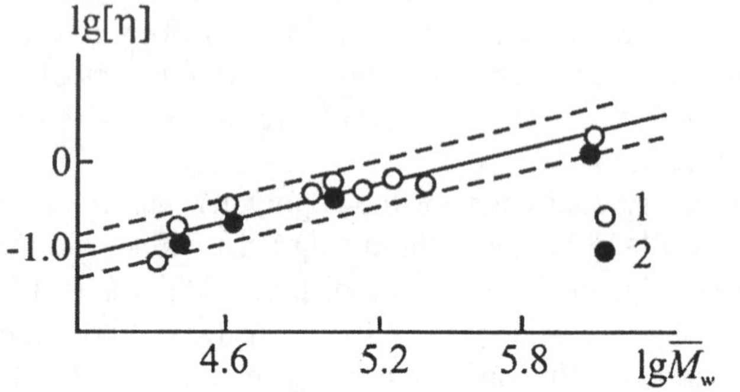

Figure 2.35. Dependence of lg[η] on lg $\bar{M}_\omega$ for silarylenecarborganosiloxane in toluene at 25°C (1) and 20°C (2)

The dependence of [η] and $\bar{M}_\omega$ values in lg[η] – lg $\bar{M}_\omega$ coordinates, shown in Figure 2.35, indicates that the following Mark-Kuhn-Hauvink equations are corresponded to silarylenecarborganosiloxane copolymer:

$$[\eta] = 0.908\times10^{-3}\ M^{0.506 \pm 0.049}\ \text{(toluene, 20°C)};$$
$$[\eta] = 0.383\times10^{-3}\ M^{0.588 \pm 0.084}\ \text{(toluene, 25°C)}.$$

Values of parameter α in the Mark-Kuhn-Hauvink equation, close to 0.5, which were obtained for silarylenecarborganosiloxane fractions, allow correspondence of the polymer molecules to the coil-like type and toluene at 20 and 25°C to θ-solvents. As a consequence, extrapolation data of the Mark-Kuhn-Hauvink equation, silarylenecarborganosiloxane molecules display the coil-like conformation, which is slightly disturbed in toluene at 20 and 25°C by interaction with the solvent.

Information about the shape of macromolecular coil may also be obtained from dependence of the second virial coefficient, A_2, on MM and [η] of fractions (Table 2.23). Considering variation of parameter, A_2, of excluded volume, $\bar{M}_\omega/[\eta]$, on $\bar{M}_\omega$ of silarylenecarborganosiloxane fractions (Figure 2.36), one may observe that A_2 value of $\bar{M}_\omega/[\eta]$ decreases monotonously from 1 to 0.5 with $\bar{M}_\omega$.

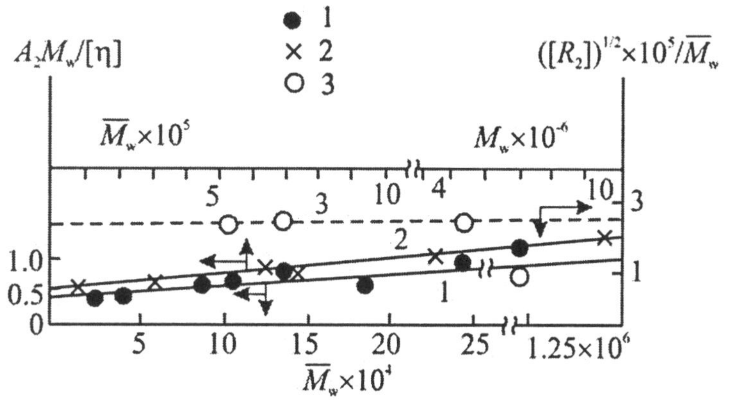

Figure 2.36. Variation of A_2 of $\bar{M}_\omega/[\eta]$ with $\bar{M}_\omega$ increase for 1 - silarylenecarborganosiloxane and 2 - polysilane [116]; 3 - $(\langle R^2\rangle_z)^{1/2}/\ \bar{M}_\omega$ – $\bar{M}_\omega$ dependence silarylenecarborganosiloxane fraction

It is known [127, 128] that theoretical values of the relation $\bar{M}_\omega/[\eta]$ of A_2 are independent of molecular mass, if the macromolecule forms the Gaussian impermeable coil in the solution. For flexible-chain macromolecules and good solvents, for A_2 $\bar{M}_\omega/[\eta]$ = 1.0 - 1.4 and decreases significantly with deterioration of the solvent quality; for rod-shaped molecules it equals 0.4 [129].

The author of the work [130], who has observed the decrease of A_2 $\bar{M}_\omega/[\eta]$ on MM in the range of molecular mass below 8,000 and 9,600,000 [116] (Figure 2.36), explains this effect by partial permeability of coils caused by insufficient quantity of segments for forming statistical coil. In the present case, occurrence of the effect of A_2 $\bar{M}_\omega/[\eta]$ decrease at MM of 2.0×10^5 magnitude may also be explained by partial permeability of the coil, which is stipulated by chemical structure of the unit of macromolecule (Table 2.24). For determining numerical value of the Kuhn segment, a computer experiment was performed. In this case, skeletal rigidity of macromolecule in the absence of substituting agents (free rotation) was estimated.

True value of macromolecule flexibility was calculated from $[\eta]$, A_2 and $\bar{M}_\omega$ of the fractions using some theoretical equations. Calculation of flexibility parameters at free rotation by the Monte-Carlo method indicates that

Table 2.24

Conformational parameters of several organosilicon polymers

No.	Repeated unit of the polymer	$(\langle h^2\rangle_0/M)\times 10^{-16}$	A, Å	A_{CB}, Å	$S_{CB} = A_{CB}/l_0$	$S = A/l_0$	σ	M_0/l_0 (l_0, Å)	Ref.
1	Silarylenecarborganocyclosiloxane	0.546[1] 0.575[2] $(\langle h^2\rangle_{CB}/M)$ = 0.555	18.9[1] 19.9[2]	19.2	1.04	1.03 1.08	1.02	638/18.4	-
2	~Me_2SiO~	0.42 - 0.43 $(\langle h^2\rangle_{CB}/M)$ = 0.24	9.4-10	6.0	2	3.2-3.4	1.26	74/2.96	129
3	-CH_2-CH_2-$SiMe_2SiMe_2$-	1.17[3] 0.96[2]	26.5 21.8	5.44	1	4.2 4.8	2.21 2.0	144/6.34	129
4	—Si— with $(CH_2)_5CH_3$ above and $(CH_2)_5CH_3$ below	5.43 nm^2	54	-	-	-	-	198/2.35	116
5	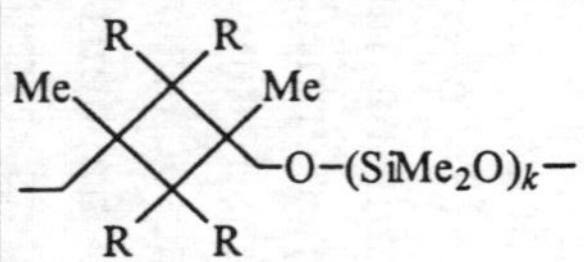R = Ph, k = 1	$(\langle h^2\rangle_{CB}/M)$ = 0.28	-	15	1.3	-	-	618/11.5	112
6	R = Ph, k = 5	0.303	16	10	-	1	1.12	914/21.9	112
7	R = Me, k = 0	0.603	23[3]	-	-	4.5	-	283/6.4	58

Note: Calculations have been executed by the following equations: 1 – Curata-Yamakawa; 2 – Stockmayer-Fixman; 3 – Yamakawa-Fudgi.

free rotation of the molecule is possible in four cases only (l_2, l_4, l_5, l_6); in three cases it is hindered (l_1, l_3, l_7), which is observed from Table 3 of the schematic presentation of the chain unit (scheme 2.1).

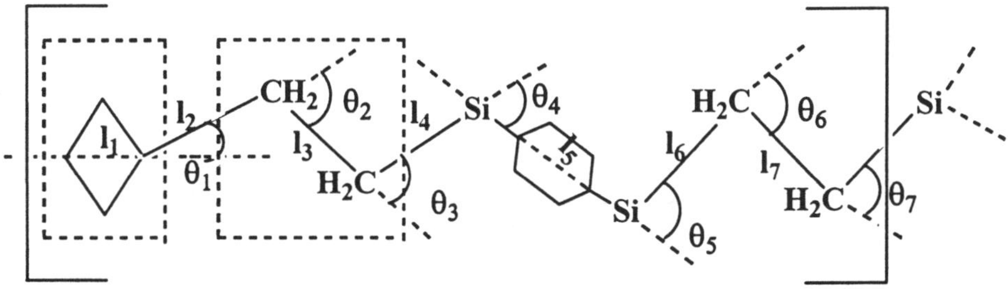

Scheme 2.1

In accordance with calculations, current flexibility parameter ($\langle h_2\rangle_{fr}/M$), where $\langle h_2\rangle_{fr}$ is the mean-square distance between macromolecule chain ends at free rotation, equals 0.555×10^{-6}. If molecular mass of the chain unit equals M_0 = 638 and length of the unit $l_0 = 18.41\times10^{-8}$ cm, the flexibility parameter in terms of the Kuhn segment, A, at free rotation equals: $A_{fr} = (\langle h^2\rangle_{fr}/M)M_l = 19.2$ Å.

Specified data show (Table 2.24) that the Kuhn segment, A_{fr}, and A of silarylenecarborganosiloxane are comparable with the molecule unit length ($l_0 = 18.4$ Å). Table 2.23 shows that for the sequence of silicon-containing high-molecular compounds, this takes place for A_{fr} of poly(ethylene tetramethyldisilane) (compound 3) and polyorganocyclosiloxane (compound 5), the repeated units of which contain -CH_2CH_2- groups [129] and cyclic fragment, respectively. For the studied silarylenecarborganosiloxane rotation is prohibited both around -CH_2CH_2- (l_3 and l_7 bond lengths) and cyclic fragments (l_1) (Table 2.25). However, all the fractions studied displayed sufficient quantity (31 - 1,880) of segments (L/A_{fr}) to provide for macromolecule the coil shape.

For estimation of silarylenecarborganosiloxane macromolecules under natural conditions, when the coil is "perturbed" by interaction with the solvent, authors of the work [125] have used the following prerequisites observed in the current case: $L >> \alpha$, i.e. the macromolecule is of the coil-shaped type ($\alpha = A/2$); low concentrations of solutions, in which the Einstein equation is active; the selected solvent is close to the θ-solvent.

Table 2.25

Bond lengths and unit rotation angles in silarylenecarborganocyclosiloxane chain

Bond length, Å	Angles between bonds (θ), deg	Type of rotation
$l_1 = 4.25$	54.5	Prohibited (*trans*)
$l_2 = 1.86$	66.0	Free
$l_3 = 1.54$	66.0	Prohibited (*trans*)
$l_4 = 1.87$	68.0	Free
$l_5 = 6.51$	68.0	Free
$l_6 = 1.87$	66.0	Free
$l_7 = 1.54$	66.0	Prohibited

Numerical values of $\langle h^2 \rangle_0/M$ and A are estimated by the Shtockmayer-Fixman equation [130], which is based on the Gaussian impermeable coil model, perturbed by interaction with the solvent, and the Kirkwood-Riseman equation [131, 132] for partly permeable coils, which may possibly take place in the current case at $M = 2.0 \times 10^5$ or lower (Figure 2.37, lines 1 and 2, respectively).

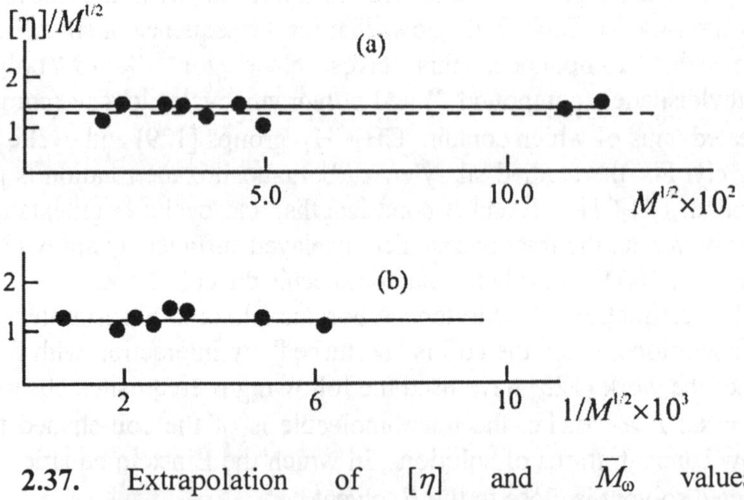

Figure 2.37. Extrapolation of $[\eta]$ and $\bar{M}_\omega$ values for silarylenecarborganocyclosiloxane fraction in accordance with Stockmayer-Fixman (a) and Kirkwood-Riseman (b) equations. Dotted line represents calculation results for free rotation

According to extrapolation by the Stockmayer-Fixman equation as follows:

$$[\eta]/M^{1/2} = F_0(\langle h^2\rangle_0/M)^{3/2} + 0.51F_0BM^{1/2},$$

the section cut at the axis of ordinates by the straight line, composed in $[\eta]/M^{1/2}$-$M_\omega^{1/2}$ coordinates (Figure 2.38), equals to K_θ; hence, $(K_\theta/F_0)^{2/3}$ = $(\langle h^2\rangle_0/M)$. For silarylenecarborganocyclosiloxane, $(\langle h^2\rangle_0/M)$ = 0.575×10^{-16} (Table 2.23). Hence, the Kuhn segment A equals K_θ = 19.9 Å. Figure 2.37 shows also that $(\langle h^2\rangle_0/\mathrm{M})$ value, obtained by $[\eta]/M^{1/2}$ extrapolation, coincides with $(\langle h^2\rangle_{fr}/M)$, which is typical of Gaussian impermeable coil.

For polymers of every type, estimation of the structural parameter, M_L, is the most difficult stage of calculation of conformational parameters: the Kuhn segment, A, and persistent length, $\alpha = A/2$. By the present time, techniques for estimating this parameter for persistent chains have been considered in the works by Bogdanetzki [133] and Shan and Nike [134], using results of $[\eta]$ and molecular mass determination. Concerning macromolecules, which form a coil in solutions, both permeable and impermeable for the solvent, no approach to estimation of M_L of these molecules was suggested. On the example of silarylenecarborganocyclosiloxane, the authors of the present monograph have made an attempt to estimate the value of M_L basing on the data on $[\eta]$ and molecular mass of the studied polymer fraction.

For determining the swelling coefficient from characteristic viscosities, let us use the known ratio [130]: $\alpha\eta \approx [\eta]/[\eta]_\theta = \alpha_{st}\alpha_t$; these viscosities were obtained based on the scaling approach, true in the considered in the locality of the θ-point. Here α_{st} is the swelling coefficient of the macromolecule by the inertia radius, R_{st}; α_t is the swelling coefficient of the macromolecule by the hydrodynamic radius, R_t, which was calculated by the Kirkwood-Riseman approach [131]. Under θ-conditions:

$$R_{st(\theta)} = (L\alpha/3)^{1/2};\quad R_{(\theta)} = 3/8[(\pi L\alpha)/3]^{1/2}.$$

Standard correction data to these values for low (as in the current case) deviations from the θ-point can be easily determined within the frames of the perturbation theory. However, as estimated in [125], they exceed the experimental accuracy and are neglected in this discussion. As a result, it was obtained that:

$$[\eta]_\theta = 2.5(\pi/3)^{3/2} N_A (L\alpha)^{3/2}/2M.$$

Table 2.25 shows assessment data on conformational parameters, persistent length (α), the Kuhn segment (A) and structural parameter (M_L) for silarylenecarborganocyclosiloxane fractions. Average values of α, A and M_L, deduced from $[\eta]$ and $\bar{M}_\omega$ of silarylenecarborganocyclosiloxane fractions, equal 9.6 Å, 19.2 Å and 34.45×10^8, respectively. Thus, the structural parameter, M_L, calculated from $\bar{M}_\omega$ and $[\eta]$ values, and the one given by chemical structure (M_L = 34.65×10^8) of molecules are practically coincident. As a consequence, the above considered A_{fr} really equals 19.2 Å. Under natural conditions, the value of A in solution is close to A_{fr}; that is why rotation dormancy factor $\sigma = (A/A_{fr})^{1/2} = (19.9/19.2)^{1/2} = 1.02$.

Thus, according to data of mathematical modeling and hydrodynamic studies, obtained by the Stockmayer-Fixman equation [130], rotation around valence bonds in silarylenecarborganocyclosiloxane chains is practically free (rotation is possible around l_2, l_4, l_5 and l_6 bonds (Table 2.24)). This conclusion was proved by calculations of flexibility parameters of silarylenecarborganocyclosiloxane molecules, executed by the authors of the present monograph using $\bar{M}_\omega$ and A_2 values, measured by the light scattering method. The calculations were performed by Krigbaum [135] and Kurata-Yamakava [136] equations.

Table 2.25 and Figure 2.38 show data on $xF_0(x)$ dependence on molecular mass for silarylenecarborganocyclosiloxane fraction. It follows that for eight fractions, except for the most low-molecular one, the value of $xF_0(x)$ parameter is close to $[xF_0(x)]_{av} = 1.18$ within the limits of the experiment error. In accordance with Kirkwood and Riseman [131], for completely permeable coils in θ-solvent $\lim xF_0(x) = 1$. For infinitely large chain, $\lim xF_0(x) = 1.259$ that is corresponded to complete screening. In this case, $xF_0(x) = 1.18$ vary between 1 and 1.259.

Independent measurements of $\langle R^2\rangle_z^{1/2}$ by the light scattering method have given values close to the results of hydrodynamic studies (Figure 2.38): $\langle h^2\rangle_0/LA$ ratio is on the verge of one. For the basis of comparative estimation of the size of coil-shaped "unperturbed" macromolecules, Flory [137] has suggested to use characteristic relation C_∞, the value of which for a broad range of studied polymers falls within the limits from 4 to 10. This is much above the values corresponded to free rotation of macromolecular chain. In accordance with Flory, parameter $C_\infty = 1$ for flexible chain at all values $n = M/M_0$. Table 2.26 indicates that for silarylenecarborganocyclosiloxane macromolecules, C_∞ is close to one.

Table 2.25

Conformational parameters (α and A), structural parameter (M_L) in accordance with the data on $[\eta]$ and $\bar{M}_\omega$ for silarylenecarborganocyclosiloxane fraction

Fraction No.	α, Å	A, Å	$[\eta]$ toluene 25°C, dl/g	M_L*×10^{-8}, mol/cm		
				(Sch-F)	(K-R)	$xF_0(x)$
1	5.3	10.6	0.070	18.8	19.5	0.49
2	8.95	17.9	0.168	31.1	33.0	1.07
3	10.24	20.5	0.250	35.6	37.7	1.31
4	10.6	21.2	0.406	36.9	39.1	1.38
5	10.8	21.6	0.458	37.5	39.8	1.41
6	9.21	18.4	0.411	32.0	33.9	1.11
7	9.75	19.5	0.528	33.9	35.9	1.23
8	8.3	16.3	0.468	28.8	30.5	0.94
9	9.3	18.5	1.247	32.2	34.1	1.12
Average	9.6	19.2	-	33.5	35.4	-

Note: * The M_L value, calculated by the formula, equals 34.65×10^{-8}. Sch-F is the Stockmayer-Fixman equation; K-R is the Kirkwood-Riseman equation.

Table 2.26

Conformational parameters of silarylenecarborganocyclosiloxane macromolecules*

No.	Equation	$\langle h^2\rangle_0/M$×10^{-16}	C_∞
1	Stockmayer-Fixman	0.575	1.08
2	Kirkwood-Riseman	0.546	1.03
3	Kurata-Yamakawa	0.543	1.03
4	Krigbaum	0.547	1.03

Note: * At M_0 = 638 and l_0 = 18.41 Å.

A wide complex of investigations for assessment of conformational parameters of silarylenecarborganocyclosiloxane macromolecules using hydrodynamic and molecular parameters allowed the authors of the present monograph to make some conclusions. The length of silarylenecarborganocyclosiloxane macromolecule segment is comparable with the unit length, l_0. At free rotation, the Kuhn segment value equals A_{fr} = 19.2 Å. Under natural conditions, value of the Kuhn segment in solution is practically coincident with A_{fr}, and the dormancy parameter of rotation around virtual bond is close to one, i.e. rotation is practically free. This is also proved by the value of characteristic ratio C_∞, which is close to one. The chain rotates due to the presence of connective junctions. Thus, as follows from the above-said, the "flexibility mechanism" gives silarylenecarborganosiloxane molecules the opportunity to form a dense coil, which approaches the Gaussian impermeable one for high molecular masses and semi-permeable as molecular mass is reduced.

References

1. Hurd C.B., *J. Am. Chem. Soc.*, 1946, vol. **68**, p. 364.
2. Hunter M.J., Warrick E.L, and Hyde J.F., *J. Am. Chem. Soc.*, 1946, vol. **68**, p. 2284.
3. Fletcher H.J. and Hunter M.J., *J. Am. Chem. Soc.*, 1949, vol. **71**, p. 2918.
4. Wilcock D.F., *J. Am. Chem. Soc.*, 1946, vol. **68**, p. 691.
5. Andrianov K.A., Vinogradov G.V., Ditsent V.E., Titkova L.V., Makarova N.N., Chernyavski A.I., and Terentieva N.A., *Vysokomol. Soedin.*, 1979, vol. **21A**(7), p. 1475. (Rus)
6. *Patent No. 3,297,632* (USA), 1967; *C.A.*, 1967, vol. **66**, 46790x.
7. *Patent No. 3,264,259*, 1966.
8. *A.C. No. 794,029* (USSR); *Byull. Izobr.*, No. 1, 1981. (Rus)
9. Gilman H. and Oita K., *J. Amer. Chem. Soc.*, 1955, vol. **77**, p. 3386.
10. Dolgoplosk S.B., Kagan E.G., and Skomorovskaya A.D., *Vysikomol. Soedin.*, 1969, vol. **11B**(2), p. 635. (Rus)
11. Mileshkevich V.P. and Yuzhelevski Yu.A., *Properties of Oxygen-Containing Silicon Compounds. Synthetic Rubber Industry*, TsNIITEneftekhim, Moscow, 1977, pp. 51 – 60. (Rus)
12. Khananashvili L.M. and Mukbaniani O.V., *Macromol. Chem. Suppl.*, 1984, vol. **6**, p. 77.
13. Mukbaniani O.V., Koyava N.A., Meladze S.M., and Khananashvili L.M., *Abstr. Commun. 6th Intern. Symp. Organosilicon Chemistry*, 1981, Budapest, p. 62.
14. Mukbaniani O.V., Achelashvili V.A., Koyava N.A., and Khanashvili L.M., *Soobshch. AN GSSR*, 1986, vol. **122**(3), p. 537. (Rus)
15. Mukbaniani O.V., Achelashvili V.A., Meladze S.M., Koyava N.A., and Khananashvili L.M., *Soobshch. AN GSSR*, 1986, vol. **122**(1), p. 105. (Rus)
16. Khananashvili L.M., Mukbaniani O.V., Inaridze I.A., Porchkhidze G.V., and Koberidze Kh.E., *Intern. J. Polym. Mater.*, 1992, vol. **24**, p. 111.
17. Andrianov K.A., Nogaideli A.I., Khananashvili L.M., and Nakaidze L.I., *Thes. Rep. IV All-Union Conference on Chemistry and Application of Organosilicon Compounds*, Tbilisi, 1968, p. 77. (Rus)
18. Andrianov K.A., Nogaideli A.I., Khananashvili L.M., and Nakaidze L.I., *Izv. AN SSSR, Ser. Khim.*, 1968, vol. **???**, p. 2146. (Rus)
19. Andrianov K.A., Nogaideli A.I., Khananashvili L.M., and Nakaidze L.I., *Soobshch. AN GSSR*, 1970, No. 3, p. 581. (Rus)
20. Andrianov K.A., Nogaideli A.I., Nakaidze L.I., and Khananashvili L.M., *Soobshch. AN GSSR*, 1972, vol. **65**, p. 589. (Rus)

21. Nogaideli A.I., Tkeshelashvili R.SH., Nakaidze L.I., and Mukbaniani O.V., *Trudy Tbilisskogo Gos. Universiteta*, 1976, vol. **167**, p. 69. (Rus)
22. Andrianov K.A., Tikhonov V.S., and Astapov B.A., *Vysokomol. Soedin.*, 1970, vol. **12B**(8), p. 577. (Rus)
23. Lukevits E.Ya., Romadan Yu.P., Giller S.A., and Voronkov M.G., *Doklady AN SSSR*, 1962, vol. **145**, p. 806. (Rus)
24. Orlov N.F. and Dolgov B.N., *Doklady AN SSSR*, 1959, vol. **125**, p. 817. (Rus)
25. Khananashvili L.M., Mukbaniani O.V., Inaridze I.A., Porchkhidze G.V., and Koberidze Kh.E., *European Polymeric Journal*, 1998, vol. **34**(3/4), p. 581.
26. Makarova N.N., Godovski Yu.K., and Lavrukhin B.D., *Vysokomol. Soedin.*, 1995, vol. **37A**(3), p. 375. (Rus)
27. Makarova N.N., Petrova I.M., Godovski Yu.K., Lavrukhin B.D., and Zhdanov A.A., *Doklady AN SSSR*, 1983, vol. **269**(6), p. 1368. (Rus)
28. Makarova N.N., Petrova I.M., and Zhdanov A.A., *Doklady AN SSSR*, 1982, vol. **262**(3), p. 619. (Rus)
29. Makarova N.N., Kuzmin N.N., Lavrukhin B.D., Godovski Yu.K., Mamaeva I.I., Matukhina E.V., and Petrova I.M., *Vysokomol. Soedin.*, 1989, vol. **31B**(9), p. 708. (Rus)
30. Mukbaniani O.V., Achelashvili V.A., Meladze S.M., Koyava N.A., Khananashvili L.M., and Sturua G.I., *Soobshch. AN GSSR*, 1986, vol. **122**(1), p. 105. (Rus)
31. Achelashvili V.A., Mukbaniani O.V., Meladze S.M., and Khananashvili L.M., *Abstr. Commun. VII Intern. Symp. Organosilicon Chemistry*, 1984, Kyoto, Japan, p. 216.
32. Godovski Yu.K., Makarova N.N., Petrova I.M., and Zhdanov A.A., *Makromol. Chem. Rapid Commun.*, 1984, No. 5, p. 427.
33. Corriu R.J.P., Guerin C., and Moreau J.J.E., In: *The Chemistry of Organic Silicon Compounds*, London: Wiley, 1989. p. 305.
34. Bassindaller A.R. and Tayler P.G., In: *The Chemistry of Organic Silicon Compounds*, London: Wiley, 1989, p. 840.
35. Lavrukhin B.D., Makarova N.N., and Petrova I.M., *Izv. AN SSSR, Ser. Khim.*, 1993, No. 8, p. 1460. (Rus)
36. Makarova N.N., Lavrukhin B.D., Turkeltaub G.N., Kuzmin N.N., Petrova I.M., and Matukhina E.V., *Izv. AN SSSR*, 1989, No. 6, p. 1351. (Rus)
37. Sommer L.H. and Bennet D.F., *J. Am. Chem. Soc.*, 1959, vol. **81**, p. 251.
38. Matukhina E.V., Godovski Yu.K., Makarova N.N., and Petrova I.M., *Vysokomol. Soedin.*, 1995, vol. **37A**(10), p. 1680. (Rus)

39. Makarova N.N., Kuzmin N.N., Lavrukhin B.D., Godovski Yu.K., Mamaeva I.I., Matukhina E.V., and Petrova I.M., *Vysokomol. Soedin.*, 1989, vol. **31B**(9), p. 708. (Rus)
40. Beatty C.L. and Karasz F.E., *Bull. Am. Phys. Soc.*, 1973, vol. **18**(3), p. 461.
41. Godovski Yu.K., Mamaeva I.I., Papkov V.S., and Kuzmin N.N., *Macromol. Chem. Rapid Commun.*, 1985, No. 6, p. 797.
42. Godovski Yu.K. and Papkov V.S., *Adv. Polymer Sci.*, 1989, vol. **88**, p. 129.
43. Godovski Yu.K., Makarova N.N., and Kuzmin N.N., *Macromol. Chem., Macromol. Symp.*, 1989, vol. **26**, p. 91.
44. Makarova N.N., Kuzmin N.N., Lavrukhin B.D., Godovski Yu.K., Mamaeva I.I., Matukhina E.V., and Petrova I.M., *Vysokomol. Soedin.*, 1989, vol. **31B**(9), p. 708. (Rus)
45. Timofeeva T.V., Boda E.E., Polischuk A.P., Antipin M.Yu., Matukhina E.V., Petrova I.M., Makarova N.N., and Struchkov Yu.T., *Mol. Cryst. Liq., Cryst.*, 1994, vol. **248**, p. 125.
46. Blumshtain A., *Polym. J.*, 1985, vol. **17**, p. 277.
47. Wunderlich B. und Orobowicz J., *Fortshritte Hochpolym-Forsch.*, 1984.
48. Makarova N.N., Matukhina E.V., Godovski Yu.K., and Lavrukhin B.D., *Vysokomol. Soedin.*, 1992, vol. **34B**(2), p. 56. (Rus)
49. Makarova N.N., Godovski Yu.K., and Lavrukhin B.D., *Vysokomol. Soedin.*, 1995, vol. **37A**(3), p. 375. (Rus)
50. Mukbaniani O.V., Achelashvili V.A., Meladze S.M., Koyava N.A., Khananashvili L.M., and Sturua G.I., *Soobshch. AN GSSR*, 1986, vol. **122**(1), p. 105. (Rus)
51. Makarova N.N. and Lavrukhin B.D., *Izv. AN SSSR, Ser. Khim.*, 1986, No. 3, p. 652. (Rus)
52. Makarova N.N., Godovski Yu.K., Matukhina E.V., Volkova L.M., Lavrukhin B.D., and Yakubovich O.V., *Vysokomol. Soedin.*, 1993, vol. **35A**(2), p. 136. (Rus)
53. Godovski Yu.K., Makarova N.N., and Kuzmin N.N., *Vysokomol. Soedin.*, 1988, vol. **30A**(2), p. 370. (Rus)
54. Buzin M.I., Kvachev Yu.L., Svistunov V.S., and Papkov V.S., *Vysokomol. Soedin.*, 1992, vol. **34B**(1), p. 66. (Rus)
55. Makarova N.N., Godovski Yu.K., and Kuzmin N.N. *Macromol. Chem.*, 1987, vol. **188**, p. 119.
56. Godovski Yu.K., Makarova N.N., and Mamaeva I.I., *Macromol. Chem. Rapid Commun.*, 1986. vol. **7**(7), p. 325.
57. Ebert M., Herrmann-Schoherr O., Wendorff J.H., Ringsdorf R., and Tschirner P., *Liq. Cryst.*, 1990, vol. **7**(1), p. 63.

58. Mamaeva I.I., Makarova N.N., Petrova I.M., Tverdokhlebova I.I., and Pavlova S.A., *Vysokomol. Soedin.*, 1987, vol. **29A**(7), p. 1507. (Rus)
59. Stockmayer W.H. and Fixman M., *J. Polym. Sci.*, 1963, vol. **1C**(1), p. 137.
60. Tsvetkov V.N., Frisman E.V., and Boitsova N.M., *Vysokomol. Soedin.*, 1960, vol. **2**(7), p. 1001. (Rus)
61. Tsvetkov V.N., Andrianov K.A., Ryumtsev V.I., Shtennikova I.N., Vitovskaya M.G., Makarova N.N., and Kurasheva N.A., *Vysokomol. Soedin.*, 1973, vol. **15A**(2), p. 400. (Rus)
62. Vitovskaya M.G., Astapenko E.P., Bushin S.V., Skazka V.S., Yamshchikov V.M., Makarova N.N., Andrianov K.A., and Tsvetkov V.N., *Vysokomol. Soedin.*, 1973, vol. **15A**(11), p. 2549. (Rus)
63. Tsvetkov V.N. and Andrianov K.A., *Eur. Polym. J.* 1973, vol. **9**, p.27.
64. Aleksandrova Yu.A., Nikitina T.S., and Pravednikov A.N., *Vysokomol. Soedin.*, 1968, vol. **10A**, p. 1078. (Rus)
65. Tomas T.H. and Kendrik T.C., *J. Polym. Sci.*, 1969, vol. **27A**, p. 537.
66. Plate N.A. and Shibaev V.P., *Comb-Like Polymers and Liquid Crystals*, Khimia, Moscow, 1980, p. 303. (Rus)
67. *Liquid Crystalline Order in Polymers*, Ed. Blumshtein A., Academic Press, New York, 1978.
68. Beatty C.L., Pochan J.M., Froix M.F., and Hinman P.P., *Macromolecules*, 1975, vol. **8**, p. 547.
69. Papkov V.S., Godovski Yu.K., Litvinov V.M., Svistunov V.S., and Zhdanov A.A. *International Conference of Socialist Countries on Liquid Crystals*, Tbilisi, 1981, vol. **2**, p. 201. (Rus)
70. *Liquid Crystalline Order in Polymers*, Ed. Blumshtein A., Academic Press, New York, 1978, Chapter 9.
71. Andrianov K.A., Souchek I., and Khananashvili L.M., *Uspekhi Khimii*, 1979, vol. **48**, p. 1233. (Rus)
72. Pomerantseva M.G., Belyakova Z.V., and Golubtsov S.A., *Synthesis of Carbofunctional Organosilanes by Polyaddition Reaction*, Moscow, 1971. (Rus)
73. Reikhsfeld V.O., Vinogradova V.N., and Fillipov N.A., *Zh. Obshch. Khim.*, 1973, vol. **43**(10), p. 2216. (Rus)
74. Fillipov N.A., Reikhsfeld V.O., Zaslavskaya T.N., and Kuzmina T.A., *Zh. Obshch. Khim.*, 1977, vol. **47**(6), p. 1374. (Rus)
75. Andrianov K.A. and Magomedov G.K., *Doklady AN SSSR*, 1973, vol. **228**(5), p. 1094. (Rus)
76. Andrianov K.A., Gavrikova L.A., and Radionov E.F., *Vysokomol. Soedin.*, 1971, vol. **13A**(4), p. 937. (Rus)

77. Watanabe H., Kitahara T., Motegi X., and Nagai H., *J. Organomet. Chem.*, 1977, vol. **139**(2), p. 215.
78. Souchek I., Andrianov K.A., Khananashvili L.M., and Myasina V.M., *Doklady AN SSSR*, 1975, vol. **222**(1), p. 128. (Rus)
79. Andrianov K.A., Petrashko A.N., Asnovich L.Z., and Gashnikova N.P., *Izv. AN SSSR, Ser. Khim.*, 1967, No. 6, p. 1267. (Rus)
80. Andrianov K.A., Zhdanov A.A., Rodionova E.F., and Vasilenko N.G., *Zh. Obshch. Khim.*, 1975, vol. **45**(11), p. 2444. (Rus)
81. Andrianov K.A., Sidorov V.I., Zaitseva M.G., and Khananashvili L.M., *Khim. Geterotsicl. Soed.*, 1967, No. 1, p. 32. (Rus)
82. Zagradnik R., *Uspekhi Khimii*, 1961, vol. **10**, p. 1291. (Rus)
83. Benson S., *Chemical Kinetics Grounds*, Moscow, 1964, p. 66. (Rus)
84. Zhdanov A.A. and Astapova T.V., *Vysokomol. Soedin.*, 1981, vol. **23A**(3), p. 626. (Rus)
85. Akhrem I.S., Chistovalova N.I., Myisov E.I., and Volpin M.E., *Zh. Obshch. Khim.*, 1972, vol. **42**(8), p. 1868. (Rus)
86. Akhrem I.S., Chistovalova N.I., Myisov E.I., and Volpin M.E., *Izv. AN SSSR, Ser. Khim.*, 1976, No. 9, p. 2069. (Rus)
87. Stober M.R., Musole M.C., and Speier J.L. *J. Org. Chem. Soc.*, 1965, vol. **30**(5), p. 1651.
88. Bulkin A.F., Papkov V.S., Zhdanov A.A., and Andrianov K.A., *Vysokomol. Soedin.*, 1978, vol. **20B**(1), p. 70. (Rus)
89. Mukbaniani O.V., Khananashvili L.M., Koyava N.A., Porchkhidze G.V., and Tolchinski Yu.I., *Intern. J. Polym. Mater.*, 1992, vol. **17**(3-4), p. 113.
90. Mukbaniani O.V., *Doctor's Dissertation on chemical sciences*, I. Dzhavakhishvili Tbilisi State University, 1993, Tbilisi, Georgia. (Rus)
91. Astapova T.V., *Candidate's Dissertation on Chemistry*, I. Dzhavakhishvili Tbilisi State University, 1983, Tbilisi, Georgia. (Rus)
92. Sobolevski M.V., Skorokhodov I.I., Ditsent V.E., Sobolevskaya A.V., and Efimov V.M., *Vysokomol. Soedin.*, 1969, vol. **11A**(5), p. 1109. (Rus)
93. Andrianov K.A., Pavlova S.A., Zhuravleva I.V., Tolchinski Yu.I., and Astapov B.D., *Vysokomol. Soedin.*, 1977, vol. **19A**(4), p. 896. (Rus)
94. Andrianov K.A., Pavlova S.A., Zhuravleva I.V., Tolchinski Yu.I., Makarova N.N., and Mukbaniani O.V., *Vysokomol. Soedin.*, 1977, vol. **19A**(4), p. 1387. (Rus)
95. Andrianov K.A., Papkov V.S., Zhdanov A.A., and Yakushina S.E., *Vysokomol. Soedin.*, 1969, vol. **11A**(9), p. 2030. (Rus)
96. Pathnode W. and Wilcock D.F., *J. Amer. Chem. Soc.*, 1946, vol. **68**, p. 358.

97. Astapova T.V., Matukhina E.V., Petrovski P.V., Blagodatskikh I.V., Makarova N.N., and Godovski Yu.K., *Vysokomol. Soedin.*, 1999, vol. **41A**(4), p. 581. (Rus)
98. He Y., Lapp A., and Herz J., *Makromol. Chem.*, 1986, Bd. **189**(5), S. 1061.
99. Astapova T.V. and Makarova N.N., *Vysokomol. Soedin.*, 1996, vol. **38B**(8), p. 1442. (Rus)
100. Levis L.N. and Lewis N., *J. Amer. Chem. Soc.*, 1986, vol. **108**(23), p. 7228.
101. Matukhina E.V., Boda E.E., Timofeeva T.V., Godovski Yu.K., Makarova N.N., Petrova I.M., and Lavrukhin B.D.,
102. Makarova N.N., Petrova I.M., Lavrukhin B.D., Matukhina E.V., and Godovski Yu.K., *Vysokomol. Soedin.*, 1997, vol. **39A**(10), p. 1616. (Rus)
103. Makarova N.N., Astapova T.V., Godovski Yu.K., Matukhina E.V., Lavrukhin B.D., and Yakubovich O.Ya., *Vysokomol. Soedin.*, 1993, vol. **35A**(2), p. 190. (Rus)
104. . Makarova N.N., Astapova T.V., and Lavrukhin B.D., *Izv. RAN, Ser. Khim.*, 1996, No. 4, p. 958. (Rus)
105. Achelashvili V.A., Mukbaniani O.V., Meladze S.M., and Khananashvili L.M., *Abstr. Commun. VII Intern. Symp. Organosilicon Chemistry*, Kyoto, Japan, 1984, p. 216.
106. Inaridze I.A., *Candidate's Dissertation on Chemistry*, I. Dzhavakhishvili Tbilisi State University, 1993, Tbilisi, Georgia. (Rus)
107. Chizhova N.V., Makarova N.N., Godovski Yu.K., and Buzin A.I., *Vysokomol. Soedin.*, 2000, vol. **42A**(11), p. 1797. (Rus)
108. Belousov S.I., Sauter E., Godovski Yu.K., Makarova N.N., Rechold W., *Vysokomol. Soedin.*, 1996, vol. **38A**(10), p. 1722. (Rus)
109. Belousov S.I., Sauter E., Godovski Yu.K., Makarova N.N., Rechold W., *Vysokomol. Soedin.*, 1996, vol. **38A**(9), p. 1532. (Rus)
110. Buzin A.I., Sauter E., Godovski Yu.K., Makarova N.N., Rechold W., *Vysokomol. Soedin.*, 1998, vol. **40A**(5), p. 782. (Rus)
111. Tverdokhlebova I.I., Kyrginyan P.A., Larina T.A., Makarova N.N., and Mukbaniani O.V., *Vysokomol. Soedin.*, 1981, vol. **23A**(5), p. 995. (Rus)
112. Damewood I.K. and West I.K., *Macromolecules*, 1985, vol. **18**(2), p. 159.
113. Sundarajian P.R., *Macromolecules*, 1988, vol. **21**(5), p. 1256.
114. Shukla P., Cotts P.M., Miller Ro D., Kussel T.K., Smith B.A., Wallraff G.M., Baier M., and Thiyagarajan P., *Macromolecules*, 1991, vol. **24**(20), p. 5607.
115. Cotts P.M., Ferline S., Dagli G., and Pearson S., *Macromolecules*, 1991, vol. **24**(25), p. 6731.

116. Mukbaniani O.V., Achelashvili V.A., and Sturua G.I., *Thes. Rep. XXII All-Union Conf. High-Molecular Compounds*, Alma-Ata, 1985, p. 89. (Rus)
117. Mukbaniani O.V., Karchkhadze M.G., Matsaberidze M.G., Achelashvili V.A., Kha-nanashvili L.M., and Kvelashvili N.G., *Intern. J. Polym. Mater.*, 1998, vol. **41**, p. 103.
118. Matsaberidze M.G., *Candidate's Dissertation on Chemistry*, I. Dzhavakhishvili Tbilisi State University, 2001, Tbilisi, Georgia. (Rus)
119. Mukbaniani O., Scherf U., Matsaberidze M., Karchkhadze M., and Khananashvili L., *Proceeding of the Georgian Academy of Science*, 1999, vol. **25**(1-2), p. 54.
120. Nogaideli A.I., Tkeshelashvili R.Sh., Nakaidze L.I., and Mukbaniani O.V., *Trudy Tbilisskogo Gos. Universiteta*, 1976, vol. **167**, p. 69. (Rus)
121. Mukbaniani O.V., Khananashvili L.M., Inarisdze I.A., Porchkhidze G.V., and Koberidze Kh.E., *Eur. Polym. J.*, 1998, vol. **34**(3/4), p. 581.
122. Zhdanov A.A., Phyakhina T.A., Afonina R.I., and Kotov V.M., *Vysokomol. Soedin.*, 1993, vol. **35**(5), p. 475. (Rus)
123. Zhdanov A.A., Phyakhina T.A., Gritsenko O.T., Kotov V.M., Zhukov V.P., Afonina R.I., and Levin V.Yu., *Vysokomol. Soedin.*, 1992, vol. **33**(8), p. 45. (Rus)
124. Tverdokhlebova I.I., Kuznetsov D.V., Pertsova N.V., Kotov V.M., Pryakhina T.A., and Bragina T.P., *Vysokomol. Soedin.*, 1993, vol. **35**(8), p. 1278. (Rus)
125. Eskin V.E., In Coll.: *Light Scattering by Polymer Solutions*, Moscow, Nauka, 1973. (Rus)
126. Orofino I.A. and Flory P.I., *J. Chem. Phys.*, 1957, vol. **26**(3), p. 1067.
127. Yamakawa H., In Coll.: *Modern Theory of Polymer Solutions*, 1971, p. 406.
128. Tverdokhlebova I.I., Sutkevich O.I., Polyakov Yu.P., Guselnikov L.E., Matveichev P.M., and Bashkirova S.A., *Vysokomol. Soedin.*, 1991, vol. **33A**(2), p. 362. (Rus)
129. Stocmayer W.H. and Fixman M., *J. Polym. Sci.*, 1963, No. 1, p. 137.
130. Kirkwood I. and Riseman I., *J. Chem. Phys.*, 1955, vol. **23**(1), p. 213.
131. Kirkwood I. and Riseman I., *J. Chem. Phys.*, 1948, vol. **23**(2), p. 565.
132. Bohdanecky M., Macromolecules, 1983, vol. **16**(9), p. 1483.
133. Shah P.P. and Naik H.B., *Eur. Polym. J.*, 1987, vol. **23**(1), p. 73.
134. Krigbaum W.R., *J. Polym. Sci.*, 1958, vol. **28**(116), p. 213.
135. Kurata M. and Yamakava H., *J. Chem. Phys.*, 1958, vol. **29**(1), p. 311.
136. Flory P., *Statistical Mechanics of Chain Molecules*, Moscow, Mir, 1971. (Rus)

Chapter 3. Copolymers with cyclic fragments in dimethylsiloxane backbone

It is known [1, 2] that introduction of different components or groups of different chemical origin or structure into the backbone is one of the effective methods for modifying linear organosiloxane polymers. Introduction of alien fragments is resulted in variation of the spiral-shaped structure of dimethylsiloxane polymers, which causes variation of their physical and chemical properties [3].

In particular, introduction of cyclic fragments to linear poly(organosiloxanes) [4, 5] hampers the chain transfer reaction accompanied by ring separation which, in its turn, increases stability of the mentioned polymers.

There is some information in the literature about synthesis of polymers with alternating diorganosiloxane and organosilsesquioxane units [6 – 10] in cyclolinear double-stranded macromolecule of the polymer [11]. The first series of works has used co-hydrolysis products of dimethyldichlorosilane with phenyltrichlorosilane [6, 7], dimethyldichlorosilane with methyltrichlorosilane [8] or methylphenyldichlorosilane with phenyltrichlorosilane [9] for synthesis of the polymers.

As a result of anionic polymerization of co-hydrolysis products at equimolar ratio of diorgaosiloxane and organosilsesquioxane units, 3D-polymers were synthesized. Polymerization of bicyclodimethylsiloxanes with various lengths of dimethylsiloxane chain between two cyclotetrasiloxane rings has given spatially cross-linked polymers [10]; copolymerization of octamethylcyclotetrasiloxane with polyphenylsilsesquioxane leads to formation of soluble low-molecular polymers [11].

However, all above-mentioned co-hydrolysis reactions lead to formation of low-molecular compounds with statistical disposition of linear and cyclic fragments, in which statistical disposition of siloxane (*D*) and silsesquioxane (*T*) units is preserved.

Discussed in this Chapter is synthesis of cyclolinear copolymers with regular disposition of monocyclic fragments in dimethylsiloxane backbone using HFC and hydride polyaddition reaction as the methods for synthesis of polymers.

3.1. Siloxane cyclolinear copolymers with organocyclotetra(penta,hexa)-siloxane fragments in dimethylsiloxane backbone

HFC reaction of difunctional organocyclosiloxanes with α,ω–dichloro(dihydroxy, dimethylamino)dimethylsiloxanes of different length (n) proceeding with formation of cyclolinear copolymers possessing regular disposition of cyclic fragments in dimethylsiloxane chain is discussed in the current Section.

In HFC reaction study, organocyclotetrasiloxanes with different disposition (1,3- and 1,5-positions by edge and diagonal) of functional groups in cyclotetrasiloxane were used [12 – 17], namely:

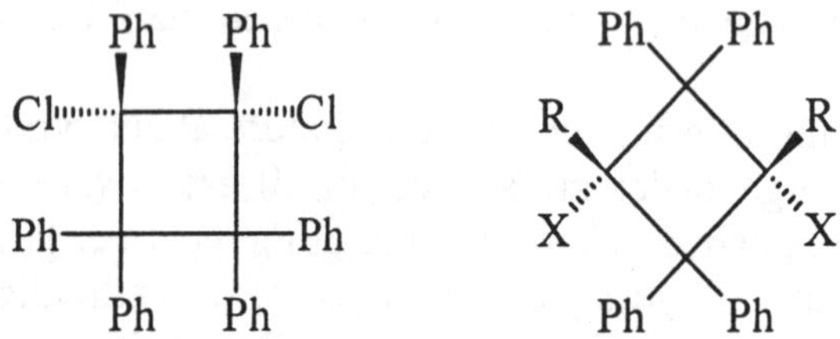

where R = Cl, OH; R = Me, Ph.

The effect of substituting agents at silicon atom in difunctional organocyclotetrasiloxanes on reactivity of haloid and hydroxyl groups interacting with α,ω–dichloro(dihydroxy)dimethylsiloxanes was studied. Polycondensation of dichlorohexaorganocyclotetrasiloxanes were performed at room temperature in 70% solution of anhydrous toluene or benzene both with acceptor and without it.

When HFC reaction proceeded without acceptor, the run of hydrogen chloride liberation was searched for. Figure 3.1 shows dependence of hydrogen chloride liberation on time at 20°C, whence it follows that as the length of dimethylsiloxane unit increases (n = 2 – 6), conversion by hydrogen chloride decreases from 45% (n = 2, R = Me) to 33% (n = 6, R = Me) (Figure 3.1, curves 1 – 3).

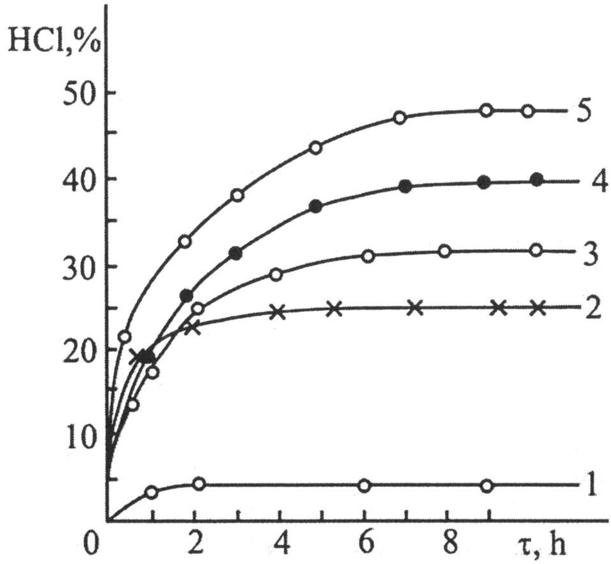

Figure 3.1. Dependence of polycondensation degree between 1,5-dichloro-1,5-diorganotetraphenylcyclotetrasilocane and α,ω–dihydroxy-dimethylsiloxanes: curves 1, 2, 3 – at R = Me, n = 2, 4, 6, respectively; curves 4, 5 – at R = Ph, n = 2, 4, respectively

Substitution of methyl side group by phenyl one at silsesquioxane atom of silicon induces abrupt decrease of hydrogen chloride liberation rate from 20 – 25% (n = 2, R = Ph) to 5% (n = 4, R = Ph), which is displayed by curves 4 and 5 in Figure 3.1.

Gas-liquid analysis of the product obtained by polycondensation of 1,5-dichlorohexaphenylcyclotetrasiloxane with 1,3-dihydroxytetramethyldisiloxane has shown that initial compounds are absent in it and octamethylcyclotetrasiloxane, which would be formed by homocondensation of disiloxanediol in acidic medium is also absent, but products with higher boiling points, i.e. the products of partial intramolecular condensation are present. In the presence of pyridine, HFC reaction of 1,5-dichlorohexaphenylcyclotetrasiloxane with α,ω-dihydroxydimethylsiloxanes proceeds by analogy (at low values of n).

Thus basing on data obtained in the study of HFC reaction between dichloro(dihydroxy)organocyclotetrasiloxanes with α,ω–dihydroxy(dichloro)-dimethylsiloxanes, it has been shown [14] that the reaction is both intermolecular forming cyclolinear copolymer and intramolecular giving bicyclic structures (at low values of *n*). As a consequence, the current reaction proceeds in accordance with the following scheme:

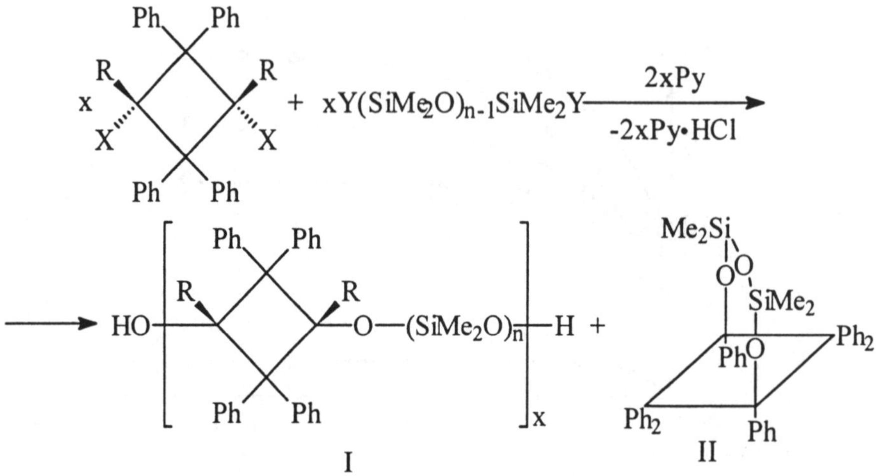

(3.1)

where X = Cl, OH; Y = OH, Cl; R = Me, Ph; *n* = 2, 4, 5, 6, 12, 25, 51, 70, 101.

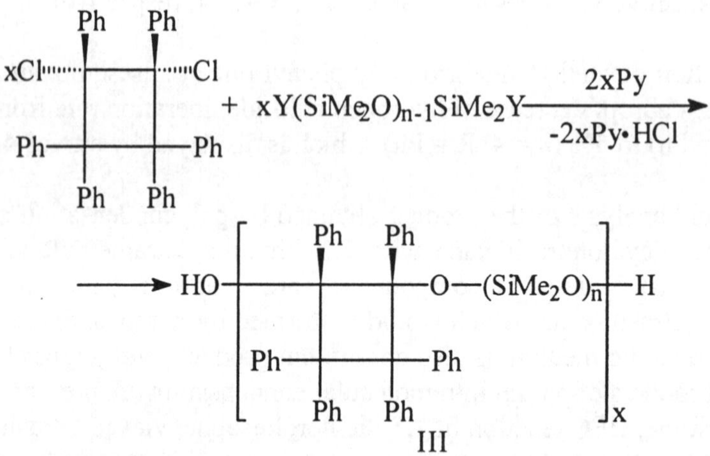

(3.2)

where *n* = 2, 4, 6, 12, 25, 70.

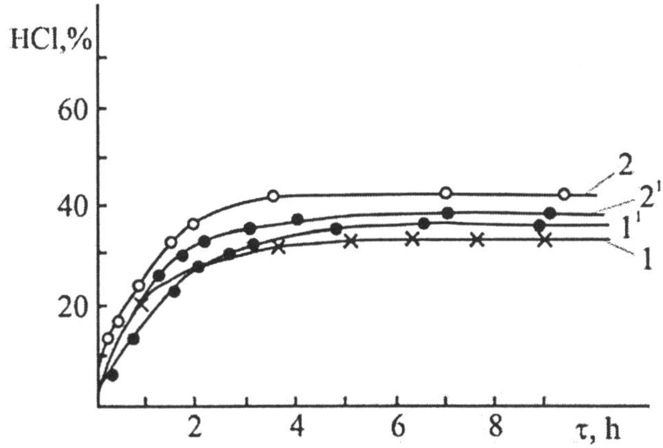

Figure 3.2. Dependence of hydrogen chloride liberation rate in polycondensation reaction of dichlorohexaphenylcyclotetrasiloxane with α,ω–dihydroxydimethylsiloxanes: curves 1 and 2 – with 1,5-disposition of chlorine atoms and at n = 25, 70, respectively; curves 1′ and 2′ - with 1,3-disposition of chlorine atoms and at n = 25, 70, respectively

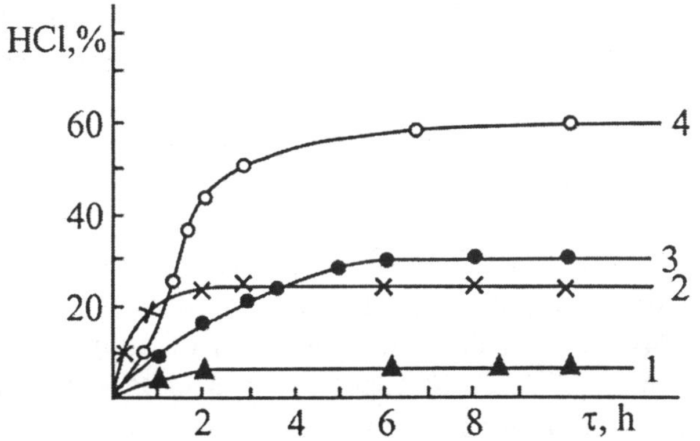

Figure 3.3. Hydrogen chloride conversion in polycondensation of 1,5-dichloro- and 1,5-dihydroxyhexaphenylcyclotetrasiloxanes with α,ω-dihydroxy(dichloro)dimethylsiloxanes (n = 2, 4): curves 1 and 2 - for 1,5-dichlorohexaphenylcyclotetrasiloxane with n = 4, 2, respectively; curves 3 and 4 – for 1,5-dihydroxyhexaphenylcyclotetrasiloxane with n = 4, 2, respectively

Reactivity of functional groups at their different location in the ring (1,5- or 1,3-position) was estimated in the reaction with α,ω-dihydroxydimethylsiloxane with the polymerization level $n = 25, 70$ at 180°C in block. From experimental data a conclusion was made [18] that 1,3- and 1,5-positions of chlorine atoms at silicon in dichlorohexaphenylcyclotetrasiloxane interacting with α,ω-dihydroxydimethylsiloxanes cause a negligible effect on their reactivity during polycondensation (Figure 3.2), which runs counter to the conclusion in the ref. [19].

At room temperature, polycondensation of 1,5-dihydroxyhexaphenylcyclotetrasiloxane with α,ω–dichlorodimethylsiloxanes in 70% solution of anhydrous toluene proceeds at deeper level, and conversion by hydrogen chloride reaches 60%, whereas in HFC reaction of 1,5-dichlorohexaphenylcyclotetrasiloxane with α,ω–dichlorodimethylsiloxanes it is below 30% (Figure 3.3).

As the screening effect of substituting agents at silicon atoms is considered separately, they display equal values of it. Thus, for estimating reactivity of functional groups in current reactions [18], both the screening effect of substituting agents at silicon atoms linked to reactive groups and the inductive action of this radical on the reactive group are considered. Because maximal conversion by hydrogen chloride did not exceed 50 – 60%, the acceptor (pyridine) was introduced for the purpose of increasing the depth of HFC proceeding.

It is shown that under experimental conditions at 23 – 25°C condensation of α,ω-dihydroxydimethylsiloxanes by hydroxyl groups does not proceed, which correlates with the previous conclusions [20]. In this work it is indicated that homofunctional condensation of 1,3-dihydroxytetramethyldisiloxane does not proceed even in the presence of more basic amines.

As HFC proceeds in 60 – 70% solution of anhydrous toluene in the presence of pyridine, the yield of copolymers is increased to 72 – 93%, and copolymers synthesized after reprecipitation represent colorless or light-yellow transparent solids of viscous substances, well soluble in usual organic solvents with $\eta_{spec} = 0.1 - 0.5$.

Study of the reaction (3.1) has indicated that at short length of linear dimethylsiloxane unit the yield of copolymers after reprecipitation is much lower. After eliminating solvent from the mother solution, a viscous product with molecular mass of ~830 units was obtained. Such molecular mass can be displayed by the product of intramolecular condensation with the structure II only. To prove the possibility of forming compound with the structure II, direct synthesis of this compound was performed [21]. Results of the current synthesis were also used for estimation of *cis*- and *trans*-isomers' contents in the initial 1,5-dichlorohexaphenylcyclotetrasiloxane. Namely, HFC reaction of a mixture of *cis*- and *trans*-isomers of 1,5-dichlorohexaphenylcyclotetrasiloxane with 1,3-dihydroxytetramethyldisiloxane was studied (equimolar ratio of initial reagents, 5% anhydrous toluene solution, 0°C, in the presence of acceptor – pyridine). In the case of *cis*-isomer, formation of polycyclic structured compound is the most probable, whereas *trans*-isomers give polymeric products. The reaction proceeds in accordance with the scheme as follows:

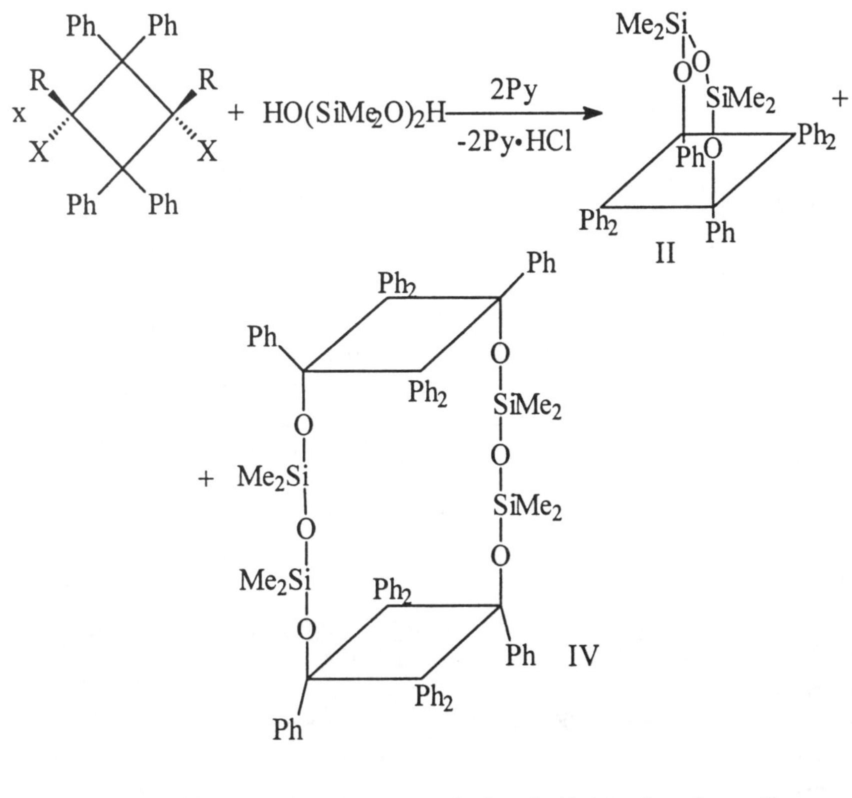

(3.3)

Composition and structure of the individual polycyclic compounds, synthesized by fractionation in high vacuum, were determined on the basis of the ultimate analysis by determination of the molecular mass, IR and ^{1}H NMR spectral data. Based on this reaction, the authors have concluded that 1,5-dichlorohexaphenylcyclotetrasiloxane used in HFC reactions represents a mixture of *cis*- and *trans*-isomers. The *cis*-form promotes for structures II and

Table 3.1

Some physical and chemical parameters and yield of structure I cyclolinear copolymers with 1,5-disposition of cyclotetrasiloxane fragment in dimethylsiloxane backbone

No.	Copolymer	R	n_{SiO}	Yield, %	T_g, °C	η_{spec}*
1	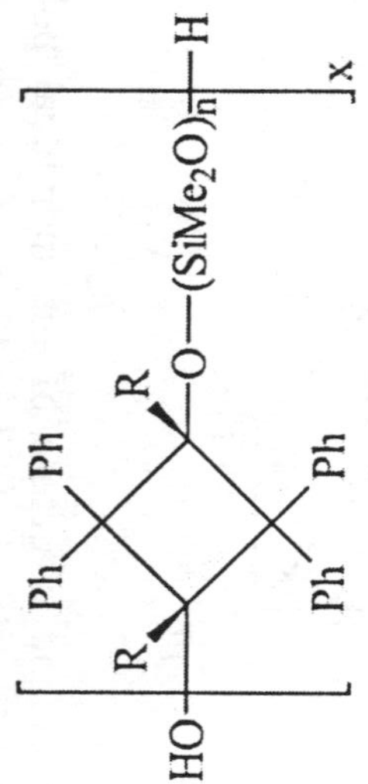	Me	2	72	-54	0.15 - 0.16
2		Me	4	83	-57	0.09 - 0.16
3**		Me	5	86	-	0.20
4		Me	6	86	-71	0.10 - 0.22
5		Me	12	90	-94	0.24 - 0.27
6**		Me	25	90	-117	- 0.33
7		Me	51	91	-123	- 0.40
8		Me	70	93	-123	0.42 -
9		Me	101	96	-123	- 0.49
10		Ph	2	72	-10÷-20	0.06 - 0.11
11		Ph	4	80	-34	0.10 - 0.14
12		Ph	6	93	-55	0.10 - 0.21
13		Ph	12	93	-83	0.19 - 0.26
14		Ph	25	95	-110	0.19 - 0.34
15		Ph	51	96	-123	- 0.40
16		Ph	70	92	-123	0.15 -
17		Ph	101	97	-123	- 0.42

Note: * Hereinafter, second values of viscosity are obtained by high-temperature polycondensation of 1,5-dihydroxyorganocyclohexasiloxanes with α,ω–bis(dimethylamino)dimethylsiloxanes.

** For these samples, viscosities in benzene are determined.

IV formation, total yield of which equals ~50%; the rest 50% are represented by oligomeric structures, formed from *trans*-forms of 1,5-dichlorohexa-phenylcyclotetrasiloxane. Spectral data of ^{1}H NMR analysis have indicated the regular structure of cyclolinear copolymers, which correlates well with the data from the literature [22]. Some physical and chemical parameters of copolymers with cyclotetrasiloxane fragments in the backbone are shown in Tables 3.1 – 3.3.

Table 3.2

Some physical and chemical parameters and yield of cyclolinear copolymers of the structure II with 1,3-disposition of cyclotetrasiloxane fragment in dimethylsiloxane backbone

No.	Copolymer	n_{SiO}	Yield, %	T_g, °C	η_{spec}
1	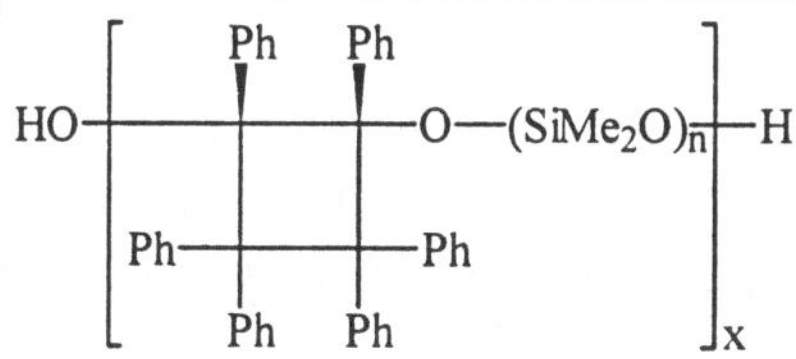	2	83	-23	0.06
2		4	90	-35	0.10
3		6	91	-	0.11
4		12	92	-83	0.31
5		25	93	-107	0.37
6		70	92	-123	0.49

For copolymers, thermomechanical studies were carried out in accordance with the technique, described in ref. [23]. Expectedly, based on the data from the literature [24, 25] the studies have shown that the presence of inclusions of any chemical origin in the polydimethylsiloxane backbone (framing groups different from methyl ones, rings of different structure, branching points) causes the loss of the ability of copolymers to crystallize, if the distance between neighboring inclusions equals ~30 siloxane units.

Copolymers with $n = 2 - 25$ studied do not crystallize, and the ones with $n = 70$ behave as crystallizing poly(dimethylsiloxanes) (PDMS). The temperature range corresponded to glass transition in copolymers regularly expands towards higher temperatures with decrease of n. Of special attention is the fact that disposition of cyclotetrasiloxane fragment in the poly(dimethylsiloxane) backbone (1,5- and 1,3-positions) causes a negligible effect on T_g of the copolymers (Figure 3.4).

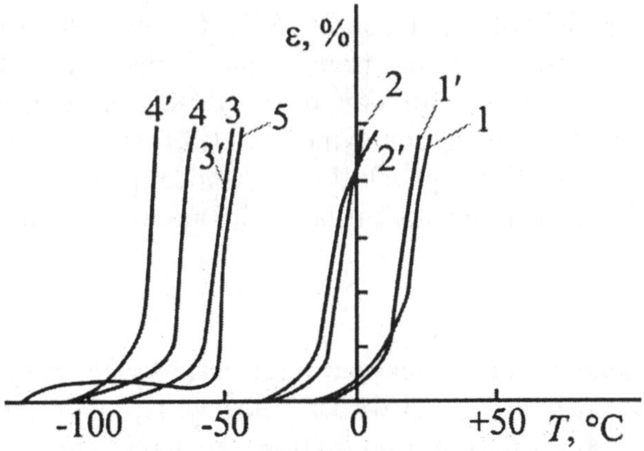

Figure 3.4. Thermomechanical curves for copolymers with 1,3- and 1,5-disposition of cyclotetrasiloxane fragments in PDMS backbone: 1, 2, 3, 4, 5 – for copolymers with 1,3-disposition of hexaphenylcyclotetrasiloxane fragment in the backbone at $n = 2, 4, 12, 25, 70$, respectively; 1′, 2′, 3′, 4′ - for copolymers with 1,5-disposition of hexaphenylcyclotetrasiloxane fragment in the backbone at $n = 2, 4, 12, 25$, respectively (100 g load, 5 deg/min heating rate)

Of special importance is the fact that introduction of rings containing phenyl framing groups into the linear chain is accompanied by T_g increase, compared with cyclolinear carbosiloxane polymers containing methyl framing groups [26]. In the case of cyclotetrasiloxanes with phenyl groups, T_g increase is higher than under the action of network points, which do not change T_g of PDMS networks at $n = 20$ [27].

The latter is determined by a significant effect of bulky phenyl groups, the presence of which at any place of PDMS backbone significantly increases T_g [28]. Substitution of phenyl side group by the methyl one in organosilsesquioxane fragment of organocyclotetrasiloxanes decreases the glass transition temperature by ~15 - 20°C. The effect of cyclotetrasiloxane fragment is detected at dimethylsiloxane unit $n = 25$ long and lies above T_g of PDMS.

Thermogravimetric studies of cyclolinear copolymers indicate that 1,3- and 1,5-dispositions of hexaphenylcyclotetrasiloxane fragment cause no difference in mass losses of copolymers (refer to Figure 3.5).

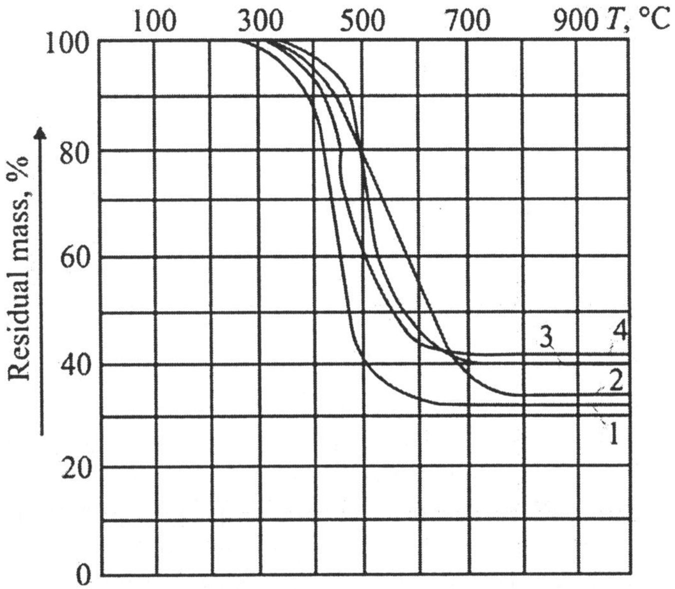

Figure 3.5. Thermogravimetric curves of copolymers with 1,3- and 1,5-disposition of hexaphenylcyclotetrasiloxane fragments in PDMS backbone: curves 1 and 3 – for copolymer with 1,5-disposition at n = 70, 6, respectively; curves 2 and 4 – for copolymer with 1,3-disposition at n = 70, 6, respectively (in air, 5 deg/min heating rate)

As a result, by HFC reaction of 1,5-dihydroxy-1,5-bistrimethyl-siloxytetraphenylcyclotetrasiloxane with α,ω-dichloromethyl-siloxanes in the presence of pyridine and in 60% anhydrous toluene solution at 20 - 25°C cyclolinear copolymers with regular disposition of 1,5-bistrimethylsiloxy-tetraphenylcyclotetrasiloxane in the dimethylsiloxane backbone, completely dissoluble in organic solvents, were synthesized [29]. The reaction proceeds in accordance with the general scheme as follows:

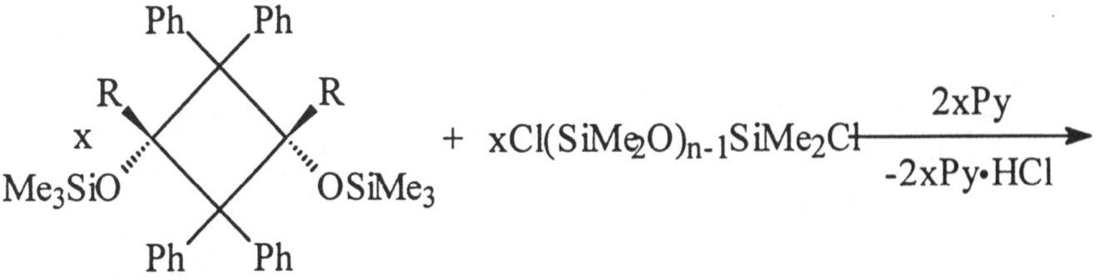

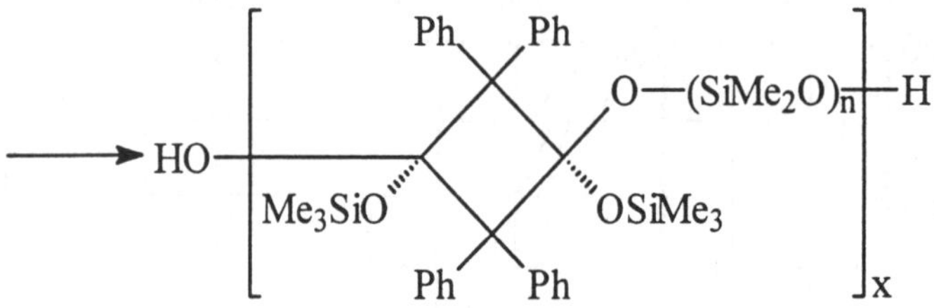

(3.4)

I

where $m = 1, 2, 3, 10$.

Depending on the length of dimethylsiloxane unit, copolymers represent crystalline (T_g = 100°C, n = 1), rubbery (T_g = 30°C, ε_{elong} = 400%) and quite strong (δ_{break} = 40 kg/cm^2 at 24°C) products. Of interest is that dielectric constant of this polymer decreases linearly with the temperature increase.

A suitable method has been suggested [12] for obtaining cyclolinear copolymers, which concludes in high-temperature HFC reaction of dihydroxyorganocyclotetra(penta, hexa)siloxanes with α,ω–disdiamino-(dimethylamino)dimethylsiloxanes. The method suggested is simpler than the above-discussed reactions (3.1) and (3.2), because synthesized copolymers possess low molecular mass. Moreover, obtaining of pure copolymers in accordance with the above-shown schemes (3.1) and (3.2) requires application of an additional stage of polymer purification from secondary product which is amine hydrochloride.

For the purpose of synthesizing cyclolinear copolymers with regular disposition of organocyclotetra-, organocyclopenta- and organocyclohexasiloxane fragments in linear the dimethylsiloxane backbone, HFC reaction of 1,5-dihydroxy-1,5-dimethyl(diphenyl)phenylcyclotetra(penta, hexa)siloxanes and 1,7-dihydroxy-1,7-dimethyl(diphenyl)octaphenylcyclohexasiloxane with α,ω–bis(dimethylamino)(dichloro)dimethylsiloxanes was studied [15 – 17].

In this case, the reaction proceeds in accordance with the general scheme as follows:

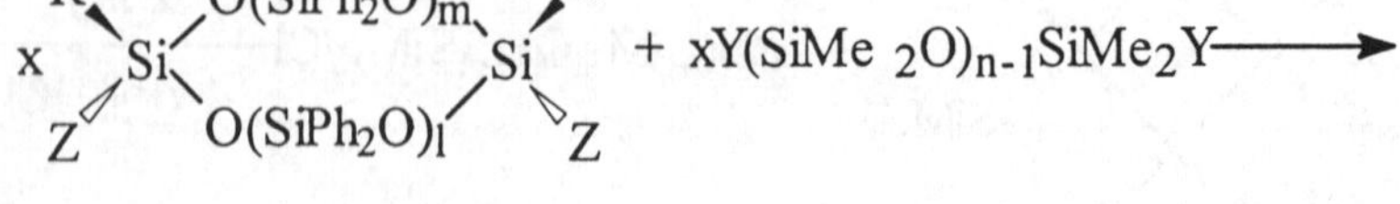

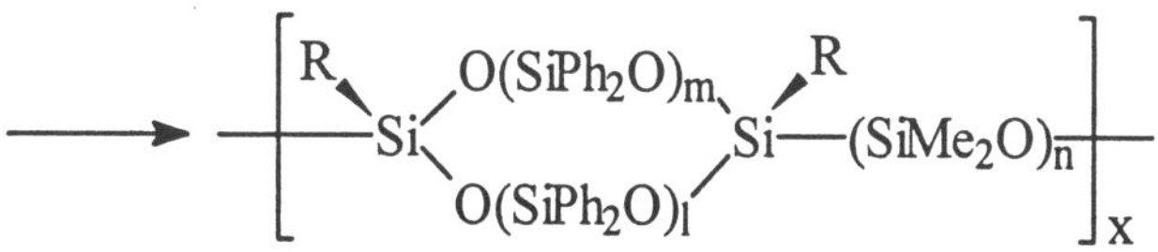

(3.5)

where Z = OH, Cl; Y = Me_2N, OH*; m = l = 1 - structure I; m = 1, l = 2 - structure V; m = 1, l = 3 - structure VI; m = l = 2 - structure VII; R = Me, Ph.

HFC was studied at 20 - 150°C in anhydrous nitrogen or argon flow both in block and in solution. As the reaction proceeds in block, the reaction mixture is heated up to 50 - 60°C until a homogeneous mixture is formed; thereafter, the reaction is continued in vacuum at P = 3 – 5 mmHg and temperature range of 120 - 150°C up to constant viscosity. Application of reactive α,ω–bis(dimethylamino)dimethylsiloxanes allowed performance of the initial stage of the reaction in solution at 20 - 50°C until a homogeneous mixture is formed. For more complete elimination of amine, liberated in the reaction, the mixture is aerated by dry inert gas. Polymers synthesized by this technique require no purification, because amines obtained as secondary reaction products possess very low boiling point and are completely removed from the reaction mixture. In as much as the mixture of dichloro(dihydroxy)-organocyclosiloxanes' isomers was used in HFC, atactic copolymers were synthesized.

In the case of HFC of 1,5-dihydroxyhexaphenylcyclotetrasiloxane with α,ω–bis(diethylamino)dimethylsiloxanes, gel formation proceeds after diethylamine conversion reaching ~5%. Previously, it was informed about a possibility of siloxane bond break in trimethyltriphenylcyclotrisiloxane by diethylamine and formation of a complex with the charge transfer [30]. In accordance with the authors' opinion [15 – 17], immediate removal of diethylamine produced in the reaction with α,ω–bis(diethylamino)dimethylsiloxanes could be hardly guaranteed. In its turn, dimethylamine interacts with the siloxane bond in the cyclic fragment and forms bipolar zwitter-ion [31]. The HFC reaction proceeding with α,ω–bis(diethylamino)(diamino)dimethylsiloxanes [12] forms copolymers completely dissoluble in organic solvents, because dimethylamine and ammonium liberated during synthesis cause no breakage of siloxane bonds in cyclic fragments of the polymeric backbone.

It is proved that at short lengths of dimethylsiloxane unit, HFC reaction proceeds in two directions: intramolecular ring formation giving polycyclic products and intermolecular formation of cyclolinear copolymers. Formation of polycyclic products is proved by the direct synthesis.

To prove formation of copolymers with regular disposition of cyclic fragments in macromolecular backbone, some copolymers were fractionated into several fractions. Results of the ultimate analysis have indicated that the values detected for fractions coincide with the calculated ones, which represents direct proof of the regular structure of copolymers.

Many investigators have studied dissolved solutions of poly(organosiloxanes) with different side groups at silicon atoms in the macromolecule backbones [32 - 35]. These works show results of the studies of the effect of the side groups' origin, their disposition and the influence of hydrodynamic and conformation parameters of macromolecules.

Table 3.3

Bond lengths and angles between them in poly(organocyclosiloxanes) of the structure I ($n = 1$)

Bond	Bond length, Å	θ_1^*, deg	Rotation conditions
l_1	4.25	54.5	Prohibited
l_2	1.63	37.0	Free
l_3	1.63	70.0	Free
l_4	1.63	37.0	Free
l_5	1.63	54.5	Free

Note: * For the polymer backbone, this coefficient is exclusively associated with the structure of macromolecules and potentials of internal rotation around internal bonds of the backbone. This dependence gives information about energy of the chain conformation and mutual transitions.

The effect of introduction of regularly disposed cyclic fragments into the macromolecular backbone on conformational and hydrodynamic parameters is also studied [30]. For this purpose, copolymers 3 and 4 (Table 3.1) were fractionated into twelve fractions from the benzene (solvent) – methanol (precipitator) system. The influence of cyclic groups, introduced into the polymer backbone, on rigidity parameters was determined by direct computer simulation of macromolecular coil with the help of the Monte-Carlo method. The process was simulated for the copolymer of the structure I with $n = 1$, 5 and 10 in the structural unit of dimethylsiloxane chain, and the Kuhn segment A and $\langle h^2 f \rangle / nM_0$ were calculated, where M_0 is the molecular mass of the structural unit, $<h^2f>$ is the mean-square distance between chain ends at free rotation around virtual bonds[1], n is the number of structural units.

Structural unit of the copolymer was simulated on the basis of the data from the literature on the simplest molecules' structure, close by the composition and structure to monomeric units. Shown below is the geometrical structure of repeated structural unit in poly(organosiloxane) at $n = 1$. Table 3.3 shows values of bond lengths and angles between them.

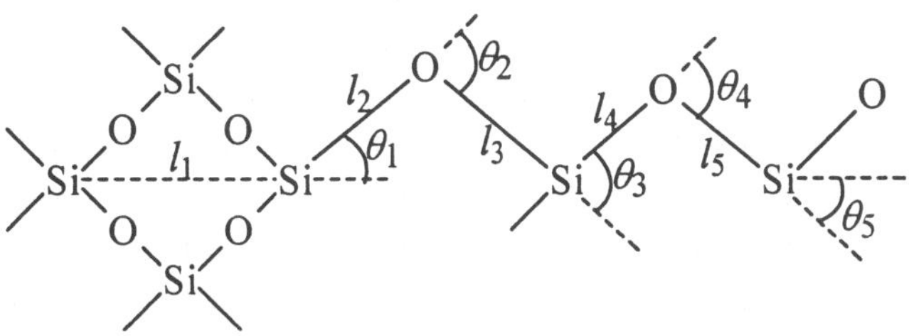

To avoid the ring twisting, rotation around virtual Si-------Si (l_1) bond is prohibited. Table 3.4 shows calculation results, which represent theoretical values of the Kuhn segment (A) and $<h^2f>/nM_0$ for polymers of the structure I ($n = 1$; $n = 5$; $n = 10$) and, for comparison, analogous values for PDMS [37].

Table 3.4

Calculation values of the Kuhn segment (*A*) and $<h^2f>/nM_0$

Polymer of structure I	n_{SiO}	$A = <h^2f>/nl_0A$	$<h^2f>/nM_0$
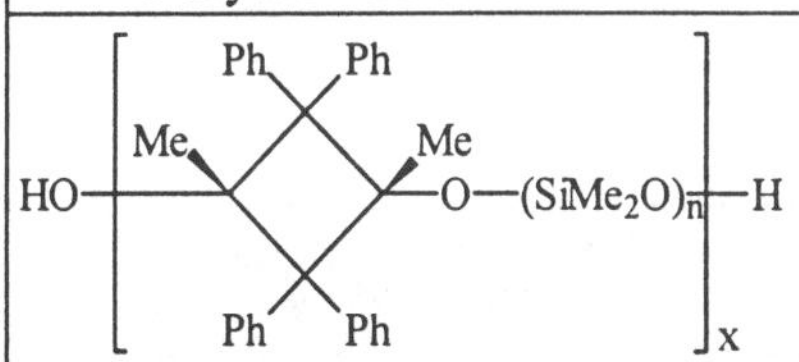	1	15	0.28
	5	10	0.24
	10	10	0.22
PDMS	10	10	0.24

These theoretical data show that the influence of the octatomic siloxane ring is observed at $n = 1$ only. Saturation is observed already at transition from n

[1] By the term of virtual bond the area of the chain is meant, rotation around which is possible. In particular case, it may be a simple valence bond –(l_2-l_5); in the general case, it may contain not only valence bonds, but also rings, or may determine the distance between atoms in the ring as (l_1) does.

= 5 to n = 10, and increase of the quantity of ≡Si-O- groups causes no effect on the coil size.

For the fraction of copolymer 3 of the structure I and n = 5, experimental values of the Kuhn segment and $<h^2 f>/nM_0$ are shown in Table 3.1.

One of the methods for estimating experimental thermodynamic flexibility of the polymer backbone is determination of parameters, associated with the size of isolated macromolecule coil. At the present time, methods of macromolecular coil parameter estimation under ideal and non-ideal conditions. In the first case, characteristic viscosity of the fraction, $[\eta]$, with the known molecular mass in ideal θ-solvent was measured. This method is based on the known Flory-Fox relation [38]. The second method represents measurements of the fraction $[\eta]$ in good thermodynamic solvents and extrapolation of experimental data in accordance with the known techniques [39 - 41]. Because in the current work [36] all measurements were performed in good thermodynamic solvent, unperturbed dimensions of macromolecules, $<h_\theta^2>$, were determined by graphical extrapolation in accordance with the Stockmayer-Fixman, suggested by the authors for flexible macromolecules, in $[\eta]M^{1/2}$ - $M^{1/2}$ coordinates (Figure 3.6).

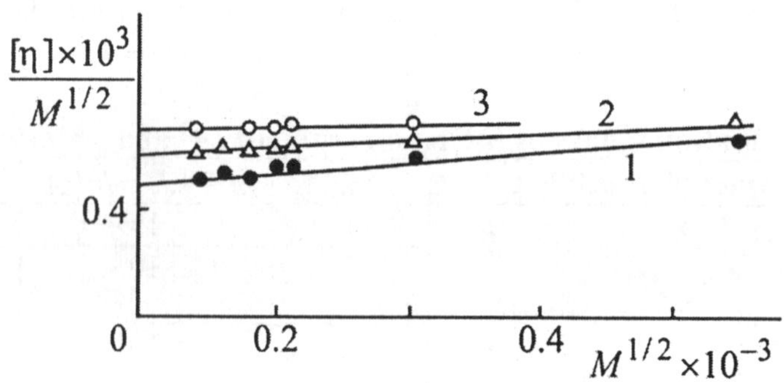

Figure 3.6. Determination of unperturbed dimensions by extrapolation in accordance with the Stockmayer-Fixman method for copolymers 3 (Table 3.1): 1 – at 25°C; 2 – at 40°C; 3 – at 50°C

Table 3.5 shows mean-weight molecular masses, $\bar{M}_\omega$, and $[\eta]$ for the fraction of copolymer 3 (n = 5) (Table 3.1) of the structure I at 25, 40 and 50°C in toluene.

Table 3.6 shows experimental values of conformational parameters for the Kuhn segment, $A = <h_0^2>/nl_0$ (where l_0 is the structural unit length), the hindrance factor of rotation around the single bond, $\sigma = (\langle h_0^2\rangle/\langle h^2 f\rangle)^{1/2}$, and constants of the Mark-Kuhn-Hauvink equation, $K_\theta = [\eta]/M^{1/2}$, at different temperatures.

Table 3.5

Molecular masses and characteristic viscosities of copolymer 3 fractions (Table 3.1, structure I) at different temperatures

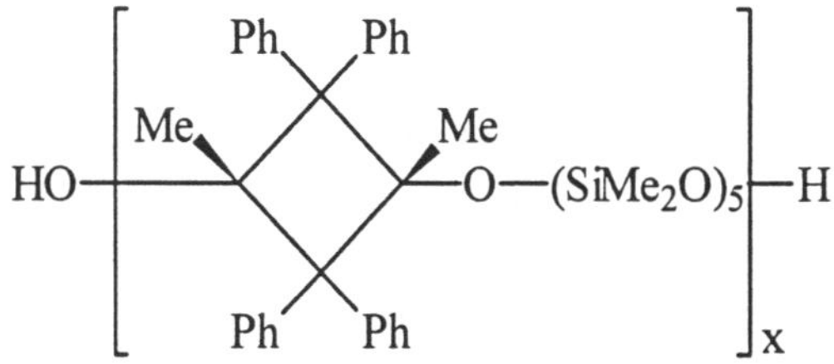

Fraction No.	$\bar{M}_\omega \times 10^{-3}$	[η] in toluene (dl/g) at temperature, °C		
		25	40	50
1	9.0	0.04	-	-
2	17.0	0.06	0.07	0.13
3	22.0	0.07	0.09	0.11
4	26.0	0.08	0.10	0.15
5	33.0	0.09	0.11	0.12
6	42.0	0.12	0.13	0.14
7	44.0	0.11	0.13	0.15
8	72.0	0.15	0.15	0.12
9	85.0	0.16	0.17	0.18
10	92.0	0.19	0.20	0.21
11	141.0	0.26	0.23	0.38
12	239.0	0.33	0.38	-

Table 3.6

Experimental conformational parameters of copolymer 3 (structure I, Table 3.1)

No.	T, °C	$K_\theta \times 10^3$	$\langle h_0^2\rangle/nM_0$	A, Å	$\sigma = (\langle h_0^2\rangle/\langle h^2 f\rangle)^{1/2}$
1	25	0.46	0.303	16.0	1.12
2	40	0.58	0.340	18.0	1.18
3	50	0.67	0.380	20.0	1.25

Experimental data on conformational parameters of copolymer 3 (Table 3.1), shown in Table 3.6, exceed these calculated in the supposition that rotation around virtual and valence bonds is free.

Comparison of experimental and calculated values of the above-mentioned parameters indicates the hindrance of rotation around valence bonds ≡Si-O- and ≡Si-Ph due to their short lengths, because distances between neighboring atoms Si----Si and O----O are shorter or close to the sum of Van-der-Waals radii of these atoms. Data from Table 3.6 indicate also that experimental values of the hindrance factors are low.

As shown by Flory, these low values of σ, i.e. a negligible increase of PDMS molecule dimension compared with dimensions, calculated in the supposition of free rotation, are associated with predominant containing of the plane *trans*-chain type conformation rather than the absence of rotation hindrance, compared with coiled conformations. Hence, low σ values for these chains may be realized at both low and high differences of energies of rotational isomers under the condition of energy gain of the *trans*-shape [42, 43]. On the other hand, low σ must be combined with positive temperature coefficient of unperturbed dimensions of macromolecules[2], because relative content of the *trans*-shape must decrease with temperature growth [42, 43]. For linear PDMS, studies of the temperature coefficient of unperturbed dimensions of macromolecules were carried out in the work [44], and the value obtained equaled $(0.78 \pm 0.06)\times10^{-3}$ deg^{-1}.

For the purpose of determination of the temperature coefficient for unperturbed dimensions of copolymer 3 (Table 3.1), $[\eta]$ values were measured in the same solvent (toluene) at different temperatures. In accordance with the Stockmayer-Fixman method, $K_\theta = (\langle h_0^2\rangle/M)^{1/2}\times F_0$ values was determined using the least-square technique. The temperature coefficient of unperturbed dimensions was calculated from the values obtained at different temperatures (Table 3.6) using the relation, suggested in the work [45]: $d \ln\langle h_0^2\rangle/dT = 2/3\times \ln K_\theta/dT$.

The coefficient of unperturbed coil dimension ($d\ln\langle h_0^2\rangle/dT$), determined for copolymer 3 (Table 3.1), equals 0.85×10^{-3} deg^{-1}. Figure 3.7 displays this

---

[2] For polymeric chain, this coefficient is associated with the structure of macromolecules and potentials of internal rotation around internal bonds in the backbone. Such dependence gives information about energetic conformation of the backbone at mutual transitions.

value is positive and quite close to the temperature coefficient of unperturbed PDMS dimensions [44].

Thus, data shown in Table 3.6 indicate that thermodynamic rigidity, mean-square dimensions and hindrance of copolymer 3 chains (Table 3.1) are increased with temperature. These facts as well as positive temperature coefficient of unperturbed dimensions testify about predominance of coiled *trans*-shape, more energetically preferable for copolymer 3 chains, similar to PDMS, compared with isomers. As temperature increases, the fracture of plane *trans*-chains decreases, which, as a result of transition from *trans*- to *gauche*-state, proceeds with an increase of unperturbed dimensions [41].

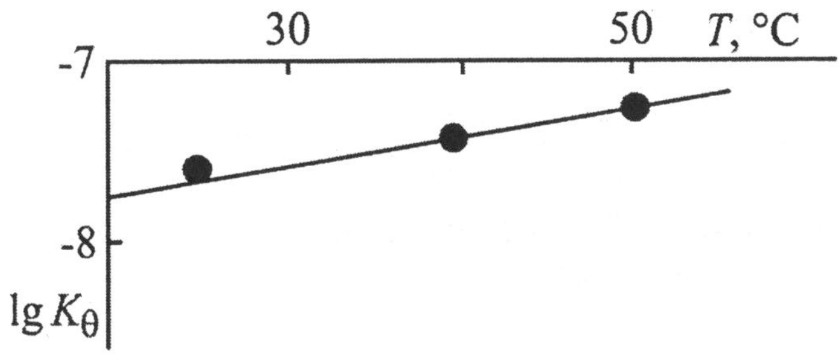

Figure 3.7. Temperature dependence of $\lg K_\theta$ for copolymer 3 (Table 3.1)

Summing up all above-discussed, one may conclude that, compared with PDMS, introduction of octatomic cyclic fragments with symmetrically disposed phenyl side groups causes no change of internal rotation conditions, as it takes place for poly(methylphenylsiloxane) with asymmetrical side phenyl groups [43]. However, already for 1:5 ratio (1 cyclic group per 5 siloxane units), i.e. at the decrease of the distance between cyclic fragments in the backbone, their introduction into dimethylsiloxane backbone induces a significant increase of thermodynamic rigidity of macromolecules. The rigidity increase of copolymer 3 is also displayed in thermomechanical properties of copolymers. For example, T_g of copolymer 3 ($T_g = -50$°C) is much higher than T_g of PDMS.

The study of hydrodynamic behavior of copolymer 6 (Table 3.1) indicates that, apparently, macromolecules of this copolymer possess branched backbones.

In the Mark-Kuhn-Hauvink equation, parameter α for copolymer 6 solution in toluene at 25°C equals 0.30, which is typical of branched macromolecules. For cyclolinear polymer 3 under the same conditions, this parameter equals 0.62 (Figure 3.8). Molecular masses vary within the range from 4×10^3 to 565×10^3.

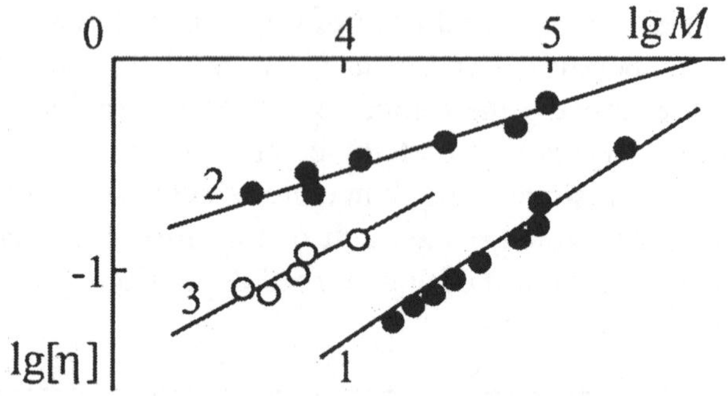

Figure 3.8. Dependence of lg[η] on lg*M*: 1 – for copolymer 3 in toluene at 25°C, $[\eta] = 1.394\times10^{-4}M^{0.62}$; 2 – for copolymer 6 at 25°C, $[\eta] = 1.79\times10^{-2}M^{0..30}$; 3 – for copolymer 6 in isopropyl alcohol at 22°C, $[\eta] = 9.86\times10^{-4}M^{0..53}$

Mean-weight molecular masses, $\bar{M}_\omega$, were measured on Sofica device. Moreover, molecular masses below 10,000 were also determined by the ebullioscopy method ($\bar{M}_n$). Good coincidence between $\bar{M}_\omega$ and $\bar{M}_n$ was obtained. The reason for branching may be the fact that initial oligomers (α,ω–dichlorodimethylsiloxanes) for synthesizing cyclolinear copolymers (poly(organocyclotetrasiloxanes)) with n = 5 and 10 were produced by partial hydrolysis reaction of dimethyldichlorosilane, and oligomers with n = 25 were telomerized in autoclave [46], which may cause formation of copolymers containing units capable of branching. Additional proof of branching is the loss of copolymer 6 solubility with time. However, behavior of copolymer 6 under θ-conditions, different from the behavior of branched macromolecules was observed. Parameter α equaled 0.53, which exceeded α = 0.25 typical of branched copolymers under θ-conditions. For copolymer 6, θ-conditions were set on a device of temperature polymer precipitation. Isopropyl alcohol at 22°C was used as θ-solvent (Figure 3.9).

The attempt to prove correctness of θ–conditions selection for copolymer 6 by checking equality to zero of the virial coefficient, A_2, has given interesting results. It has been found that purification of the polymer solution from dust by filtration via dense Shott filter (No. 5) causes a sharp increase of light scattering intensity, which does not allow measurement of molecular mass and A_2. Repeated filtering induces much more sharp increase of light scattering intensity. The precipitation temperature of filtered solution is increased. This fact contradicts to the supposition that copolymer 6 represents a branched structure. Possible reason for this effect may be solution structuring during long-term filtration. To check this suggesting [36], copolymer solutions (filtered and non-filtered) were studied on optical microscope. Films obtained from 1% filtered solution displayed oriented chains and aggregates from them, which were not observed in the films from non-filtered solution. Studies carried out with the help of electron microscope have not allowed definite conclusions, because the technique requires dilution to 0.01 g/ml concentration of the solution, i.e. an order of magnitude lower than that, at which molecular mass is determined.

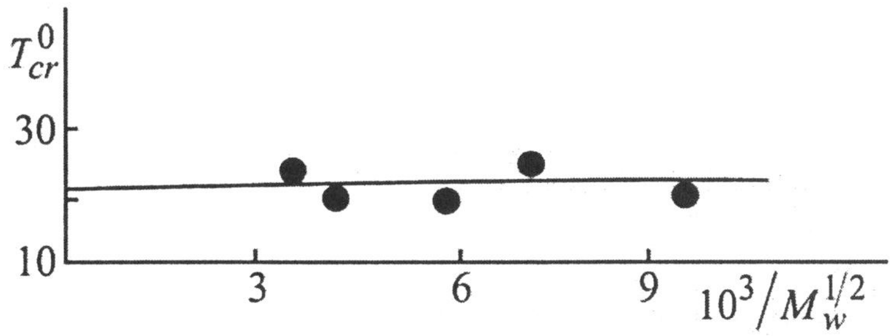

Figure 3.9. Dependence of critical precipitation temperature, T^0_{cr}, of copolymer 6 fraction (Table 3.1) on molecular mass

Thermogravimetric studies of copolymers 2 and 4 (Table 3.1) have determined the influence of regularly disposed organocyclotetrasiloxane fragments in linear dimethylsiloxane chain on thermal stability of the studied copolymers. Pyrolytic spectrum (Figure 3.10) indicates that thermal degradation of polymers 2 and 4 below 300 and 350°C, respectively, displays full absence of organosilicon and organic compounds in the pyrolysis products.

At 350 - 400°C, thermal degradation rate is considerably increased, and the reaction proceeds with benzene isolation. In 380 - 440°C temperature range, degradation products display cyclosiloxane components of D_n composition, D_4 < 1.5%. Note also that pyrolysis temperature increase induces sharp increase of CH_4, D_3 and D_4 compounds extracted. Liberation of benzene and methane is due to ≡Si-C and C-H bond break, which induces cross-linking at the expense of methyl and phenyl groups [47].

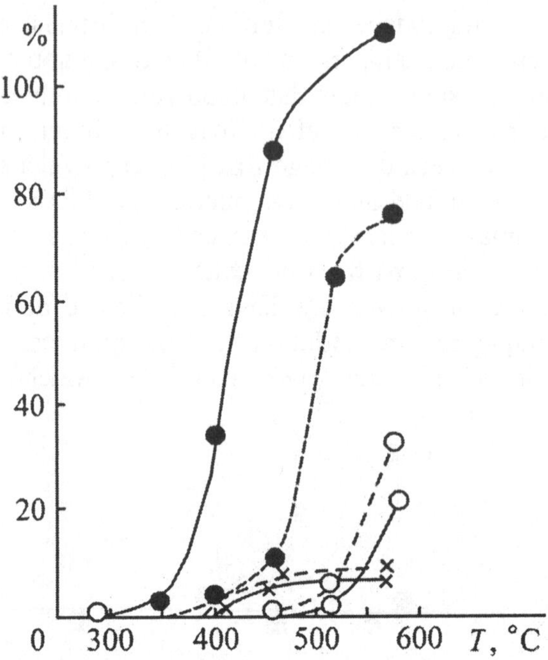

Figure 3.10. CH_4, C_6H_6, D_3 and D_4 liberation curves for cyclolinear copolymers 2 and 4 (structure I, Table 3.1): • - C_6H_6, ○ - CH_4, × - D_3 and D_4 (<1.5%). Here — is corresponded to copolymer 2 and ---- to copolymer 4

It has been shown before [47 – 51] that degradation proceeds by the radical mechanism, mainly forming oligomeric products instead of organocyclosiloxanes. Such phenomenon was observed in studies of thermal degradation of organocyclosiloxane cyclolinear copolymers [49]. This gives an opportunity to suggest that thermal pyrolysis of copolymers 2 and 4 is guided by the backbone break in compliance with the law of randomness with formation of oligomeric products. Simultaneously with the backbone degradation, radical cross-linking reactions by methyl and phenyl groups proceed, accompanied by hydrogen, methane and benzene liberation. Introduction of organosiloxane cyclic fragments into linear dimethylsiloxane backbone induces a considerable decrease of the degradation process of such copolymers by depolymerization mechanism at 320 - 340°C, whereas unblocked PDMS is polymerized by almost 95% at these temperatures.

The general feature of all copolymers is decrease of their thermal oxidative stability and increase of final mass losses with dimethylsiloxane unit length increase. Figure 3.11 gives comparative estimation of thermal oxidative stability of copolymer, which proves negligible effect of 1,5- or 1,7-disposition of cyclohexasiloxane fragments (similar to organocyclotetrasiloxane fragments) on thermal oxidative stability of copolymers.

Table 3.7

Some physical and chemical parameters of copolymers with organocyclopentasiloxane (structure V) fragments in the backbone

No.	Copolymer	R	n_{SiO}	Yield, %	η_{spec}	T_g, °C
1	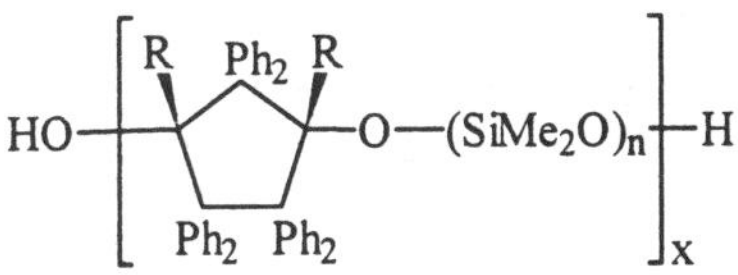	Me	2	77	0.08	+15
2		Me	4	81	0.13	-14
3		Me	6	93	0.14	-38
4		Me	8	96	0.18	-
5		Me	12	96	0.20	-65
6		Me	40	94	0.25	-123
7		Me	51	97	0.32	-123
8		Ph	2	79	0.06	+15
9		Ph	4	83	0.08	0
10		Ph	6	93	0.11	-25
11		Ph	8	93	0.14	-
12		Ph	12	93	0.18	-50
13		Ph	40	96	0.23	-123
14		Ph	51	98	0.29	-123

Furthermore, at the increase of cyclic fragment volume, i.e. at transition from organocyclotetrasiloxane fragment to organocyclopenta- and organocyclohexasiloxane ones, at the same length of dimethylsiloxane unit induces a decrease of thermal oxidative stability of the copolymers, which, in turn, is explained by increase of the mass part of phenyl groups.

Comparative thermomechanical studies of copolymer with organocyclotetra-, organocyclopenta- and organocyclohexasiloxane (1,5- and 1,7-positions) fragments in the chain have been implemented. Data on T_g of copolymer, shown in Table 3.7, indicate that the increase of dimethylsiloxane unit length leads to obligate decrease of T_g of the copolymers. It has been found that both cases of hexaphenylcyclotetrasiloxanes (1,3- and 1,5-positions) and organocyclohexasiloxanes 1,5- or 1,7-disposition of cyclic fragments with the same framing groups at atom of silicon in silsesquioxane and equal length of the linear dimethylsiloxane unit causes negligible effect on T_g of copolymers.

Table 3.8

Some physical and chemical parameters of copolymers with 1,5-disposition of organocyclohexasiloxane (structure VI) fragments in the backbone

No.	Copolymer	R	n_{SiO}	Yield, %	η_{spec}	T_g, °C
1	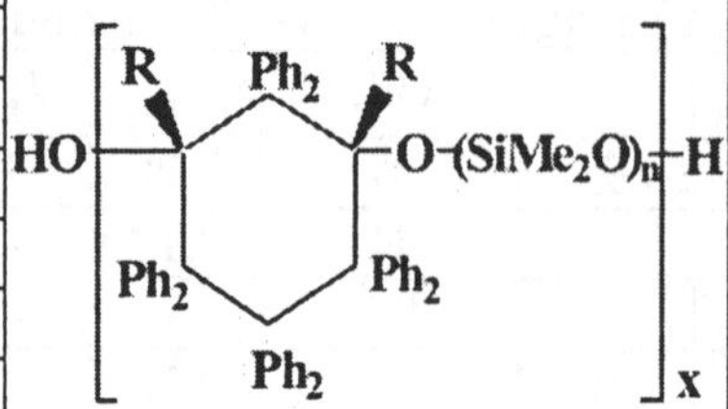	Me	2	75	0.09	+10
2		Me	4	80	0.14	-1
3		Me	8	83	0.17	-47
4		Me	34	88	0.26	-123
5		Me	51	90	0.30	-123
6		Ph	2	76	0.10	+21
7		Ph	4	81	0.13	+8
8		Ph	8	82	0.18	-38
9		Ph	34	87	0.24	-123
10		Ph	51	91	0.29	-123

Table 3.9

Some physical and chemical parameters of copolymers with 1,7-disposition of organocyclohexasiloxane (structure VII) fragments in the backbone

No.	Copolymer	R	n_{SiO}	Yield, %	η_{spec}	T_g, °C
1	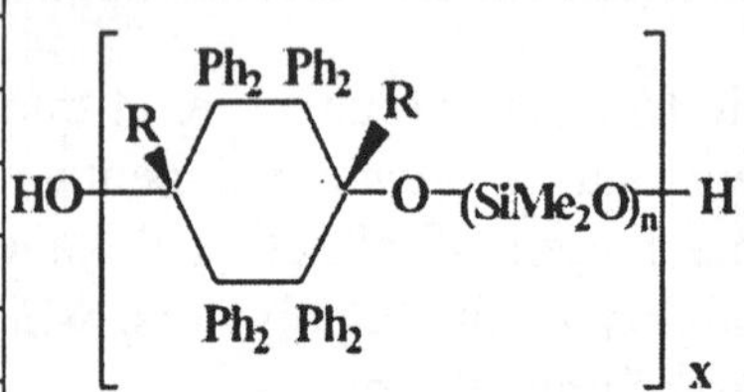	Me	2	85	0.06	+11
2		Me	4	92	0.08	0
3		Me	6	93	0.17	-
4		Me	12	94	0.21	-50
5		Me	51	94	0.26	-123
6		Me	101	96	0.37	-123
7		Ph	2	82	0.05	+24
8		Ph	4	90	0.07	+14
9		Ph	6	93	0.13	-35
10		Ph	12	93	0.19	-54
11		Ph	51	94	0.22	-123
12		Ph	101	96	0.35	-123

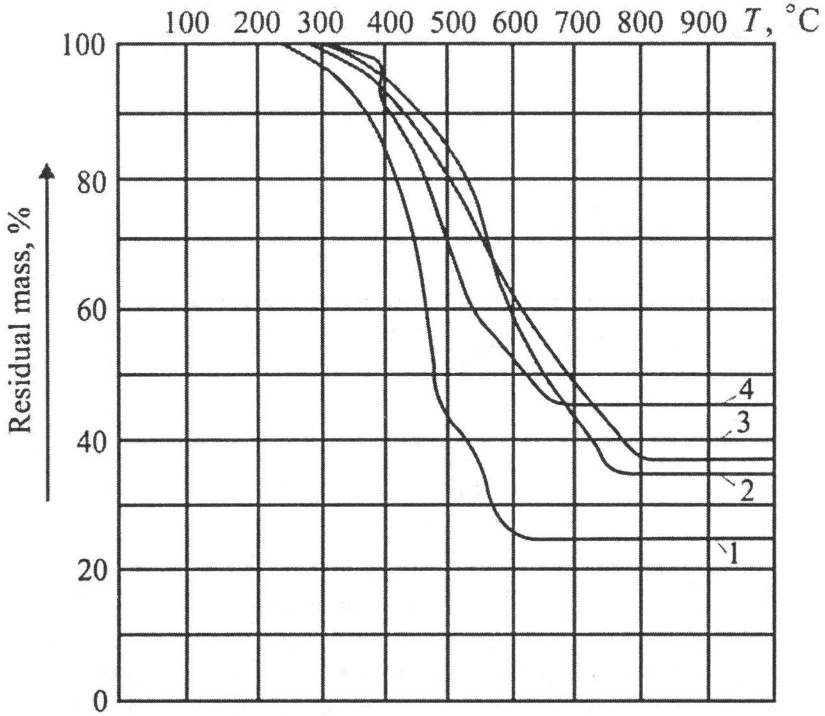

Figure 3.11. Thermogravimetric curves of cyclolinear copolymers in air: 1 – copolymer with 1,5-disposition of octaphenylcyclopentasiloxane fragment, n = 8; 2 - copolymer with 1,5-disposition of decaphenylcyclohexasiloxane fragment, n = 4; 3 - copolymer with 1,7-disposition of decaphenylcyclohexasiloxane fragment in dimethylsiloxane backbone, n = 4; 4 - copolymer with 1,5-disposition of octaphenylcyclopentasiloxane fragment in dimethylsiloxane backbone, n = 4

Figure 3.12 shows T_g dependence on dimethylsiloxane unit length, which indicates that the increase of T_g of copolymers with the number of phenyl groups and volume of cyclic fragment. Partial substitution of methyl groups at silicon atom in silsesquioxane fragment by phenyl ones will increase T_g by ~10°C. Similar dependence was detected for all copolymers with cyclic fragments in dimethylsiloxane backbone.

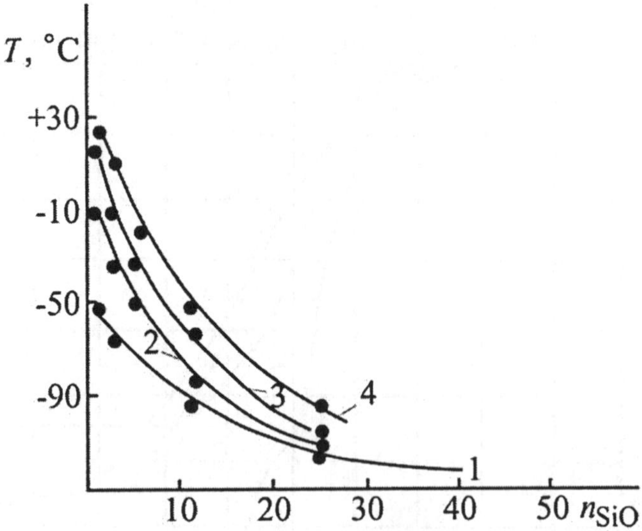

Figure 3.12. Dependence of T_g on the number of dimethylsiloxane units for cyclolinear copolymers: 1 – copolymers with 1,5-disposition of dimethyltetraphenylcyclotetrasiloxane fragments; 2 – copolymers with 1,5-disposition of hexaphenylcyclotetrasiloxane fragments; 3 - copolymers with 1,5-disposition of dimethylhexaphenylcyclopentasiloxane fragments; 4 - copolymers with 1,7-disposition of decaphenylcyclohexasiloxane fragments

X-ray diffraction studies were carried out for cyclolinear copolymers. Figure 3.13 displays copolymers to be amorphous systems. Independently of the length of single-unit dimethylsiloxane chain, diffraction patterns display two amorphous halos. It is known that the first amorphous halo, d_1, characterizes the average interchain distance in amorphous polymer [52], whereas d_2 is more complicated and corresponds to both intrachain and interchain or interatomic distances.

For copolymers with cyclopentasiloxane and cyclohexasiloxane fragments in dimethylsiloxane backbone, difference in d_1 values is negligible.

Maximal interchain distance (d_1 = 9.60 - 9.54 Å) is observed at short lengths of dimethylsiloxane chain (n = 2, 4), and d_1 is decreased to 7.20 Å, i.e. to PDMS interchain distance, with increase of the length, n.

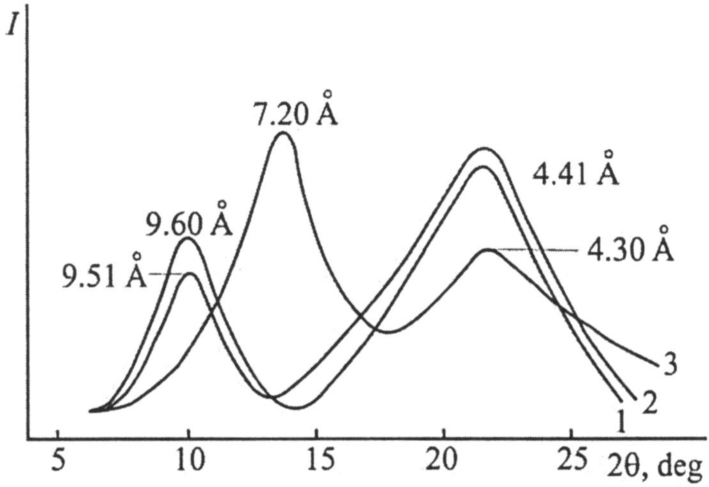

Figure 3.13. Diffraction patterns for cyclolinear copolymers: 1 – copolymer with 1,5-octaphenylcyclopentasiloxane fragment in dimethylsiloxane backbone (n = 2); 2 – copolymer with 1,5-decaphenylcyclohexasiloxane fragment in dimethylsiloxane backbone (n = 2); 3 – copolymer with 1,5-dimethyloctaphenylcyclohexasiloxane fragment in dimethylsiloxane backbone (n = 51)

It is known from the literature that occurrence of the mesomorphous state in cyclolinear siloxane polymers (CLSP) is mainly determined by the size of cyclosiloxane fragment, i.e. by the number of diorganosiloxane groups in the structure of cyclosiloxane fragment, and its symmetrical disposition in the polymeric backbone [53], molecular mass [54] and tacticity of the polymer backbone [55]. Mesomorphous state of CLSP occurs at sufficient length of symmetrical alkyl substituting agents ($R = C_nH_{2n+1}$, $n \geq 2$) irrespective of the backbone structure. Kuhn segment increase [56] and considerable increase of thermal stability of the mesophase for cyclolinear atactic hexalkylcyclotetrasiloxanes (ACTS) [57, 58] happen in the consequence of minimal increase of chain rigidity at transition from linear polyorganosiloxanes to atactic poly(hexaorganocyclotetrasiloxanes). Atactic methyl-substituted ACTS, as well as its linear analogue (PDMS) does not possess mesomorphous properties [57 – 59]. Further increase of poly(organosiloxane) chain rigidity by increasing the fracture of *trans*-units in ACTS macromolecules induces qualitative changes: occurrence of mesomorphous properties for methyl-substituted ACTS [59 – 61].

For the purpose of broadening conceptualization of mesomorphous properties display in polymers with tetrasiloxane rings in the backbone [62] copolymers, derived from ACTS, with alkyl substituting agents in the ring and different length of dialkylsiloxane junction were synthesized by HFC reaction of 1,5-dihydroxyhexaalkylcyclotetrasiloxanes with α,ω–dichloropolydialkylsiloxanes in the presence of hydrogen chloride acceptors in various solvents in accordance with the scheme as follows:

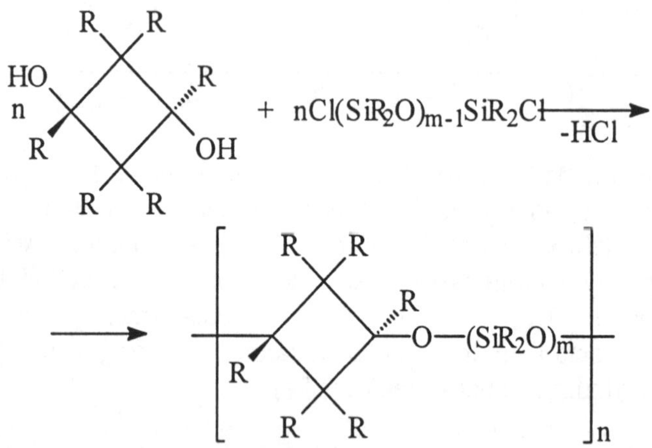

(3.6)

I

where R = Me, Et, Pr; m = 1, 2, 3.

α,ω–Dichloropolydialkylsiloxanes were synthesized by partial hydrolysis of diethyldichlorosilanes and di-*n*-propyldichlorosilanes. Physical and chemical parameters of initial poly(dialkylsiloxanes) and abbreviations and characteristics of synthesized copolymer are shown in Tables 3.10 and 3.11, respectively.

It is found that cyclolinear hexapropylcyclotetrasiloxane copolymers with di-*n*-propylsiloxane units are not formed under current reaction conditions. In accordance with the results of gas-liquid chromatography studies, primary compounds enter the reaction completely, but the reaction conditions (temperature and diluter concentration) do not provide for sufficient chain propagation.

Table 3.10

Physical and chemical parameters of α,ω–dichloropolydialkylsiloxanes, structural formula $Cl(SiR_2O)_{m-1}SiR_2Cl$

No.	R	*m*	T_{boil}, °C *P* (Pa)	n_D^{20}	^{29}Si chemical shifts, ppm
1	Et	2	215-217 (10^5)	1.4370	8.40
2	Et	3	139-141 (700-900)	1.4375	5.57 -17.20
3	*n*-Pr	2	100-102 (133)	1.4420	5.87
4	*n*-Pr	3	164-166 (133)	1.4412	2.92 19.32

All synthesized ACTS copolymers are completely soluble in usual organic solvents. The structure of copolymers was proved by the data of ultimate analysis, NMR and ^{29}Si NMR spectra, chemical shifts of which are shown in Table 3.11.

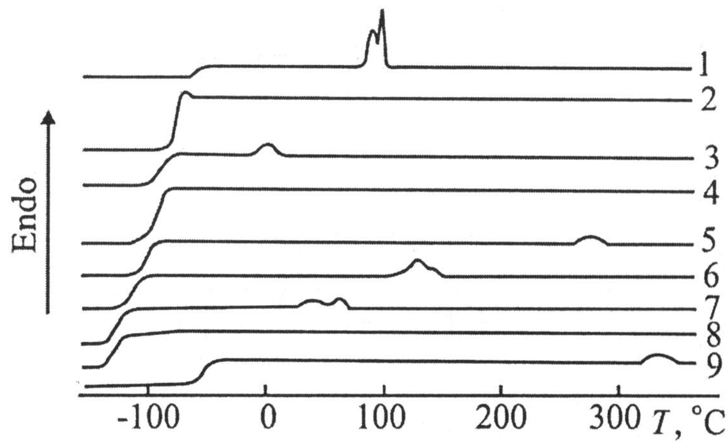

Figure 3.14. DSC curves for MCTS (1), MCTSS-1 (2), MCTSS-2 (3), MCTSS-3 (4), ECTS (5), ECTSS-1 (6), ECTSS-2 (7), ECTSS-3 (8), and PCTS (9)

ACTS copolymers were studied by differential scanning calorimetry (DSC) method. Figure 3.14 shows DSC curves for copolymers, which indicate

Table 3.11

Physical and chemical characteristics of PACS

No.	Copolymer	R	m	$[\eta]$*, dl/g	Yield, %	*Cis*:*trans* units relation in copolymer	^{29}Si chemical shifts δ, ppm (I_{rel})**		
							D_c	D_l	T
1	2	3	4	5	6	7	8	9	10
1	MCTS***	Me	0	0.17	80	88:12	-19.27 -19.31	-	-65.74 (0.11) -65.53 (0.78) -65.77 (0.10)
2	MCTSS-1	Me	1	0.15	81	93:7	-19.51 (0.83) -19.58 (0.17)	-21.15 (0.03) -21.23 (0.26) -21.33 (0.70)	-65.12 (0.85) -65.30 (0.15)
3	MCTSS-2	Me	2	0.16	75	95:5	-19.64	-21.56	-65.30
4	MCTSS-3	Me	3	0.13	70	95:5	-19.67	-21.64	-65.39
5	ECTS***	Et	0	0.26	-	70:30	-19.66 -19.65	-	-66.61 -66.65 -66.97 -66.99

1	2	3	4	5	6	7	8	9	10
7	ECTS-1	Et	1	0.08	75	70:30	-20.20 (0.65) -20.23 (0.35)	-21.92 (0.10) -21.95 (0.40) -21.95 (0.50)	-66.63 (0.70) -66.89 (0.30)
8	ECTS-2	Et	2	0.09	80	70:30	-20.16	-22.07	-66.71 (0.70) -67.01 (0.30)
9	ECTS-3	Et	3	0.09	47	70:30	-20.28	-22.23 (0.33) -22.37 (0.67)	-66.91 (0.70) -67.17 (0.30)
10	PCTS***	*n*-Pr	0	0/23	60	70:30	-21.86 (0.85) -21.98 (0.15)	-	-68.14 (0.15) -68.33 (0.85)

Notes: * In toluene at 25°C.

** D_c and D_l are chemical shifts for R_2Si groups in the structure of rings (cyclic fragments) and linear fragments, respectively; *T* is chemical shift for $RSiO_{1.5}$ groups.

*** Corresponded polyhexaalkylcyclotetrasiloxane copolymers are shown for a comparison.

that MCTSS-1 and MCTSS-3 are incapable of regulation. MCTSS-2 crystallizes and transits into an isotropic melt above T_{melt} = +5°C. As concerns ECTSS-1 and ECTSS-2, they are in the mesomorphous state below the isotropization temperature, T_{iso}, transiting into mesomorphous glass below glass transition temperature, T_g.

Flexible junction length increase, *m*, decreases thermal stability of the mesophase, and in the case of $m = 3$ (ECTSS-3), to suppression of copolymer ability to transit into the mesophase, i.e. ability of copolymer to the long-range intermolecular regularity is completely suppressed due to the loss of ability to intramolecular regulation. However, sufficient length of flexible junction and considerable intermolecular overlapping make possible occurrence of close regularity between cyclic fragments both at intramolecular and intermolecular levels. This conclusion is based on occurrence of amorphous dispersion in the angular area, close to $2\theta_1$ value for poly(ethylcyclotetrasiloxane), on diffraction patterns of ECTSS-3 at low temperatures. To put it differently, a tendency to close regulation by type of nematic diskophase, N_D, is observed for ECTSS-3 [63].

MCTSS-1 and MCTSS-3 diffraction patterns, obtained in the temperature range of -130 - +40°C, and MCTSS-2 one, obtained above +25°C, are qualitatively identical to the diffraction pattern of homopolymeric atactic MCTS. They contain two amorphous halos: the first is intensive one in the angular area of $2\theta = 9 - 12$ ($1/2\Delta = 1.2$) with the maximum at $2\theta_1$ and the second weak diffuse one in the area of $2\theta = 16 - 25$. It is common knowledge [64] that for polyorganosiloxanes $2\theta_1$ value characterizes the average intermolecular distance in amorphous component or isotropic melt. It is shown that interplane distances, d_m, in copolymers decreases with increase of flexible junction length, *m*, and increases with the length of side framing, *n*. As flexible junction is introduced and its length is increased, angular location of the maximum of the first amorphous halo ($2\theta_1$) is shifted to the area of larger angles, and the glass transition temperature (T_g) of MCTSS decreases, approaching T_g of its linear analogue (PDMS) at $m = 3$. This means that decrease of polymeric chain rigidity with *m* increase proceeds simultaneously with decrease of average intermolecular distances.

As suggested in ref. [62], the tendency of ACTS copolymers to regulation is displayed on two interconnected levels: intramolecular and intermolecular ones. On the one hand, cylindrical fragments tend to realize a conformation with the shortest intramolecular distances. On the other hand, this conformation provides for a disposition of side radicals, which may induce rather strong interaction between them to affect intramolecular regularity and stabilize a definite revolute configuration. Moreover, if isomers of organic polymers are stabilized by van-der-Waals detachment of valence unbonded atoms or radicals only, in the case of considered compounds, this mechanism is added by a factor, stipulated by the following feature of ACTS copolymer macromolecules: different nature of the backbone and side framing. Similar to linear analogues [65 – 70], side radicals in ACTS copolymers will tend to form organic cover for coiled siloxane chain.

Thus, lack of anisodiametry of the cyclic fragment is compensated by two interconnected factors – configuration features of the backbone and dephility of the backbone and side framing. These factors promote for formation and stabilization of configurative *trans*-tactic isomers, large sequences of which may be considered as peculiar conformational mesogens.

For previously synthesized poly(methylcyclosiloxanes) with cyclic pentasiloxane and heptasiloxane fragments, no mesophase was observed. However, it should be noted that cyclolinear poly(organocyclotetrasiloxanes) with aliphatic substituting agents form a mesomorphous state, substituting agent length increase from methyl to propyl fragment inducing mesophase isotropization temperature increase, and at sufficient molecular mass poly(hexa-*n*-propylcyclotetrasiloxane) is present in the mesophase in the range from glass transition to decomposition temperature transiting into the mesophase glass below the glass transition temperature [57].

These results gave grounds [72] to suggest that the introduction of aliphatic substituting agents (ethyl, propyl, etc.) into poly(organosiloxanes) with cyclic pentasiloxane fragments in the linear backbone promotes mesophase formation and increases thermal stability. For this purpose, HFC reaction of 1,5-dichlorooctaethylcyclopenta siloxane with 1,3-dihydroxytetraethyldisiloxane in the presence of pyridine, proceeding in argon flow in the temperature range from 25 to 50°C in accordance with the scheme shown below, was studied:

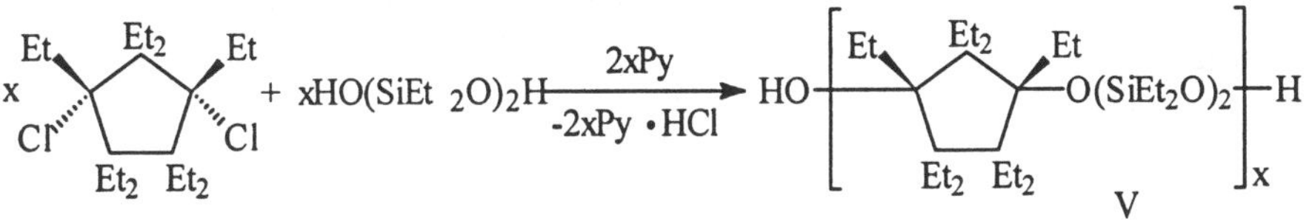

(3.7)

After precipitation, polymer with [η] = 0.07 with 36% yield was obtained. The structure of synthesized polymer was detected based on the spectral data.

^{29}Si NMR spectra of polymer solution display D^{Et} (diethylsiloxane group) and T^{Et} (ethylsesquioxane group) reflexes: reflexes in the *D*-range at -22.39, -22.47, -22.52, and -22.56 ppm and 3:4:2:1 ratio and in the *T*-range at –68.50 and -68.56 ppm and 1 : 5 ratio. Complexity of these reflexes is stipulated by spatial nonequivalence and for D^{Et}, in addition, by positional nonequivalence of silicon atoms.

X-ray and calorimetric studies of the polymer shows T_g = -127°C, i.e. introduction of a flexible junction decreases T_g compared with poly(hexa-*n*-propylcyclotetrasiloxane) bead-shaped polymer [57] and approaches it to T_g of linear poly(diethylsiloxane) (PDES) [73]. Above T_g, diffraction patterns of the polymer display a single sharp peak at 50°C: $2\theta = 10.13°$ ($\Delta \sim 30'$). Dispersion remains unchanged up to isotropization at 35°C. The data obtained correlate with the results of DSC studies and thermal analysis. Taking into consideration temperature dependencies of angular dispositions of observed reflexes and atactic structure of the backbone, one may suggest that for synthesized polymer long-range positional order in the packing of gravity centers of macromolecules is realized in two directions only in the plane perpendicular to chain axes: in the whole temperature range below the isotropization point.

The effect of methylcyclosiloxane fragments in dimethylsiloxane backbone on properties of copolymers and formation of the mesophase state was studied [74] by HFC reaction of 1,7-dichloro(dihydroxy)-1,7-diorganooctamethyl cyclohexasiloxane with α,ω–dihydroxy(dichloro)-dimethylsiloxanes at equimolar ratio of the initial components in the presence of amines. The reaction proceeds as follows:

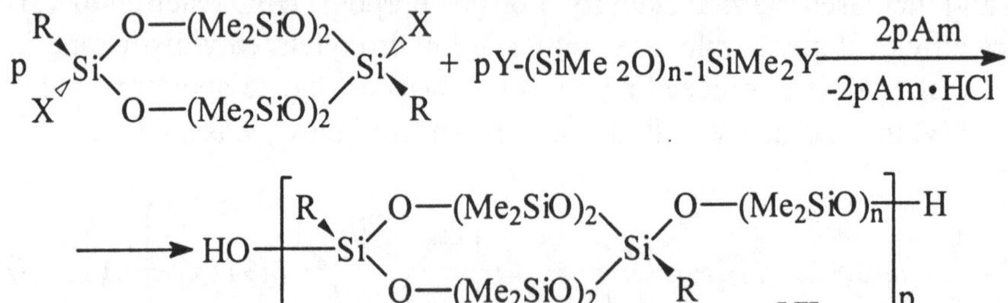

VII

(3.8)

where R = Me, Ph; *n* = 0, 1, 2, 3.

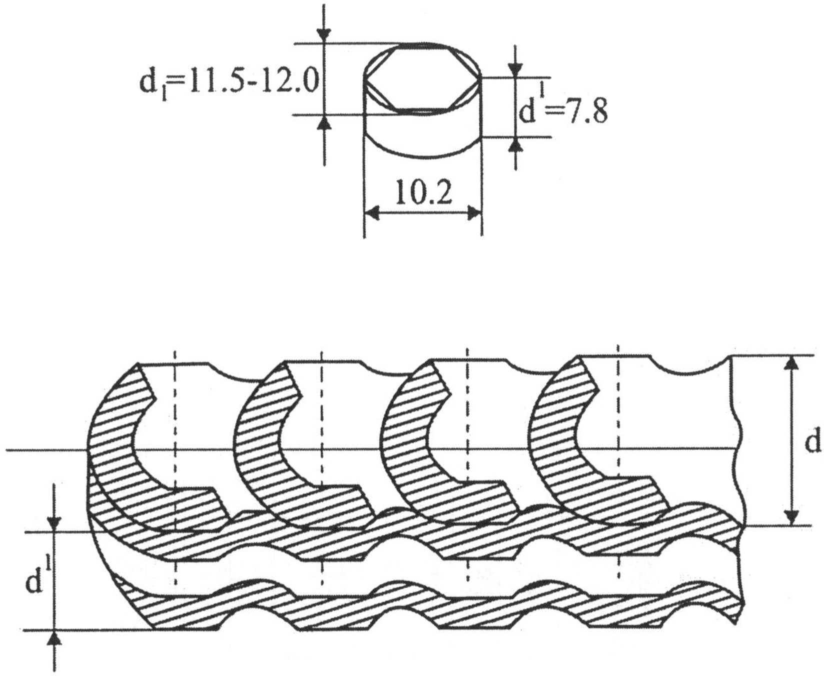

Figure 3.15. Hypothetical model of decaorganocyclohexasiloxane polymer chain

In synthesized cyclolinear siloxane polymers (CLSP) the dimethylsiloxane unit was taken as a flexible spacer [74]. Table 3.10 shows temperatures and heat transitions in relation to the flexible spacers' lengths. These polymers were found in the crystalline and thermotropic states, like *trans*-tactic poly(decaorganocyclohexasiloxanes), but temperature range of the mesomorphous state is considerably narrower; *n* increase to 3 gives completely amorphous polymers. This sharp decrease of isotropization temperature in CLSP with flexible spacers is not unexpected, since addition of dimethylsiloxane spacer instead of the hydrocarbon one has similarly led to a sharp decrease of isotropization temperature in liquid-crystalline polymers with classic mesogenic groups [75]. The authors studied the structure of decaorganocyclohexasiloxane polymer unit with the help of Stuart-Briegleb models and compared reflex variation with temperature at $2\theta = 9 - 11°$. Figure 3.15 shows CLSP unit resembling a disc with $d_l = 11.5$ Å and height $h = 7.8$ Å, introduced into the backbone as disc-shaped thin plates similar by thickness to

Table 3.10

Temperatures and heat transitions of cyclolinear poly(methylphenylsiloxanes) with flexible spacers

No.	Backbone	R	n_{SiO}	[η] in toluene, dl/g	T_g, °C	T_{melt}, °C	H_m, kJ	T_{iso}, °C	H_i, J/g
1	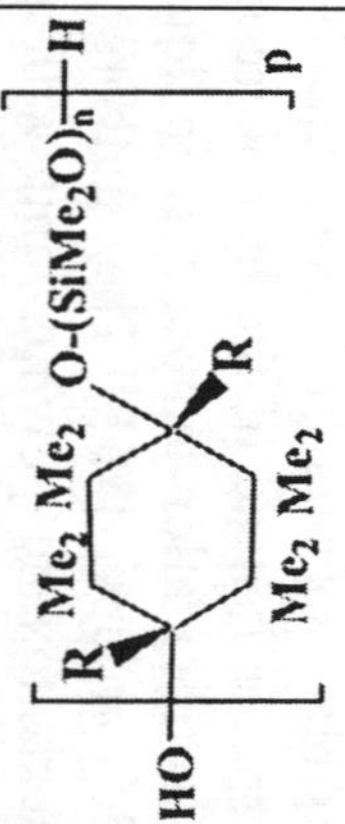	Me	0	0.15	-91	-53	5.5	49	10.0
2			1	0.13	-88	-7 - -12	17.6	67-97	3.6
3			2	0.14	-91	-	-	7-57	1.3
4			3	0.15	-90	-	-	-	-
5		Ph	0	0.20	-41	187	13.8	427	3.0
6			1	0.19	-59	82-107	16.8	137-167	1.3
7			2	0.15	-67	52	14.7	-	-
8			3	0.14	-80	-	-	-	-

interplanar distance $d = 8.4$ Å, which was calculated from experimental data on the maximum position (refer to Figure 3.16).

Existence of the long-range order in the packing with the lower limit of the coherence area from $0.5h \geq 400$ Å indicates that the packing shown in Figure 3.15 only is possible for these intermolecular distances.

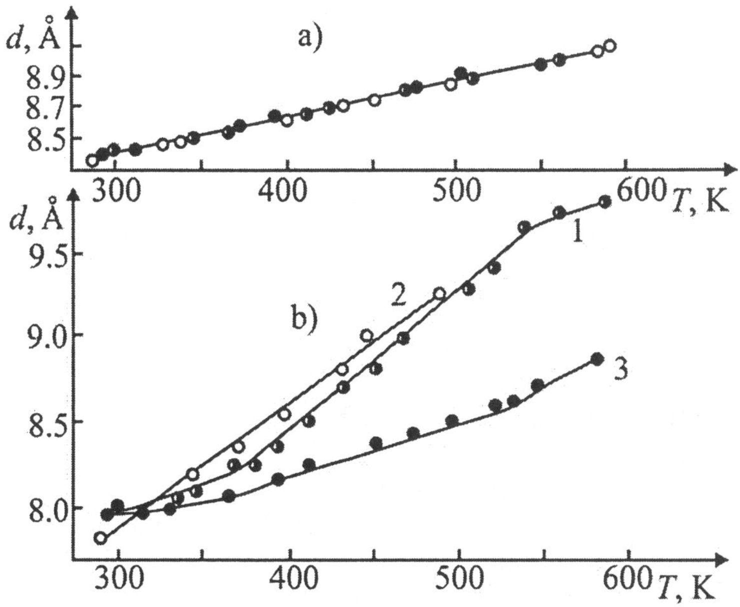

Figure 3.16. Temperature dependence of interplanar distance, *d*, for decamethylcyclohexasiloxane polymer: a – mesomorphous component; b – amorphous component, depending on the polymer chain tacticity: 1 – atactic, 2 – microtactic, 3 – *trans*-microtactic

It should be noted that distances of close values are realized in the ring itself; in this case, however, existence of so high long-range order is impossible, since intra- and intermolecular distances would alternate: first, they are different and secondly, differ. As a consequence, it is impossible to retain an unchanged long-range order in the broad temperature range of the mesophase. Existence of the single peak only follows from the suggested model. Thus, the order in the basic plane is realized in one direction only – perpendicular to the plane of the ring that is polymer chain unit, which is simultaneously perpendicular to the main axis of the polymer macromolecule.

Comparison of the data shown in Figure 3.15 with *d* values from the diffraction pattern in Figure 3.16 indicate (Figure 3.16a) that the mesomorphous component (a narrow reflex), *d*, is independent of poly(decamethylcyclohexasiloxane) chain tacticity. Hence, irrespective of *trans*-tactic polymer ability to crystallization, the mesophase temperature range changes insignificantly. At lower temperature of –73°C only, a bending on the temperature dependence curve, *d*(*T*), of the mesophase peak occurs.

Variation of *d*(*T*) dependence shape at the bending is accompanied by a considerable increase of the narrow reflex integral half-width and decrease of the integral intensity. In this temperature range, the interplanar distance, *d*, is comparable with the disc – decamethylcyclohexasiloxane unit – height (Figure 3.15) due to close packing, leading to decreased long-range order areas.

Using the methods of amorphous halo separation from the narrow reflex, Figure 3.16b shows *d*(*T*) dependenceis for poly(decamethylcyclohexasiloxanes) of different tacticity. Figure 3.16b indicates that tacticity variation affects the values and shape of *d*(*T*) dependence of amorphous halo. At 20°C, the difference of studied samples in *d* is insignificant, but at 197 - 247°C difference in the amorphous halo approaches 1 Å, approximately. The bending observed at the beginning of active melting of the mesophase is indicated by abrupt intensity drop of the mesophase peak.

Thus, as follows from the analysis of *d*(*T*) dependence of amorphous dispersion, the polymer chain tacticity breakdown causes changes of packing in two directions only, where the long-range order is not realized, but the interplanar spacing is unaffected.

As follows from *d*(*T*) dependence (Figure 3.16) of the mesophase peak and amorphous halo, introduction of phenyl substituting agents to organosilsequioxane fragments in the organocyclohexasiloxane unit leads to the differences depending on the polymer chain tacticity, not only in the amorphous dispersion, but also in the mesophase peak position.

Thus, the above data show the main factors responsible for mesomorphous state formation, which are, first, the increased conformational parameters and the Kuhn segment (A) of CLSP as a consequence of certain conformation of organocyclosiloxane unit, most likely, the folded ring conformation or any other ensuring, approximately, co-planar conformation of these cyclosiloxanes; secondly, inclusion of organocyclosiloxane unit with symmetrical position into the polymer chain by *para*-position type; thirdly, the use of organic substituting agents ensuring amplification of both intra- and intermolecular interactions simultaneously capable of increasing the rigidity of the backbone.

The factors responsible for the mesomorphous state temperature range are: first, the polymer chain tacticity, perfect for the mesophase emerging in CLSP is the *trans*-tactic structure enabling to achieve the highest isotropization temperatures at low degrees of polymerization; secondly, growth of CLSP molecular weight leading, irrespective of the CLSP unit structure, to extended mesomorphous state temperature range in CLSP.

Thermal decomposition of branched chain methylsiloxane polymers has been studied before [5, 76-79]. Methylsiloxane rings and polycycles were detected as volatile products of the methylsiloxane resin pyrolysis [5, 76-79]. The results indicate that pyrolysis decomposition of methylsiloxane resins is similar to that observed for methylsiloxane chain polymers [4]. The relation between the polymer structure and thermal degradation products can be interpreted in terms of this decomposition mechanism, but a resin of irregular structure does not provide proper system for studying this relation.

Thermal decomposition products [80] of two methylsiloxane polymers of strictly regular and known structure were examined to clarify the mentioned relations. The polymers studied include di- and tri-functional structural units (i.e. $(CH_3)_2SiO_{2/2}$ denoted as -D- and $CH_3SiO_{3/2}$ denoted as -T<) and involve cyclolinear chain structures, as follows:

$(T_2D_6)_n$: 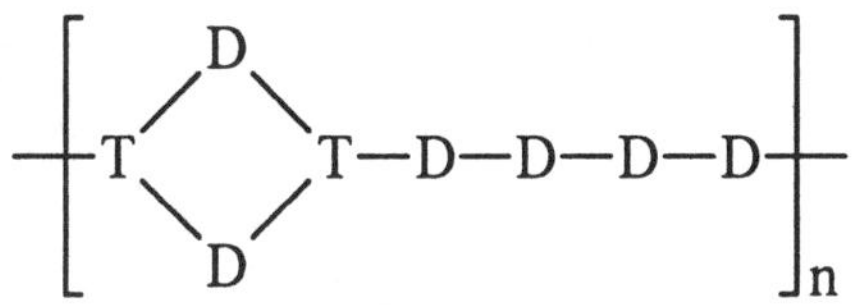

and

$(T_2D_9)_n$: 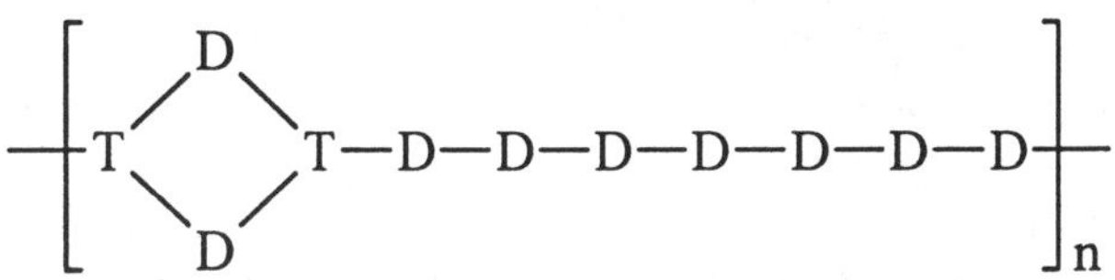

These structures are ensured by the method of synthesis used.

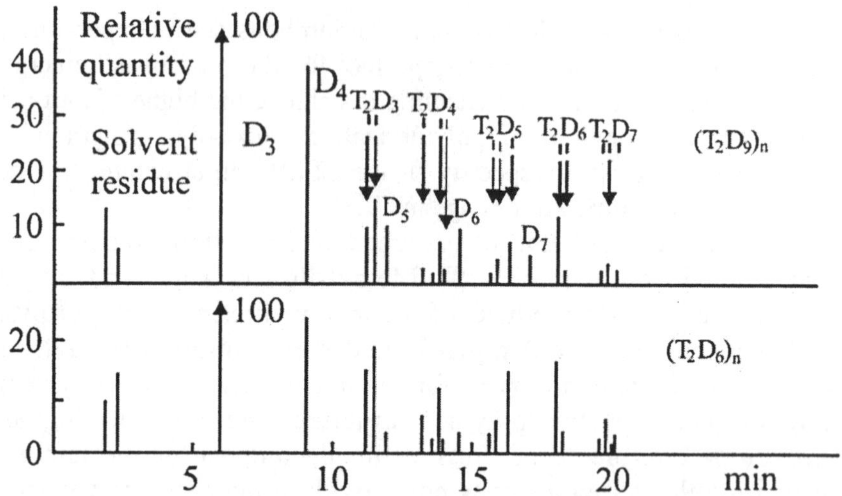

Figure 3.17. Cyclolinear methylsiloxane polymers' pyrolysis products separated in OV-1 gas chromatographic phase

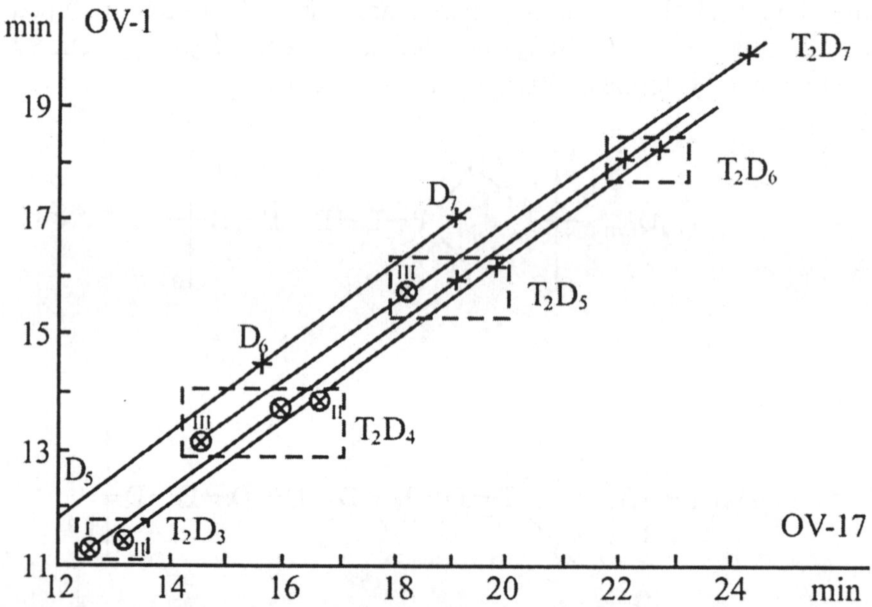

Figure 3.18. A plot of retention time against retention time for pyrolysis products on temperature programmed columns coated with OV-1 and OV-17 phase, respectively

Thermal decomposition of investigated cyclolinear methylsiloxane polymers at 350°C gives cyclic methylsiloxane oligomers. Generally, the polymer $(T_2D_6)_n$ decomposes to D_3, and polymer $(T_2D_9)_n$ to D_3, D_4, D_5 etc., similar to poly(dimethylsiloxane), but the degree of conversion is very low at this temperature. At higher temperatures several polycyclic compounds are also produced from both polymers. Figure 3.17 shows relative amounts of products, produced by pyrolysis at 550°C due to GLC results. The gas chromatographic peak areas are related to the peak area of the most abundant product. The product distribution in the temperature range of 450 - 600°C does not vary significantly with pyrolysis duration or pyrolysis temperature. The author identified ten of fifteen main products by comparing gas chromatographic retention times with these pure standard substances on three different GC stationary phases. Standard polycyclic siloxanes were obtained from methylsiloxane resin by pyrolysis followed by preparative gas chromatography.

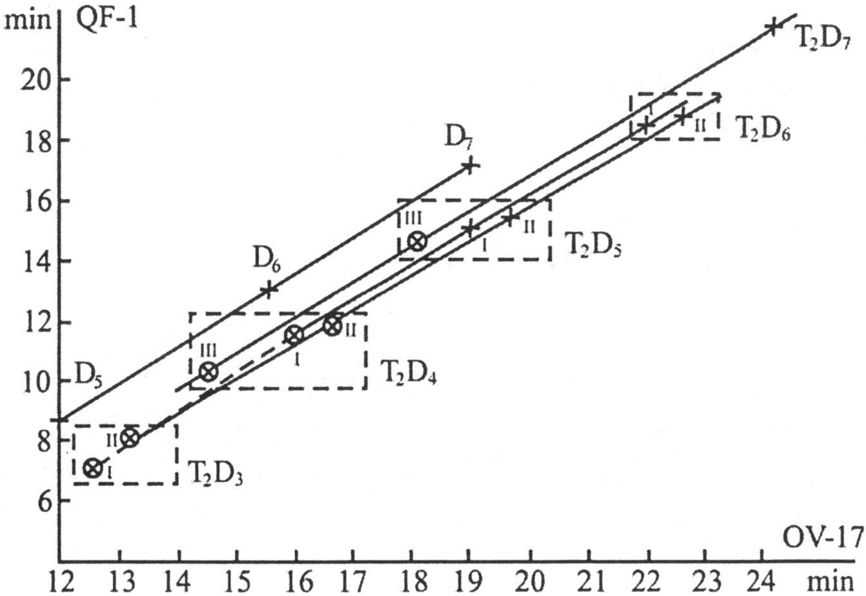

Figure 3.19. A plot of retention time against retention time for the pyrolysis products on temperature programmed columns coated with QF-1 and OV-17 phase, respectively

Their identity was proved by mass spectrometry, ^{29}Si NMR spectroscopy and X-ray diffraction patterns [76-79]. Standards are not available for some important products, but these seem to be homologous with the products identified. This is indicated by straight lines which join the points given by the retention time values of the products measured on two different stationary phases with a linear temperature program (Figures 3.18 and 3.19). The products identified are represented by circled points in the Figures.

Examining the structures and amounts of the pyrolysis products, one may observe some correlation between the polymer structure and the pyrolysis products.

Pyrolysis products are either cyclic or bi-cyclic methylsiloxanes ranging from 3 to 9 siloxane units. Pyrolizate of the polymer $(T_2D_9)_n$ displays higher quantity of cyclic products than in that of $(T_2D_6)_n$. Relative quantities of cyclic products D_3, D_4 and D_5 in the pyrogram of $(T_2D_9)_n$ are similar to these found in the pyrogram of poly(dimethylsiloxane), but smaller quantity of D_4 and D_5 from $(T_2D_6)_n$ was observed.

Table 3.11

Homolog series of cyclolinear methylsiloxane copolymer pyrolysis

Composition	Series I	Series II	Series III
T_2D_3	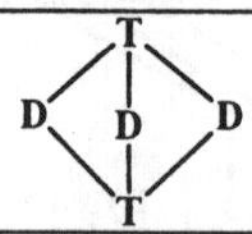	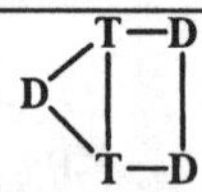	–
T_2D_4	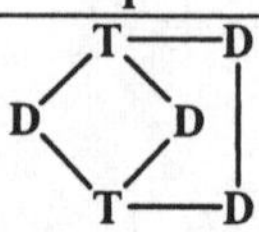	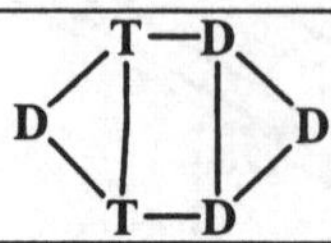	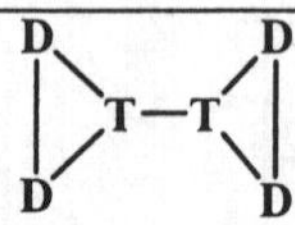
T_2D_5	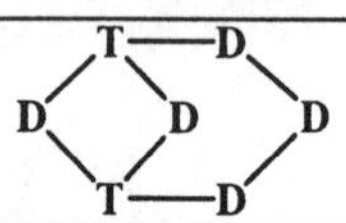	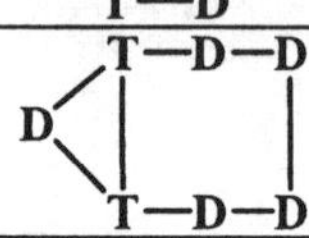	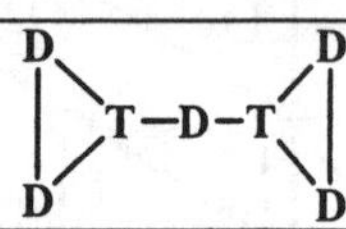
T_2D_6	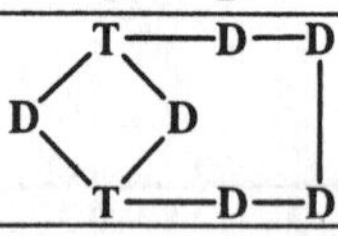	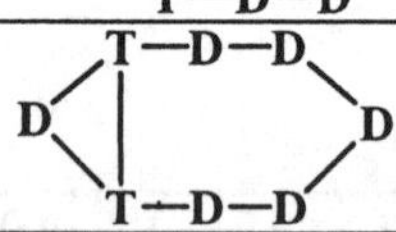	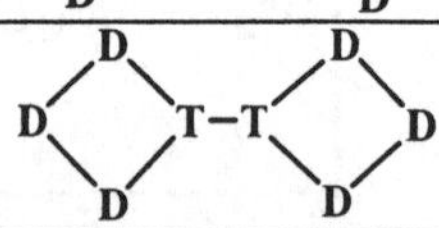
T_2D_7	–	–	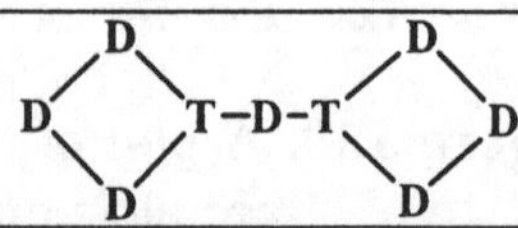

All important pyrolytic products belong to a homolog series of siloxane units (see Figures 3.18 and 3.19). Three series were found for double rings shown in Table 3.11. None of the products include more than two T units.

Quantitative evaluation of the total amounts of T and D units, found in the pyrolyzate, gives a T/D ratio corresponded to that in the original polymer. For the conversion of peak area values into weights authors used reflex factors for the Flame Ionization Detector [81].

Structures of volatile degradation products show that all these molecules were detached from the cyclolinear polymer by the mechanism first suggested for poly(dimethylsiloxane) by Thomas [4]. In this mechanism, degradation starts with formation of an intermediate four-centered structure, and then two siloxane bonds involved are rearranged. For poly(dimethylsiloxane) chain, resulting volatile product represents an oligomer ring. Similarly, D_3, D_4, D_5 cyclic oligomers are detached from the linear part of studied cyclolinear methylsiloxanes. When the siloxane bond rearrangement between two linear segments, bonded to the same ring of the polymer takes place, double ring of the Series I can be detached as follows:

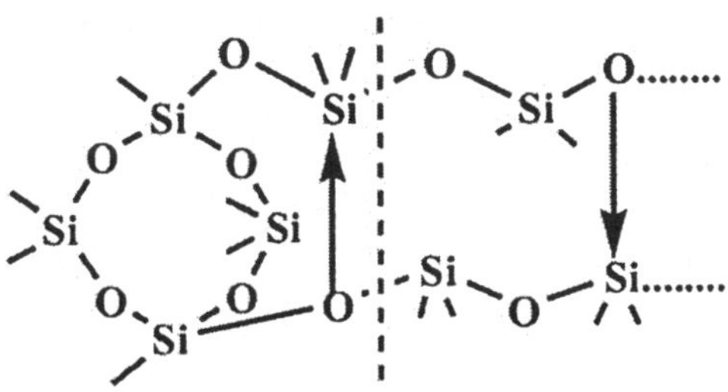

(3.9)

In T and D symbols:

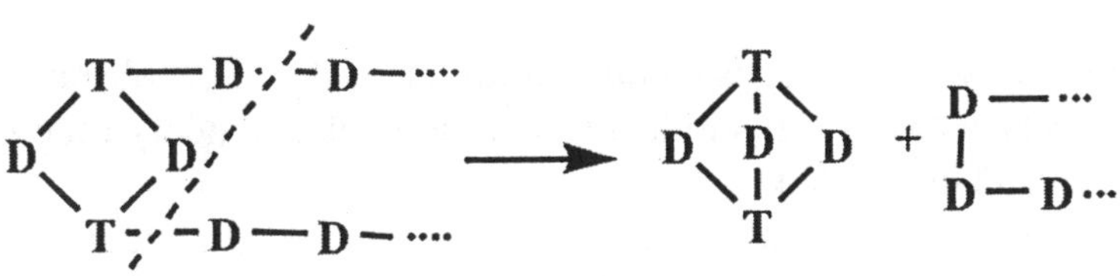

(3.9)

may result from such inter- or intramolecular bicycles of Series II may form similarly, but the macromolecular segment, in which the siloxane bond rearrangement takes place has to include a triatomic siloxane ring:

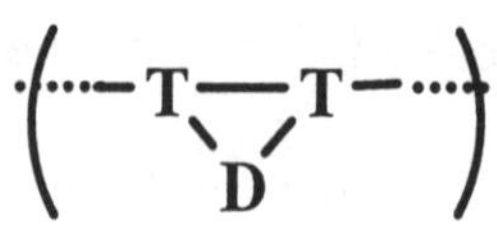

This modified molecular segment may result from such inter- or intramolecular rearrangement, as follows:

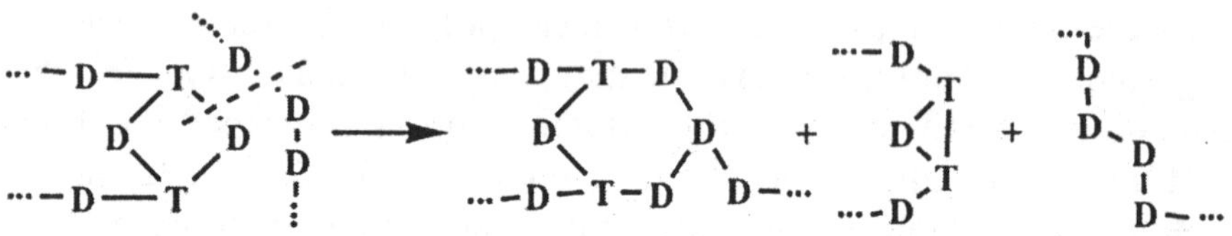

(3.10)

As a consequence, Series II products are assumed to be the products of three consecutive rearrangement steps.

Distinct double rings were also found among the pyrolysis products. Their formation can be described in terms of bond rearrangement between the linear part and the adjacent ring of the polymer, as follows:

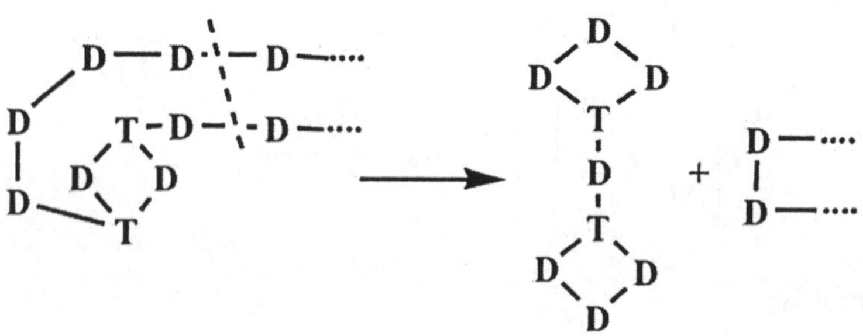

(3.11)

In reaction 3 two cleavages occur simultaneously. The absence of products with more than two T units indicates that none of secondary reactions of the primary products takes place.

Under the pyrolysis conditions, the authors used stoichiometric composition of total pyrolyzate similar to that of the polymer. In addition, no change in functionality of siloxane units takes place, which correlates with the authors' earlier results [77].

Accordingly, the structure of degradation products is closely connected with the macromolecular one. The original linear segments of the macromolecule appear in the pyrolyzate as cyclic oligomers, and the branching points are detached with adjacent fragments as polycyclic rings.

The ratio of mono- and double ring products' quantities characterizes the average frequency of branching points in the polymer network. The exact description of structures of double ring products may help in distinguishing distribution of the branching points, and thus the microstructure of methylsiloxanes.

The above discussion indicates that synthesized CLSP with phenyl framing groups are characterized by high thermal oxidative stability and easy processing to articles. That is why these substances were suggested for production of specific rubbers [12].

For the purpose of increasing selectivity of immovable phase for complete and selective separation of organic compounds and complex mixtures of organosilicon compounds by gas-liquid chromatography method, CLSP of the following general formula was suggested [82]:

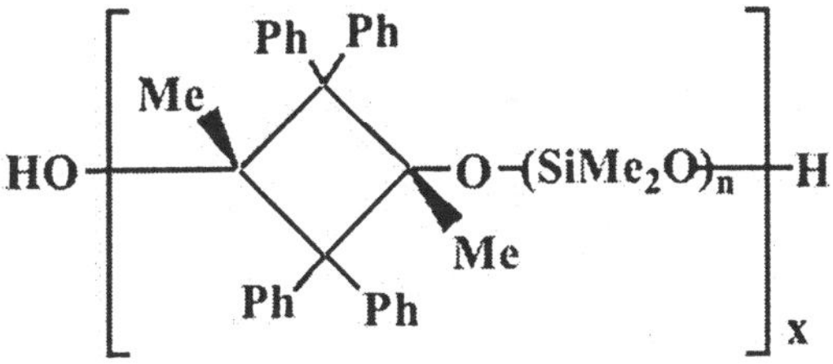

I

where $n = 3 - 20$; $x = 10 - 100$.

As the immovable phase, CLSP was applied to N-AW chromaton carrier in 20% quantity of its mass. The column used was 2 m long and 2 mm of internal diameter. Figure 3.20 shows the chromatogram of difficultly separable compound separation: cyclohexanol (1), cyclohexylamine (2), phenol (3), and aniline (4) at 100°C. Moreover, the possibility of suggested phase to separate compounds from the naphthene sequence from compounds of the standard structure is illustrated on the example of hexane (5) separation from cyclohexane (6) (Figure 3.20).

The phase peculiarity to alcohols of the fatty sequence is shown on the example of synthetic alcohols $C_{10} - C_{21}$ fraction separation in the temperature range from 100 to 270°C (at 4 deg/min rate) without preliminary transfer to volatile derivatives.

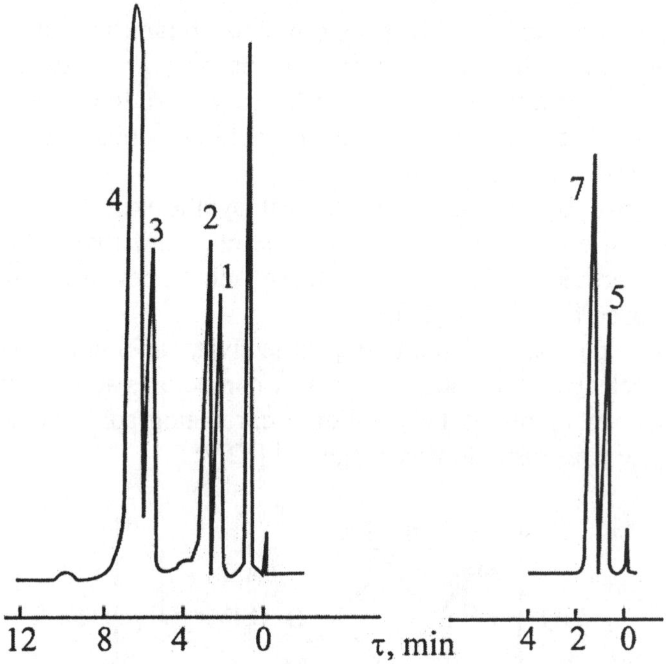

Figure 3.20. Chromatographic curves for compounds

Table 3.12

Comparative data on Mac-Reynolds constants

Immovable phase	Mac-Reynolds constants				
	X'	*Y'*	*Z'*	*I'*	*S'*
SE-30	15	53	44	64	41
SKT	17	57	46	67	45
SE-54	33	72	66	99	67
DPOC	86	130	134	181	154
Suggested CLSP naphthene phase ($n = 3$)	95	135	120	185	145

In spite of the known phases, the advantage of suggested selective immovable phase is the possibility of selective analysis of various classes of organic and organosilicon compounds. Of special attention should be peculiarity of the phase to separation of cyclic and naphthene hydrocarbons, amines and alcohols with preservation of high thermal stability. Table 3.12 shows comparative data on Mac-Reynolds constants for a series of universal non-polar phases and high parameters of suggested phase.

Note also that in spite of the suggested phase, immovable phases considered give no opportunity to separate many classes of chemical compounds. In particular, amines and alcohols can be separated on these immovable phases, if transferred to volatile derivatives only; cyclic and naphthene amines and alcohols cannot be separated, and complete separation of complex mixtures of organosilicon compounds cannot be performed.

3.2. Cyclolinear copolymers with cyclic carbosiloxane fragments in dimethylsiloxane backbone

Cyclolinear copolymers containing cyclic carbosiloxane fragments in dimethylsiloxane backbone were synthesized by HFC reaction of dichlorine-containing organocyclocarbosiloxanes (described in Chapter 1) with α,ω-dihydroxymethylsiloxanes. Copolymers were synthesized in 60 – 70% solution of anhydrous toluene both in the presence and in the absence of pyridine (hydrogen chloride acceptor) at 20 - 25°C. It is found that at HFC proceeding without acceptor, increase of dimethylsiloxane unit length and volume of the cyclic fragment decrease conversion by hydrogen chloride. In both cases, the reaction proceeds by the scheme as follows [14, 83]:

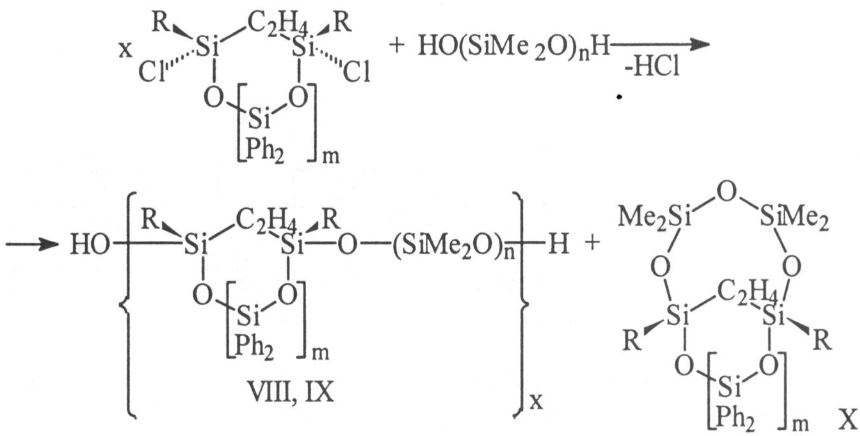

(3.12)

where m = 1 (VIII), 2 (IX); n = 2, 4, 10, 20, 37.

When precipitated from toluene solution with methanol, synthesized copolymers are solid or viscous substances with $\eta_{spec} = 0.07 - 0.30$, well soluble in usual organic diluters, with regard to dimethylsiloxane unit value.

Table 3.13

Physical and chemical properties for copolymer with structure VIII ($m = 1$, $n = 2$)

$[\eta]$, dl/g	Diffusion coefficient (D), cm^2/s	Sedimentation parameter, S	Specific volume (γ), cm^3/g	Refraction index increment (dn/dc), cm^3/g	$\bar{M}_{SD}$
0.07	28×10^{-7}	2.6	0.87	0.28	16,500

Table 3.14 shows the yield, T_g and viscosity parameters of cyclolinear carbosiloxane copolymers.

Table 3.14

Physical and chemical properties of cyclolinear carbosiloxane copolymers of structures VIII and IX

No.	Copolymer	m	n_{SiO}	Yield, %	η_{spec} in toluene at 25°C	T_g, °C	d_1, Å
1	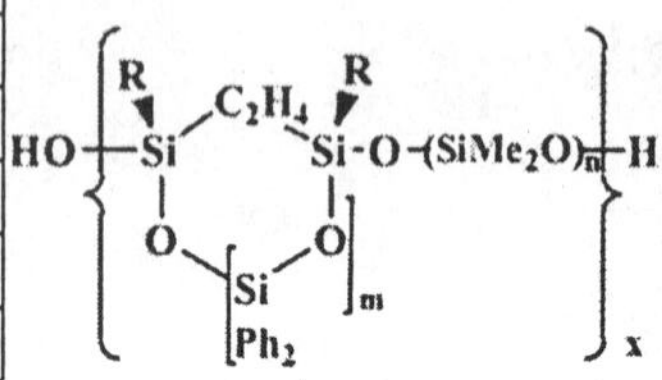	1	2	78	0.08	-25	8.82
2		1	4	79	0.11	-60	-
3		1	10	86	0.18	-90	-
4		1	20	87	0.25	-105	7.24
5		1	37	90	0.29	-123	7.21
6		2	2	77	0.09	-10	9.80
7		2	4	80	0.12	-40	-
8		2	10	85	0.18	-75	-
9		2	20	87	0.26	-90	7.25
10		2	37	89	0.30	-123	7/24

Comparatively reduced yield at low lengths of dimethylsiloxane unit ($n = 2, 4$) in reaction (3.12) is explained by partial proceeding of HFC reaction by intramolecular cyclization mechanism with formation of polycyclic structures. Similar to HFC reaction (3.1) a product of intramolecular condensation was extracted from the mother solution and additionally derived by direct synthesis.

Table 3.13 shows inherent viscosity, molecular mass and some physical and chemical properties for copolymer of the structure VIII (m = 1, n = 2), determined in benzene solution.

Thermogravimetric studies of cyclolinear copolymers have displayed their sufficiently high thermal oxidative stability. For short lengths of dimethylsiloxane unit, initial mass loss is observed at 260 – 280°C and 5% loss at 300°C. Above 750°C no mass loss variation is observed, which is apparently associated with formation of secondary structures. Thermal oxidative stability of copolymers with carbocyclosiloxane fragments in the backbone is similar to that of their pure siloxane analogues. Moreover, thermal oxidative stability of copolymers is decreased with increase of the volume of cyclic carbosiloxane fragment.

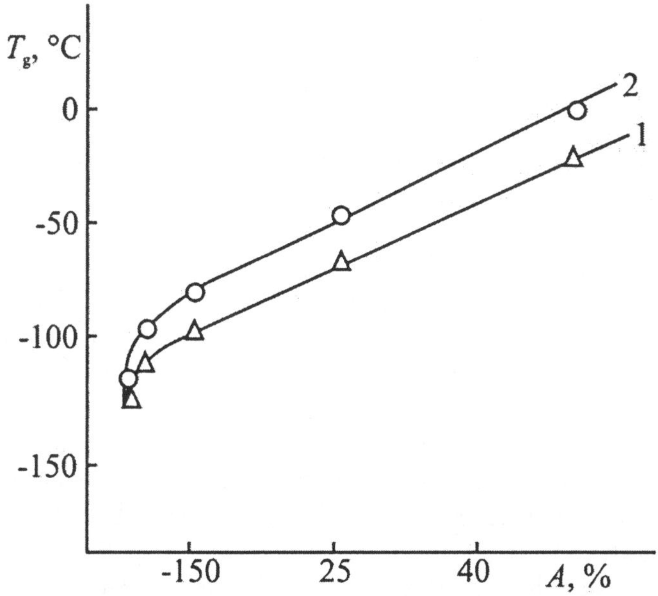

Figure 3.21. Dependence of glass transition temperature of copolymers on mol% concentration of cyclic carbosiloxane (A) fragments: 1 – copolymers with m = 1; 2 – copolymers with m = 2

Figure 3.21 shows dependence of the glass transition temperature of copolymers on cyclic carbosiloxane fragment content in dimethylsiloxane backbone. Similar to copolymers from poly(dimethylphenylsiloxane) sequence [53], the effect of non-dimethylsiloxane units on T_g of copolymers with cyclic carbosiloxane fragments in the backbone is primarily observed at ~3% concentration of the latter. The Figure shows also that the effect of cyclic carbosiloxane fragments with $m = 2$ on T_g of studied copolymers is higher rather than for copolymers with $m = 1$; T_g of copolymers is increased with mol% concentration of cyclic carbosiloxane fragments.

Diffraction patterns of copolymers with cyclic carbosiloxane fragments in the backbone are characterized by the presence of two diffraction maximums typical of amorphous polymers. Data in Table 3.15 show that interchain distances, d_1, decrease with the increase of dimethylsiloxane backbone length and at $n = 37$ reach the value typical of PDMS. Simultaneously, the increase of cyclic fragment volume in copolymers induces a considerable increase of d_1.

One more interesting feature of these copolymers is that initial difunctional cyclic organocarbosiloxanes can be derived easily, and copolymers themselves contain reactive cyclic carbosiloxane fragments capable of cross-linking without liberating volatile compounds [84].

3.3. Organosiloxane copolymers with heterocyclic fragments in the dimethylsiloxane backbone

It has been shown in the literature [85] that introduction of arylene fragments into the dimethylsiloxane backbone increases thermal oxidative stability of copolymers. For the purpose of synthesizing and studying properties of copolymers with 1,3-diorgano-1,3-disila-2-oxaindane fragments in the backbone [85, 86], HFC reaction of 1,3-dichloro-1,3-diphenyl(dimethyl)-1,3-disila-2-oxaindane with α,ω–dihydroxydimethylsiloxanes, dihydroxydiphenylsilane and 1,4-bis(hydroxydimethylsilyl)benzene both in the presence and in the absence of pyridine (hydrogen chloride acceptor) was investigated [87].

Initial 1,3-dichloro-1,3-diphenyl(dimethyl)-1,3-disila-2-oxaindanes were synthesized by high-temperature pyrolysis of 1,3-dichloro-1,3-dimethyl(diphenyl)-1,3-diphenyldisiloxane [88] in the inert medium in accordance with the scheme (1.45). HFC reaction of 1,3-dichloro-1,3-diphenyl-1,3-disila-2-oxaindane with dihydroxydiphenylsilane or 1,4-bis(hydroxydimethylsilyl)benzene proceeded with different ratios of reagents, which gave dihydroxy-containing oligomers with different transformation level (m = 6 - 18). Several parameters of these oligomers are shown in Table 3.14. Further on, they were used for deriving block-copolymers of the $(AB)_mC_n$ type.

HFC reaction of 1,3-dichloro-1,3-diorgano-1,3-disila-2-oxaindane with α,ω–dihydroxydimethylsiloxanes proceeded at equimolar ratio of initial components. In both cases, it proceeds in accordance with the scheme as follows:

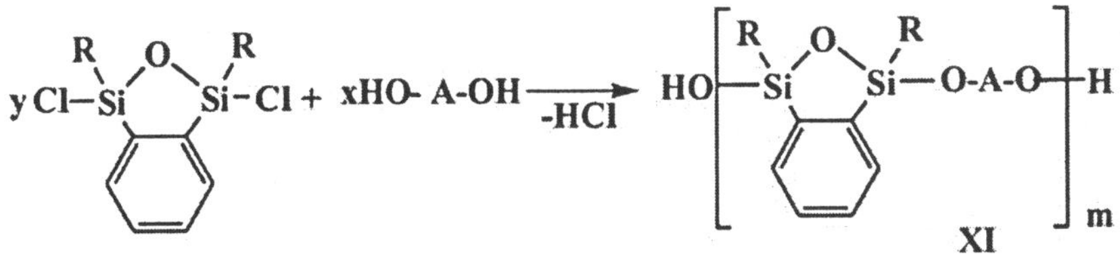

(3.13)

where A = -Me_2Si-C_6H_4-$SiMe_2$-: R = Ph, $x = y$, m = 18 (XI[1]); A = (R'_2SiO): R = R' = Ph; $x \neq y$; m = 6, 9 (XI[2]); A = (R'_2SiO): R = Me, Ph; R' = Me, ($x = y$), n = 2, 4, 8, 20, 37 (XI[3]).

If the reaction proceeds in 60 – 79% toluene solution at 20 – 25°C in the absence of acceptor, final conversion by hydrogen chloride decreases from 70% (n = 2) to 54% (n = 4) with increase of the length of dihydroxydimethylsiloxane unit.

Methyl radical substitution by phenyl one at the atom of silicon in silaindane ring also decreases conversion by hydrogen chloride from 62% (n = 2) to 43% (n = 4). High conversion by hydrogen chloride at low n of dimethylsiloxane backbone can be explained by increased reactivity of ≡Si-Cl bond, which is apparently associated [89] with a decrease of the valence angle of ≡Si-O-Si≡ bond in 1,3-disila-2-oxaindane. Lower reactivity of silaindane fragment with phenyl framing at atoms of silicon is explained by both steric effect and inductive influence of the framing groups. That is why for synthesizing high-molecular copolymers, HFC reactions were studied in the presence of pyridine. Re-precipitated copolymers are transparent or slightly opalescent products, well soluble in different organic solvents.

Several physical and chemical parameters and yields of synthesized copolymers are shown in Table 3.15.

Data shown in Table 3.15 indicate that the yields of copolymers – the products of HFC reaction without acceptor, are rather low. Moreover, the experimental values, detected by the ultimate analysis, are somewhat different from theoretical ones. This fact proves that besides basic HFC process, side homofunctional condensation of initial α,ω–dihydroxydimethylsiloxanes proceeds under the reaction conditions, which disturbs the ratio and decreases molecular mass. Slight underestimation of the yields of copolymers with disilaoxaindane fragments in the backbone at short lengths of dimethylsiloxane unit (n = 2, 4) is caused by simultaneous proceeding of intermolecular and intramolecular cyclization during HFC, the latter synthesizing two-ring compounds. To prove regular structure of obtained copolymers, copolymer 11 (Table 3.15) was fractionated into four fractions. The ultimate analysis of all four fractions gave values similar to calculated ones. ^1H NMR spectrum of the copolymer displays a singlet reflex for methyl protons with chemical shift at δ= 0.15 ppm, and the ratio of methyl and phenyl protons found coincides well with the calculated one. In turn, this proves the reaction proceeding with the formation of copolymers of regular structure. These data coincide with the experimental results [90].

Thermogravimetric analysis of the copolymers has displayed 5% mass loss at 330 - 360°C. Basic degradation process proceeds in the temperature range of 420 - 550°C, and above 620°C mass change curves become supersaturated. Thermal oxidative stability of copolymers with disilaindane fragments in the dimethylsiloxane backbone is higher than for copolymers with cyclic carbosiloxane fragments in the backbone.

Thermomechanical studies have shown that the effect of disilaindane fragment on the dimethylsiloxane backbone is negligible, and T_g of the copolymer, even in the presence of tetramethylsiloxane units, remains below zero at the methyl framing, whereas phenyl framing in copolymer 9 (Table 3.16) gives T_g = +11°C. Expansion of the linear fragment length decreases T_g down to T_g of PDMS (-123°C). Figure 3.22 shows dependence of T_g of copolymers on the content of disilaindane fragments in the backbone. Similar to the case of poly(dimethylphenylsiloxane) copolymers [91], T_g is linearly increased with concentration of non-dimethylsiloxane units in them. Note also that for copolymers 4 – 8 (Table 3.15), T_g dependence on disilaindane fragment content in them is close to that on the content of diphenylsiloxy units in poly(dimethylphenylsiloxane). For copolymers 9 – 13 (Table 3.15), increase of mol% concentration of disilaindane fragments induces sharper increase of T_g.

X-ray diffraction studies have shown that diffraction patterns of all studied copolymers possessing disilaindane fragments in the backbone display two diffraction maximums: d_1 = 7.24 - 8.81 Å and d_2 = 4.45 Å, typical of amorphous polymers. Substitution of methyl framing by phenyl one in silaindane ring is accompanied by an increase of interchain distance. Analysis of diffraction patterns obtained proves the increase of interchain distance in copolymers with concentration of disilaindane fragments in them.

Table 3.15

Physical and chemical parameters and yields of (oligomers) copolymers with 1,3-diorgano-1,3-disila-2-oxaindane fragments in the backbone of the structure XI

No.	Copolymer (oligomer) unit	R	n (m)	Yield, %	η_{spec} at 25°C	T_g, °C	d_1, Å	$M_\omega \times 10^{-3}$
1	XI^1	Ph	(19)	89	0.21	215-226	10.42	4.75
2	XI^2	Ph	(6)	88	0.07	83-88	10.41	2.99
3	XI^2	Ph	(9)	85	0.09	91-96	10.42	4.49
4	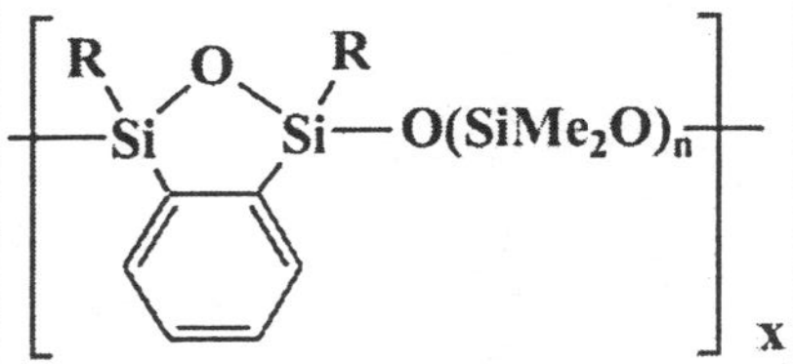	Me	2	76 (53)	0.14 (0.06)	-43 -	8.60	36.4
5		Me	4	79 (56)	0.16 (0.07)	-78 -	-	-
6		Me	8	86	0.18	-93	-	-
7		Me	20	90	0.26	-118	-	-
8		Me	37	90	0.31	-123	7.24	144.1
9		Ph	2	78 (53)	0.13 (0.05)	+11 -	8.81	29.8
10		Ph	4	80 (57)	0.15 (0.06)	-40	-	-
11		Ph	8	82	0.18	-57		87.9
12		Ph	20	86	0.29	-87	-	-
13		Ph	37	90	0.39	-123	-	-

Note: Shown in brackets are parameters of copolymers derived in the absence of acceptor.

3.4. Poly(organocarbosiloxanes) with cyclic siloxane fragments in the backbone

Besides HFC reaction of preliminarily prepared cyclic organosiloxanes with functional groups and difunctional organosilicon compounds, which give an opportunity to preserve cyclic groups in the polymeric backbone, hydride polyaddition is also widely used, which proceeds under soft conditions and does not involve cyclic structures, introduced into the backbone.

For the purpose of synthesizing poly(carbosiloxanes) with cyclic tetrasiloxane fragments in the dimethylsiloxane backbone, hydride polyaddition of divinylorganocyclosiloxane by dihydrodimethylsiloxane was studied [92]. Polymers were synthesized in argon at 1:1 molar ratio of the initial reagents in the absence of diluter or in inert organic solvent (toluene) at 100 - 110°C. The reaction temperature was selected at the level causing no scission of organosiloxane rings. Platinum-hydrochloric acid, double added to the reaction mixture in amount $1 - 1.5\times10^{-5}$ g of $H_2PtCl_6\cdot6H_2O$ per 1 g of the initial mixture, was used as the catalyst. A half of this amount was added before the reaction initiation, and the second half 25 – 140 hours after beginning of heating.

Platinum-hydrochloric acid was added in the form of 0.01 M solution in tetrahydrofurane. Isopropyl alcohol, used as diluter for $H_2PtCl_6\cdot6H_2O$, decreased relative viscosity of synthesized polymers, apparently, due to proceeding of side alkoxylation reaction:

$$\equiv Si\text{-}H + HOC_3H_7 \rightarrow \equiv Si\text{-}O\text{-}C_3H_7 + H_2 \qquad (3.14)$$

Linear poly(organocarbosiloxanes) with cyclic structures in the backbone were synthesized in accordance with the following scheme [92]:

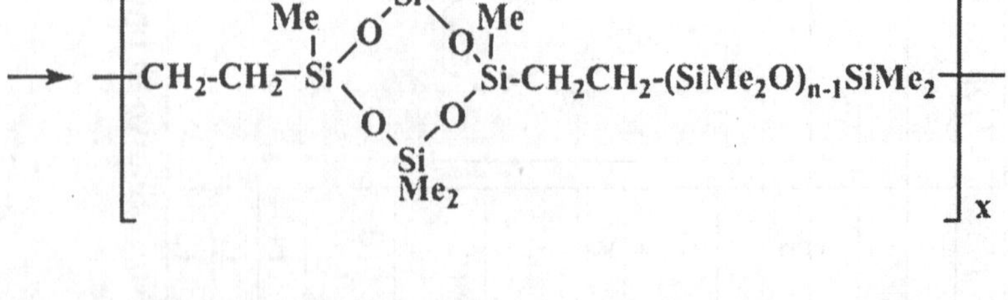

(3.15)

XII

where n = 0, 1, 4, 5, 6, 10, 20, 27, 34, 57, 94, 150, 200.

Synthesized polymers represent viscous and highly viscous colorless transparent liquids, soluble in cyclic hydrocarbons and lower eaters.

The effect of reaction proceeding in an inert organic solvent (for example, toluene) on inherent viscosity of derived polymer is negligible. Obtaining of high viscosity values in the presence of solvent requires just longer-term heating up of the reaction mixture.

To the authors' point of view, polymers of such structure, possessing cyclic fragments in the backbone, are of interest due to their high reactivity. For example, these polymers are capable of easy formation of cross-linked structures in anionic catalysts.

Initial divinylhexamethylcyclotetrasiloxane was synthesized by combined hydrolysis of dimethyldichlorosilane and methylvinyldichlorosilane. Despite the application of efficient rectification columns and analytical chromatograph with preparative add-on device, the attempts of the authors to separate isomeric 1,3- and 1,5-divinylhexamethylcyclotetrasiloxanes, which may be formed in cooperative hydrolysis, have failed. That is why isomeric structural groups as follows (XIII) may also be present in synthesized polymers:

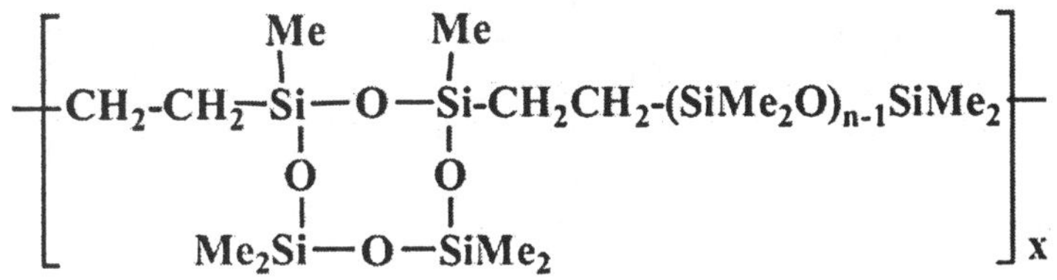

XIII

Semiquantitative assessment of the ratio of isomeric 1,3- and 1,5-cyclic structures in synthesized polymers with the help of NMR spectra was performed, which was found 1:1.

In a series of processes variation of functional groups' content (≡Si-H due to IE-spectroscopy data) during reaction proceeding and type of the increase of reaction mixture specific viscosity were studied. Maximal viscosities of polymers ($[\eta]$ = 0.17 - 0.97 dl/g) are reached after 50 – 160 hours of heating and in majority of cases depend on the length of α,ω–dihydropolydimethylsiloxane chain and purity of initial compounds used.

Studies of IR spectra of synthesized poly(organocyclocarbosiloxanes) and preliminary experiments on long-term heating of the mixture of initial hexamethyldivinylcyclotetrasiloxane isomers under polyaddition conditions allow a suggestion that polymers are synthesized due to hydride polyaddition proceeding with preservation of structures of initial compounds, but not polymerization of cyclic hexamethyldivinylcyclotetrasiloxane. The presence of organocyclotetrasiloxane fragments in the structure of synthesized poly(organocyclocarbosiloxanes) may be proved by their transition into non-fusible, insoluble state due to polymerization of organosiloxane cycles existing in the polymer structure. As re-precipitated polymers are heated at 100 - 110°C in the presence of 0.001 – 0.01 wt.% of anionic polymerization catalysts, viscosity is considerably increased and gel is formed. Varying length of alkylenesiloxane bridge between organocyclotetrasiloxane fragments of poly(organocyclocarbosiloxane) backbone, one may change the average distance between cross-link points and, consequently, properties of cross-linked polymers formed.

Hydride polyaddition between 1,5-divinyl-1,5-dimethyl-3,3,7,7-tetraorganocyclotetrasiloxane and methylphenylsilane has been studied [93]. All attempts to separate initial divinylorganocyclotetrasiloxanes into *cis*- and *trans*-isomers have failed. Thus, according to NMR data, initial divinylorganocyclotetrasiloxanes represent a mixture of *cis*- and *trans*-isomers. The reaction proceeds as follows:

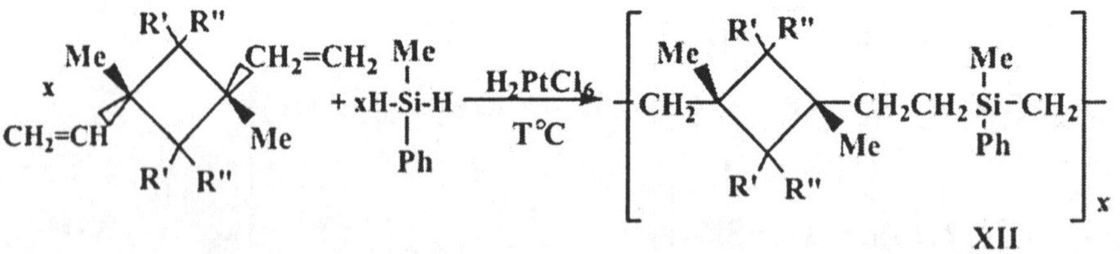

(3.16)

where R′ = R″ = Me = Ph; R′ ≠ R″.

Polyaddition was carried out at 60 – 70°C, and at the final stage the mixture was heated up to 100°C. The catalyst in amount 5×10^{-4} Pt mol/mol was added to vinylcyclosiloxane, heated up to 50°C. Some parameters of synthesized copolymers are shown in Table 3.16.

For the purpose of synthesizing carbosiloxane copolymers with organocyclopentasiloxane fragments in the dimethylsiloxane backbone, hydride polyaddition of α,ω-dihydridedimethylsiloxanes to 1,5-divinyl-1,5-dimethylhexaphenylcyclopentasiloxane in the presence of platinum hydrochloric acid as a catalyst was studied at temperatures below 100°C: 75°C, 80°C and 85°C. Forasmuch as the initial 1,5-divinyl-1,5-dimethylhexaphenylcyclopentasiloxane represents a mixture of *cis*- and *trans*-isomers (at the ratio 52:48), which cannot be separated by preparative methods, copolymers derived from them are atactic. Preliminary heating of initial compounds within the temperature range of 80 - 95°C in the presence of the catalyst indicated that under these conditions organocyclopentasiloxane fragments are not polymerized.

Table 3.16

Physical and chemical parameters of poly(organocarbosiloxane) copolymers of cyclolinear structure

No.	Copolymer	R′	R″	[η], dl/g	T_{degr}* of 5% mass loss	Coke residue, (800°C), %	T_g, °C
1	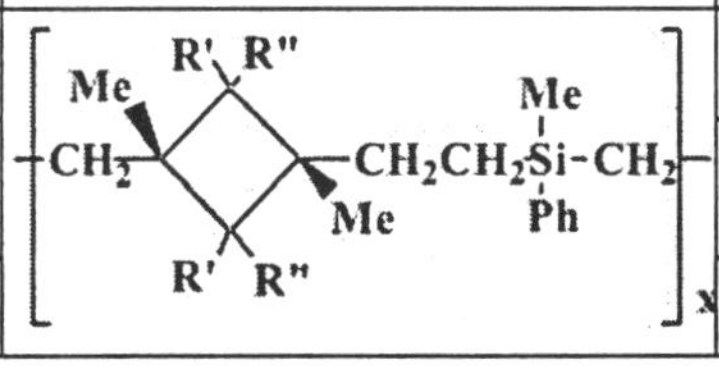	Me	Me	0.08	240	52	-7
2		Me	Ph	0.06	320	45	26
3		Ph	Ph	0.04	370	41	13

Note: * TGA data on polymers processed by heptamethylvinylcyclotetrasiloxane.
Copolymer structure was determined from ^{29}Si NMR spectral data.

The reaction proceeding was detected by a decrease of amount of active ≡Si-H groups. It was observed that the rate and depth of polyaddition decrease with the increase of α,ω-dihydridedimethylsiloxane chain length. Hydride polyaddition proceeds in accordance with the following scheme [94, 95]:

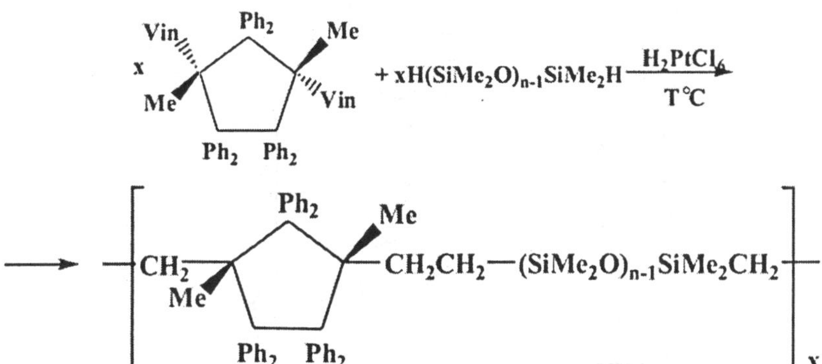

XIV (3.17)

where $n = 2 - 23$.

Table 3.17

Physical and chemical parameters of structure XIV carbosiloxane copolymers containing cyclopentasiloxane fragments

No.	Copolymer	n_{SiO}	Yield, %	T, °C	η_{spec}*	T_g, °C	d_1, Å	5% mass loss	$\bar{M}_\omega \times 10^{-3}$
1	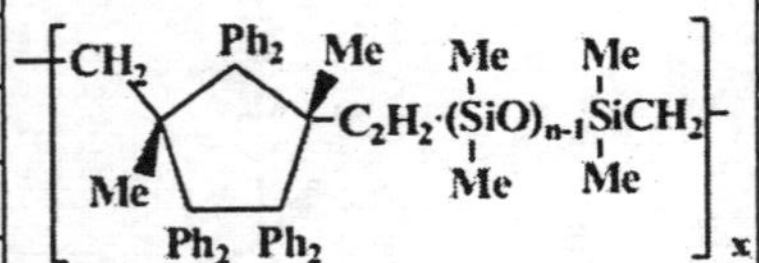	2	75	85	0.09	0÷-2	9.20	320	189
2		4	80	85	0.14	-22	-	-	-
3		6	92	75	0.15	-	-	-	-
3′		6	93	80	0.18	-	-	-	-
3″		6	95	85	0.20	-53	-	295	211
4		12	95	85	0.24	-82	-	-	-
5		23	96	85	0.31	-123	7.21	285	236

Note: * In toluene at 25°C.

As a result of the reaction, copolymers with $\eta_{spec} = 0.09 - 0.26$ are obtained, which are liquid or glassy light-yellow products, soluble in ordinary organic solvents. Some physical and chemical parameters and the yield of copolymers are listed in Table 3.17. As indicated by the data in the Table, in the case of short lengths of the dimethylsiloxane backbone, *n*, the yield of copolymers is low. This may be explained by the fact that besides intermolecular reaction, intramolecular cyclization proceeds forming a polycyclic structure. This conclusion is in agreement with data from the literature [12 - 17].

The amount of active ≡Si-H groups was decreased during proceeding of hydride polyaddition. Figure 3.22 shows that the rate of hydride polyaddition increases with temperature (at one and the same values of dimethylsiloxane units, *n*), but on the other hand, with an increase of the length of dimethylsiloxane links (*n*) at the same temperatures, the rate of hydride polyaddition decreases. Figure 3.22 shows that conversion of active ≡Si-H groups is not complete and decreases from 20% ($n = 6$) to 15% ($n = 12$).

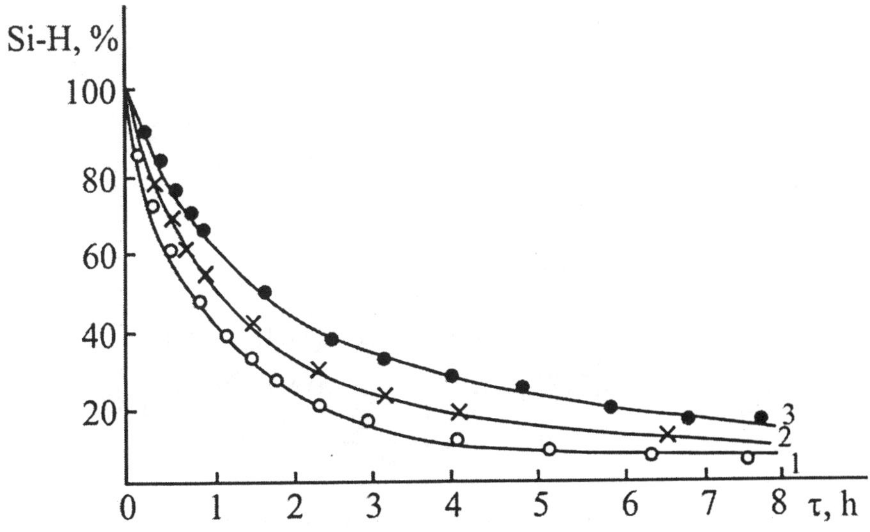

Figure 3.22. Time dependence of changes in active ≡Si-H% groups during polyaddition of α,ω-dihydridedimethylsiloxane ($n = 6$) with 1,5-divinyl-1,5-dimethylhexaphenylcyclopentasiloxane: 1 - at 85°C; 2 - at 80°C; 3 - at 75°C

It was found that polyaddition is the second order reaction. The reaction rate constants and the activation energy were calculated: $k_{75°C} = 1.4004\times10^{-2}$, $k_{80°C} = 1.965\times10^{-2}$, $k_{85°C} = 2.559\times10^{-2}$; $E_a = 62.1$ kJ/mol, respectively.

^1H NMR spectra of copolymers indicate that catalytic hydride polyaddition mainly proceeds by the Farmer rule with formation of dimethylenic bridges. In these spectra a reflex of $-CH_2-CH_2-$ group with chemical shift δ = 0.34 ppm is observed; it is indicated that hydride polyaddition partly (about 6 - 7%) proceed by the Markovnikov rule.

Cyclolinear carbosiloxane copolymers with 1,7- and 1,5-disposition of dimethyloctaphenylcyclohexasiloxane fragments in the dimethylsiloxane backbone were synthesized by hydride polyaddition of α,ω-dihydridedimethylsiloxane to 1,7-divinyl-1,7-dimethyloctaphenylcyclo-hexasiloxane and 1,5-divinyl-1,5-dimethyloctaphenylcyclohexasiloxane in the presence of a catalyst. Polyaddition reactions were studied below 100°C. It was also indicated that under these conditions polymerization or polycondensation of initial compounds does not take place. Polyaddition proceeds in accordance with the following scheme [95 - 97]:

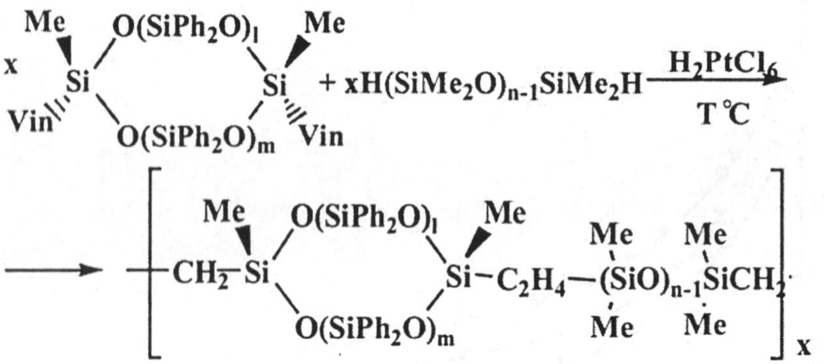

(3.18)

XV, XVI

where $m = l = 2$ (XV); $n = 2 - 23$; $l = 1$, $m = 3$ (XVI); $n = 2$ - 23.

Forasmuch as 1,7- and 1,5-divinylcyclohexasiloxanes, used in polyaddition, represent mixtures of *cis*- and *trans*-isomers of the approximate 52:48 ratio, synthesized copolymers are atactic.

Re-precipitation of copolymers from toluene solution by methyl alcohol has given viscous or solid (with regard to the value of flexible junction) transparent products with η_{spec} = 0.09 - 0.29, well soluble in different organic solvents. It is found that at short length of dimethylsiloxane unit ($n \leq 4$), copolymer yields are slightly decreased that may be explained by partial proceeding of hydride polyaddition by intramolecular cyclization mechanism (see Tables 3.18 and 3.19).

After solvent removal from the mother solution of re-precipitated copolymer 1 (Table 3.18), a semi-crystalline compound with the molecular mass equal ~1,100 was obtained [96, 97]. The product of intramolecular cyclization of 1,7-divinyl-1,7-dimethyloctaphenylcyclohexasiloxane and 1,3-dihydridetetramethyldisiloxane of the following structure only may display the current molecular mass:

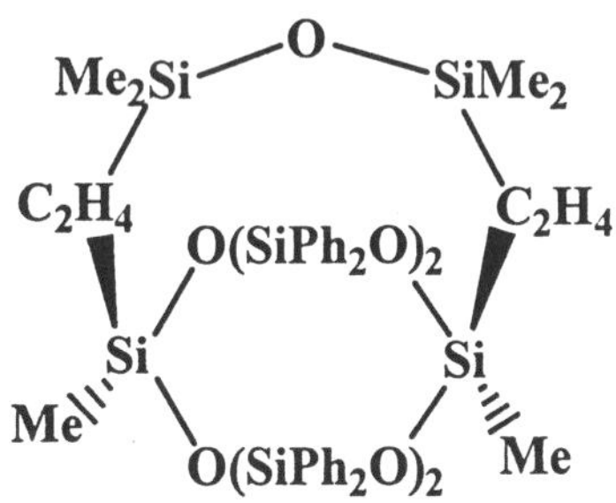

XVII

,

because divinylorganocyclohexasiloxane of the *trans*-structure participates in formation of macromolecular chain.

Structure and composition of synthesized cyclolinear carbosiloxane copolymers were determined by functional and ultimate analysis, IR and NMR spectral data. Some parameters of copolymers are shown in Tables 3.18 and 3.19.

A reflex with chemical shift at δ = 0.35 ppm typical of $-CH_2-CH_2-$ group is observed in ^1H NMR spectrum of copolymer 1 (Table 3.18). This indicates that polyaddition proceeds pursuant to the Farmer rule. A duplet centered at δ= 1.06 ppm, corresponded to methyl protons in $=CH-CH_3$ group, is also observed in the spectrum. Based on the ratio of intensities, it was concluded [96, 97] that polyaddition partly proceeds by the Markovnikov mechanism (6 – 8%). A complex multiplet with chemical shift at δ = 5.6 - 6.2 ppm typical of vinyl protons not entered polyaddition reaction, and a singlet for ≡Si-H protons with chemical shift at δ = 4.4 ppm, not participated in the reaction, too, were observed in the spectra.

Hydride polyaddition proceeded at different temperatures. Figures 3.23 and 3.24 show variations of ≡Si-H bond concentration during polyaddition of

α,ω–dihydridedimethylsiloxane (n = 6) to 1,7-divinyl-1,7-dimethyloctaphenylcyclohexasiloxane and 1,5-divinyl-1,5-dimethyloctaphenylcyclohexasiloxane. It is observed that hydride polyaddition depth is increased with the reaction temperature. Moreover, the effect of 1,7- or 1,5-disposition of vinyl groups in cyclohexasiloxane fragments is the negligible factor for their reactivity.

It is found that at the initial stages, polyaddition represents the second order reaction. In the case of 1,7-divinyl-1,7-dimethyloctaphenylcyclohexasiloxane, polyaddition rate constants for different temperatures were determined as follows: $k_{90°C}$ = 3.0797×10^{-2}; $k_{85°C}$ = 2.3007×10^{-2}; $k_{80°C}$ = 1.6781×10^{-2}. Activation energies for 1,7-divinyl-1,7-dimethyloctaphenylcyclohexasiloxane and 1,5-divinyl-1,5-dimethyloctaphenylcyclohexasiloxane were also calculated: E_a = 66.7 and E_a = 69.7 kJ/mol, respectively. Obviously, these values are very close.

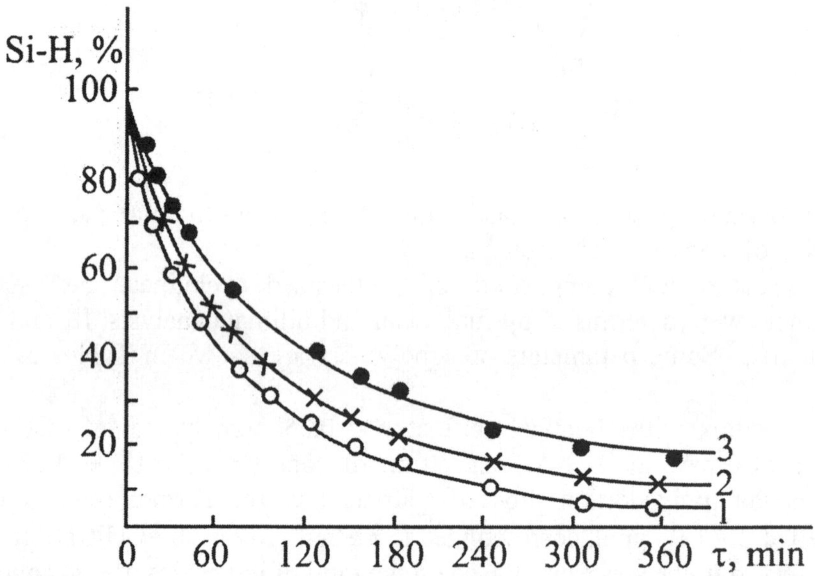

Figure 3.23. Decrease of ≡Si-H bond concentration during hydride polyaddition of α,ω–dihydridedimethylsiloxane (n = 6) to 1,7-divinyl-1,7-dimethyloctaphenylcyclohexasiloxane: 1 - 90°C; 2 - 85°C; 3 - 80°C

X-ray diffraction studies have indicated that copolymers are single-phase amorphous systems, and maximal interchain distance is observed for short dimethylsiloxane unit length ($n = 2$); hence, for copolymer 1 (Table 3.19), $d_1 = 9.60$ Å. This value is slightly greater than the interchain distance of carbosiloxane copolymer 1 (Table 3.18) with 1,7-disposition of cyclohexasiloxane fragment in the dimethylsiloxane backbone ($n = 2$). As flexible junction length is increased, d_1 decreases and approaches the interchain distance in PDMS; it increases with the volume of cyclic fragment at the same lengths of flexible dimethylsiloxane unit, i.e. at transition from cyclopentasiloxane to cyclohexasiloxane fragment.

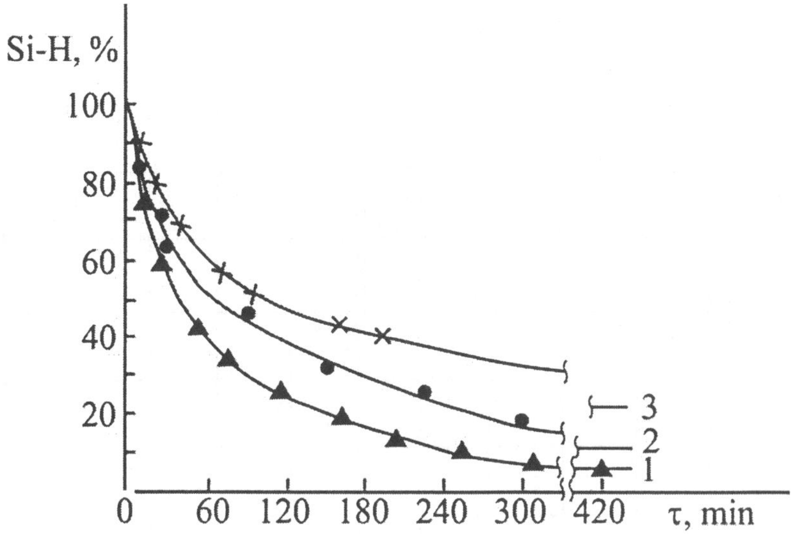

Figure 3.24. Decrease of ≡Si-H bond concentration during hydride polyaddition of α,ω-dihydridedimethylsiloxane ($n = 6$) to 1,5-divinyl-1,5-dimethyloctaphenylcyclohexasiloxane: 1 - 100°C; 2 - 90°C; 3 - 80°C

Thermogravimetric studies of carbosiloxane copolymers have indicated 5% mass loss of the compounds in the temperature range of 250 - 260°C. The main degradation process proceeds in the range of 380 - 630°C, and above 700°C the mass loss is not observed. It is found that thermal oxidative stability of copolymers is decreased with increase of the cyclic fragment volume, i.e. at the transition from cyclic pentasiloxane to hexasiloxane fragments in cyclolinear carbosiloxane copolymer. It is also found that carbosiloxane copolymers with 1,7- and 1,5-disposition of cyclic hexasiloxane fragments in the backbone are characterized by almost identical thermal oxidative stability. Thus, is was concluded [96, 97] that the effect of 1,7- or 1,5-disposition of cyclic hexasiloxane fragment in carbosiloxane copolymer is negligible for thermal oxidative stability of copolymers (Figure 3.25). On the other hand, compared with pure siloxane analogues, thermal oxidative stability of carbosiloxane copolymers is lower.

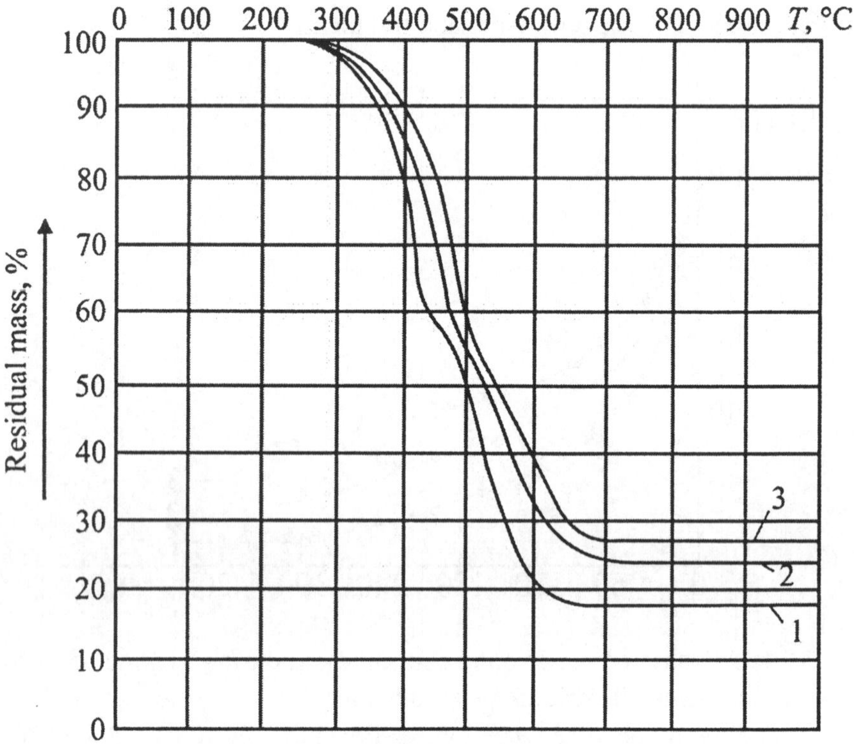

Figure 3.25. Thermogravimetric curves of carbosiloxane copolymers: 1 – copolymer 4 (Table 3.19) with 1,5-disposition of cyclic hexasiloxane fragment in the backbone; 2 – copolymer 1 (Table 3.18) with 1,7-disposition of cyclic hexasiloxane fragment in the backbone; 3 – copolymer 1 (Table 3.17) with cyclic pentasiloxane fragment in the backbone

Thermogravimetric studies have displayed that the cyclic fragment causes a considerable effect on carbosiloxane copolymer at $n = 12$ only, and at $n = 23$ no effect of cyclic fragment on the glass transition temperature of the copolymer is observed.

Figure 3.26 shows dependence of T_g on the length of dimethylsiloxane unit for cyclolinear carbosiloxane copolymers.

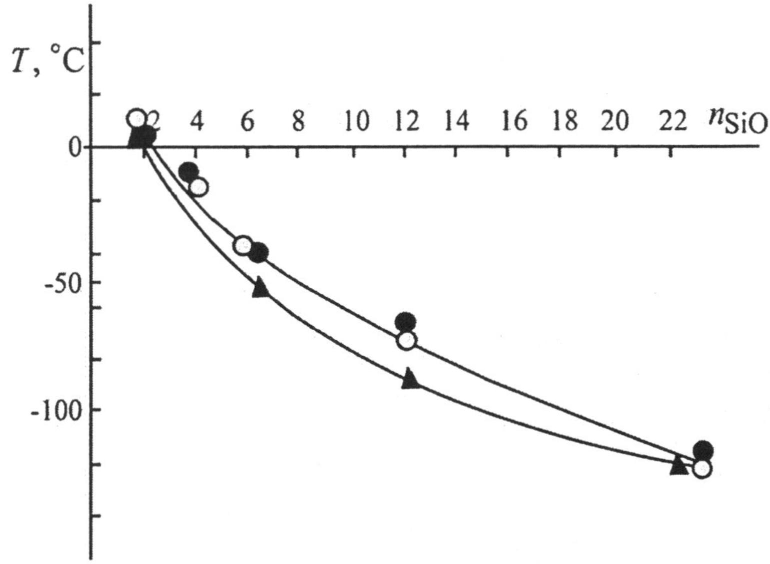

Figure 3.26. Dependence of T_g for cuclolinear carbosiloxane sopolymers on the length of dimethylsiloxane unit: 1 – copolymer with 1,7-position of cyclic hexasiloxane fragment; 2 – copolymer with 1,5-position of cyclic hexasiloxane fragment

It has been found that expansion of the cyclic fragment volume at the same length of dimethylsiloxane unit, i.e. introduction of a single diphenylsiloxane unit, T_g of the copolymer is increased by ~10°C. It is also shown that the effect of 1,7- or 1,5-disposition of cyclic hexasiloxane fragment on T_g of the copolymer is negligible, which conform to the previous results on pure siloxane copolymers.

Table 3.18

Physical and chemical parameters of carbosiloxane copolymers with 1,7-disposition of cyclic hexasiloxane fragments in the dimethylsiloxane backbone (structure XV)

No.	Copolymer	n_{SiO}	Yield, %	T, °C	η_{spec}*	T_g, °C	d_1, Å	5% mass loss	$\bar{M}_\omega \times 10^{-3}$
1	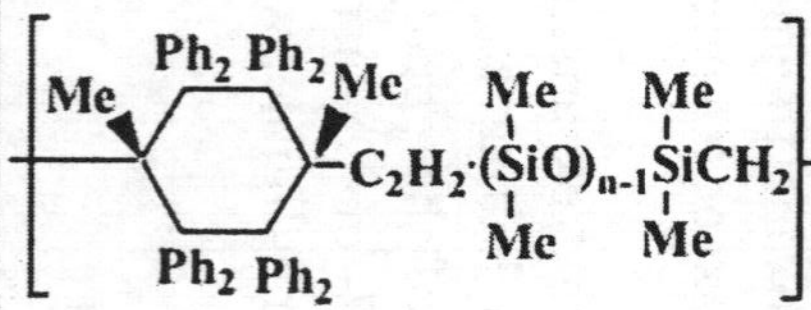	2	74	90	0.10	+5	9.31	280	174
2		4	80	90	0.12	-10	-	-	-
3		6	94	80	0.17	-	-	-	-
3′		6	94	80	0.17	-	-	-	-
3″		6	95	85	0.18	-40	8.81	280	194
4		12	96	90	0.23	-68	8.40	-	-
5		23	95	90	0.29	-123	7.24	260	231

Note: * In toluene at 25°C.

Table 3.19

Physical and chemical parameters of carbosiloxane copolymers with 1,5–position of cyclic hexasiloxane fragments in the dimethylsiloxane backbone (structure XVI)

No.	Copolymer	n_{SiO}	Yield, %	T, °C	η_{spec}*	T_g, °C	d_1, Å	5% mass loss	$\bar{M}_{\omega}\times10^{-3}$
1	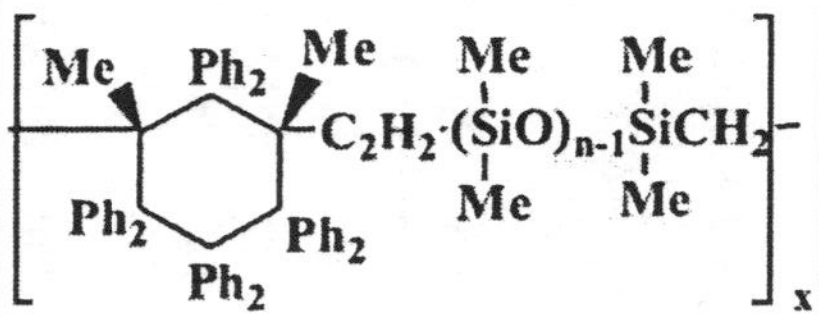	2	72	10	0.09	+8	9.60	270	159
2		4	84	85	0.11	-12	-	-	-
3		6	86	80	0.15	-	-	-	-
3′		6	89	90	0.18	-	-	-	-
3″		6	94	100	0.15	-38	8.90	265	180
4		12	95	100	0.22	-72	8.34	-	210
5		23	95	100	0.28	-123	-	260	-

Note: * In toluene at 25°C.

3.5. Cyclolinear copolymers derived by polymerization of organobicyclo- and organotricyclosiloxanes

The above-mentioned materials indicate that cyclolinear organosiloxane copolymers are usually synthesized by homopolycondensation and HFC reactions of difunctional organocyclosiloxanes or organopoly(cyclosiloxanes) with α,ω–dihydroxy(diamino, dichloro)diorganosiloxanes [99 – 102]. As known from the literature, polymerization and copolymerization of organopoly(cyclosiloxanes) not always give soluble polymers and their yield is usually low [103].

3.5.1. Synthesis of organodi- and organotricyclosiloxanes with labile cyclic fragments

To synthesize organodi- and organotricyclosiloxanes containing thermodynamically and sterically labile rings, step HFC reaction of cis-1,3,5,7-tetrahydroxy-1,3,5,7-tetraphenylcyclotetrasiloxane with dimethyldichlorosilane and trimethylchlorosilane at 1:1:2 ratio of initial components was studied in the presence pyridine in 5% absolute ester. The reaction was performed in two stages. At the first stage, tetrole solution was added by dimethyldichlorosilane and a part pyridine, and at the second stage by trimethylchlorosilane and the rest of pyridine; at the final stage, the mixture was boiled and continuously mixed, simultaneously. The reaction proceeded in accordance with the scheme as follows:

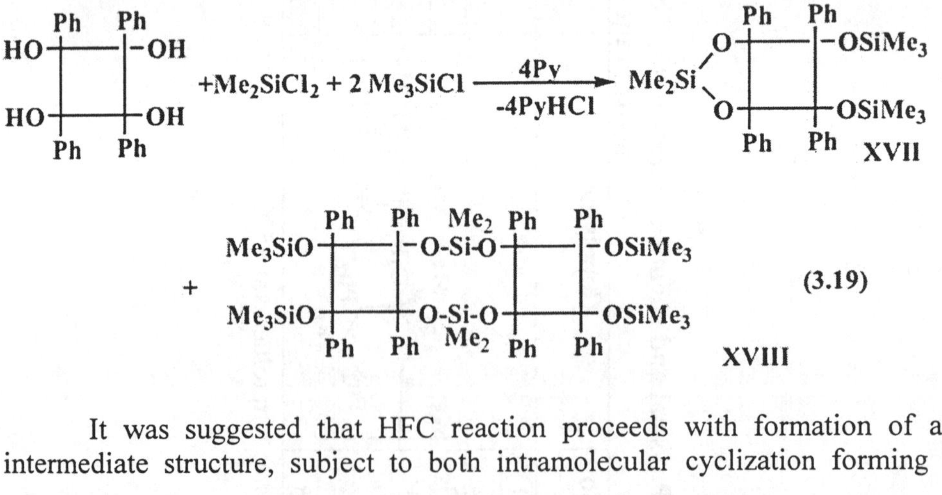

(3.19)

It was suggested that HFC reaction proceeds with formation of an intermediate structure, subject to both intramolecular cyclization forming a dicyclic compound of the structure XVII and intermolecular condensation forming a dicyclic compound of the structure XVIII.

Similarly, the step HFC reaction of tetrole with 1,3-dichlorotetramethylsiciloxane and trimethylchlorosilane (1:1:2 ratio of initial reagents) under the same reaction conditions was studied. In this case, the reaction is supposed proceeding via formation of a transitional structure, and as a result of intra- and intermolecular cyclization both di- and tricyclic compounds are formed in accordance with the scheme as follows [104]:

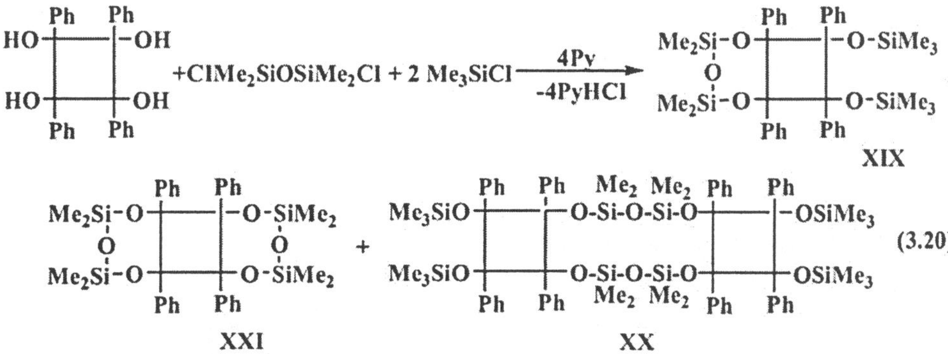

(3.20)

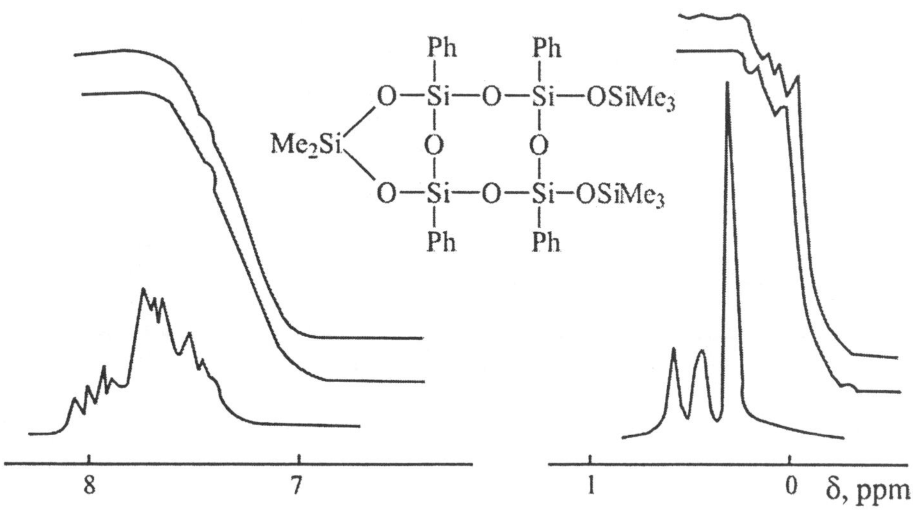

Figure 3.27. ^{1}H NMR spectrum of dicyclic compound 1 with structure XVII (Table 3.19)

All synthesized organodi- and organotricyclic compounds represent dense transparent oil-like liquids, crystallized in calm. Tricyclic compound 5 of the structure XXI was separated during re-crystallization of compound 3 of the structure XIX (Table 3.20).

Composition and structure of obtained compounds was detected by the ultimate analysis, determination of the molecular mass and from IR and ^1H NMR spectral data. Yields and some physical and chemical parameters of synthesized di- and tricyclic organosiloxanes are shown in Table 3.20. Figure 3.27 shows ^1H NMR spectrum of dicyclic compound 1 of the structure XVII (Table 3.20) displaying a triplet with the chemical shifts at $\delta = 0.20$, 0.33 and 0.46 ppm, and a complex multiplet in the range of 7.2 – 8.4 ppm.

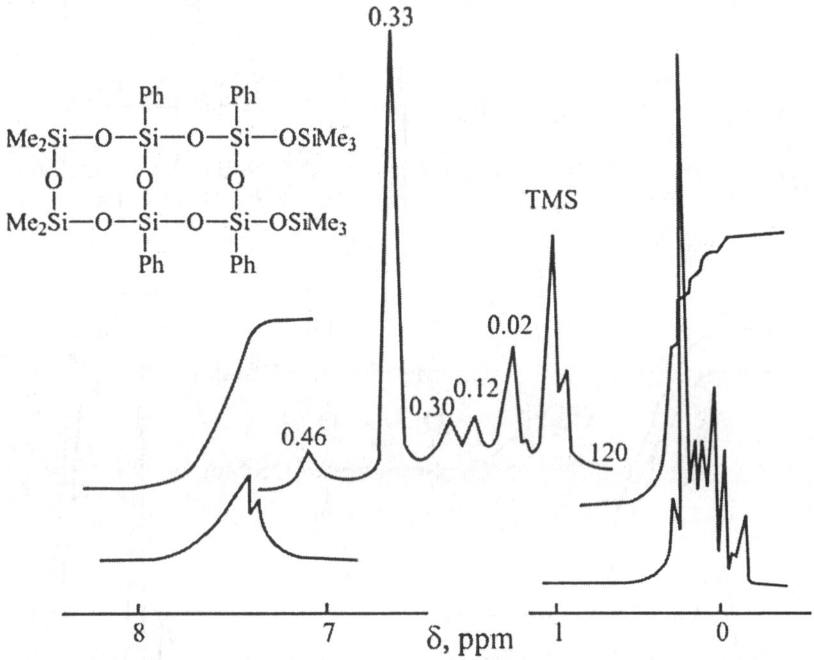

Figure 3.28. ^1H NMR spectrum of dicyclic compound 3 with structure XIX (Table 3.20)

Figure 3.28 shows ^1H NMR spectrum of dicyclic compound 3 with structure XIX (Table 3.20) containing a multiplet with chemical shifts at $\delta =$ 0.02, 0.12, 0.30, 0.33 and 0.46 ppm, corresponded to methyl protons.

Synthesized organodi- and organotricyclosiloxanes contain a single labile cycle each. A hexatomic cyclic fragment is more labile in dicyclic compound 1 (structure XVII), because it is more stressed than octatomic one, and the more so, the latter is framed by trimethylsiloxy- and phenyl groups, which naturally makes steric hindrances to polymerization. The compounds contain octa-, dodeca- and hexatomic cyclic fragments, framed by dimethyl radicals, which makes them accessible for impact of nucleophilic catalysts during polymerization.

Table 3.20

Physical and chemical parameters of di- and tricyclic organosiloxanes

No.	Structure	Yield, %	T_{boil}, °C (P, mm Hg)	T_{melt}, °C
1	XVII	43	260 - 268 (6×10^{-3})	100 - 102
2	XVIII	21	320 - 327 (6×10^{-3})	-
3	XIX	50	262 - 270 (6×10^{-3})	116 - 117
4	XX	24	330 - 335 (6×10^{-3})	-
5	XXI	5	-	128 - 129

3.5.2. Polymerization of cyclolinear organosiloxane copolymers with cyclic fragments in the backbone

Synthesized bi- and tricyclic organosiloxane compounds were used in the polymerization reaction, which was performed in the presence of α,ω–tetramethylammonium oxydimethylsiloxanolate (n = 11) and 0.1 H KOH alcohol solution in the temperature range of 50 - 200°C. It was found that the reaction proceeding at low temperature (< 100°C) gives low-molecular oligomers (η_{spec} = 0.04). The reaction temperature increase over 140°C induces decomposition of the catalyst mentioned. That is why future studies of polymerization of tricyclic organodi- and organotrisiloxanes were carried out in 0.1 H alcohol solution of potassium hydroxide.

Di- and tricyclic compounds were polymerized with formation of low-molecular cyclolinear copolymers:

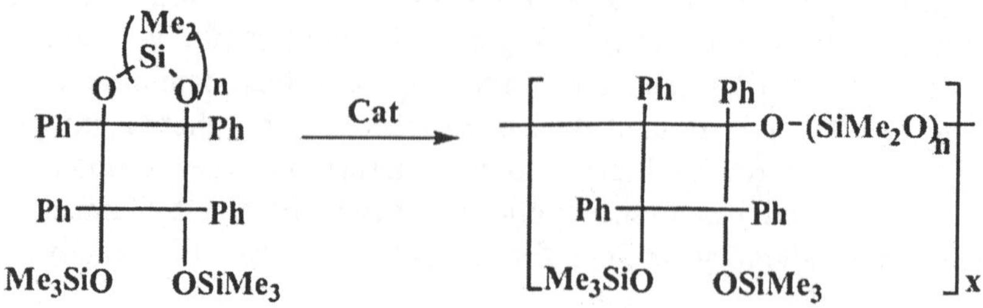

(3.21)

XXII

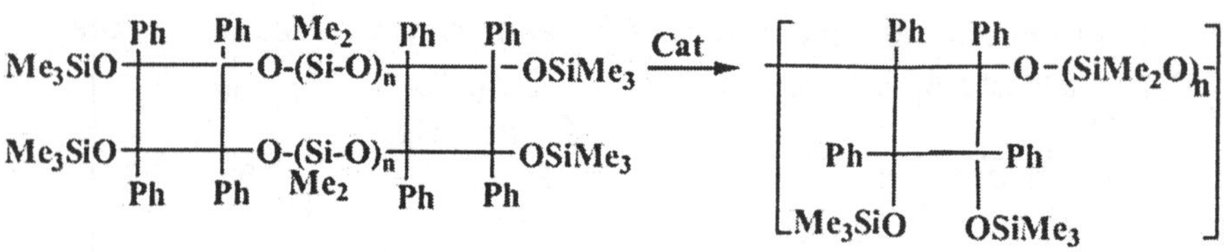

(3.22)

XXII

where $n = 1, 2$.

Synthesized regular copolymers represent light-yellow solids with $\eta_{spec} = 0.4 - 0.7$, well soluble in various organic solvents. It is found that the effect of increasing catalyst concentration (0.1 – 0.5 wt.% of initial compound) and reaction temperature on the molecular mass of low-molecular compounds obtained is negligible.

Thermogravimetric analysis has indicated mass losses of copolymers at 260 - 270°C not exceeding 2 – 3%. At 300 – 350°C thermal degradation rate is considerably increased, and at 550°C mass losses are maximal. The mass loss is regularly increased with the length of dimethylsiloxane backbone, i.e. at transition from dimethylsiloxane to disiloxane bridge.

Thus, synthesis of cyclolinear organosiloxane and carbosiloxane copolymers with monocyclic fragments in the backbone, obtained in HFC, hydride polyaddition and polymerization reactions, are considered in this Chapter. The above-shown data on high-molecular polymers with regular disposition of cyclic fragment in the backbone are mainly synthesized by HFC and hydride polyaddition reactions. As for polymerization of bicyclosiloxanes, they are almost useless for high polymer synthesis.

References

1. Andrianov K.A., *Doklady AN SSSR*, 1963, vol. **151**(5), p. 1093. (Rus)
2. Andrianov K.A., Talanov V.N., and Khananashvili L.M., *Izv. AN SSSR, Ser. Khim.*, 1969, No. 9, p. 2004. (Rus)
3. Mileshkevich V.P., *Kauchuk i Resina*, 1978, No. 6, p. 4. (Rus)
4. Thomas T.H. and Kendric T.C., *J. Polym. Sci.*, 1969, vol. **2A**(7), p. 537.
5. Alexander G. and Garzo G., *Chromatography*, 1974, vol. **7**(4), p. 225.
6. Andrianov K.A., Jenchelskaya S.M., and Petrashko Yu.K., *Zh. Obshch. Khim.*, 1958, vol. **28**(3), p. 685. (Rus)
7. Sokolov N.N., *Methods of Poly(organosiloxane) Synthesis*, Moscow – Leningrad, Gosnergoizdat, 1959, p. 103. (Rus)
8. Andrianov K.A., *Polymers With Inorganic Backbones of Macromolecules*, Moscow, Izd. AN SSSR, 1962, p. 144. (Rus)
9. Andrianov K.A., Jenchelskaya S.M., and Petrashko Yu.K., *Plast. Massy*, 1960, No. 3, p. 20. (Rus)
10. Volkova L.M. and Andrianov K.A., *Vysokomol. Soedin.*, 1964, vol. **6**(9), p. 1662. (Rus)
11. Andrianov K.A., Khanashvili L.M., and Zavalny V.G., *Plast. Massy*, 1967, No. 6, p. 48. (Rus)
12. Andrianov K.A., Khanashvili L.M., Makarova N.N., Mukbaniani O.V., and Meladze S.M., *A.C. No. 791,757* (USSR), *Bull. Izobr.*, 1981, No. 4. (Rus)
13. Andrianov K.A., Nogaideli A.I., Slonimski G.L., Levin V.Yu., Makarova N.N., Kvachev Yu.P., and Mukbaniani O.V., *Vysokomol. Soedin.*, 1976, vol. **13B**(1), p. 359. (Rus)
14. Mukbaniani O.V., *Organosiloxane Copolymers and Block-copolymers With Different Cyclic Structure of Macromolecules*, Doctor's Dissertation on Chemistry, 1993, Tbilisi State University, Tbilisi, Georgia. (Rus)
15. Koiava N.A., Mukbaniani O.V., and Khanashvili L.M., *Vysokomol. Soedin.*, 1985, vol. **27A**(11), p. 2261. (Rus)
16. Koiava N.A., Mukbaniani O.V., and Khananashvili L.M., *Abstr. Commun. 27th Intern. Symp. Macromol.*, Strasburg, 1981, p. 18.
17. Mukbaniani O.V., Meladze S.M., and Khananashvili L.M., *Vysokomol. Soedin.*, 1984, vol. **24B**(4), p. 250. (Rus)
18. Andrianov K.A., Nogaideli A.I., Makarova N.N., and Mukbaniani O.V., *Izv. AN SSSR, Ser. Khim.*, 1977, No. 6, p. 1388. (Rus)
19. Zhdanov A.A. and Astapova T.V., *Vysokomol. Soedin.*, 1981, vol. **23A**(3), p. 626. (Rus)

20. Yuzhelevski Yu.A., Kurlova T.V., Kogan B.G., and Suvorova M.V., *Zh. Obshch. Khim.*, 1972, vol. **42**(9), p. 2006. (Rus)
21. Mukbaniani O.V., *Synthesis of Polycyclic Organosilicon Compounds and Block-Copolymers on Their Basis*, Candidate's Dissertation on Chemistry, 1976, Tbilisi State University, Tbilisi, Georgia. (Rus)
22. Bostick E.E., *Polymer Prep.*, 1969, vol. **10**(2), p. 877.
23. Tsetlin B.L., Gavrilov V.I., Velikovskaya N.A., and Kochkin V.V., *Zavodskaya Laboratoria*, 1956, vol. **22**(3), p. 352. (Rus)
24. Andrianov K.A., Slonimsky G.L., Levin V.Yu., Zhdanov A.A., and Godovsky Yu.K., *J. Polym. Sci.*, 1972, vol. **10**(A-1), p. 1.
25. Andrianov K.A., Slonimsky G.L., Levin V.Yu., Zhdanov A.A., and Godovsky Yu.K., *J. Polym. Sci.*, 1972, vol. **10**(A-1), p. 23.
26. Zhdanov A.A., Andrianov K.A., Malyikhin A.P., and Emelianov V.N., *Izv. AN SSSR, Ser. Khim.*, 1973, No. 10, p. 2299. (Rus)
27. Andrianov K.A., Slonimski G.L., Zhdanov A.A., and Levin V.Yu., *Vysokomol. Soedin.*, 1968, vol. **10A**(9), p. 2102. (Rus)
28. Andrianov K.A., Slonimski G.L., Levin V.Yu., Godovski Yu.K., Kuznetsova N.K., Tsvankin D.Ya., Moskalenko V.A., and Kuteinikova L.N., *Vysokomol. Soedin.*, 1970, vol. **12A**(6), p. 1268. (Rus)
29. Khyinku E.S., Zhukov V.P., Gerasimov M.V., Il'ina M.N., Kvachev Yu.P., and Papkov V.S., *Thes. Proc. VII All-Union Conference on Chemistry, Production Technology and Application of Organosilicon Compounds*, 1990, November 20 – 23, Tbilisi, Part 1, p. 183. (Rus)
30. Andrianov K.A., Zhdanov A.A., Zavin B.G., and Evdokimov A.M., *Doklady AN SSSR*, 1971, vol. **199**(3), p. 597. (Rus)
31. Andrianov K.A., Zhdanov A.A., Zavin B.G., Evdokimov A.M., and Biryukova T.V., *Vysokomol. Soedin.*, 1972, vol. **14B**(1), p. 327. (Rus)
32. Buch R., Klimisch H., and Johnson O., *J. Polym. Sci.*, 1970, vol. **8(A-2)**, No. 4, p. 541.
33. Erenburg V.E., Kartasheva G.G., Eremina M.A., and Poddubni I.Ya., *Vysokomol. Soedin.*, 1967, vol. **9A**(12), p. 2709. (Rus)
34. Skazka V.S. and Shaltyiko L.G., *Vysokomol. Soedin.*, 1960, vol. **2**(4), p. 572. (Rus)
35. Andrianov K.A., Pavlova S.A., Tverdovhlebova I.I., Pertsova N.V., and Larina T.A., *Abstr. Commun., Prague IUPAC Microsimposium on Macromolecules. Thermodinamics of Interactions in Polymer Solutions*, 1971, September 6 - 9, p. 116.

36. Tverdokhlebova I.M., Kurginyan P.A., Larina T.A., Makarova N.N., Ronova I.A., Pavlova S.S., and Mukbaniani O.V., *Vysokomol. Soedin.*, 1981, vol. **23A**(5), p. 995. (Rus)
37. Tverdokhlebova I.M., *Studies in the Sphere of Molecular and Conformational Transformations of Poly(organosiloxanes)*, Doctor's Dissertation on Chemistry, Moscow, 1978, INEOS, AS USSR. (Rus)
38. Flory P.J., Fox T.G., and Shaefgen J.R., *J. Amer. Chem. Soc.*, 1951, vol. **73**(5), p. 1904.
39. Inagaki H., Suzuki H., and Kurata M., *J. Polym. Sci.*, 1966, vol. **15**, p. 409.
40. Stockmayer W.H. and Fixmann M., *J. Polym. Sci.*, 1963, vol. **1**(1), p. 137.
41. Flory P.J., Gescenzi V., and Mark J., *J. Amer. Chem. Soc.*, 1964, vol. **86**(3), p. 146.
42. Bohdanecky M., *Collect Czech. Chem. Commun.*, 1968, vol. **33**(12), p. 4397.
43. Kartasheva G.G., Erenburg V.E., and Poddubni I.Ya., *Vysokomol. Soedin.*, 1972, vol. **14B**, p. 665. (Rus)
44. Mark J.E. and Flory P.J., *J. Amer. Chem. Soc.*, 1964, vol. **86**(1), p. 138.
45. Tarasova G.I., *Fractionation of Cyclic Polyesters. Chemistry and Technology of High-molecular Compounds*, Coll., Comp. Teplyakov M.S., Ed. V.V. Korshak, Moscow, VINITI, 1973, vol. **4**, p. 109. (Rus)
46. Severni V.V., *Organosiloxanes' Reaction with Alkyl(aryl)chlorosilanes*, Doctor's Dissertation on Chemistry, 1962, Moscow, INEOS AS USSR. (Rus)
47. Sobolevski M.V., Skorokhodov I.I., Ditsent V.E., Sobolevskaya A.V., and Efimov V.M., *Vysokomol. Soedin.*, 1969, vol. **11A**(5), p. 1109. (Rus)
48. Andrianov K.A., Pavlova S.A., Zhuravleva I.V., Tolchinski Yu.I., and Astapova B.A., *Vysokomol. Soedin.*, 1977, vol. **19A**(4), p. 896. (Rus)
49. Andrianov K.A., Pavlova S.A., Zhuravleva I.V., Tolchinski Yu.I., Makarova N.N., and Mukbaniani O.V., *Vysokomol. Soedin.*, 1977, vol. **19A**(4), p. 1387. (Rus)
50. Andrianov K.A., Papkov V.S., Zhdanov A.A., and Yakushina S.E., *Vysokomol. Soedin.*, 1969, vol. **11A**(9), p. 2030. (Rus)
51. Pathnode W. and Wilcock D.F., *J. Amer. Chem. Soc.*, 1946, vol. **46**, p. 358.
52. Andrianov K.A., Kononov A.M., and Tsvankin D.Ya., *Vysokomol. Soedin.*, 1968, vol. **10B**(5), p. 320. (Rus)
53. Makarova N.N., Petrova I.M., Godovski Yu.K., Lavrukhin B.D., and Zhdanov A.A., *Doklady AN SSSR*, 1983, vol. **209**(6), p. 1368. (Rus)
54. Godovsky Yu.K., Makarova N.N., and Mamaeva I.I., *Macromol. Chem. Rapid Commun.*, 1986, vol. **7**(1), p. 32.

55. Makarova N.N., Godovsky Yu.K., and Kuzmin N.N., *Makromol. Chem.*, 1987, Bd. **188**(1), S. 119.
56. Mamaeva I.I., Makarova N.N., Petrova I.M., Tverdokhlebova I.I., and Pavlova S.A., *Vysokomol. Soedin.*, 1987, vol. **29A**(7), p. 1507. (Rus)
57. Makarova N.N., Kuzmin N.N., Lavrukhin B.D., Godovski Yu.K., Mamaeva I.I., Matukhina E.V., and Petrova I.I., *Vysokomol. Soedin.*, 1989, vol. **31B**(9), p. 708. (Rus)
58. Makarova N.N., Godovski Yu.K., and Lavrukhin B.D., *Vysokomol. Soedin.*, 1995, vol. **37A**(3), p. 1. (Rus)
59. Matukhina E.V., Boda E.E., Timofeeva T.V., Godovski Yu.K., Makarova N.N., Petrova I.I., and Lavrukhin B.D., *Vysokomol. Soedin.*, 1996, vol. **38A**(9), p. 1545. (Rus)
60. Godovsky Yu.K., Makarova N.N., Petrova I.M., and Zhdanov A.A., *Macromol. Chem. Rapid Commun.*, 1984, vol. **5**(3), p. 427.
61. Timofeeva T.V., Boda E.E., Polishchuk A.P., Antipin M.Yu., Matukhina E.V., Petrova I.M., Makarova N.N., and Struchkov Yu.T., *Mol. Cryst. Liq. Cryst.*, 1994, vol. **248**, p. 125.
62. Makarova N.N., Petrova I.M., Matukhina E.V., Godovski Yu.K., and Lavrukhin B.D., *Vysokomol. Soedin.*, 1997, vol. **39A**(10), p. 1616. (Rus)
63. Destrade C., Tinh N.H., Gasparoux H., Malthete J., and Levelut A.M., *Mol. Cryst. Liq. Cryst.*, 1981, vol. **71**, p. 111.
64. Kuzmin N.N. and Matukhina E.V., *Vysokomol. Soedin.*, 1991, vol. **33B**(7), p. 547. (Rus)
65. Beatty C.L. and Karasz F.E., *Bull. Am. Phys. Soc.*, 1973, vol. **18**(3), p. 461.
66. Papkov V.S., Godovsky Yu.K., Svistunov V.S., Litvinov V.M., and Zhdanov A.A., *J. Polym. Sci., Polym. Chem. Ed.*, 1984, vol. **22**(12), p. 3617.
67. Godovsky Yu.K. and Papkov V.S., *Adv. Polym. Sci.*, 1989, vol. **88**, p. 129.
68. Tsvankin D.Ya., Papkov V.S., Zhukov V.P., Godovsky Yu.K., Svistunov V.S., and Zhdanov A.A., *J. Polym. Sci., Polym. Chem. Ed.*, 1985, vol. **23**(4), p. 1043.
69. Godovsky Yu.K., Makarova N.N., Papkov V.S., and Kuzmin N.N., *Macromol. Chem. Rapid Commun.*, 1985, vol. **6**, p. 443.
70. Out G., Turetski A., Moeller M., and Oelfin D., *Macromolecules*, 1994, vol. **27**(12), p. 3310.
71. Makarova N.N., Petrova I.M., Godovski Yu.K., Lavrukhin B.D., and Zhdanov A.A., *Doklady AN SSSR*, 1983, vol. **209**(6), p. 1368. (Rus)
72. Makarova N.N., Matukhina E.V., Godovski Yu.K., and Lavrukhin B.D., *Vysokomol. Soedin.*, 1992, vol. **34B**(2), p. 56. (Rus)

73. Beatty C.L. and Karasz F.E., *Bull. Amer. Phys. Soc.*, 1973, vol. **18**(3), p. 461.
74. Makarova N.N. and Godovski Yu.K., *Vysokomol. Soedin.*, 1986, vol. **28B**(4), p. 243. (Rus)
75. Ringsdorf H. and Scheller A., *British Polym. J.*, 1981, vol. **13**, p. 43.
76. Garzo G. and Perhsson K., *J. Organometal. Chem.*, 1971, vol. **30**, p. 187.
77. Garzo G., Tamas J., Szekely T., and Ujszaszi K., *Acta Chim., Acad. Sci. Hung.*, 1971, vol. **69**, p. 273.
78. Garzo G. and Alexander G., *Chromatogr.*, 1971, vol. **4**, p. 554.
79. Alexander G. and Garzo G., *Chromatogr.*, 1974, vol. **7**, p. 190.
80. Blazso M. and Garzo G., *J. Organomet. Chem.*, 1979, vol. **7**, p. 273.
81. Lengyel B., Garzo G., Fritz D., and Till F., *J. Chromatogr.*, 1966, vol. **24**, p. 8.
82. *A.C. No. 842,572* (USSR), 1981; *Bull. Izobr.*, No. 24, 1981. (Rus)
83. Mukbaniani O.V., Achelashvili V.A., Levin V.Yu., Meladze S.M., Inaridze I.A., and Khananashvili L.M., *Izv. AN GSSR, Ser. Khim.*, 1990, vol. **16**(1), p. 20. (Rus)
84. Inaridze I.A., *Synthesis and Investigation of Organocyclosiloxane and Orga-nocarbocyclosiloxane Fragments Containing Polyorganosiloxanes*, Candidate Dissertation on Chemistry, 1993, Tbilisi State University, Tbilisi, Georgia. (Rus)
85. Papkov V.S., Il'ina M.N., Makarova N.N., Zhdanov A.A., Slonimski G.L., and Andrianov K.A., *Vysokomol. Soedin.*, 1975, vol. **12A**, p. 2700. (Rus)
86. Achelashvili V.A., Mukbaniani O.V., Koiava N.A., and Komalenkova N.G., *Abstr. Commun. 31th IUPAC Macromolecular Symposium*, GDR, Merseburg, 1987, p. 155.
87. Achelashvili V.A., Mukbaniani O.V., Khananashvili L.M., Levin V.Yu., Komalenkova N.G., and Chernyishev E.A., *Vysokomol. Soedin.*, 1990, vol. **32A**(3), p. 480. (Rus)
88. Chernyishev E.A., Komalenkova N.G., Klochkova T.A., Kuzmina G.M., and Bochkarev V.N., *Zh. Obshch. Khim.*, 1975, vol. **45**(10), p. 2229. (Rus)
89. Chernyishev E.A., Belkina T.V., Nikitin V.S., Komalenkova N.G., Shchapatin A.S., and Zhinkin D.Ya., *Zh. Obshch. Khim.*, 1978, vol. **48**(3), p. 630. (Rus)
90. Khananashvili L.M. and Mukbaniani O.V., *Intern. J. Polym. Mater.*, 1994, vol. **27**, p. 31.
91. Andrianov K.A., Slonimski G.L., Levin V.Yu., Podovski Yu.K., Kuteinikova N.K., Tsvankin D.Ya., Moskalenko V.A., and Kuteinikova L.I., *Vysokomol. Soedin.*, 1970, vol. **12A**(6), p. 1268. (Rus)

92. Zhdanov A.A., Andrianov K.A., and Malyikhin A.P., *Doklady AN SSSR*, 1973, vol. **211**(5), p. 1104. (Rus)
93. Zhdanov A.A., Pryakhina T.A., Strelkova T.V., Afonina R.I., and Kotov V.M., *Vysokomol. Soedin.*, 1993, vol. **35**(5), p. 475. (Rus)
94. Karchkhadze M.G., Mukbaniani N.O., Samsonia A.Sh., Tkeshelashvili R.Sh., Kvelashvili N.G., Chogovadze T.V., and Khananashvili L.M., *Bull. Georg. Acad. Sci.*, 1998, vol. **158**(1), p. 75.
95. Mukbaniani N.O., *Synthesis and Investigation of Properties of Carbosiloxane Cyclolinear Copolymers*, Candidate's Dissertation on Chemistry, 2001, Tbilisi State University, Tbilisi, Georgia.
96. Karchkhadze M.G., Mukbaniani N.O., Khananashvili L.M., Meladze S.M., Kvelashvili N.G., and Doksopulo T.P., *Intern. J. Polym. Mater.*, 1998, vol. **41**, p. 89.
97. Mukbaniani N.O., Karchkhadze M.G., Samsonia A.Sh., Tkeshelashvili R.Sh., and Khananashvili L.M., *Bull. Georg. Acad. Sci.*, 1999, vol. **160**(1), p. 84.
98. *Patent No. 3,297,632* (USA), 1967; Wu T.C., *C.A.*, 1967, vol. 66, 46790x.
99. *Patent No. 3,264,259* (USA), 1966.
100. *A.C. No. 794,029* (USSR); *Bull. Izobr.*, No. 1, 1981. (Rus)
101. *A.C. No. 791,758* (USSR); *Bull. Izobr.*, No. 48, 1980. (Rus)
102. Andrianov K.A. and Zachernyuk A.B., *Vysokomol. Soedin.*, 1974, vol. **16A**(8), p. 1435. (Rus)
103. Mukbaniani O.V., Meladze S.M., Makarova N.N., and Khananashvili L.M., *Soobshch. AN GSSR*, 1980, vol. **99**(4), p. 109. (Rus)
104. Meladze S.M., *Candidate Dissertation on Chemistry*, 1980, Tbilisi State University, Tbilisi, Georgia.

Chapter 4. Organosiloxane copolymers with cyclic fragments in the side chain

It is known from the literature [1] that the main difference between branched oligo(dimethyl)methylsiloxanes and equal viscosity linear oligodimethylsiloxanes is the absence of tendency to crystallize at definite concentration of branched methylsilsesquioxane units. The main reason for suppression of crystallization in oligo(dimethyl)methylsiloxanes is disturbance of the regular structure of linear siloxane chain at the introduction of methylsilsesquioxane units into them.

The most effective method of modification of linear organosiloxane polymers is introduction of various elements or groups different by chemical origin and structure into the macromolecular chain. Introduction of different fragments into the dimethylsiloxane backbone induces changes of its physical and chemical properties [2].

At thermal degradation, introduction of cyclic fragments into the linear dimethylsiloxane backbone hinder the chain transfer reactions proceeding with liberation of the D_n-type (where D = Me_2SiO) cyclic products that increases thermal oxidative stability of the mentioned polymers [3].

4.1. Synthesis of monomeric difunctional organocyclosiloxanes

As mentioned in Chapter 1, synthesis of organocyclosiloxanes with the known disposition of functional groups at silicon atoms is realized by two ways: (1) formation of a cyclosiloxane skeleton and (2) modification of functional groups contained in organocyclosiloxanes.

For the purpose of synthesizing organocyclosiloxanes with dichlororganosiloxy functional groups at silicon atoms, the HFC reaction of hydroxyorganocyclotri(tetra, penta)siloxanes with organochlorosilanes in the presence of amines was investigated [4 - 6]. For the case of HFC of hydroxyorganocyclosiloxanes with trimethylchlorosilane at 1:1 ratio of initial compounds in the presence of amines, it was proved that both HFC and homofunctional condensation with formation of dicyclic structure proceed [7].

For increasing the yield of chlororganosiloxyorganocyclosiloxanes and decreasing the yields of side reaction products, HFC performed (with organochlorosilanes excess) at 1:2 ratio of initial compounds, in 10% anhydrous toluene solution (in the case of 1-hydroxy-1-organo-tetraphenylcyclotrisiloxanes and 1-hydroxy-1-organohexaphenylcyclotetrasiloxanes) or anhydrous ether (in the case of hydroxymethylcyclosiloxanes), in the presence of pyridine and the temperature range -5 ÷-10°C. However, under these conditions HFC with formation of side products proceeds in accordance with the following scheme [7, 8]:

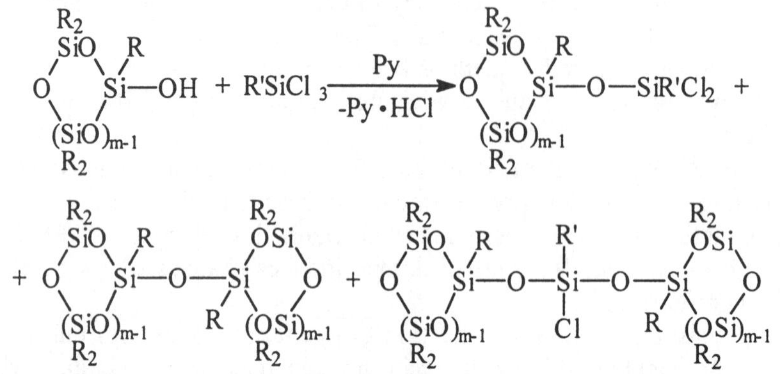

(4.1)

where for the structures 1 and 3: $m = 2, 3, 4$; R = R' = Ph, Me; R ≠ R'.

Under the above-mentioned conditions the reaction proceeds ambiguously, and besides organochlorosiloxycyclosiloxanes of the structure I, bis(cyclosiloxy)organochlorosilane of the structure III are obtained. GLC analysis has indicated that besides the compounds of the structure III, about 3-5% of bis(organocyclosila)oxides of the structure II are produced.

Synthesized organosilicon compounds were identified by functional and ultimate analysis, determination of molecular masses, IR and NMR spectral data [9]. Yields and some physical and chemical parameters of synthesized organocyclosiloxanes are shown in Table 4.1.

^{1}H NMR spectra of compounds 1 and 3 (Figure 4.1, Table 4.1) display singlet reflexes for methyl protons with chemical shifts of $\delta = 0.57$ ppm and $\delta = 0.58$ ppm, respectively, and multiplets in the range of $\delta = 7.1 - 7.7$ ppm for phenyl ones.

Table 4.1

Physical and chemical parameters and yields of organodichlorosiloxyorganocyclotetra(penta)siloxanes

No.	Compound	Yield, %	T_{boil}, °C (P, mmHg)	T_{melt}, °C	d_4^{20}	n_D^{20}	M_{RD}*
1	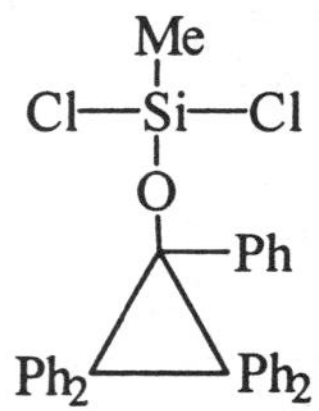	66	230-232 (0.03)	95-96	-	-	-
2	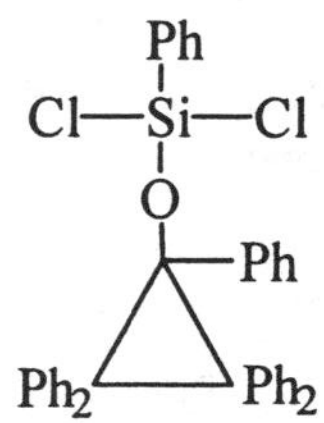	77	268-270 (0.02)	120-121	-	-	-
3	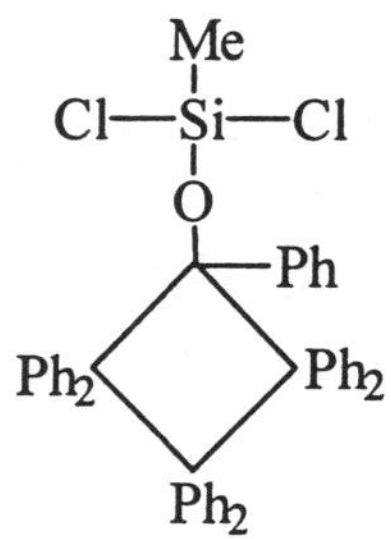	79	252-254 (0.01)	141-142	-	-	-

4	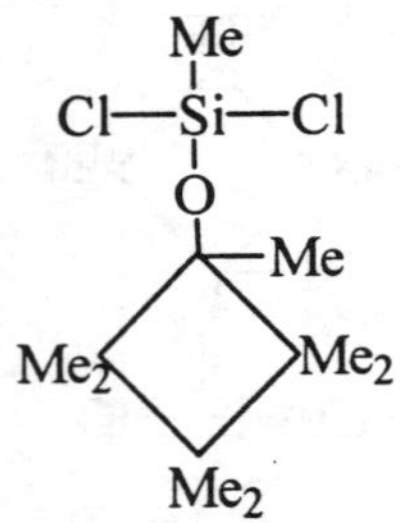	85	54 (0.2)	-	1.4061	1.0826	410/411
6	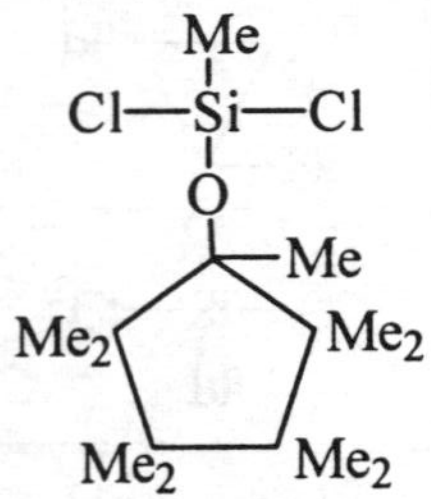	86	70-71 (0.2)	-	1.4058	1.0643	484/485
7	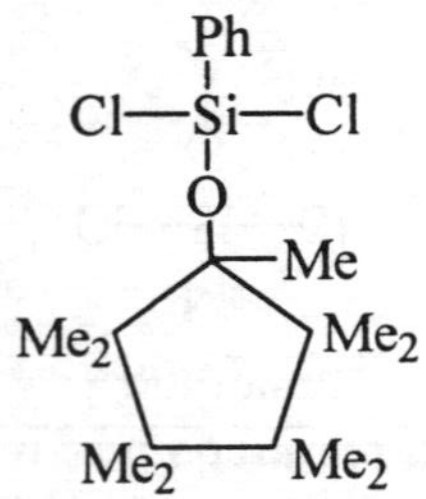	82	104-106 (0.2)	-	1.4444	1.0990	546/547

8	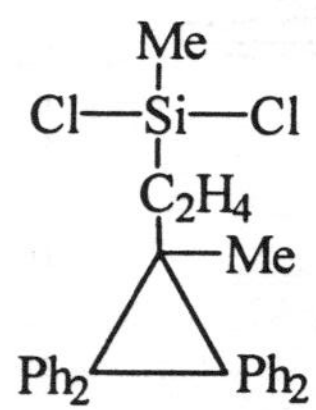	84	224-228 (0.02)	148-149	-	-	-

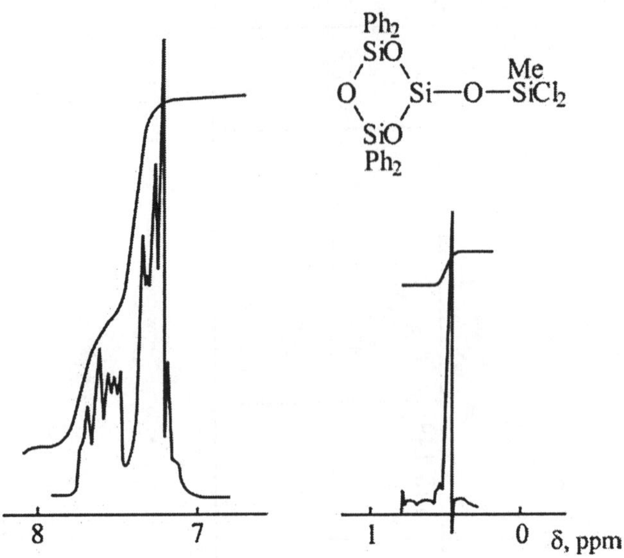

Figure 4.1. ^{1}H NMR spectra of compound 1 (Table 4.1)

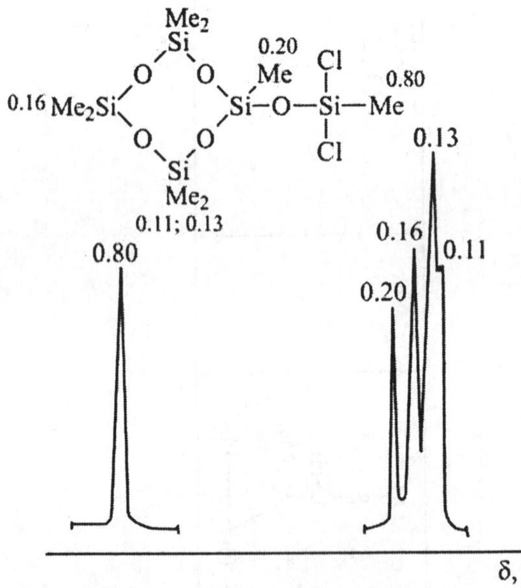

Figure 4.2. ^{1}H NMR spectra of compound 4 (Table 4.1)

^1H NMR spectrum of compound 4 (Table 4.1) contains two singlet reflexes for methyl groups at atom of silicon in silsesquioxane with the chemical shift of $\delta = 0.18$ ppm and methyl groups in linear fragment with the chemical shift of $\delta = 0.50$ ppm (3H). Reflexes for two pairs of enantiotropic methyl groups, distinguished from each other by spatial disposition in relation to cyclosiloxane fragment with chemical shifts of $\delta = 0.10$ and 0.14 ppm, and for methyl groups (these groups are not exchanged) at $\delta = 0.12$ ppm (Figure 4.2).

Note also that ^1H and ^{13}C NMR spectra of organodichlorosiloxymethylcyclotetra(penta)siloxanes were mainly studied at room temperature. NMR spectra of the cyclic part of methylcyclosiloxanes were also studied [10].

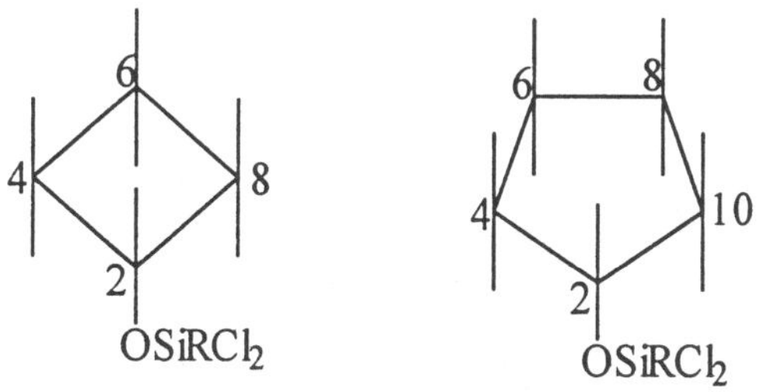

R=Me (4) R=Me (6), Ph (7)

Scheme 4.1

^1H NMR spectrum of compound 4 (refer to Table 4.2) displays two pairs enantiotropic methyl groups (Me^{4a}, Me^{8a} and Me^{4b}, Me^{8b}) differing by spatial disposition with respect to methylsiloxy-substituting agent, and Me^{6a} and Me^{6b} groups (these groups are not exchanged in symmetry operations), which leads to occurrence of spin-spin interaction between protons of geminal methyl groups ($4JHCSiCH_{gem} = 0.4$ Hz) [9]. Thus ^1H NMR spectrum of the cyclic part of the compounds occurrence of a singlet reflex, corresponded to Me^2 protons, and four quartet reflexes of $Me^{4a,8a}$, $Me^{4b,8b}$, Me^{6a} and Me^{6b} protons should be expected. But it should be noted that occurrence of a small difference in chemical shifts of quartet reflexes for the above-mentioned enantiotropic methyl protons can only be recorded by spectrometers with the resonance frequency above 200 MHz. While overlapping of resonance bands and results takes place at spectrum recording on a spectrometer with lower resonance frequency (99.54 Hz), no fine splitting of the spin-spin interaction of protons in geminal methyl groups is observed, but broadening only happens. Resonance reflexes of the rest of methyl groups in the cyclic fragment are grouped in three bands with 2:3:1 ratio of integral intensity (for the unit the integral intensity of the reflex from protons of Me^2 group is accepted).

^{13}C NMR spectra of methyldichlorosiloxymethylcyclosiloxanes (refer to Table 4.3 and Figures 4.3 and 4.4) clearly display the absorption bands of carbon nuclei in Me^2 groups in strong fields. The resonance reflexes from carbon nuclei of rest methyl groups in the cyclic part as grouped in two bands with 2:4 ratio of integral intensities (hereinafter, weak-field band is first mentioned).

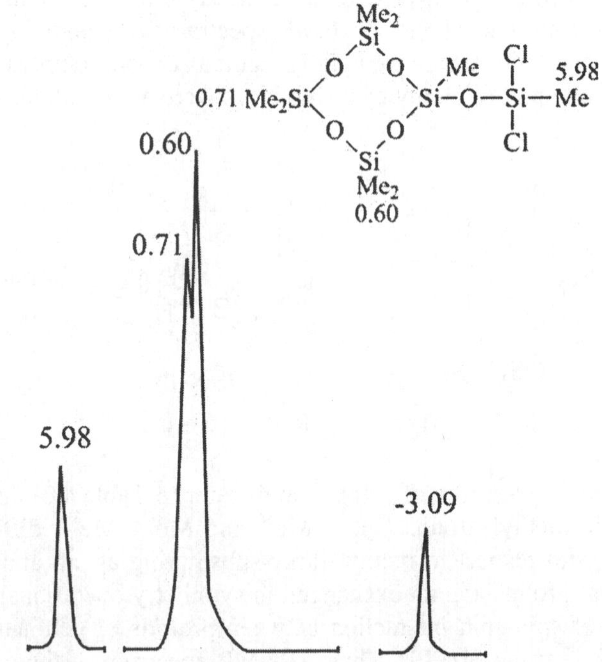

Figure 4.3. ^{13}C NMR spectrum of compound 4 (Table 4.1)

The synthesized chlororganosiloxymethylcyclosiloxanes was studied by the mass spectrometry method. The mass spectra of the compounds indicate the presence of fragments in the molecules, corresponded to definite typical structural components and do not contradict with the suggested structures.

Table 4.2

^{1}H NMR spectral data and chemical shifts of methyl groups in organodichlorosiloxymethylcyclotetra-(penta)siloxanes (δ, ppm)

No.	Group position	Me^{2}	$Me^{4a, 8a}$	$Me^{4b, 8b}$	6^{a}	6^{b}	R
4		0.20	0.13	0.16	0.11	0.13	0.80
	Group position	Me^{2}	$4^{a}, 10^{a}$	$4^{b}, 10^{b}$	$6^{a}, 8^{a}$	$6^{b}, 8^{b}$	
6		0.21	0.11	0.16	0.11	0.13	0.79
7		0.40	0.25	0.29	0.25	0.27	-

Table 4.3

^{13}C NMR spectral data and chemical shifts of methyl groups in organodichlorosiloxymethylcyclotetra-(penta)siloxanes (δ, ppm)

No.	Group position	Me^{2}	$Me^{4a, 8a}$	$Me^{4b, 8b}$	6^{a}	6^{b}	R
4		-0.09	0.60	0.60	0.71	0.71	5.98
	Group position	Me^{2}	$4^{a}, 10^{a}$	$4^{b}, 10^{b}$	$6^{a}, 8^{a}$	$6^{b}, 8^{b}$	
6		-2.80	0.77	0.68	0.89	0.89	6.00
7		-2.65	0.77	0.77	0.92	0.92	-

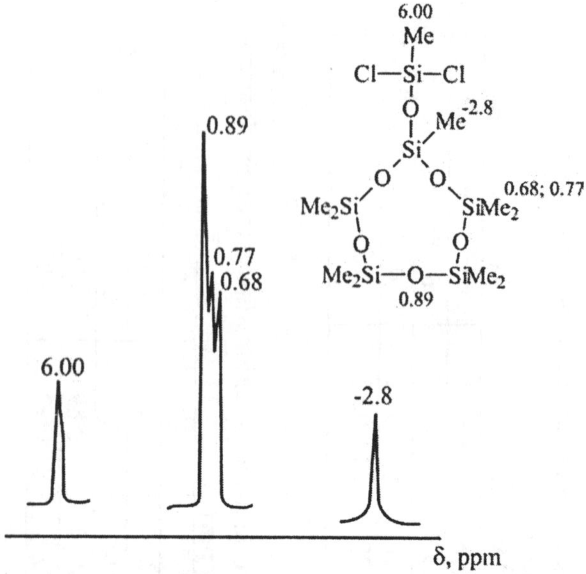

Figure 4.4. ^{13}C NMR spectrum of compound 6 (Table 4.1)

Hydride addition of methyldichlorosilane to 1-vinyl-1-methyltetraphenylcyclotrisiloxane proceeding at 1:1 ratio of initial components, in the presence of 0.1 M platinum-hydrochloric acid solution in tetrahydrofurane and $Co_2(CO)_8$ as the catalysts and in temperature range of 50 - 60°C and has produced dichloromethylsilylcarbocyclotrisiloxane 8 [11,12]. The reaction scheme is as follows:

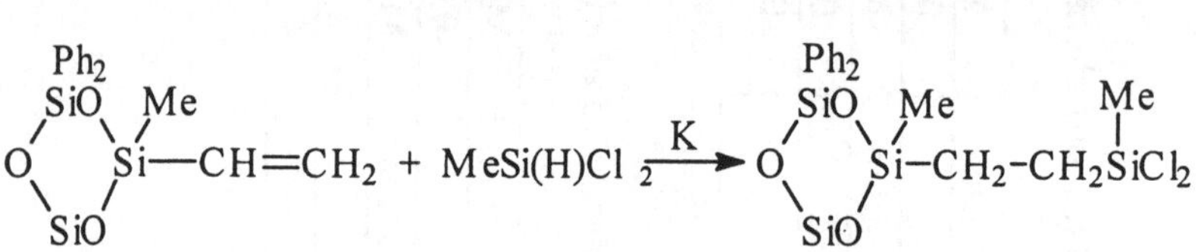

IV (8) (4.2)

The yields and some physical and chemical parameters of synthesized compound 8 of the structure IV are shown in Table 4.1. ^{1}H NMR spectrum of compound 7 displays a broadened singlet reflex (4H) with the chemical shift $\delta =$ 0.77 ppm for $-CH_2$-CH_2- group, a singlet reflex (3H) with the chemical shift $\delta =$

0.49 ppm for –O-SiMe-O- group, a singlet reflex (3H) with δ = 0.28 ppm for methyl protons in –SiMeCl$_2$ group, and a complex multiplet with the chemical shift within δ = 6.90 – 7.60 ppm for phenyl protons (refer to Figure 4.5). As a consequence, one may conclude about hydride addition of methyldichlorosilane to 1-vinyl-1-methyltetraphenylcyclotrisiloxane proceeding with formation of β-adduct.

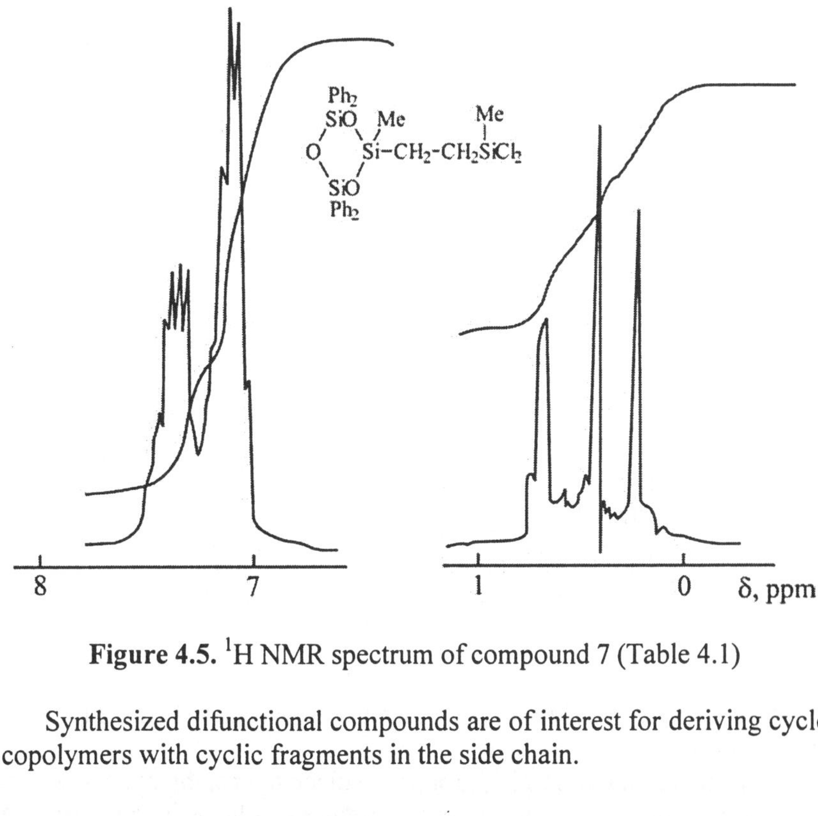

Figure 4.5. ^{1}H NMR spectrum of compound 7 (Table 4.1)

Synthesized difunctional compounds are of interest for deriving cyclolinear copolymers with cyclic fragments in the side chain.

4.2. Synthesis of organosiloxane copolymers with monocyclic (cyclotri-, cyclotetra-, cyclopentasiloxane) fragments in the side chain

The synthesis of linear organosiloxane copolymers with regular disposition of organocyclosiloxane fragments in the side chain was implemented using HFC of dichlororganosiloxyorganocyclosiloxanes with α,ω-dihydroxydimethylsiloxanes and dihydroxydimethylsilane in the presence of pyridine (the acceptor of hydrogen chloride). It is known that application of the acceptor increases of depth of the reaction [13, 14]. The reaction proceeded at room temperature, and at the final stage the reaction mixture was heated up to the boiling point of the solvent used. It was found that at short lengths of linear dimethylsiloxane unit the yields of copolymers are low. Studying the heterofunctional polycondensation of dichloroorganosiloxydiorganocyclosiloxanes with 1.5-dihydroxyhexamethyltrisiloxane after polymer reprecipitation from the mother solution, low-molecular products, corresponded to dicyclic compound by structure and composition, were obtained. As a consequence, the above-mentioned HFC reaction proceeds in accordance with the following scheme [15 - 21]:

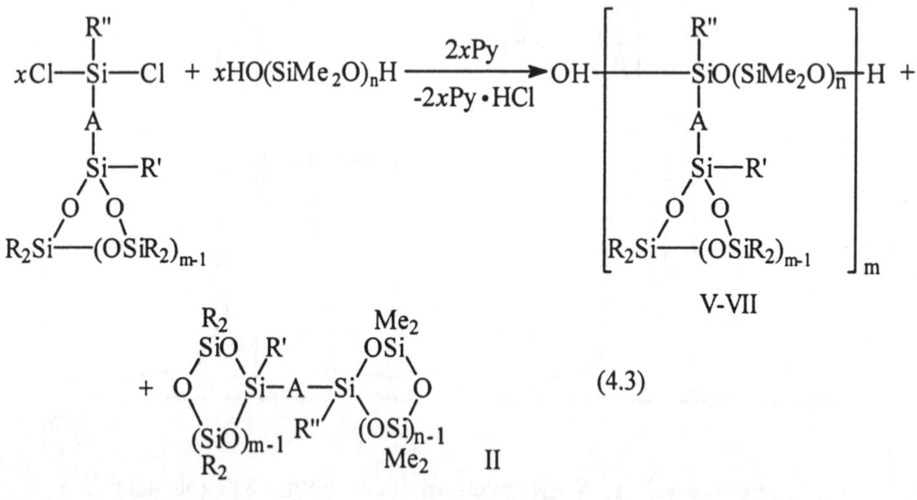

(4.3)

where R = R' = R" = Me, Ph; R' ≠ R"; A =-O-, $-C_2H_4-$; m = 1 (V), 2 (VI), 3 (VII); m = 1 - 51.

Depending on the framing groups at silicon atoms, the copolymers represent transparent or slightly opalescent substances, soluble in ordinary organic solvents with η_{spec} = 0.04 – 0.4. The yield of synthesized copolymers equals 72 - 95%. Some physical and chemical parameters and yields of copolymers are shown in Tables 4.4 - 4.8.

Table 4.4

Some physical and chemical parameters and yields of copolymers with phenylcyclotrisiloxane fragments in the side chain

No.	Copolymer structure	R	n_{SiO}	Yield, %	η_{spec}*	T_g, °C	d_1, Å
1	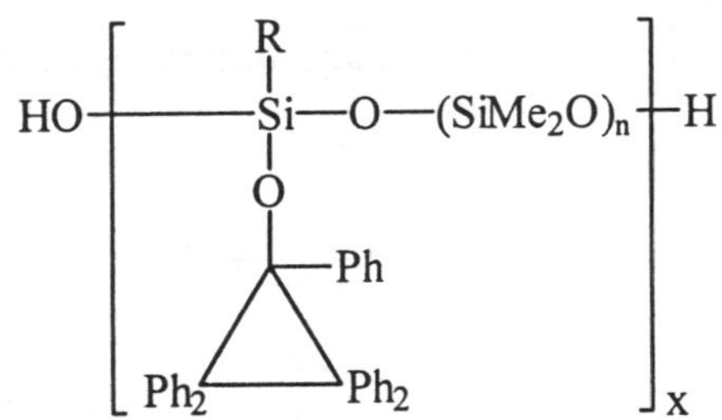	Ph	2	73	0.08	+18 - +20	10.05
2		Ph	3	73	0.09	+8- +10	-
3		Ph	4	78	0.11	0 - +5	-
4		Ph	8	85	0.14	-38	-
5		Ph	34	94	0.20	-115	-
6		Ph	51	95	0.24	-123	7.54
7		Me	2	76	0.06	+5 - +10	8.64
8		Me	3	77	0.08	0	-
9		Me	4	80	0.12	-12	-
10		Me	8	85	0.19	-50	7.59
11		Me	34	94	0.28	-108	-
12		Me	51	94	0.37	-123	7.31

Table 4.5

Some physical and chemical properties and yields of copolymers with carbocyclotrisiloxane fragments in the side chain

No.	Copolymer structure	R	n_{SiO}	Yield, %	η_{spec}*	T_g, °C	d_1, Å
1	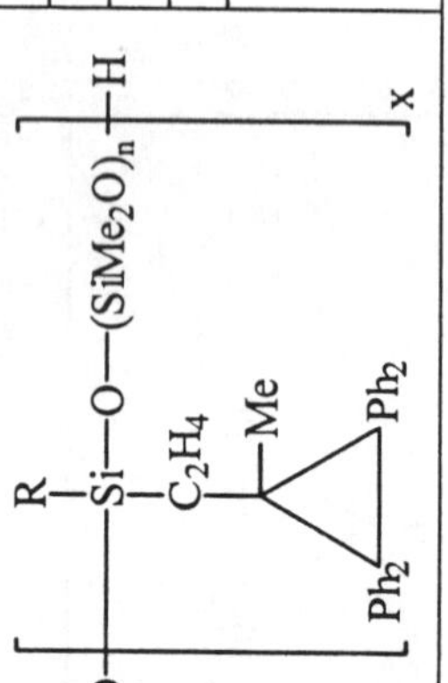	Me	2	72	0.07	0÷-5	10.05
2		Me	4	74	0.10	-20	-
3		Me	8	89	0.13	-65	-
4		Me	34	94	0.20	-123	7.42
5		Me	51	95	0.26	-123	-

Table 4.6

Some physical and chemical properties and yields of copolymers with phenylcyclotetrasiloxane fragments in the side chain

No.	Copolymer structure	R	n_{SiO}	Yield, %	η_{spec}*	T_g, °C	d_1, Å
1	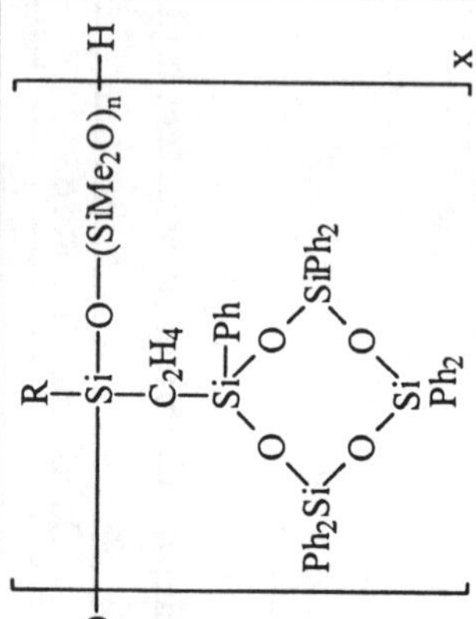	Me	2	73	0.06	+16	10.40
2		Me	4	80	0.08	-2	-
3		Me	8	88	0.10	-40	-
4		Me	34	92	0.15	-	7.50
5		Me	51	95	0.21	-123	-

Table 4.7

Some physical and chemical properties and yields of copolymers with methylcyclotetra(penta)siloxane fragments in the side chain

No.	Copolymer structure	M	n_{SiO}	Yield, %	η_{spec}*	T_g, °C	d_1, Å
1	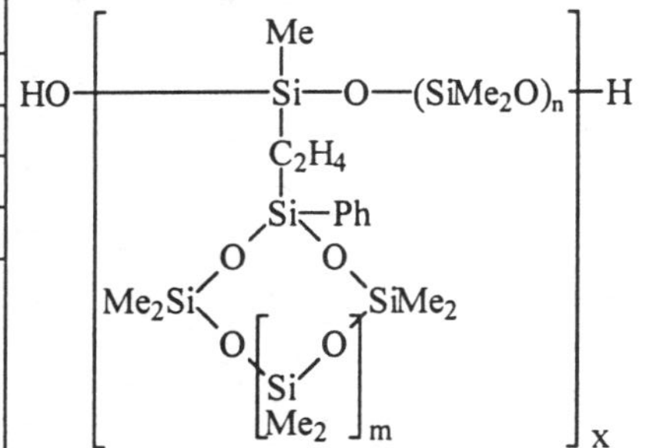	1	1	82	0.03	-90	7.83
2		1	2	76	0.10	-95	-
3		1	3	76	0.06	-100	-
4		2	1	82	0.03	-105	8.19
5		2	2	78	0.06	-110	-
6		2	3	76	0.08	-118	-

Table 4.8

Some physical and chemical properties and yields of copolymers with methylcyclotetrasiloxane fragments in the side chain

No.	Copolymer structure	n_{SiO}	Yield, %	η_{spec}*	T_g, °C	d_1, Å
1	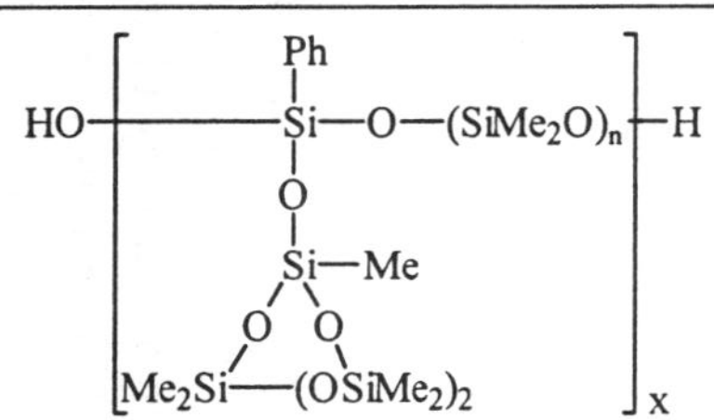	1	82	0.05	-83	8.76
2		2	75	0.08	-88	-
3		3	81	0.10	-96	-

The Tables show that at short lengths dimethylsiloxane unit (n = 2 - 4) the yields of synthesized copolymers are lower. This is explained by partial proceeding of HFC reaction by the intramolecular cyclization path with formation of dicyclic compounds with the yield of about ~8%.

After reprecipitation of synthesized copolymers 8 (Table 4.4), 3 and 6 (Table 4.7) from the mother solution and removal of the solvent (toluene), the products of the structure II were separated. For these products molecular masses were determined and IR and NMR spectra were recorded, which indicated the dicyclic structure of them.

Such dicyclic structures were obtained directly by HFC of dichloroorganosiloxyorganocyclosiloxanes with 1,5-dihydroxyhexamethyltrisiloxane at 1:1 ratio of the initial components in the presence of aniline, in 5% anhydrous ether solution and temperature range of -5 ÷ -10°C.

The reaction proceeded in accordance with the following scheme [21]:

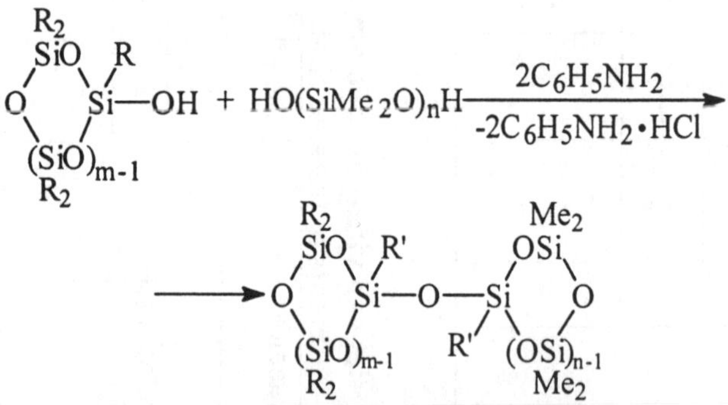

(4.4)

II

where R = R' = Ph, Me; R ≠ R'; R" =Me, m =2, 3; n = 3, 4.

The composition and structure of synthesized compounds of the structure II were detected by ultimate analyses, determination of the molecular mass, IR and NMR spectral data (refer to Table 4.9).

^{1}H NMR spectrum of compound 1 (Table 4.9) displays a singlet reflex for methyl groups (3H) with the chemical shift δ = 0.11 ppm, and two pairs of enantiotropic methyl groups (Me^{4a}, Me^{8a} and Me^{4b}, Me^{4b}), differed by spatial disposition relative to methylcyclotetrasiloxane fragments with chemical shifts δ = 0.12 and δ = 0.13 ppm, and protons of Me^{6a} and Me^{6b} methyl groups (not exchange) at δ = 0.14 ppm. Moreover, a complex multiplet for phenyl groups in the area of 6.8 – 7.4 ppm is observed. Decoding of NMR spectra for compound 3 (Table 4.9) became rather difficult problem, because the whole range of protons' absorption did not exceed 0.04 ppm.

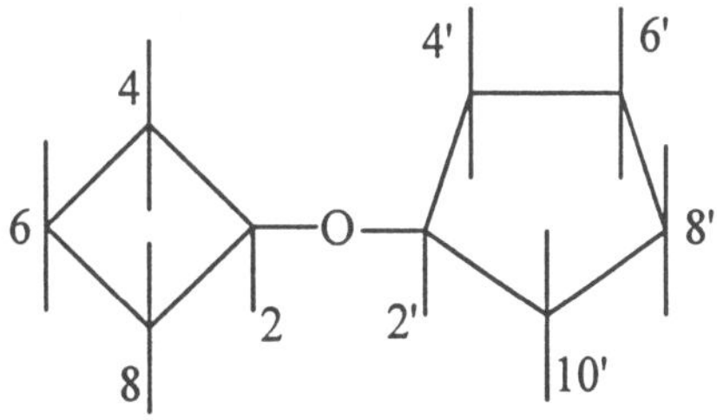

Compound 3

While in this case organocyclosiloxanes are substituting groups at Si^2 and $Si^{2'}$, the resonance bands of Me^2 и $Me^{2'}$ groups' protons will be observed within the range of chemical shifts of δ = 0.097 - 0.111 ppm. It may be indicated only that the lowest field band (δ = 0.14 ppm) includes resonance absorption reflexes of $Me^{4,\,8b}$ and $Me^{4,\,10b}$ groups' protons, and the most intensive strong field band (δ = 0.10 ppm) includes reflexes of Me^2, $Me^{2'}$, $Me^{6',8'a}$ and Me^{6a} groups' protons. Figure 4.6 shows ^1H and ^{13}C NMR spectra for compound 3 (Table 4.9).

^{13}C NMR spectrum of compound 3 in the strong-field area displays two reflexes, which may be corresponded to absorption of carbon nuclei in C^2 and $C^{2'}$ groups (the strong-field shift is stipulated by bridge oxygen atoms). Assuming that the above-mentioned dependence of the chemical shift of carbon nuclei in these methyl groups on the ring size does also occur in the considered compound, comparative strong-field reflex with the chemical shift δ = 2.71 ppm can be corresponded to C^2 nucleus, and the reflex at δ = 2.51 ppm to $C^{2'}$ nucleus. The integral intensity of absorption reflexes of the rest carbon nuclei of methyl groups is, approximately, seven times higher than total integral intensity of observed strong-field reflexes that corresponds to the compound 3 structure. The spectrum of carbon nuclei in these methyl groups their absorption shaped as three bands centered at δ = 0.92, 0.77 and 0.65 ppm, the middle band being of the highest intensity. Using data on compounds 4 and 5 (Scheme 4.1) and above-considered relations, it has been suggested [10] that the band with chemical shift δ = 0.92 ppм, corresponds to absorption of $C^{6,\,8}$ nuclei (both types) of methyl groups, and the band at δ = 0.77 ppm corresponds to C^6 nuclei (both types), and

Table 4.9

Some physical and chemical properties of bis(organocyclosiloxane)oxides

No.	R	R'	R"	*m*	*n*	Yield, %	T_{boil}, °C (*P*, mmHg)	T_{melt}, °C	n_D^{25}	d_4^{25}	M_{RD}	*M**
1	Ph	Ph	Me	2	3	57	120-123 (0.2)	62-63	-	-	-	$\frac{817}{814}$
2	Me	Me	Me	3	3	62	87-89 (0.2)	-	1.4031	1.0216	$\frac{138.1}{138.9}$	$\frac{580}{578}$
3	Me	Me	Me	3	4	67	93-95 (0.2)	-	1.4045	1.0135	$\frac{157.5}{157.7}$	$\frac{651}{652}$

one enantiotropic types of $C^{4',8'}$ or $C^{4',10'}$, and the band at $\delta = 0.65$ ppm corresponds to another type, $C^{4,8}$ or $C^{4,10}$.

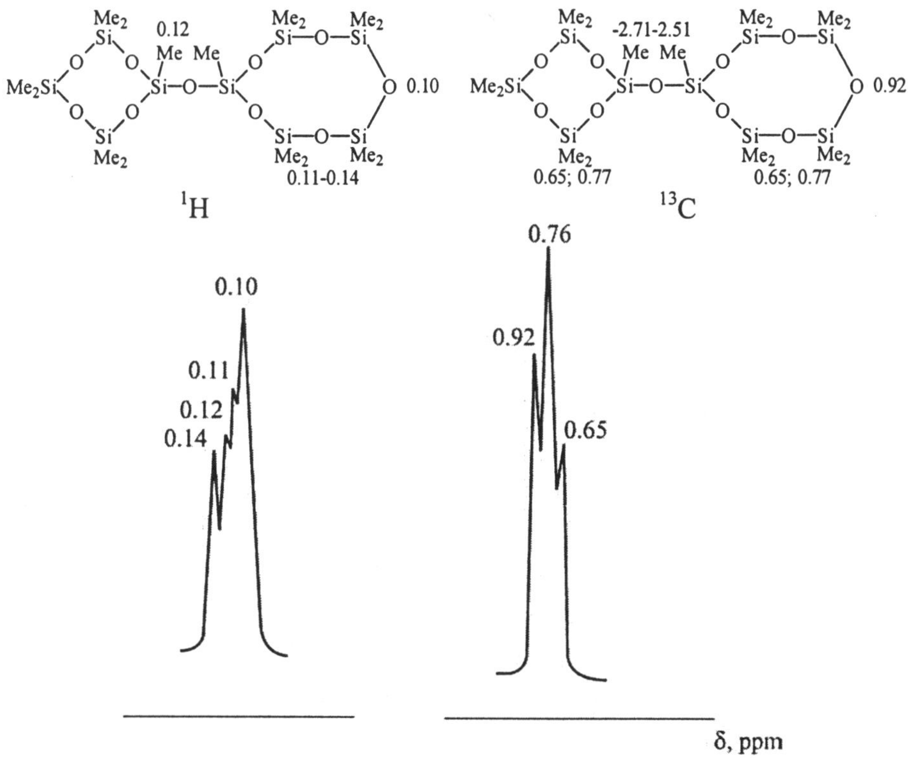

Figure 4.6. ^{1}H and ^{13}C NMR spectra of compound 3 (Table 4.9)

The structure of some synthesized copolymers was determined from IR and ^{29}Si spectral data [23]. Molecular masses, molecular mass distributions, characteristic viscosity, hydrodynamic and conformational characteristics of copolymers were found.

Figure 4.7a shows absorption bands typical of valence oscillations of ≡Si-O-Si≡ bonds in octatomic ring (D_4) at 1080 cm^{-1}. Figure 4.8a shows the absorption band of decatomic ring (D_5) at 1080 cm^{-1}. It should be noted that additional absorption bands at 1020 cm^{-1} typical of linear poly(dimethylsiloxanes) (PDMS) occur in spectra of copolymers 1 and 4 (Table 4.7). However, these bands are narrow for PDMS and rather wide for copolymers.

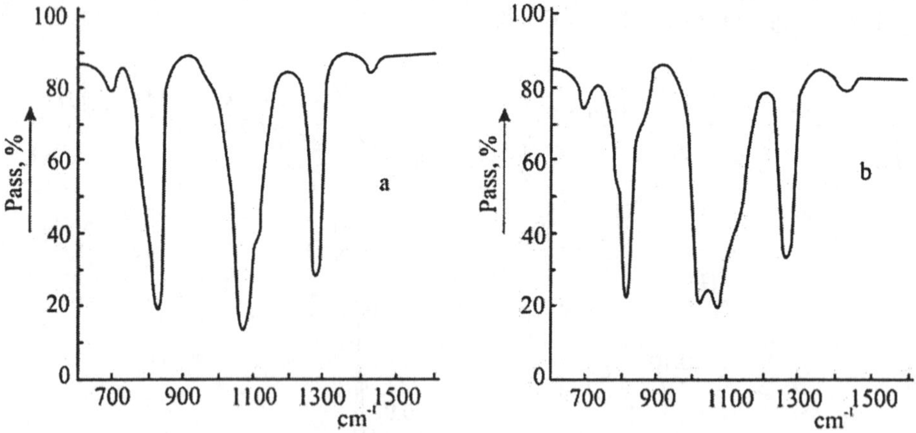

Figure 4.7. IR spectra of octamethylcyclotetrasiloxane (a) and copolymer 1 (b) (Table 4.7)

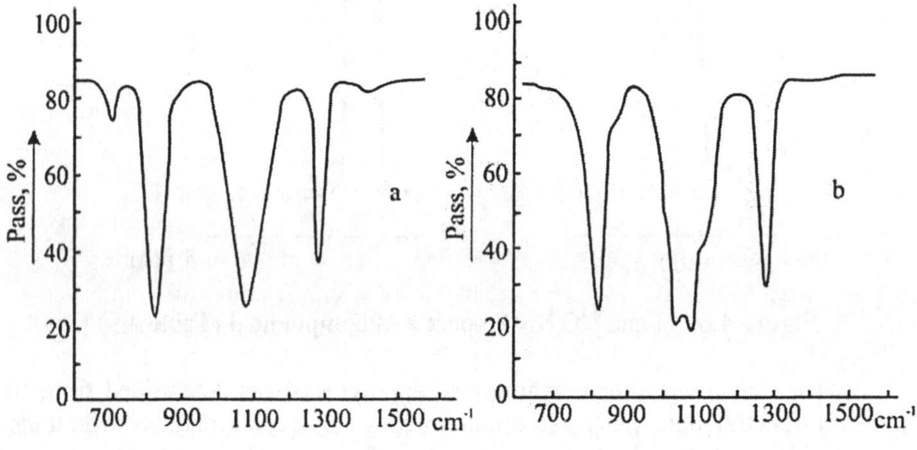

Figure 4.8. IR spectra of decamethylcyclotetrasiloxane (a) and copolymer 4 (b) (Table 4.7)

^{29}Si NMR spectrum of copolymer 1 (Table 4.7; Figure 4.9, in chloroform) possesses reflexes in the area of -19.0 ÷ -22.0 ppm (for linear D units) and -65.0 ÷ -68.0 ppm (for silsesquioxane T units). Each mentioned range contains two separate groups of reflexes. Weak-field reflexes in these ranges (-19.00 ÷ -

19.22 and -65.0 ÷ -68.5 ppm) are corresponded to a cyclic fragment of the copolymer structure. The ratio of integral intensities corresponds to anticipated structure.

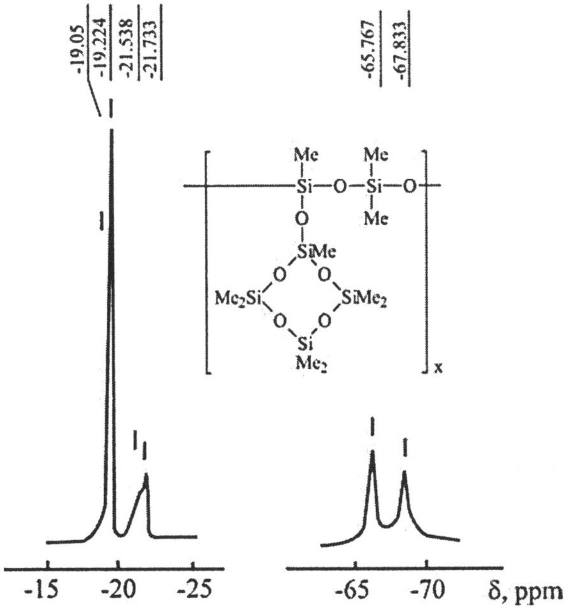

Figure 4.9. ^{29}Si NMR spectrum of copolymer 1 (Table 4.7)

The main reflexes in ^{29}Si NMR spectrum of copolymer 4 (Table 4.7, Figure 4.10) fall within the range of -21.0 ÷ -21.6 ppm (D units) and -66.5 ÷ -68.5 ppm (T units). The reflexes for linear and cyclic parts of the compound can be hardly distinguished within these ranges.

Figure 11 shows the gel chromatogram of copolymer 1 (Table 4.7). Since the copolymer 1 was not specially reprecipitated, it possesses a broad molecular mass distribution (MMD). The gel chromatographic studies were also performed for copolymer 4 (Table 4.7, Figure 4.12), which indicated narrow molecular mass distribution of it due to copolymer reprecipitation. Polydispersity coefficient was also calculated, which equaled $\varepsilon \approx \bar{M}_\omega / \bar{M}_n = 1.23$.

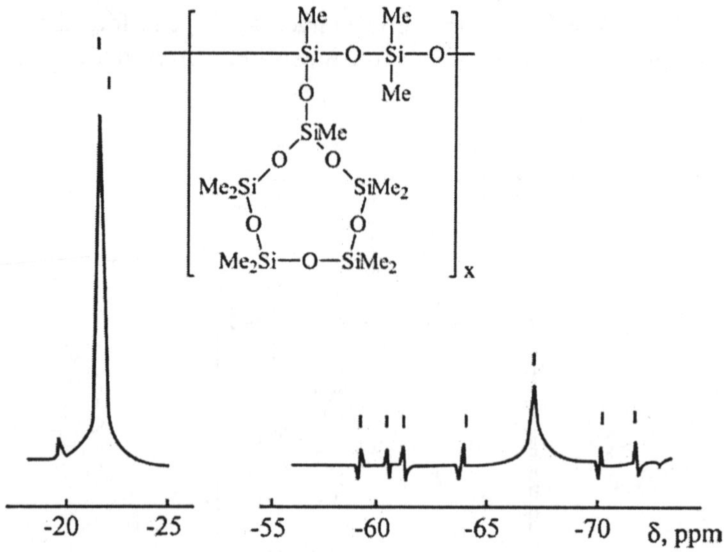

Figure 4.10. ^{29}Si NMR spectrum of copolymer 4 (Table 4.7)

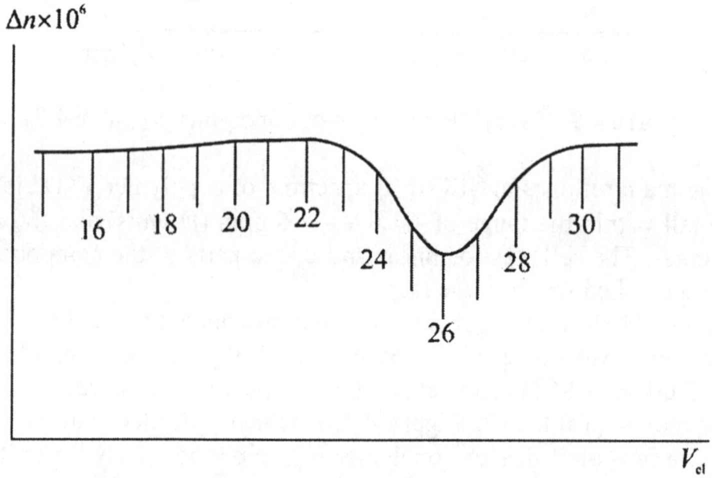

Figure 4.11. GPC of copolymer 1 (Table 4.7)

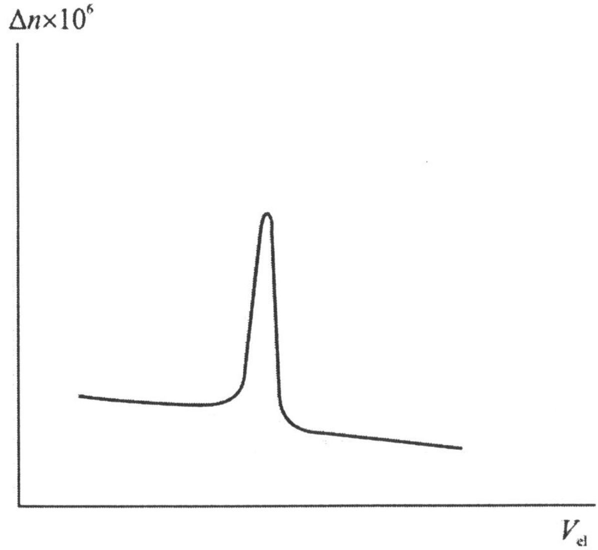

Figure 4.12. GPC data for copolymer 4 (Table 4.7)

Tables 4.10 and 4.11 show the results of fractionation of copolymers 1 and 4 (Table 4.7). Based on fractionation data and determination of $\bar{M}_n$ of the fraction and characteristic viscosity $[\eta]$, integral curves of distribution by MM and $[\eta]$ for copolymer 4 (Figure 4.13) were plotted. For the same copolymer values of the constant K in the Mark-Kuhn-Hauwink-Sakurada equation (K = $(2.38 \pm 0.01)\times10^{-4}$) parameter $\alpha = (0.5579 \pm 0.004)$ were calculated.

Comparison of the equilibrium rigidity parameter of copolymers 4 and PDMS has shown that volumetric cyclic substituting group in copolymer 4 molecule results in an increase of equilibrium rigidity of copolymer macromolecules, A_0 = 27 Å, compared with that of PDMS, A_0 = 10 Å, and poly(dimethyldiphenylsiloxanes) (PDMDPS), A_0 = 16 Å. Tables 4.10 and 4.11 show viscosities and molecular-mass parameters of copolymers 1 and 4 (Table 4.7).

Thermogravimetric study of synthesized copolymers was performed. Figures 4.14 - 4.17 show thermogravimetric curves for copolymers. Figure 4.14 indicates that the substitution of a part of methyl groups by phenyl ones both at silsesquioxane silicon atom and in cyclic fragment causes rise of decomposition temperature up to 280 - 300°C. This is associated with high stability of phenyl groups to oxidation and their inhibiting effect on oxidation of methyl groups [20].

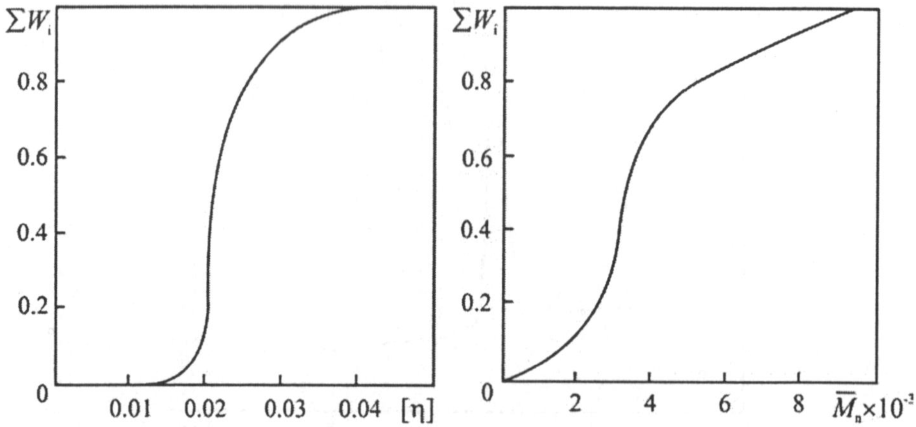

Figure 4.13. Integral curves of distribution by MM and [η] for copolymer 4 (Table 4.7)

Table 4.10

Viscosities and molecular mass parameters of copolymer 1 (Table 4.7)

Fraction No.	[η], dl/g at 25°C in toluene	$\bar{M}_n$ at 25°C in acetone	lg[η]	lg $\bar{M}_n$
1	0.322	9,585	-1.4921	3.9815
2	0.0248	3,520	-1.6055	3.5465
3	0.021	2,500	1.6778	3.3979
4	0.043	6,270	-1.3665	3.7978
5	0.05	-	-1.3011	-
6	0.0295	1,080	-	-
Non-fractionated	0.032	4,157 ÷ 5,460	-1.4979	3.6187÷ 3.732

Table 4.11

Viscosities and molecular mass parameters of copolymer 4 (Table 4.7)

Fraction No.	[η] dl/g at 25°C in toluene	$\bar{M}_n$ at 25°C in acetone	P_i, fraction mass (g)	W_i, fraction mass part	Corrected total mass part of fraction ΣW_i	lg[η]
1	0.02	3,080	0.6129	0.1127	0.0564	-1.6989
2	0.0204	1,916	2.7165	0.4994	0.5558	-1.6989
3	0.027	4,207	0.3047	0.0560	0.6118	-1.5686
4	0.021	3,450	0.7522	0.1383	0.7501	-1.6778
5	0.031	5,227	0.2956	0.0543	0.8044	-1.5086
6	0.026	5,073	0.2248	0.0413	0.8457	-1.5850
7	0.036	6,634	0.2398	0.0441	0.8898	-1.4559
8	0.0275	-	0.1565	0.0288	0.9186	-1.5607
9	0.0358	9,375	0.1365	0.0251	0.9437	-1.4461
Non-fractionated	0.021	2,116	5.4395	1.0000	-	-1.6778

Table 4.12

Some comparative physical properties of organosiloxane polymers

Copolymer, polymer unit	$M_l = M_0/\lambda_0$	Kuhn segment, Å	Free rotation	Dormancy factor
1. Poly(methylcyclosiloxane) (copolymer 4, Table 4.7)	100 Å	27.0	-	2.12
2. Poly(dimethylsiloxane) $\sim(Me_2SiO)_n\sim$	-	10.0	0.362	1.35
3. Poly(methylphenylsiloxane) (PMPS) $\sim(MePhSiO)_n\sim$	-	22.3	0.362	1.56
4. *Cis*-PMPS $\sim(MePhSiO)_n\sim$	-	21.2	0.365	1.97
5. *Trans*-PMPS $\sim(MePhSiO)_n\sim$	-	22.1	0.365	2.0
6. Poly(dimethyldiphenylsiloxane) $\sim(Me_2SiO)_5\text{-}(Ph_2SiO)_5\sim$	-	16.0	0.365	1.71

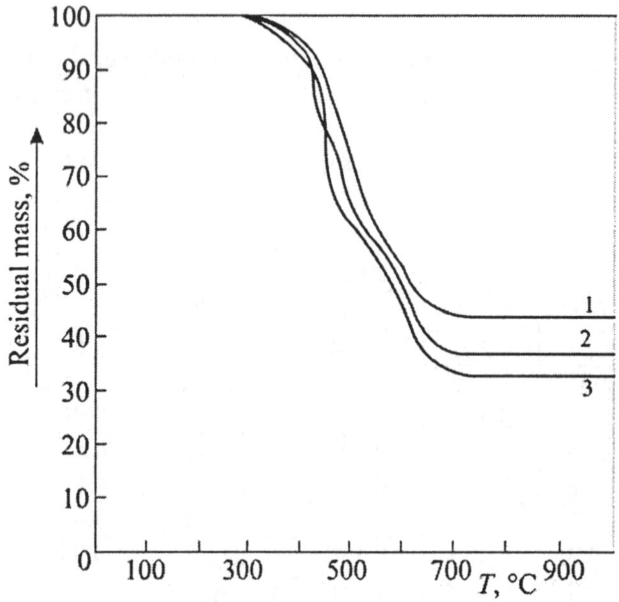

Figure 4.14. Thermogravimetric curves for copolymers: 1 - copolymer 11; 2 – copolymer 6; 3 – copolymer 1 (Table 4.4)

Oxygen bridge substitution by ethylene group causes a slight decrease of thermal oxidative stability of copolymers (Figure 4.15) [17]. As the cyclic fragment volume increases, i.e. as pentaphenylcyclotrisiloxane fragment is substituted by heptaphenylcyclotetrasiloxane one (Figure 4.16), a slight decrease of thermal oxidative stability happens. Similar phenomenon is observed at heptamethylcyclotetrasiloxane fragment substitution by nonamethylcyclopentasiloxane one (Figure 4.16).

For phenyl-containing copolymers 5% mass losses are observed in the temperature range of 300 - 350°C. The main degradation process proceeds in the temperature range of 300 - 600°C. Above 700°C the curves of mass losses come to a plateau. For copolymer with methylcyclosiloxane fragments as side groups, initial mass losses are observed in the temperature range of 230 - 250°C (Figure 4.17).

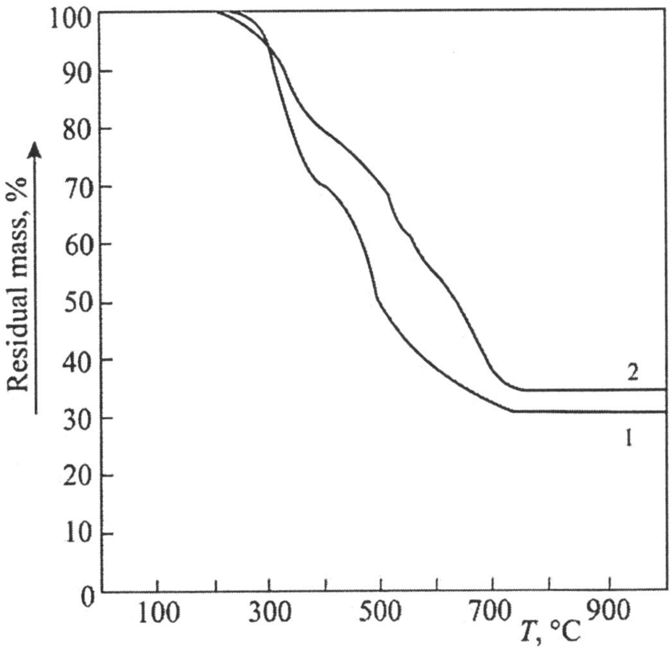

Figure 4.15. Thermogravimetric curves for copolymers: 1 - copolymer 2; 2 – copolymer 1 (Table 4.5)

The main degradation process proceeds in the temperature range of 300 - 600°C, and above 600°C the mass change curves come to a plateau.

General regularity is the decrease of thermal oxidative stability of copolymers with increase of dimethylsiloxane unit length. Comparison of data on thermal oxidative stability of copolymers with phenylcyclosiloxane fragments in the side chain and copolymers containing cyclotetra-, cyclopenta- and cyclohexasiloxane fragments in the backbone [22 - 25] indicated their approximately equal thermal oxidative stabilities, because introduction of cyclic fragments in the side chain hinders chain transfer reactions proceeding with D_n type rings liberation during degradation. Thus these data indicate that cyclic fragments in the side chain act similar to cyclic fragments, disposed along the dimethylsiloxane backbone in degradation.

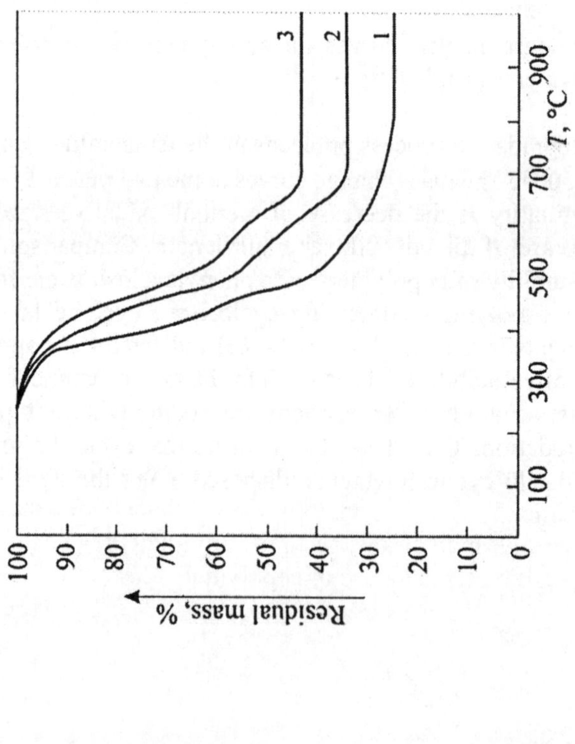

Figure 4.16. Thermogravimetric curves for copolymers: 1 - copolymer 4; 2 – copolymer 2; 3 – copolymer 1 (Table 4.6)

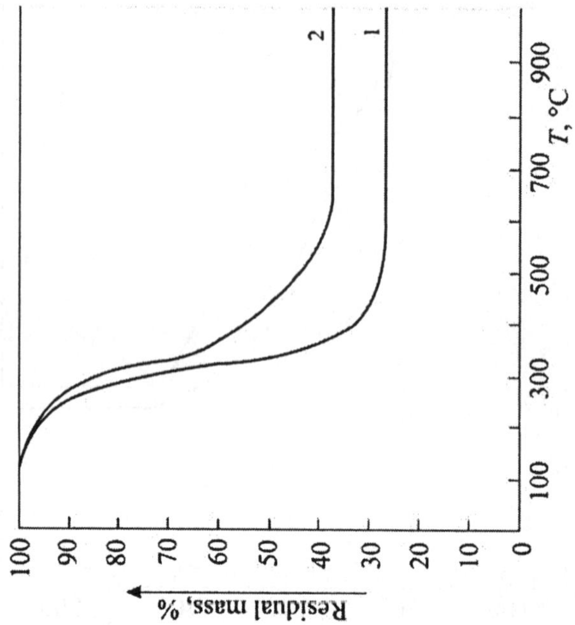

Figure 4.17. Thermogravimetric curves for copolymers: 1 - copolymer 4; 2 – copolymer 1 (Table 4.7)

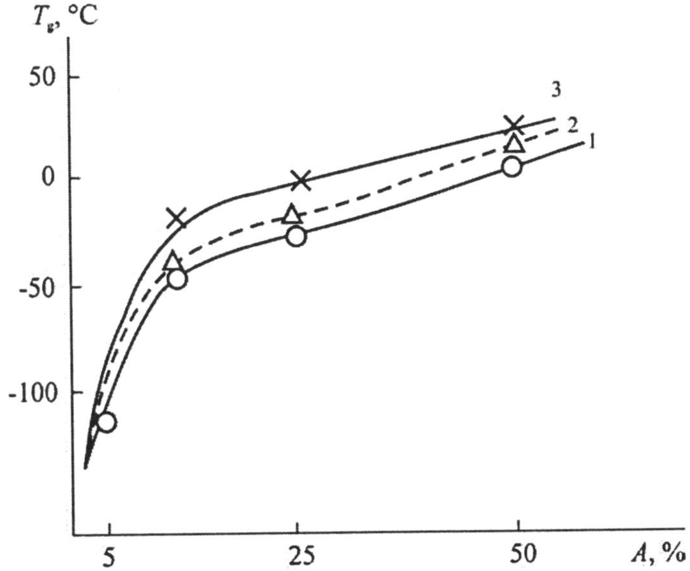

Figure 4.18. Dependence of T_g on mol% concentration of phenylcyclosiloxane fragments: 1 - copolymers 7 – 12; 2 – copolymers 1 – 6 (Table 4.4); 3 - copolymers 1-5 (Table 4.6)

Thermomechanical studies of the synthesized copolymers were performed. Figure 4.18 shows the dependence of T_g on mol% concentration of phenylcyclosiloxane units in the side chain at silicon atoms. Evidently Tables 4.4 - 5.8 indicate the increase of the glass transition temperature of copolymers with the number of framing cyclosiloxane fragments.

Substitution of a single methyl group by phenyl one in silsesquioxane unit causes an increase of T_g by ~10°C that is in agreement with previously published data [22 - 25]. Moreover, the general phenomenon for all copolymers is the decrease of their T_g down to that of linear PDMS with increase of dimethylsiloxane unit length (n).

Introduction of diphenylsiloxane units in the cyclic group increases T_g of copolymers by ~15°C. This is explained by a sufficient influence of phenyl groups, the presence of which at any place of poly(dimethylsiloxane) chain increases the glass transition temperature of copolymers [26].

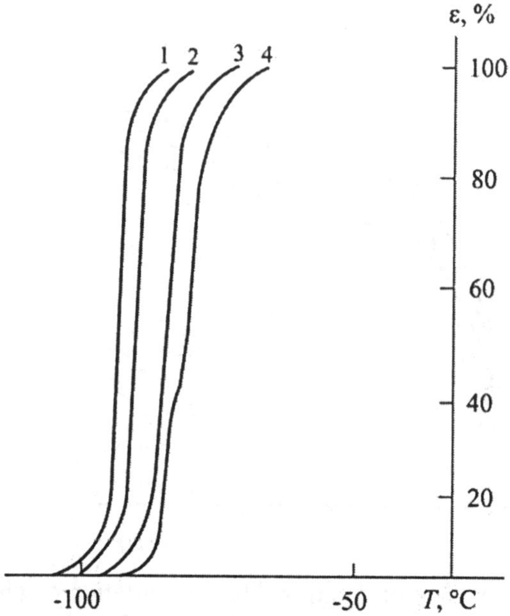

Figure 4.19. Thermomechanical curves of comb-shaped copolymers: 1 – copolymer 4; 2 – copolymer 3; 3 – copolymer 2; 4 – copolymer 1 (Table 4.7)

As oxygen is substituted by ethylene group in copolymers 1 - 5 (Table 4.5), flexibility of the copolymers increases, which decreases T_g of copolymers comparing with copolymers 1 - 6 (Table 4.4). Table 4.7 shows that introduction of methylcyclotetrasiloxane fragments as side chains to dimethylsiloxane backbone causes a negligible increase of T_g of copolymers. At $n = 1 - 3$, T_g varies from -90°C to -100°C. Glass transition temperature of copolymers with cyclopentasiloxane fragments in side chains is decreased by the increase of methylcyclopentasiloxane fragment flexibility. At the same length (n) of linear dimethylsiloxane unit, the difference between glass transition temperatures of copolymers with monocyclic (organocyclotetra-, organocyclopentasiloxane) fragments in the side chain equals 10°C (Figure 4.19). As the length of linear dimethylsiloxane unit is increased by one dimethylsiloxy-groups, T_g is decreased by 5°C (copolymers in Tables 4.7 and 4.8).

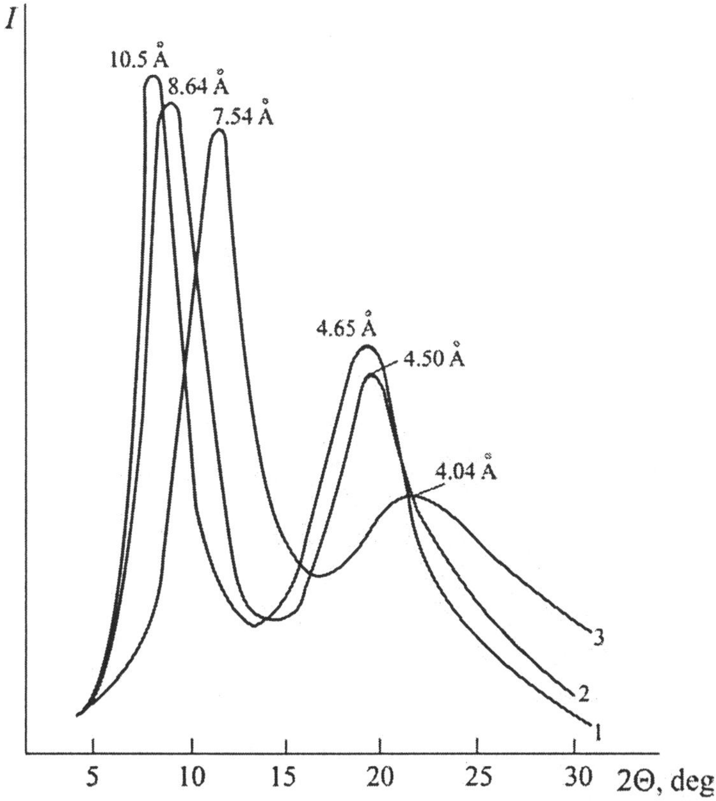

Figure 4.20. Diffraction patterns of comb-shaped methylsiloxane copolymers with phenylcyclotrisiloxane fragments in the side chain: 1 - copolymer 1; 2 - copolymer 7; 3 - copolymer 6 (Table 4.4)

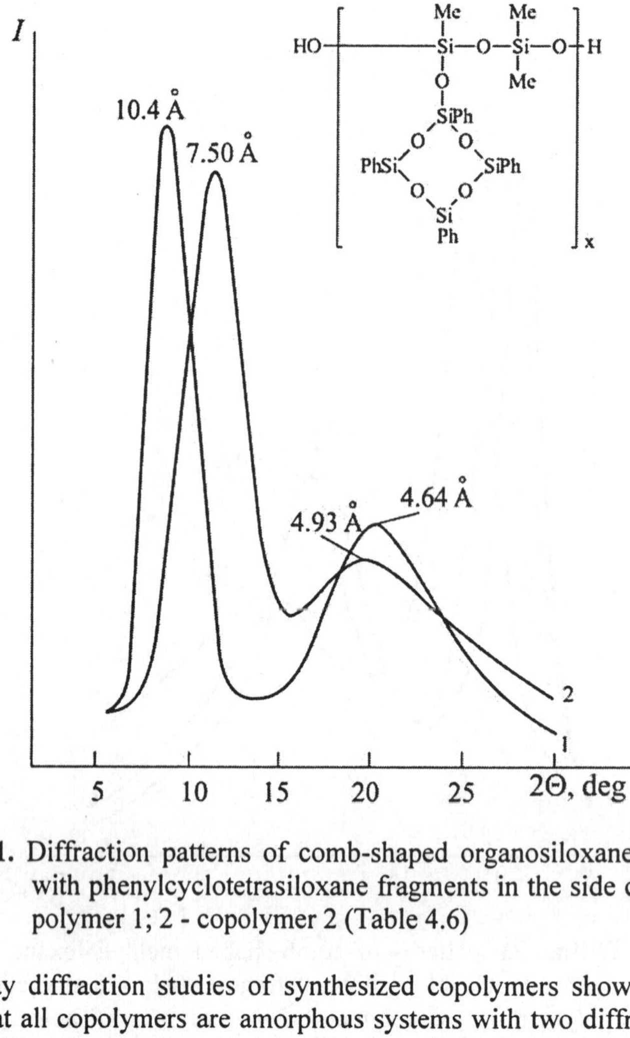

Figure 4.21. Diffraction patterns of comb-shaped organosiloxane copolymers with phenylcyclotetrasiloxane fragments in the side chain: 1 - copolymer 1; 2 - copolymer 2 (Table 4.6)

X-ray diffraction studies of synthesized copolymers show (Tables 4.4 and 4.5) that all copolymers are amorphous systems with two diffraction maximums: the first one, d_1, characterizing the interchain distance, falls within the range of 7.54 – 10.05 Å (refer to Figure 4.20); the second one, d_2, is more complicated and characterizes both interchain and interatomic interactions, and varies negligibly in all copolymers [27]. Maximal interchain distance is observed at short lengths of dimethylsiloxane unit.

Substitution of methyl radical by phenyl one at silicon atom in silsesquioxane fragment of copolymers, d_1 value is increased from 8.64 Å to 10.05 Å

(Table 4.4). In the case of heptamethylcyclotetrasiloxane fragment substitution by nonamethylcyclopentasiloxane one, d_1 is increased from 7.83 Å to 8.19 Å (Table 4.7, Figure 4.22). Similarly in the case of methyl group substitution at silsesquioxane silicon atom in copolymers with methylcyclotrisiloxane fragments in the backbone, the interchain distance d_1 is increased (Figure 4.22).

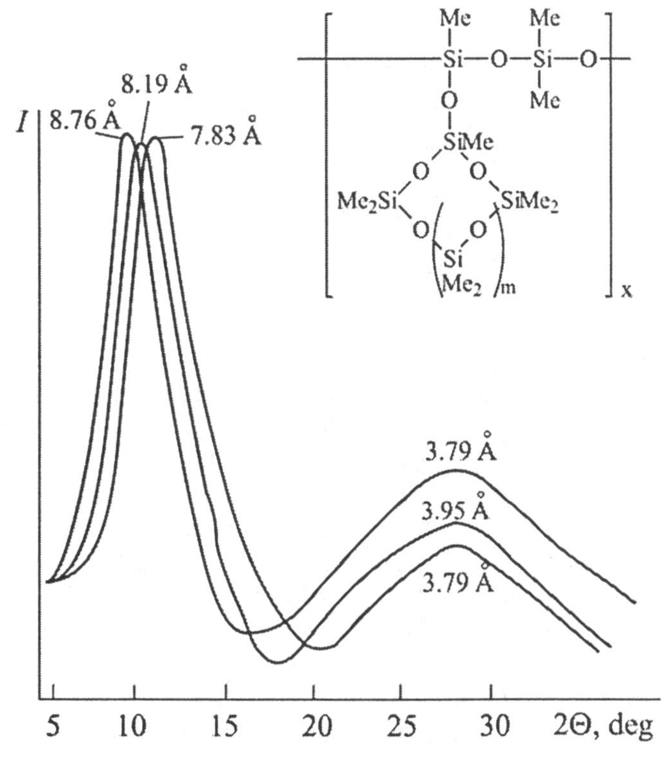

Figure 4.22. Diffraction patterns of comb-shaped methylsiloxane copolymers with methylcyclosiloxane fragments in the side chain: 1 - copolymer 1; 2 - copolymer 4 (Table 4.7); 3 - copolymer 1 (Table 4.8)

4.3. Synthesis of organosiloxane oligomers with cyclic fragments in the side chain by catalytic dehydrocondensation

Modification of functional groups containing methyl- and ethylhydridesiloxanes is one of perspective methods of synthesizing organosiloxane copolymers with cyclic fragments in the side chain. In the modification process, dehydrocondensation reactions of ≡Si-H- and ≡Si-OH-containing organosiloxanes proceeding in the presence nucleophilic catalysts (caustic potash, alkoxide) with elimination of hydrogen and formation of ≡Si-O-Si≡ bonds should be outlined.

For the purpose of synthesizing organosiloxane copolymers containing cyclosiloxane fragments in the side chain, commercially available oligoethylhydridesiloxane with end trimethylsiloxy-groups (GKZ-94, Russian product) with the polymerization degree $n = 15$ were used as initial reagents.

Catalytic dehydrocondensation of ethylhydridesiloxane with hydroxylorganocyclosiloxane proceeded in the presence of platinum-hydrochloric acid or caustic potash as a catalyst at various proportions of initial compounds at 40 - 60°C. It is shown that under these conditions polymerization or polycondensation of initial products does not take place. It is found that at 1:15 ratio of initial components dehydrocondensation is incomplete and not all active centers enter the reaction, which is explained by steric hindrances caused by organocyclosiloxane fragments during formation of the macromolecular chain.

Thorough GLC analysis of the reaction mixture has displayed the presence of unreacted D_4^{OH}. At 1:7.5 ratio of the initial components, the reaction proceeds at deeper level and conversion by hydrogen reaches 96%. It is shown that dehydrocondensation proceeding is highly dependent on the catalyst used. In the presence of platinum-hydrochloric acid, the reaction proceeds slowly at 60°C and after 6 hours the hydrogen conversion is below 50%, whereas in the presence of caustic potash dehydrocondensation proceeding is more vigorous. Dehydrocondensation proceeds in accordance with the following scheme [5, 28]:

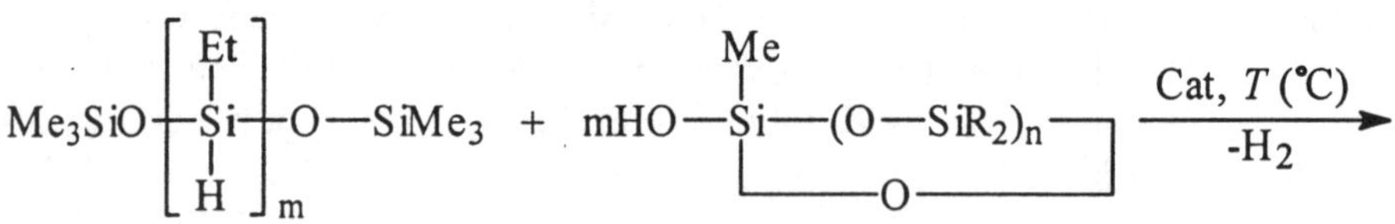

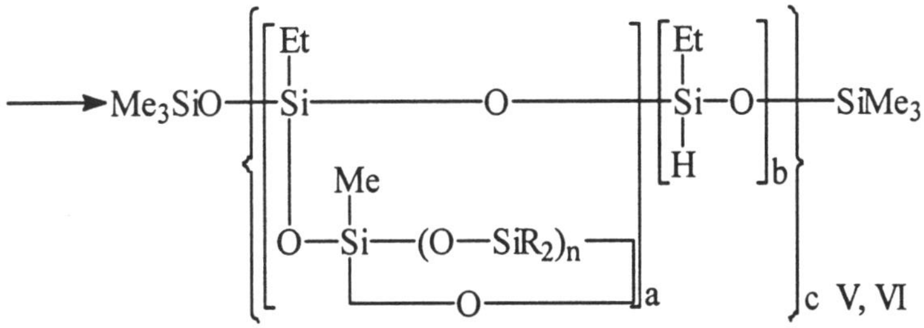

(4.5)

V, VI

where $m = 15$; $(a + b)c = m$; $n = 2$ (structure V), 3 (structure VI); R = Me, Ph.

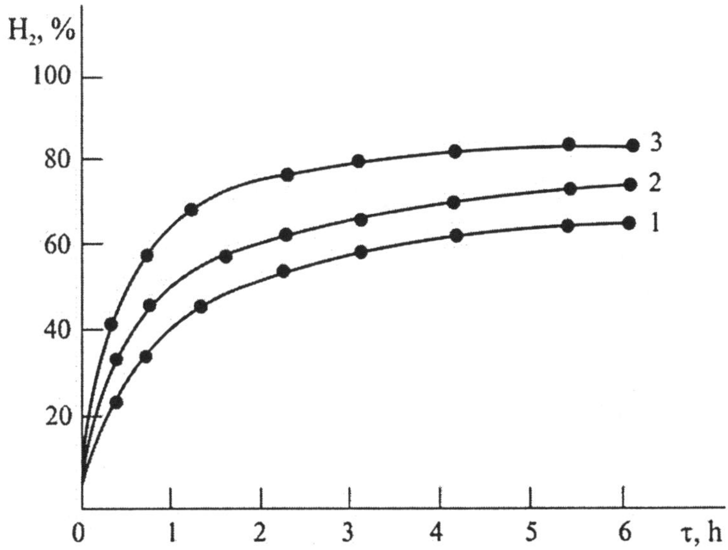

Figure 4.23. Hydrogen conversion in dehydrocondensation of oligoethylhydridesiloxane with hydroxyheptamethylcyclotetrasiloxane at the reaction temperature as follows: 1 - 40°C, 2 - 50°C, 3 - 60°C, at 1:15 ratio of initial components

Table 4.13 shows physical and chemical parameters and the yield of ethylsiloxane oligomers containing cyclic fragments in the side chain. IR spectra of these oligomers possess absorption bands of asymmetrical valence oscillations of ≡Si-O-Si≡ bonds at 1020 and 1080 cm^{-1}, typical of both linear and cyclic siloxane units. In these spectra absorption bands of unreacted ≡Si-H bonds at 2160 cm^{-1} are also observed.

Figure 4.23 shows the increase of hydrogen conversion from 65 to 80% with temperature rise from 40 to 60°C. Dehydrocondensation proceeds intensively during initial 1.5 - 2 hours, and after 5 hours the conversion is changed negligibly. Evidently, at 1:15 ratio of initial compounds the reaction is incomplete, and dehydrocondensation of all active centers is impeded by steric effects, caused by organocyclosiloxane fragments during formation of the macromolecular chain. At a 1:7.5 ratio of the initial compounds the reaction proceeds more intensively and deeper (Figure 4.24).

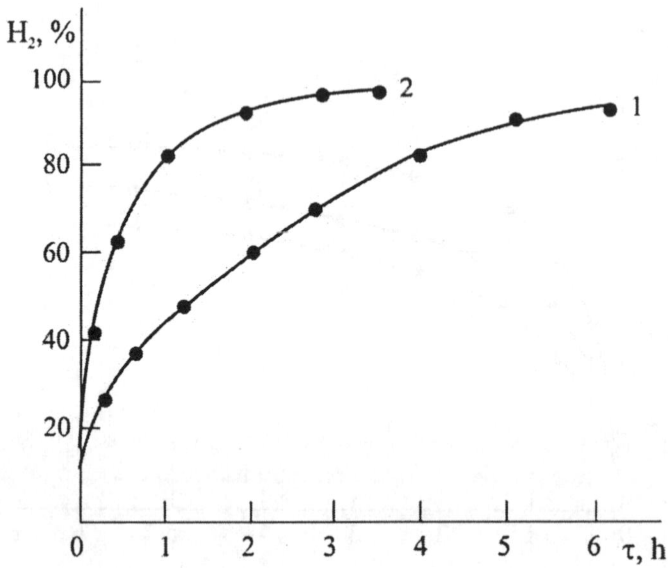

Figure 4.24. Hydrogen conversion in dehydrocondensation of oligoethylhydridesiloxane with hydroxyheptamethylcyclotetrasiloxane at reaction temperature as follows: 1 - 40°C, 2 - 60°C, at 1:7.5 ratio of initial components

The reaction was studied in the presence of anhydrous KOH (0.1 wt% of the initial components, C = 0.2271 mol/l) at 40, 50 and 60°C. Figure 4.25 shows the time dependence of the reverse concentration. It is found that dehydrocondensation is the second order reaction. The reaction rate constants at various temperatures were calculated: $k_{40°C} = 0.923\times10^{-1}$, $k_{50°C} = 1.6\times10^{-1}$, $k_{60°C} = 2.85\times10^{-1}$. From the dependence of the reaction rate constant logarithm on reverse temperature (Figure 4.26), the activation energy of dehydrocondensation was calculated, $E = 51.5$ kJ/mol. It was found that for each 10°C of temperature rise, the reaction rate constant increases by 1.7 times, approximately.

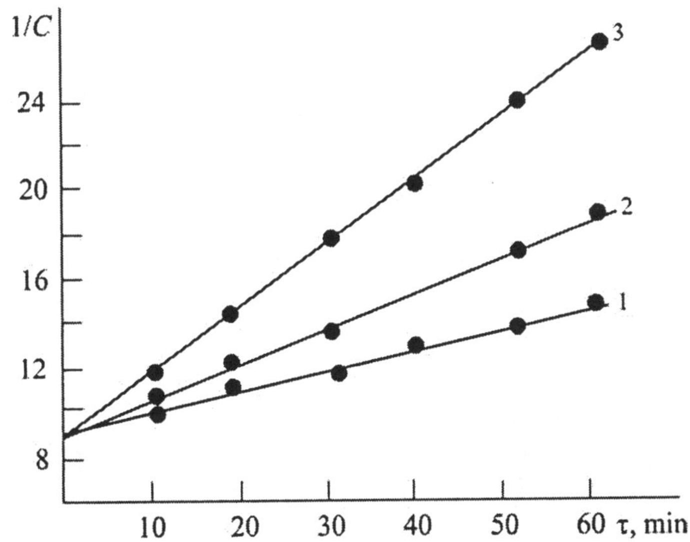

Figure 4.25. Time dependence of reverse concentration for catalytic dehydrocondensation of oligoethylhydridesiloxane with hydroxyheptamethylcyclotetrasiloxane at 1:15 ratio: 1 - at 40°C, 2 - at 50°C; 3 - at 60°C

Thermomechanical studies of the oligomers show (Table 4.13) that the glass transition temperature, T_g, of oligomer 1 equals -80°C and –95°C for oligomer 4. This indicates T_g increase with the content of methylcyclotetrasiloxane fragments in the side chain. The reason for this phenomenon is the increase oligomeric segment rigidity.

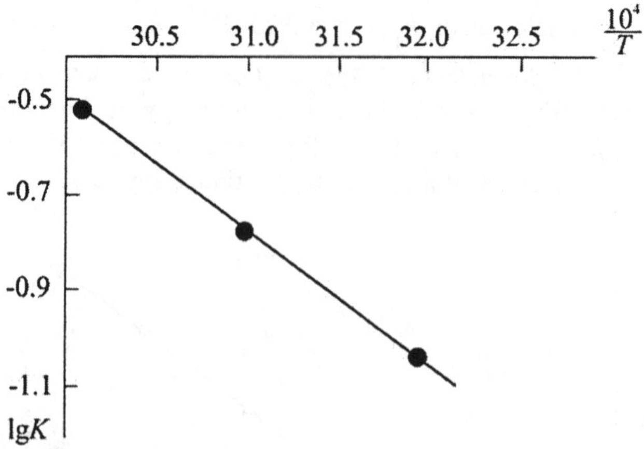

Figure 4.26. Dependence of reaction rate constant logarithm on reverse temperature

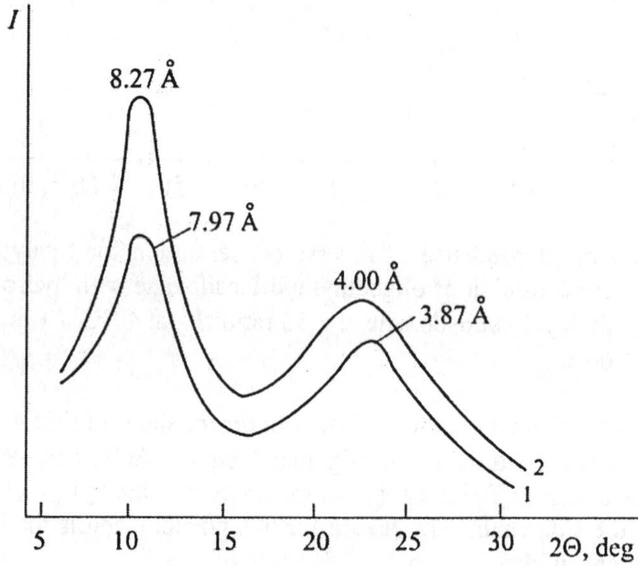

Figure 4.27. X-ray diffraction patterns for oligoethylhydridesiloxanes with methylcyclosiloxane fragments in the side chain: 1 - oligomer 4; 2 - oligomer 3 (Table 4.13)

Table 4.13

Some physical and chemical properties and yields of ethylsiloxane oligomers with organocyclotri(tetra)siloxane fragments in the side chain

Oligomer No	n	R	T_r, °C	Initial component ratio	Yield, %	η_{sp}*	d_1, Å	H_2 conversion, %	T_g, °C
1	3	Me	40	1:15	87	0.04	-	65	-
2	3	Me	50	1:15	90	0.04	8.27	72	-80
3	3	Me	60	1:15	90	0.04	8.27	80	-76
4	3	Me	60	1:7.5	91	0.03	7.97	98	-95
5	2	Ph	60	1:15	88	0.03	8.73	66	-62

Note: * In 1% toluene solution, at 25tolueneC.

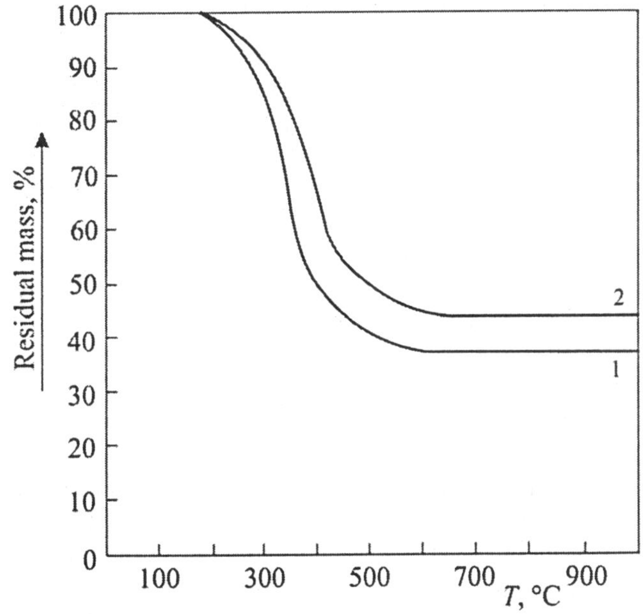

Figure 4.28. Thermogravimetric curves of oligoethylhydridesiloxane oligomers with methylcyclosiloxane fragments in the side chain: 1 - oligomer 4; 2 - oligomer 3 (in air, at 5°C/min heating rate)

X-ray diffraction studies show that the interchain distances in oligomers 3 and 5 (Table 4.13) are different, because the interchain distance increases from 7.97 Å (oligomer 4) to 8.27 Å (oligomer 3) with the number of cyclic fragments in the side chain (Figure 4.27).

Thermogravimetric studies of synthesized oligomers display (Figure 4.28) their high thermal oxidative stability and 5% mass losses at 240 - 250°C. The main destruction process proceeds in the temperature range of 300 - 550°C. Above 600°C the mass loss curves do not change. Thermal oxidative stability of oligomers is also increased with the number of cyclic fragments in the side chain that prevents chain transfer in depolymerization [6].

4.4. Synthesis of cyclolinear organosiloxane copolymers by hydrosilylation

It is known from the literature [29 - 33] that hydrosilylationis widely used for synthesis of cyclolinear copolymers. Discussed in this Chapter is synthesis of organosiloxane copolymers with carbocyclosiloxane fragments in the side chain by hydride addition of methylhydridesiloxanes and methylhydridesiloxanedimethylsiloxane copolymer to vinyl-containing organocyclosiloxanes. It is shown that hydride addition of dihydrideorganocyclosiloxanes to divinyl-containing organocyclosiloxanes in the presence of platinum-hydrochloric acid does not cause cyclosiloxane ring scission [29 - 33].

Methylhydridesiloxanes interlocked by trimethylsiloxy-groups were successfully used for modification by hydrosilylation [34, 35]. Comb-shaped methylsiloxane copolymers were synthesized by hydride addition of methylhydridesiloxanes to allyl glycidyl ester [36, 37], allyl alcohol [38, 39] or dicyclopentadiene [35, 40] in the presence of catalysts.

For purpose of synthesizing carbosiloxane copolymers with cyclic fragments in the side chain, hydride addition of methylhydridesiloxanes to vinylorganocyclosiloxanes in the presence of catalysts was studied [41]. Preliminary heating up of 1-vinyl-1-methyltetraphenylcyclotrisiloxane (at concentration C = 0.02628 mol/l) during 8 hours in the presence of platinum-hydrochloric acid (9×10^{-5} g per 1 g of reagents) in the temperature range of 80 - 100°C has not induced polymerization of vinylorganocyclosiloxane, and no changes in ^{1}H NMR spectra for methyl protons have been detected. Hydrosilylation of methylhydridesiloxane (n = 35) with 1-vinyl-1-methyltetraphenylcyclotrisiloxane at 1:35 ratio of the initial reagents at 80, 90 and 100°C was studied.

Figure 4.29 shows that hydrogen conversion is increased from 85% to 95% with temperature rise from 80 to 100°C. Hydrosilylation is intensive during initial 50 - 60 min and after 80 min becomes negligible. Note also that the reaction is incomplete, and substitution of all active ≡Si-H groups by cyclotrisiloxane fragments is not observed. This may be explained by the steric effect induced by organocyclosiloxane fragments during formation of the macromolecular chain. Complete hydrosilylation is observed at 1:17.5 ratio of the initial reagents. Hydride addition proceeds in accordance with the general scheme as follows:

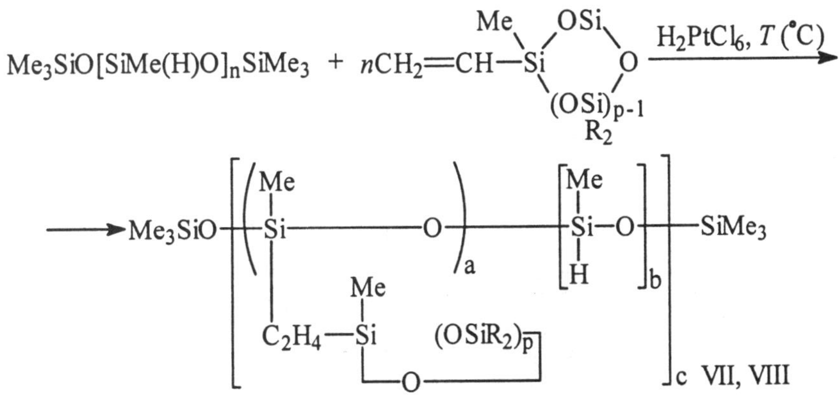

(4.6)

where $n = 35$, $(a + b)c = n$; $p = 2$ (VII), 3 (VIII); R = Ph, Me.

Synthesized copolymers represent light-yellow liquids or solids, well soluble in ordinary organic solvents, with $\eta_{sp} = 0.05 - 0.07$.

Structure and composition of synthesized copolymers were determined by ultimate analysis and from IR and NMR spectral data. Some physical and chemical parameters of the comb-shaped methylsiloxane copolymers with carbocyclosiloxane fragments in the side chain are shown in Table 4.14.

IR spectra of copolymers display absorption bands typical of unreacted ≡Si-H bond. They also display new absorption bands in the range of 2,160 – 2,170 cm^{-1}, typical of fan oscillations of CH_2 group and for asymmetrical valence oscillations of ≡Si-O-Si≡ bonds at 1,020 and 1,075 cm^{-1}, typical of both linear siloxane units and siloxane bonds in cyclotri- and –cyclotetrasiloxane rings. In copolymers with cyclotrisiloxane fragments in the side chain, absorption bands typical of asymmetrical valence oscillations of ≡Si-O-Si≡ bonds in linear dimethylsiloxane chain and cyclotrisiloxane ring overlap.

Table 4.14

Some physical and chemical parameters and yields of comb-shaped methylsiloxane copolymers with organocarbocyclosiloxane fragments in the side chain

Oligomer No.	Yield, %	p	R	T_r, °C	Initial reagent ratio	η_{spec}^*	d_1, Å	T_g, °C	H_2 conversion, %
1	93	2	Ph	80	1:35	0.05	10.30	-	85
2	94	2	Ph	90	1:35	0.06	-	-	90
3	94	2	Ph	100	1:35	0.07	10.30	+12 - +15	95
4	96	2	Ph	100	1:17.5	0.05	9.80	0-+3	100
5	96	3	Me	90	1:35	0.05	8.45	-94	92
6	95	4	Me	90	1:35	0.05	8.76	-110	90

^1H NMR spectrum of copolymer 3 (Table 4.14) displays a broadened signals for methyl protons (3H) in -O-SiMe-O- fragment with chemical shift δ= 0.27 ppm and methyl protons in the backbone with chemical shift δ= 0.10 ppm. For copolymers 5 and 6 (Table 4.14), broadened singlet reflex (4H) with chemical shift δ= 0.72 ppm, typical of $-CH_2-CH_2-$ fragment and indicating the Farmer rule proceeding of hydrosilylation, is also observed. The more so, singlet reflexes from methyl protons at 0.13 and 0.15 ppm, typical of cyclic fragments, are detected.

Figure 4.29 indicates hydrogen conversion increase from 85% to 95% with temperature rise from 80°C to 100°C. Hydrosilylation proceeds intensively during initial 2 hours. Figure 4.30 shows time dependence of reverse concentration. The second order of hydrosilylation is detected. Hydrosilylation rate constants were calculated: $k_{80^\circ C} = 5.5431\times10^{-1}$, $k_{90^\circ C} = 7.5355\times10^{-1}$, $k_{100^\circ C} = 1$. Figure 4.31 shows dependence of hydrosilylation rate constant logarithm, lg*k*, on reverse temperature, and activation energy was calculated, *E* = 31.6 kJ/mol.

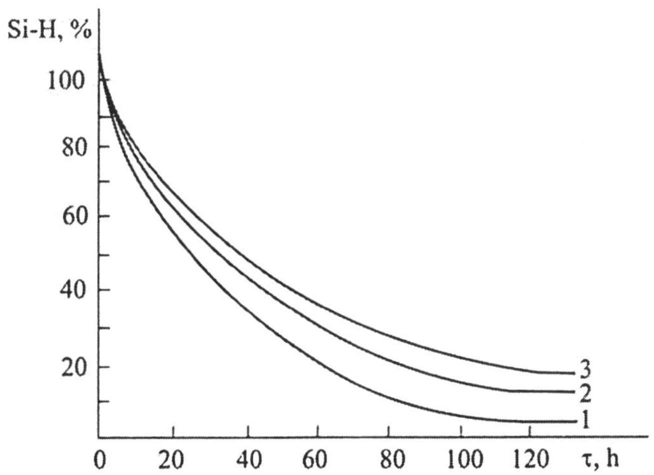

Figure 4.29. Hydrogen conversion in hydride addition of α,ω-bis-(trimethylsiloxy)methylhydridesiloxane to 1-vinyl-1-methyl-tetraphenylcyclotrisiloxane: 1 - at 100°C; 2 – at 90°C; 3 – at 80°C

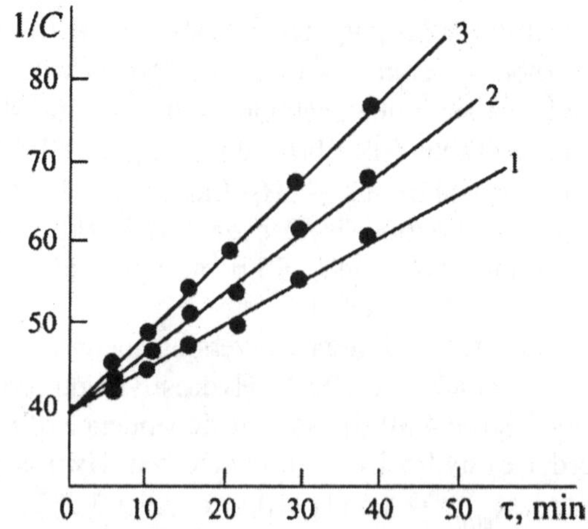

Figure 4.30. Time dependence of reverse concentration for hydride addition of methylhydridesiloxane to 1-vinyl-1-methyltetraphenylcyclotrisiloxane at 1:35 ratio of initial reagents: 1 - at 80°C; 2 - at 90°C; 3 - at 100°C

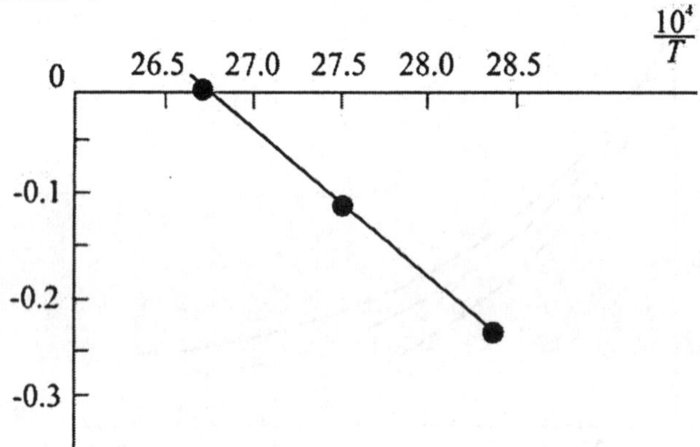

Figure 4.31. Dependence of the reaction rate constant logarithm of hydride addition of methylhydridesiloxane to 1-vinyl-1-methyltetraphenylcyclotrisiloxane on reverse temperature at 1:35 ratio pf initial reagents

X-ray diffraction studies show that synthesized copolymers are single-phase amorphous products with interchain distance for copolymer 3 (Table

4.14) equal d_1 = 10.30 Å. The interchain distance decreases with the number of cyclosiloxane fragments in the side chain (copolymer 4, Table 4.14).

Figure 32 shows differential scanning calorimetry curves of synthesized copolymers. In accordance with data from the Figure and Table 4.14 copolymer 3 possesses higher glass transition temperature, compared with copolymer 1 (Table 4.5), explained by the presence of phenyl framing groups in the cyclic fragments at silicon atoms. The glass transition temperature is decreased with the quantity of cyclic fragments (refer to Table 4.14).

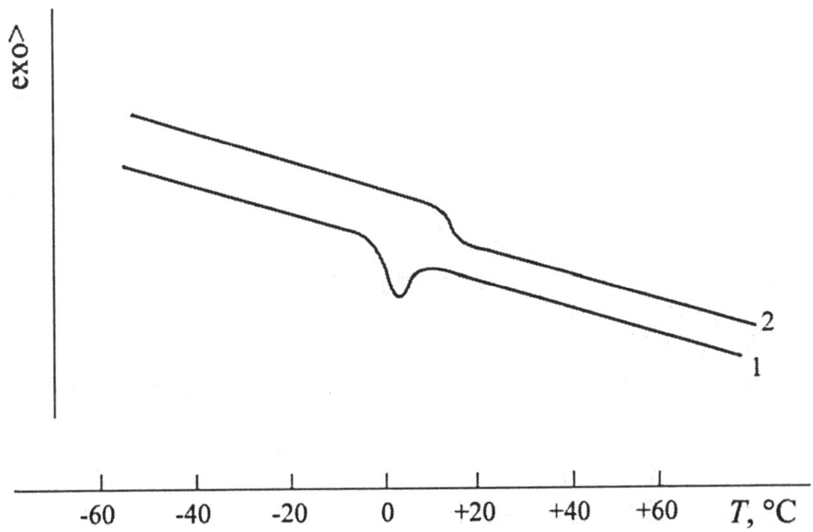

Figure 4.32. DSC patterns for oligomers with cyclic fragments in the side chain: 1 - oligomer 3; 2 - oligomer 4 (Table 4.14)

Substitution of methylcyclotetrasiloxane fragment by methylcyclopentasiloxane one decreases the glass transition temperature, T_g, by 13 - 15°C, which coincides with the previously published data [7]. Thermogravimetric studies show copolymers 3 and 4 (Table 4.14) possession of higher thermal oxidative stability compared with copolymer 1 (Table 4.5).

Thus, that was the first time, when behavior of methylsiloxane oligomers and copolymers with various cyclic fragments in the side chain similar to dimethylsiloxane copolymers with cyclic fragments in the backbone was indicated [8].

References

1. Sobolevski M.V., *Oligoorganosiloxanes: Properties, Production and Application*, Khimia, Moscow, 1985. (Rus)
2. Mileshkevich V.P., *Kauchuk i Rezina*, 1978, No. 6, p. 4. (Rus)
3. Tomas T.H. and Kendrick T.C., *J. Polym. Sci.*, 1969, vol. **2A**(7), p. 537.
4. Mukbaniani O.V., Esaptia I.G., and Khananashvili L.M., *Soobshch. AH GSSR*, 1991, vol. **142**(3), p. 537. (Rus)
5. Esartia I.G., "Linear Poly(organosiloxanes) with Cyclic Fragments in the Side Chain", *Candidate's Dissertation on Chemistry*, 1993, Tbilisi State University, Tbilisi, Georgia. (Rus)
6. Esartia I.G., Mukbaniani O.V., Khananashvili L.M., and Nizharadze N.N., *Soobshch. AN GSSR*, 1991, vol. **142**, p. 537. (Rus)
7. Mukbaniani O.V. "Organosilicon Copolymers and Block-Copolymers with Various Cyclic Structure of Macromolecules", *Doctor's Dissertation on Chemistry*, 1993, Tbilisi State University, Tbilisi, Georgia. (Rus)
8. Mukbaniani O.V., Meladze S.M., Esartia I.G., Tverdokhlebova I.I., and Khananashvili L.M., *J. Appl. Polym. Sci*, 1999, vol. **74**, p.583.
9. Esartia I.G., Tsitsishvili V.G., Mukbaniani O.V., and Khananashvili L.M., *Zh. Obshch. Khim.*, 1992, vol. **62**, p. 1120. (Rus)
10. Makarova N.N. and Lavrukhin B.D., *Izv. AN SSSR, Ser. Khim.*, 1986, No. 3, p. 652. (Rus)
11. Achelashvili V.A., "Cyclic Organosiloxanes, Organocarbosiloxanes and Copolymers Derived from Them", *Candidate's Dissertation on Chemistry*, 1989, Tbilisi State University, Tbilisi, Georgia. (Rus)
12. Mukbaniani O.V., Achelashvili V.A., Meladze S.M., and Khananashvili L.M., *Abstr. Commun., 7th International Symposium in Organosilicon Chemistry*, Japan, Kyoto, 1984, p. 216.
13. Yuzhelevski Yu.A., Kurlova T.V., Kogan B.G., and Suvorova M.V., *Zh. Obshch. Khim.*, 1972, vol. **42**, p. 2006. (Rus)
14. Andrianov K.A., Nogaideli A.I., Makarova N.N., and Mukbaniani O.V., *Izv. AN SSSR, Ser. Khim.*, 1977, No. 6, p. 1388. (Rus)
15. Nagy I., Farkas R., Mukbaniani O.V., Esartia I.G., and Khananashvili L.M., *Abstr. Commun., 7th International Symposium in Organosilicon Chemistry*, Japan, Kyoto, 1984, p. 162.
16. Mukbaniani O.V., Esartia I.G., Khananashvili L.M., Nagy I., and Farkas R., *Abstr. Commun., 31st IUPAC Macromolecular Symposium*, GDR, Merseburg, 1987, p. 93.

17. Esartia I.G., Mukbaniani O.V., Khananashvili L.M., and Meladze S.M., *Soobshch. AN GSSR*, 1986, vol. **12**, p. 234. (Rus)
18. Mukbaniani O.V., Esartia I.G., and Khananashvili L.M., *Bulletin of the Georgian Academy of Sciences*, 1991, vol. **144**, p. 277.
19. Mukbaniani O.V., Esartia I.G., and Khananashvili L.M., *Bulletin of the Georgian Academy of Sciences*, 1991, vol. **144**, p. 381.
20. Esartia I.G., Mukbaniani O.V., and Khananashvili L.M., *Proceedings of the Georgian Academy of Sciences*, 1993, vol. **19**, p. 212.
21. Mukbaniani O.V., Meladze S.M., Esartia I.G., Tverdokhlebova I.I., and Khananashvili L.M., *J. Appl. Polym. Sci.*, 1999, vol. **74**, p. 583.
22. Andrianov K.A., Nogaideli A.I., Slonimsky G.L., Levin V.Yu., Makarova N.N., Kvachev Yu.P., and Mukbaniani O.V., *Vysokomol. Soedin.*, 1976, vol. **18B**, p. 359. (Rus)
23. Meladze S.M., Mukbaniani O.V., and Khananashvili L.M., *Vysokomol. Soedin.*, 1984, vol. **26B**, p. 250. (Rus)
24. Koyava N.A., Mukbaniani O.V., and Khananashvili L.M., *Vysokomol. Soedin.*, 1985, vol. **27A**, p. 2261. (Rus)
25. Achelashvili V.A., Mukbaniani O.V., Khananashvili L.M., and Sturua G.I., Proceedings of the Georgian Academy of Sciences, Collection, 1988, p. 53. (Rus)
26. Andrianov K.A., Slonimski G.L., Levin V.Yu., Zhdanov A.A., Godovsky Yu.K., Kuznetsova N.K., Tsvankin D.Ya., Moskalenko V.A., and Kuteinikova I.I., *Vysokomol. Soedin.*, 1970, vol. **12A**, p. 1268. (Rus)
27. Andrianov K.A., Kononov L.M., and Tsvankin D.Ya., *Vysokomol. Soedin.*, 1968, vol. **10B**, p. 320. (Rus)
28. Mukbaniani O.V., Khananashvili L.M., Esartia I.G., and Khaduri S.I., *Int. J. Polym. Mater.*, 1994, vol. **24**, p. 131.
29. Khananashvili L.M., Mukbaniani O.V., Koyava N.A., Porchkhidze G.V., and Tolchinski Yu.I., *Int. J. Polym. Mater.*, 1992, vol. **17**, p. 113.
30. Karchkhadze M.G., Mukbaniani N.O., Khananashvili L.M., Meladze S.M., Kvelashvili N.G., and Doksopulo T.P., *Int. J. Polym. Mater.*, 1998, vol. **41**, p. 89.
31. Mukbaniani N.O., Karchkhadze M.G., Samsonia A.Sh., Tkeshelashvili R.Sh., and Khananashvili L.M., *Bulletin of the Georgian Academy of Sciences*, 1999, vol. **160**(1), p. 84.
32. Zhdanov A.A., Pryakhina T.A., Strelkova T.V., Afonina R.I., and Kotov V.M., *Vysokomol. Soedin.*, 1993, vol. **35**(5), p. 475. (Rus)

33. Mukbaniani O.V., Khananashvili L.M., Karchkhadze M.G., Tkeshelashvili R.Sh., and Mukbaniani N.O., In: *Synthesis and Properties of Polymers*, Ed. G.E. Zaikov, Nova Science Publ., Inc. Commack, 1996, p. 89.
34. Mukbaniani O.V., Karchkhadze M.G., and Khananashvili L.M., *Abstr. Commun., XIth International Symposium on Organosilicon Chemistry*, Montpellier, France, 1996, PB-61.
35. Mukbaniani O.V. and Koberidze Kh.E., *Thes. Rep. K.A. Andrianov 95th Anniversary Conference*, Moscow, February 1 – 4, 2000, p. 37. (Rus)
36. Mukbaniani O.V., Scherf U., Gurgenidze G.N., Karchkhadze M.G., Meladze S.M., and Khananashvili L.M., *Int. J. Polym. Mater.*, 2001, vol. **48**(3), p. 267.
37. Gurgenidze G.N., Meladze S.M., Mukbaniani O.V., and Khananashvili L.M., *Proceedings of the Georgian Academy of Sciences*, 2000, vol. **26**(1-2), p. 78.
38. Mukbaniani O.V., Gurgenidze G.N., Meladze S.M. and Khananashvili L.M., *Russian Polymer News*, 2001, vol. **6**(4), p. 18.
39. Mukbaniani O.V., Gurgenidze G.N., Meladze S.M. and Khananashvili L.M., In: *Radical and Ion Reactions, Problems and Ways of Their Solution*, Ed. G.E. Zaikov, 2001, Nova Science Publishers, Inc, p. 79.
40. Mukbaniani O.V., Scherf U., Karchkhadze M.G., Koberidze Kh.E., and Labartkava M.O., *Int. J. Polym. Mater.*, 2000, vol. **48**(2), p. 177.
41. Mukbaniani N.O., Meladze S.M., and Achelashvili V.A., *Abstr. Commun., Scientific Conference on Chemistry*, Tbilisi, Georgia, 1998, p. 45.

Chapter 5. Organosilicon copolymers with carbocyclosiloxane fragments in dimethylsiloxane backbone

Organosiloxane copolymers with carbocyclic fragments in the backbone can be synthesized by HFC reaction between primarily prepared difunctional organocyclosiloxanes and organocyclocarbosiloxanes and α,ω–dichloro(dihydroxy)dimethylsiloxanes. Another method of synthesis of copolymers with carbocyclic fragments in the backbone represents polymerization and copolymerization of organospirobicyclosiloxanes and polymerizing polycondensation of organospirobicyclosilazasiloxanes with various organosilicon diols.

Discussed in this Chapter is the synthesis of primary organosiloxane and carbosiloxane compounds with functional groups of organobicyclospirosiloxanes and organobicyclosilazasiloxanes. Methods of obtaining copolymers with carbocyclosiloxane fragments in the backbone are also discussed.

5.1. Synthesis of organospirobicyclosiloxanes and organospirobicyclosilazasiloxanes

To obtain organospirobicyclosiloxanes, at the first stage of the process initial 1,1-dichlorophenylcyclotri(tetra)siloxane was synthesized [1 – 3].

HFC reaction of 1,3-dihydroxytetraphenyldisiloxane and 1,5-dihydroxyhexaphenyltrisiloxane with silicon tetrachloride has been studied. It is shown that the reaction process and structure of the products depend on dilution and the ratio of reagents, but in the absence of hydrogen chloride acceptor, at most two chlorine atoms in a molecule of silicon tetrachloride are substituted. The reaction proceeds by the scheme as follows:

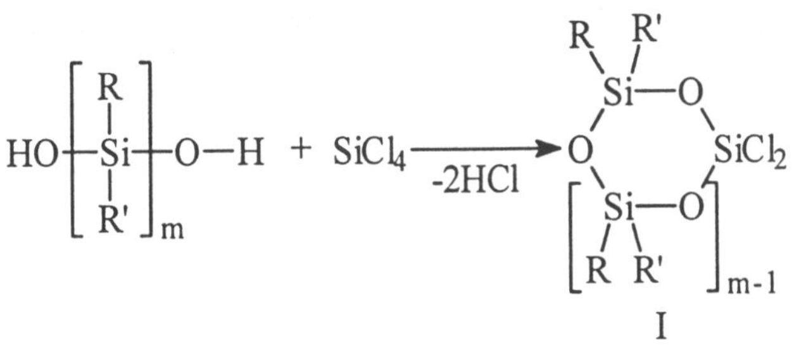

(5.1)

where R = R' = Me, Ph; R ≠ R', m = 2, 3.

It is found that affinity to homocondensation of dihydroxyorganosiloxanes increases with decrease of the framing groups' size at the atom of silicon, which results in decrease of organochlorocyclosiloxanes content. The presence of hydrogen chloride acceptor in the reaction decreases the effect of diol side homocondensation and gives an opportunity to apply these reactions to direct synthesis of mentioned organocyclosiloxanes. However, in spite of the above-mentioned references, the study of HFC reaction of silicon tetrachloride with 1,3-dihydroxytetraphenyldisiloxane [4] displayed formation of broadened cyclohexasiloxane and its structural isomers in addition to dichlorotetraphenylcyclotrisiloxane. Table 5.1 shows some physical and chemical parameters of synthesized dichlororganocyclotri(tetra)siloxanes.

HFC reaction of 1,3-dihydroxytetraorganodisiloxane with silicon tetrachloride (at 2:1 ratio of the initial components in the presence of acceptor, pyridine) proceeds with complete substitution of chlorine atoms and is accompanied by spirocyclization in accordance with the scheme as follows [5 - 8]:

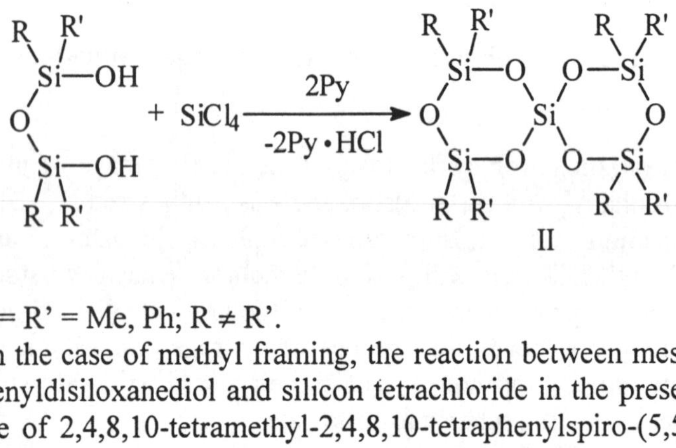

(5.2)

where R = R' = Me, Ph; R ≠ R'.

In the case of methyl framing, the reaction between meso-1,3-dimethyl-1,3-diphenyldisiloxanediol and silicon tetrachloride in the presence of pyridine gives one of 2,4,8,10-tetramethyl-2,4,8,10-tetraphenylspiro-(5,5)-pentasiloxane stereo-isomers with the yield up to 35% in accordance with the following scheme [8, 9]:

Table 5.1

Some physical and chemical properties of synthesized dichlororganocyclotri(tetra)siloxanes

No.	Compound	R	R'	*m*	Yield, %	T_{boil}, °C (P, mmHg)	T_{melt}, °C	References
1	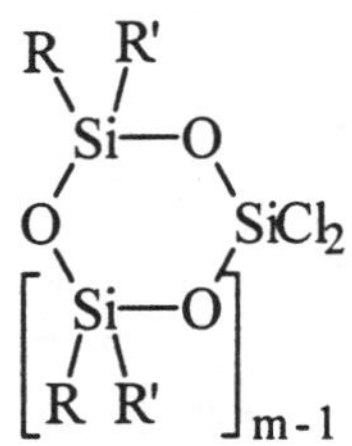	Ph	Ph	2	82	232 - 233 (1)	83-84	[3]
2		Me	Me	2	43	90 - 91 (750)	66-67	[3]
3		Me	Me	3	67	81 - 82 (2)	30-31	[3]
4		Me	Ph	2	65	170 - 171 (2)	76-77	[2]

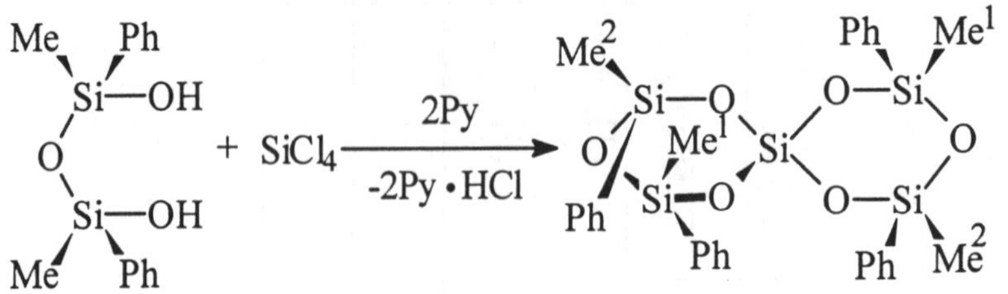

(5.3)

II

Spirobicyclosiloxanes were synthesized by HFC reaction of dichlorotetraorganocyclotrisiloxane with dihydroxydiorganosiloxanes (1:1 ratio of initial reagents, in the presence of pyridine). The reaction proceeds as follows:

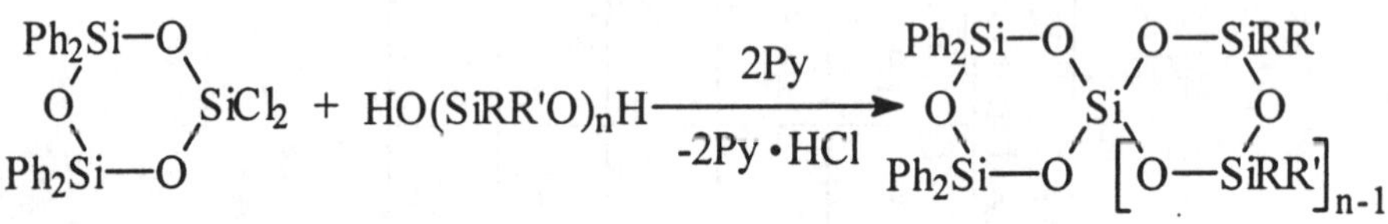

(5.4)

II

where R = R' = Me, Ph; R ≠ R'; $n = 2, 3$.

^{1}H NMR spectrum of the compound, obtained in reaction (5.7), where R = Me, R' = Ph, displays reflexes from protons of methyl groups of both isomers. Chemical shifts of *cis-* and *trans-*isomer protons differ due to different contributions to screening, provided by circular currents of phenyl groups. In such systems [10] reflexes from protons of methyl groups prove the *cis-*disposition in relation to phenyl groups. This allows relation of reflexes with $\delta =$ 0.44 and $\delta = 0.53$ ppm chemical shifts to methyl protons of *trans-* and *cis-*isomers, respectively. Table 5.2 shows some physical and chemical parameters of synthesized spirobicyclosiloxanes.

For the purpose of synthesizing silazaoxane-siloxane and silazane-siloxane dicyclic compounds with spirane atoms of silicon, HFC reaction of 1,3-diamino-1,3-dimethyl-1,3diphenyldisiloxane and 1,3-diamino-1,3-dimethyl-1,3-diphenyldisilazane with dichlorotetraphenylcyclotrisiloxane in diethyl ether (1:1 ratio of initial reagents) has been studied [11 – 13]. The reaction proceeds with formation of silazaoxane-siloxane (IIIa) and silazane-siloxane (IIIb) spirobicyclosiloxane:

Table 5.2

Some physical and chemical parameters of synthesized spirobicyclosiloxanes II

Compound, No.	R	R'	*n*	Yield, %	T_{boil}, °C (*P*, mmHg)	T_{melt}, °C	Chemical shift, δ (ppm)	References
1	Me	Me	2	22	230-231 (0.5)	150-151	0.22	[7-9]
2	Me	Ph	2	48	265-266 (0.01)	114-115	*Cis-trans* 0.53; 0.44	[7-9]
3	Me	Ph	2	36	200-201 (0.01)	70	*Cis-* 0.56; 0.48	[7-9]
4	Me	Ph	2	16	205-206 (0.01)	142-143	*Trans-* 0.42	[7-9]
5	Me	Me	3	51	214-215 (0.2)	54-55	0.05; 0.107	[7-9]
6	Ph	Ph	2	65	265-268 (0.01)	252-253	-	[1]
7*	-	-	-	36 16	200-201 (0.01) 205-206 (0.01)	70-71 142-143	*Cis-* 0.56; 0.48 *trans-* 0.42	[8]

Note: For compound 7, synthesized by reaction 5.3.

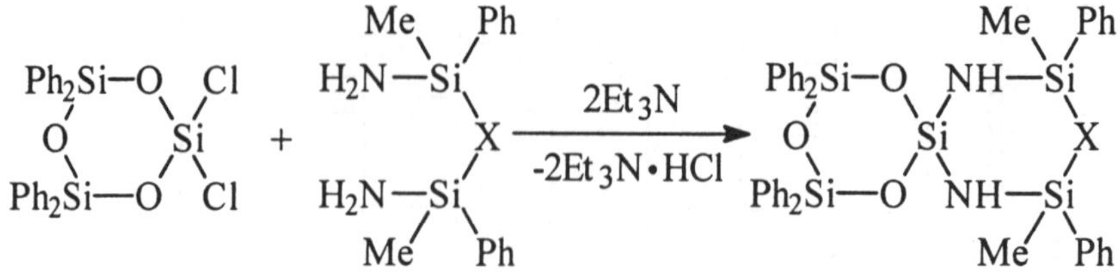 (5.5)

IIIa - IIIb

where X = O (IIIa), NH (IIIb).

The composition and structure of synthesized spirocyclosiloxanes are detected by ultimate analysis, determination of the molecular mass, IR and ^1H NMR spectroscopy. Table 5.3 shows some physical and chemical parameters of silazaoxane-siloxane (IIIa) and silazane-siloxane (IIIb) spirobicyclosiloxane.

NMR spectral analysis of spirobicyclic compounds has detected two existing isomers. ^1H NMR spectra of the compound with IIIa and IIIb structures display two reflexes of methyl protons, corresponded to *cis*- and *trans*-isomers.

It is known that for such systems [10] reflexes of protons in *cis*-oriented methyl groups in relation to phenyl ones are shifted towards stronger field, compared with reflexes of *trans*-methyl group protons. This occurs as a consequence of different contributions into shielding, stipulated by circular currents of phenyl groups at different mutual orientations of phenyl and methyl groups. This fact gave an opportunity to correspond the reflexes with chemical shifts $\delta = 0.475$ and $\delta = 0.475$ ppm to *cis*-isomers, and the reflexes $\delta = 0.350$ and $\delta = 0.340$ ppm to *trans*-isomers of compounds 1 and 2 (structures IIIa and IIIb, Table 5.4). It is of interest that spectra of compounds 1 and 2 (Table 5.4) possess equal chemical shifts of methyl proton reflexes that indicate that shielding of methyl groups in these rings depends on their orientation in relation to phenyl groups only.

Table 5.3

Some physical and chemical parameters of silazaoxane-siloxane and silazane-siloxane spirobicyclosiloxane

No.	Compound	X	Yield, % (total)	Stereo-form	T_{boil}, °C (P, mmHg)	T_{boil}, °C	δ_{Me}, ppm	References
1	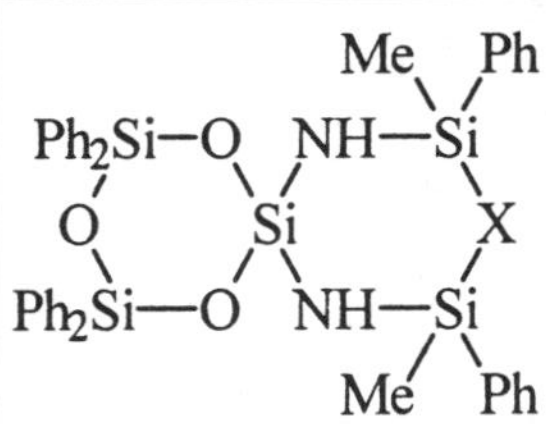 IIIa - IIIb	O	28	<u>*cis*</u> *trans*	265-270 (0.02)	<u>-</u> 134-135	<u>0.475</u> 0.350	[11]
2		NH	48	<u>*cis*</u> *trans*	299-301 (0.02)	<u>170-171</u> -	<u>0.475</u> 0.340	[11]

5.2. Synthesis of difunctional heterocyclic compounds

Difunctional heterocyclic compounds were synthesized in two stages. At the first stage, the initial compound 1,1,1,3-tetrachloro-3,3-dimethyl-1,3-disilaethylene was synthesized. Reactions of 1,1,1,3-tetrachloro-3,3-dimethyl-1,3-disilaethylene synthesis by hydride addition of dimethylchlorosilane to vinyltrichlorosilane in the presence of catalyst, platinum-hydrochloric acid [14, 15], and dicobalt octacarbonyl [16] are known, which give high yield of the mentioned product. At the second stage, HFC reaction of 1,1,1,3-tetrachloro-3,3-dimethyl-1,3-disilaethylene with dihydroxydiphenylsilane in the presence of pyridine at different temperatures and 1:1 ratio of initial reagents proceeds in accordance with the scheme as follows [16]:

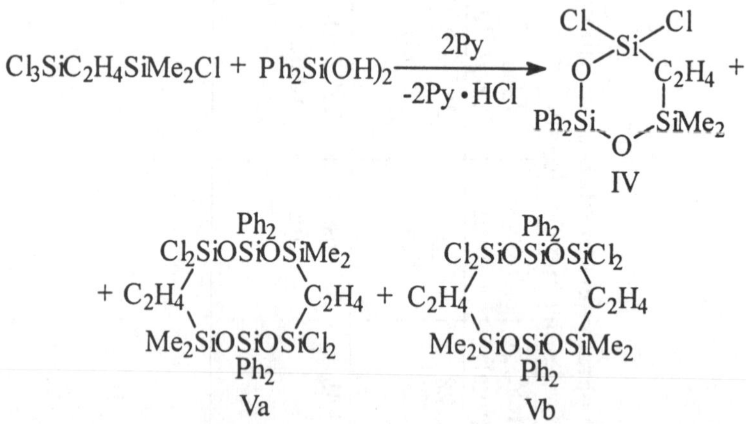

(5.6)

Table 5.4

Some physical and chemical parameters of carbocyclosiloxanes with functional groups

No.	Compound	Yield*, %	T_{boil}, °C (P, mmHg)	T_{melt}, °C	References
1	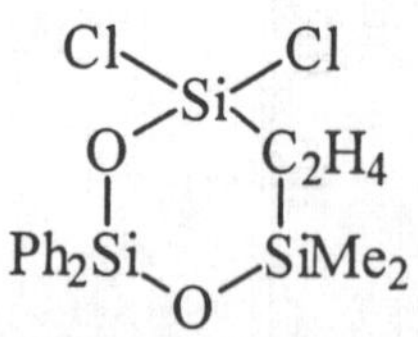	55 (68)	187-193 (0.01)	77-78	[15]
2	$[Cl_2SiC_2H_4SiMe_2OSiPh_2O]_2$	22 (17)	298-302 (0.01)	97-99	[15]

Note: * Yields at decreased temperature (-5 ÷ -10°C).

It is found that the yield of fractionated cyclic products is seriously dependent on the reaction temperature. HFC reaction proceeding at 20 - 25°C gives the yield of fractionated cyclic products below 70%, which is increased to 85% at -5 ÷ -10°C. Composition and structure of synthesized compounds were determined with the help of functional and ultimate analysis, determination of the molecular mass and from spectral data. Table 5.4 shows some physical and chemical parameters of carbocyclosiloxanes with functional groups.

^1H NMR spectra of dichlorocyclocarbosiloxane with structure IV displays a complex multiplet for protons of asymmetrical methylene groups possessing chemical shifts in the range of $\delta = 1.10 - 1.45$ ppm, centered at $\delta = 1.30$ ppm (4H), a singlet for methyl protons with chemical shift at $\delta = 0.45$ ppm (6H) and a complex multiplet with chemical shift within the range of 7.32 - 7.75 ppm. ^1H NMR spectrum of a mixture of compounds Va and Vb is more complicated, because broadened rings may also contain other isomeric structures displaying different disposition of functional groups in a cyclic fragment, which cannot be rectified.

Gas-phase pyrolytic method [17, 18] is successfully applied to synthesis of silicon-containing heterocyclic compounds. Intramolecular gas-phase pyrolysis of *o*-dichlorosiloxydiphenyl was performed [17], accompanied by hydrogen evolution and formation of a hexatomic ring of the phenanthrene type. The yield of synthesized 10,10-dichloro-10-sila-9-oxarophenanthrene[1] equals 60%. The reaction proceeds as follows:

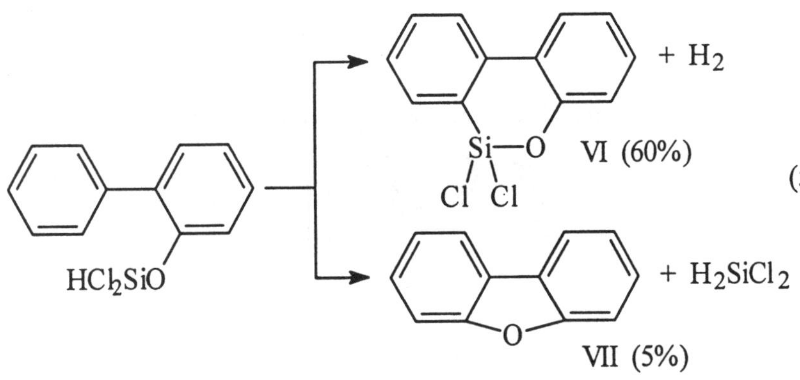

(5.7)

The above scheme indicates that the reaction partly proceeds with dibenz(*o*)furan formation. These investigators have also found that cyclization

[1] Hereinafter, 1.1-dichloro-1-sila-2-oxadihydrophenanthrene.

of *orhto*-siloxydiphenols can proceed not exclusively by ≡Si-H bond, but also by other bonds at silicon. However, this significantly decreases the yield of heterocyclic rings. For *orhto*-trichlorosiloxydiphenyl, the yield of dichlorosilaoxarophenanthrene equals 5% only. The structure of dichlorosilaoxarophenanthrene was determined by physicochemical analysis, functional and ultimate analysis, from molecular mass and molecular refraction. It was also found that dichlorosilaoxarophenanthrene is quantitatively, without breaking ≡Si-O- bond, alkylated and arylated by Grignard reagents, and reduced by lithium aluminum hydride. When treated by hydrofluoric acid, it is transformed to difluorosilaoxarophenanthrene. High chemical stability of siloxarophenanthrenes indicates a definite cyclicity of these compounds (similar to cyclicity of boroxaro-borozarophenanthrenes [19, 20]. In accordance with the Dewar classification, the name of siloxarophenanthrene [19] was given to this heterocyclic ring due to similarity of properties of the mentioned structures and possible cyclic type.

Hydrolysis of 1,1-dichloro-1-sila-2-oxadihydrophenanthrene in a neutral medium in the presence of hydrogen chloride acceptor (aniline) has mainly produced corresponded dihydroxy-derivative. The reaction proceeds as follows [21]:

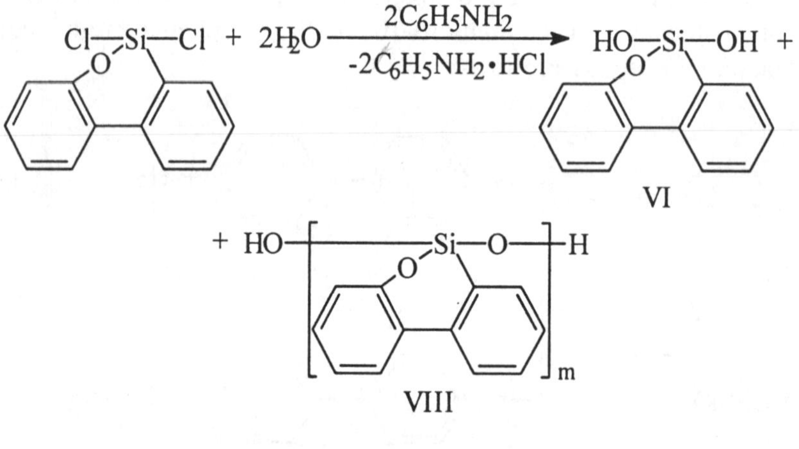

(5.8)

After recrystallization of the reaction product from toluene-heptane mixture, 1,1-dihydroxy-1-disila-2-oxadihydrophenanthrene (VI) was obtained with 86% yield. After diluter evacuation from mother solution, dixydroxy-containing oligomeric product of the structure VIII was obtained. Table 5.5 shows the yield and physical and chemical parameters of silaoxarophenanthrene with functional groups.

Table 5.5

Yield and some physical and chemical parameters of silaoxarophenanthrene with functional groups

No.	Compound	Yield, %	T_{boil}, °C (P, mmHg)	T_{melt}, °C	References
1	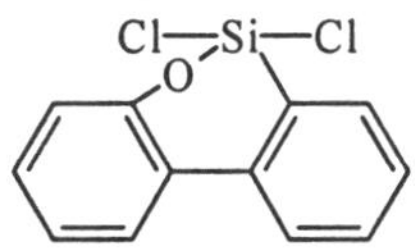	56	160 (1)	82-83	[17, 18]
2	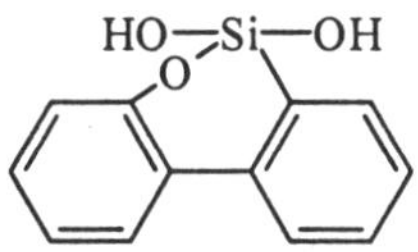	86	-	336-339	[21]

5.3. Synthesis and study of properties of organosilicon copolymers with cardocyclosiloxane fragments

Organosilicon copolymers with cardocyclosiloxane fragments in the backbone were synthesized by HFC reaction of 1,1-dichlorotetraorganocyclotrisiloxane with α,ω-dihydroxypoly(dimethylsiloxanes) [22]. Initially, condensation proceeds at room temperature, and at the final stage, for the purpose of obtaining rubbery polyorganosiloxanes, it is desirable to increase temperature to 100 - 150°C with simultaneous evacuation of diluter. Pyridine or diethylaniline were used as the acceptor of hydrogen chloride, produced during condensation. Generally, the reaction proceeds as follows:

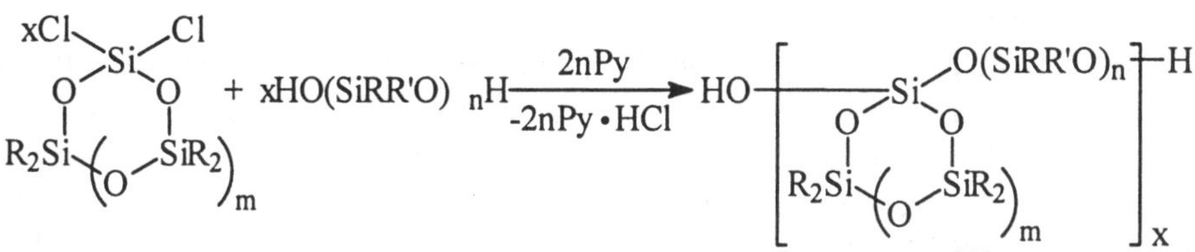

IX

(5.9)

where R = Me, Ph; n = 1 - 400.

As a consequence, organosilicon copolymers, which are transparent liquids with $[\eta] = 0.4 - 0.6$, highly soluble in organic solvents, with regular disposition of cardosiloxane fragments in the backbone were synthesized. Molecular masses of these compounds fall within the range of $1.2 - 1.9 \times 10^5$.

Cyclolinear poly(organosiloxanes) of the above-mentioned general formula represent liquids, capable of easy transformation to 3D-crosslinked elastomers at room temperature in the presence of catalytic amounts of acids and alkali (sulfuric acid, caustic alkali, ammonium and phosphonium bases), which induce regrouping of siloxane bonds. Gel formation of copolymers with structure XI in the presence of initiators, quaternary ammonium bases of the $Me_4N[SiMe_2O]_nNMe_4$ type, was studied [22]. It is shown that gel formation with elastomer networks proceeds during 15 minutes at room temperature in the presence of the above-mentioned initiator (0.02 - 0.1 wt.%). In this case, the gel-fraction concentration is 70 – 80%. By analogy, network systems are formed in the presence of sulfuric acid with 80% gel-fraction content. This property of the copolymers with cardocyclic fragments in the backbone makes them applicable as self-vulcanizing rubbers, filling compounds, and components for coatings and filler pastes.

At present, processes forming ladder-type polymers from organosilsesquioxanes are studied. In this connection, of great interest is catalytic polymerization of spirocyclic compounds. If organocyclosiloxanes are difunctional compounds in polymerization and form linear macromolecules, organospirosiloxanes may display functionality two and four and, as a matter of principle, are capable of forming spatially crosslinked polymeric structures. It has been shown [23] that octamethylspiro-(5,5)-pentasiloxane polymerization does always give branched and, finally, crosslinked systems. It is also found [24] that polymerization of phenyl- and tolylspirobicyclosiloxanes in the presence of catalytic amounts of potassium naphthalene gives soluble polymers only. The authors think that polymerization proceeds with formation of polycyclic systems.

Polymerization of organospirobicyclosiloxanes with different organic groups in the framing was studied [25], specifically, polymerization of octaphenylspiro-(5,5)-pentasiloxane (compound 1, Table 5.6) and spirobicyclic compounds with different ratio of methyl and phenyl groups in the framing (*cis*-2,4-dimethyl-2,4,8,8,10,10-hexaphenylspiro-(5,5)-pentasiloxane, 2,2,4,4-tetramethyl-8,8,10,10-tetraphenylspiro-(5,5)-pentasiloxane and *cis*-2,4,8,10-tetramethyl-2,4,8,10-tetraphenylspiro-(5,5)-pentasiloxane – compounds 2 - 4, Table 5.6, respectively) in the presence of anionic polymerization catalysts.

Polymerization of rings 2 and 3 (Table 5.6) was studied in benzene solution at 50°C in the presence of the catalyst, $Me_4NO[SiMe_2O]_{26}NMe_4$. As indicated, it proceeds differently for compounds 2 and 3 in the solution. The polymer derived from compound 3 is highly soluble and possesses specific viscosity of 1% solution in benzene equal 0.18; no gel formation was observed. IR-spectra indicate that about a half of hexatomic rings in the polymer are not disclosed, which, besides occurrence of the band at 1,090 cm^{-1}, is proved by occurrence of an intense doublet in the range of 1,025 - 1,040 cm^{-1} of the initial ring spectrum and its preservation in the spectrum of reprecipitated polymer. Simultaneously, according to data of NMR spectra, conversion of the rings approaches 100%. This fact allows an assumption of linear or low-branched structure of the polymer containing hexatomic cyclic fragments in the backbone:

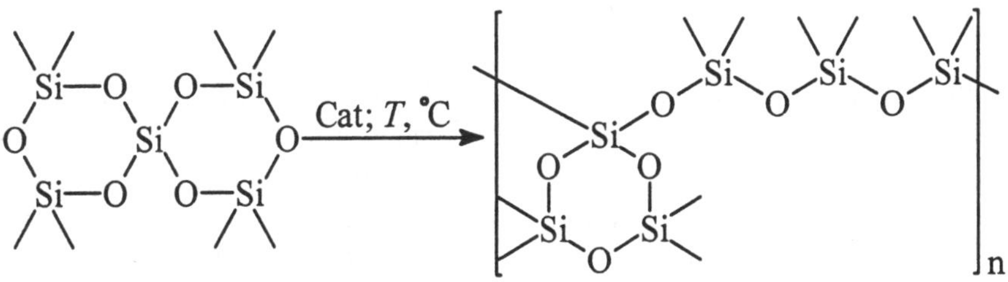

(5.10)

IX

Under the same conditions, polymerization of the ring 3 (Table 5.6) proceeds with formation of a transparent gel, the gel-fraction concentration in the reaction mixture being increased with time and approaching an equilibrium level. By analogy, compound 1 is polymerized at 100°C in the presence of ammonium catalyst in the case of preliminary solvent recovery; thus, the gel-fraction yield is somewhat higher in this case.

Polymerization of spirobicyclosiloxanes 1 – 4 (Table 5.6) in block at 150°C has given completely soluble polymer from the ring 1 only, which after reprecipitation possesses specific viscosity 0.1. For the rest spirobicyclosiloxanes, occurrence of gel-fraction accompanied formation of soluble polymers (Table 5.6). While phenyl group content in the molecule of the initial ring decreased, the equilibrium yield of the gel-fraction increased.

Table 5.6

Gel-fraction yield in polymerization of spirobicyclosiloxanes at 150°C in block, in the presence of $KO(SiMe_2O)_8K$ as the catalyst – 0.02% (recalculated per KOH)

No.	Spirobicyclosiloxane	R	R'	R"	η_{spec}**	Gel-fraction yield, %
1	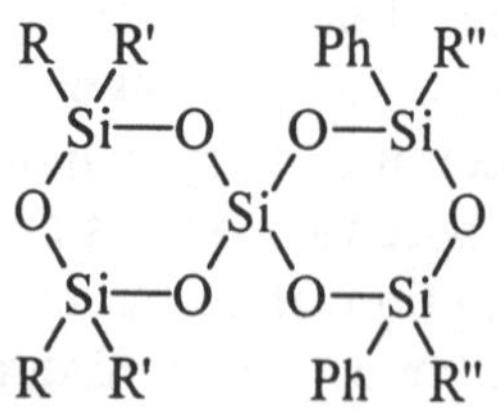	Ph	Ph	Ph	0.10	0
2		Me	Ph	Ph	0.09 (0.18)	28
3		Me	Me	Ph	-	56
4		Me	Ph	Me	-	64

Notes: * Copolymer viscosity after reprecipitation.
** 1% solution in benzene at 20°C.

Relatively equilibrium content of the gel-fraction in the polymer is the probability criterion for interchain interaction. That is why one may assume that the process of polymer synthesis proceeds in two directions, simultaneously: interchain interaction, which produces spatially crosslinked systems, and intramolecular polycyclization producing soluble polymers. If cyclic framing groups only are present in a spirocyclic molecule, their intramolecular steric detachment promotes formation of soluble polycyclic products. While phenyl groups are substituted by methyl ones, the interchain interaction inducing gel formation becomes predominant.

Spirosiloxane 2 (Table 5.6), which under soft conditions produces a soluble polymer only, is capable of forming a crosslinked product with the polymerization temperature increase. Its polymerization in block at 120 - 150°C indicates that the equilibrium yield of the gel fraction is increased with the polymerization temperature.

Temperature, °C	120	130	150
Gel fraction yield, %	13	16	28
η_{spec} of soluble polymer	0.16	0.14	0.12

This is explained by the fact that temperature increase promotes an increase of mobility of active centers in propagating chains and probability of nucleophilic attack of cyclic fragments, not disclosed by them. Viscosity of the soluble part is slightly decreased with the increase of polymerization temperature, which is probably associated with development of the chain transfer reactions with a break. Proceeding of these reactions is proved by the results of ultimate analysis of obtained polymers.

Features of spirane and cyclic siloxanes' copolymerization was studied [26] on the example hexamethylcyclotrisiloxane (HMCTS) and 2,2,4,4-tetramethyl-8,8,10,10-tetraphenylspiro-(5,5)-pentasiloxane (compound 3, Table5.6) in benzene solution in the presence of tetramethylammoniumsiloxane. It may be expected that copolymerization is initiated simultaneously with the nucleophilic attack of both HMTCS and one of siloxane rings of spironucleus 3, because simultaneous disclosure of both rings of the spironucleus is of low probability. In this case, a linear poly(organosiloxane) with cyclic fragments, included in the siloxane chain, should be formed. As free HMCTS is exhausted, the future attack of these cyclic fragments must induce formation of statistically regulated polymer networks.

Figure 5.1a shows time dependencies of polymer specific viscosity for different ratios of HMTCS and compound 3 and constant total concentration of the catalyst, $[Cat]_{tot}$, which display an induction period in copolymerization, during which no viscosity increase in the system is observed. The induction period is increased with relative concentration of compound 3: at 8 mol% concentration no viscosity increase was observed during 140 minutes.

After the end of the induction period, viscosity increases till insoluble product is formed. This happens, because rise of the compound 3 concentration decreases the fracture of active centers of the catalyst falling at it ($[Cat]_a$ is the catalyst concentration in relation to compound 3 only). Thus it may be assumed that the induction period represents duration of compound 3 homopolymerization, caused by much higher reactivity of it [27]. The rate of spironucleus restructuring, induced by the catalyst, is so higher, the higher is concentration of the latter. Hence, increasing is low, because compound 3 regrouping products apparently display the structure of globular polycyclic blocks. The viscosity increase, observed after complete depletion of spirocyclic compound 3 in the reaction mixture, happens at the expense of these blocks "overgrowth" by poly(dimethylsiloxane) chains with consequent formation of a polymer network due to the interchain exchange.

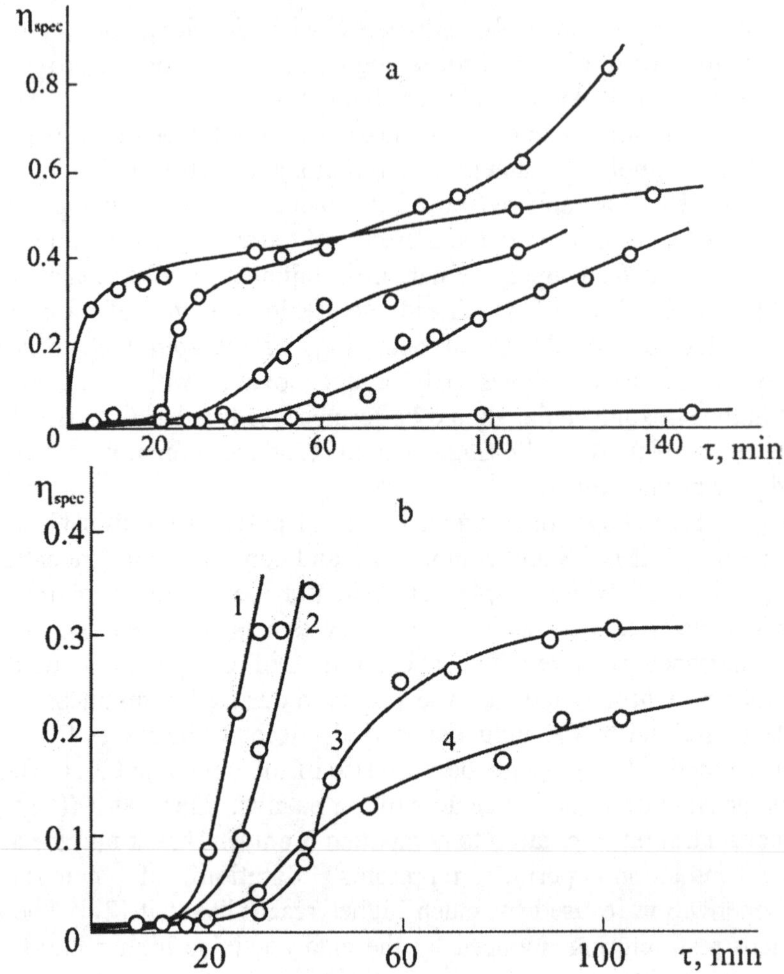

Figure 5.1. Specific viscosity variation during HMCTS and 2,2,4,4-tetramethyl-8,8,10,10-tetraphenylspiro-(5,5)-pentasiloxane (3) copolymerization: a – $[Cat]_{tot}$ = 0.015%; compound 3 concentration in the initial mixture, mol%: 1 – 0; 2 – 1; 3 – 2; 4 – 4; 5 – 8; b - $[Cat]_a$ = 0.5 wt.%; compound 3 concentration in the initial mixture, mol%: 1 – 8; 2 – 4; 3 – 2; 4 – 0.5.

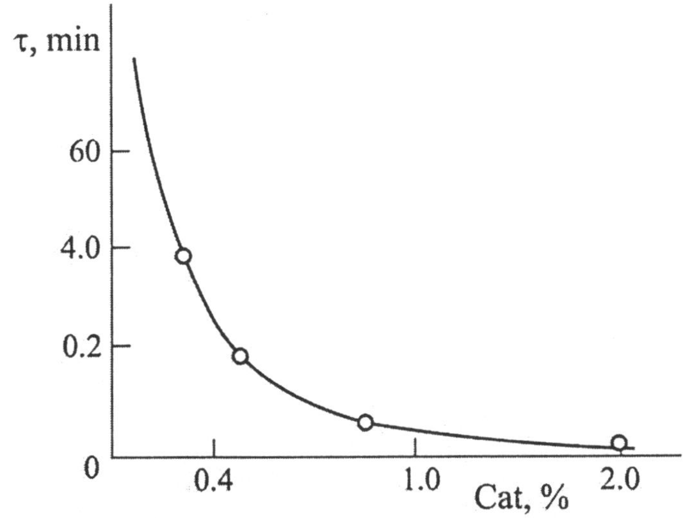

Figure 5.2. Dependence of the induction period on relative concentration of initiator, $[Cat]_a$, for HMCTS copolymerization with 2 wt.% of compound 3

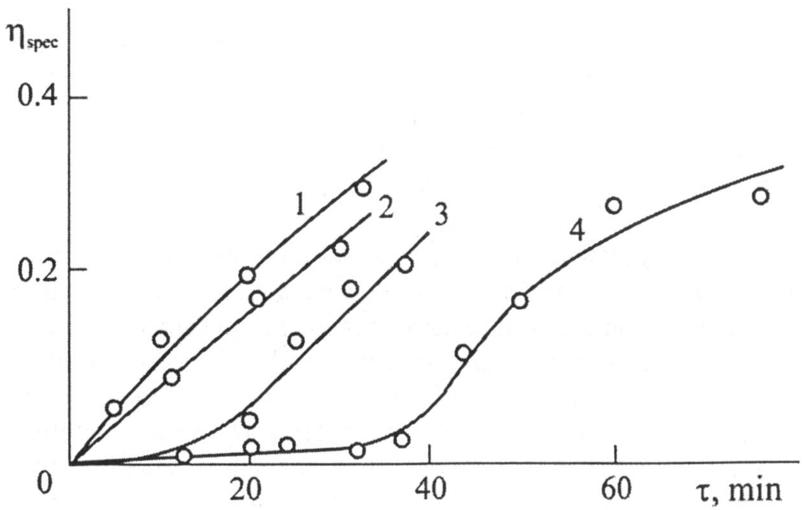

Figure 5.3. Variation of specific viscosity during HMCTS copolymerization with 2 wt.% of compound 3. $[Cat]_a$ ($[Cat]_{tot}$), wt.%: 1 – 2.0 (0.110); 2 – 1.0 (0.550); 3 – 0.50 (0.028); 4 – 0.28 (0.015)

All the above-said is proved by the fact that at constant selected catalyst concentration in relation to compound 3 ([*Cat*]), the induction period is independent of the initial ratio between HMCTS and compound 3. For this series of experiments, the induction period equaled 16 minutes, approximately (Figure 5.1b). Vice versa, at constant initial HMCTS – compound 3 ratio (2 mol% of compound 3) the induction period is decreased with the increase of the catalyst concentration (Figures 5.2 and 5.3). As the catalyst concentration is $[Cat]_a$ = 2 wt.% (that corresponds to $[Cat]_{tot}$ = 0.11 wt.%) or higher, the induction period is not observed.

Thus, experimental data give an opportunity to suggest that copolymerization of studied spirane and cyclic siloxanes is of the block-copolymerization type with polyspirocyclic blocks as the branching centers, but not via formation of linear-cyclic polymers with their consequent crosslinking.

Organosiloxane copolymers with cardocyclic fragments in the backbone, synthesized in the work [28], were used for development of artificial biocomposites, based on skeleton-free bone matrices.

High strength (up to 140 MPa) and elasticity modulus (up to 20 GPa) of the bone tissue are stipulated by both hierarchic spiral structure of its protein and mineral parts and intensive interactions between them. The mineral part forms 70% of the bone composition.

To achieve an artificial biocomposite, natural protein, the reason for occurrence of bio-incompatibility, must be removed from the bone tissue with preservation of the mineral part structure, and porous bone matrix should be filled in with a bio-compatible material, able to polymerize.

Application of IR-spectroscopy and thermal gravimetry methods enabled to develop an optimal mode of the bone treatment, which provides for complete elimination of collagen without breaking structure of the mineral part. After that the skeleton-free bone matrix was impregnated with siloxane oligomers in the presence of catalytic amounts of tetramethylammonium siloxane.

It is shown that the pore filling degree is highly dependent on molecular mass of the oligomer. The features of polymerization proceeding in pores and forming poly(organosiloxane) networks are studied. Network structures in pores of the bone matrix are formed due to splitting of cyclosiloxane fragments in the oligomer chain and their regrouping, the latter proceeding under the effect of tetramethylammonium siloxane and being not accompanied by occurrence of secondary low-molecular production. All components of synthesized composites are biocompatible and human friendly.

It is known from the literature [29 – 32] that organic and organosilicon diols interacting with silicon-nitrogen-containing heteronuclei form polymers with rather high molecular mass and specific physicochemical properties. To synthesize organosilicon copolymers with cardocyclosiloxane fragments in the linear chain, the reaction of 2,2,4,4,8,10-hexaphenyl-8,10-dimethylspiro-(5,5)-pentasila-7,11-diaza-1,3,5,9-tetraoxane (compound 1, Table 5.3) with different diols was studied [33].

Polycondensation was implemented in block or solution with 1:1 ring - diol molar ratio till ammonia evolution is ended. Ammonia conversion and properties of produced substances are highly dependent on the chemical origin of the initial diol (Table 5.7). In the case of α,ω-dihydroxypoly(dimethylsiloxane), the system became insoluble after 1.3 hour of condensation proceeding. Similar result was observed for diphenylolpropane both in block and solution. A glassy-like low-melting polymer, soluble in benzene, is derived from 1,2-bis(hydroxymethyl)carborane. N,N'-bis(hydroxydimethylsilyl)tetramethylcyclodisilazane gives soluble oligomers; however, ammonia conversion is low in this case, both in block and solution. The reaction proceeds as follows [33]:

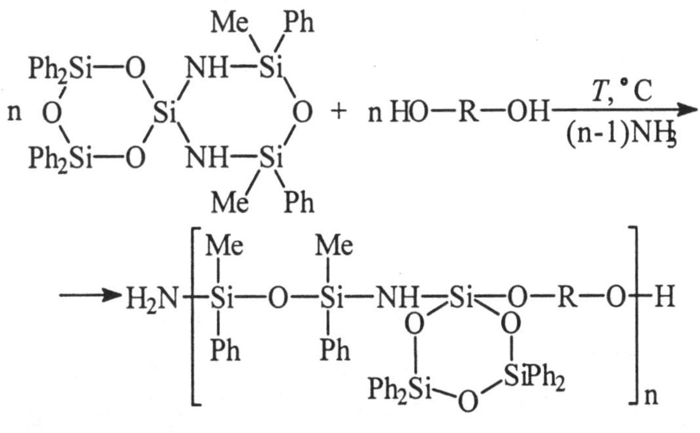

(5.11)

IX

Table 5.7 shows conditions of synthesis for copolymers with cardosiloxane fragments in the backbone.

Table 5.7

Reaction conditions for 2,2,4,4,8,10-hexaphenyl-8,10-dimethylspiro-(5,5)-pentasila-7,11-diaza-1,3,5,9-tetraoxane with diols

No.	Diol	Medium	*T*, °C	Time, h	Liberated NH_3 quantity	Gel-fraction, %
1	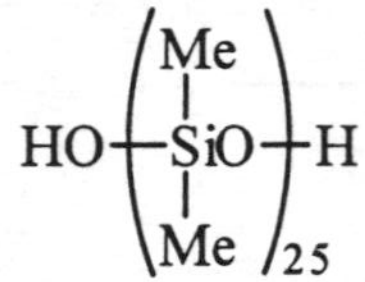	In block	180	5	70	66*
2	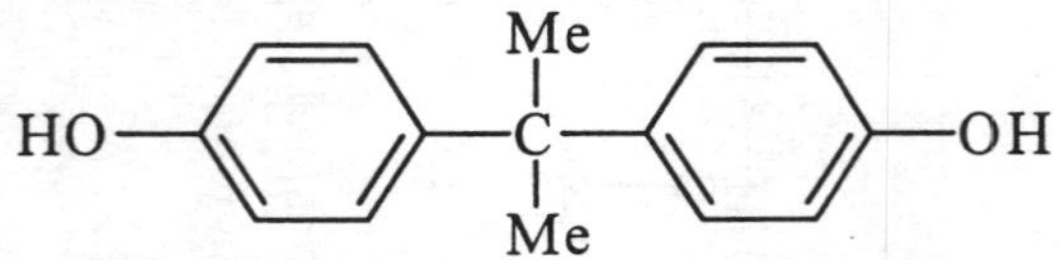	In block Toluene	150 110	6,5 3,5	86 90	73 31
3	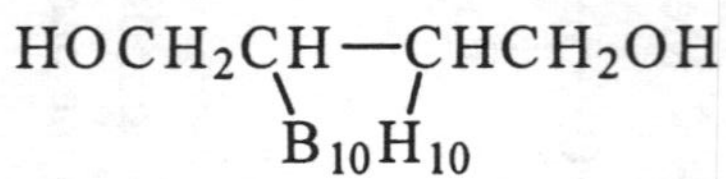	Toluene	110		99	_**
4	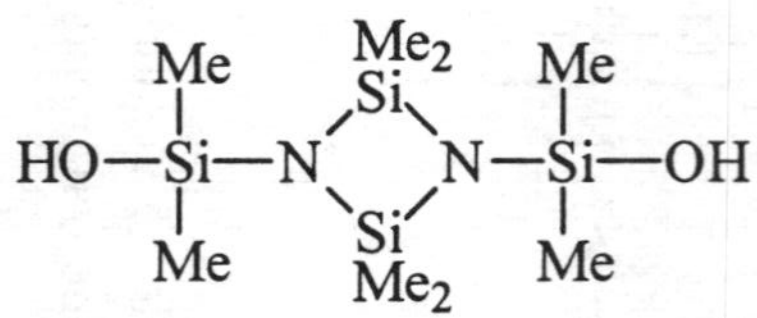	In block Toluene	140-200 110	4 9,5	48 40	- -

Notes: * 10 minutes before structuring, η_{spec} of the product in benzene equals 0.2.
** $\eta_{spec} = 0.1$.

As shown above [31], condensation of organic and organosilicon diols with organocyclosilazasiloxanes and silazanes tends to form branched polymers and even polymer networks with nitrogen in the branching points. However, the formation mechanism of trisilyl-substituted nitrogen atom is not fully studied yet. Moreover, the siloxane ring may be broken, and a branching center may be formed on the atom of silicon.

To clear up a possibility of realization of one alternative or another of branched polymer formation, model reactions have been performed, one of which between poly(dimethylsiloxane) diols (n = 25 and n = 2) with hexamethylcyclotrisilazane (HMCTS) at 150°C in block. Ammonia liberation curves are shown in Figure 5.4. No formation of insoluble products was observed.

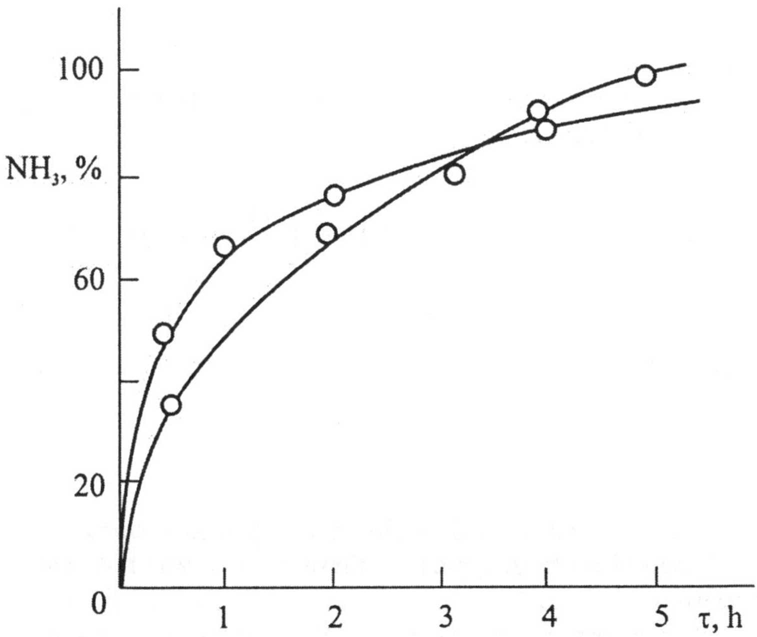

Figure 5.4. Ammonia liberation in the reaction between HMCTS and α,ω–dihydroxypoly(dimethylsiloxanes): 1 – n = 2; 2 – n = 25

Maximum viscosity of 1% polymer, derived from diol with n = 25, solution in benzene reached the value of 0.82, and for diol with n = 2 – 0.64. To clear up if the siloxane bond is broken in condensation of spirocyclic compounds with diols, HMCTS reaction with diol (n = 2) in the presence of hexamethylcyclotrisiloxane and 1:1:1 molar ratio of the reagents was performed. It was found that hexamethylcyclotrisiloxane is gradually evacuated from the reaction mixture (Figure 5.5), i.e. stressed siloxane trimer is broken under condensation conditions. Generally, the process may be described by a sequence of reactions as follows:

$$\equiv SiOH + NH(Si\equiv)_2 \rightarrow (\equiv Si)_2O + H_2NSi\equiv \quad (5.12)$$

$$\equiv SiOH + H_2NSi\equiv \rightarrow (\equiv Si)_2O + NH_3 \quad (5.13)$$

$$\equiv SiOH + NH_3 \rightleftharpoons \equiv Si\overset{-}{O}\overset{+}{N}H_4 \quad (5.14)$$

$$\equiv \overset{*}{S}i\overset{-}{O}\overset{+}{N}H_4 + (\equiv Si)_2O \longrightarrow \equiv \overset{*}{S}i{-}O{-}Si\equiv + H_4\overset{+}{N}\overset{-}{O}Si\equiv \quad (5.15)$$

Thus all three compounds, including hexamethylcyclotrisiloxane added to active end of the propagating chain by the last reaction, participate in the copolymer synthesized:

$$\equiv \overset{*}{S}i\overset{-}{O}\overset{+}{N}H_4 + HN(Si\equiv)_2O \longrightarrow \equiv SiOH + H_4\overset{+}{N}\overset{-}{N}(Si\equiv)_2 \quad (5.16)$$

$$(\equiv Si)_2\overset{-}{N}\overset{+}{N}H_4 \xrightarrow{+(\equiv Si)_2O} (\equiv Si)_3N + H_4\overset{+}{N}\overset{-}{O}Si\equiv$$
$$(\equiv Si)_2\overset{-}{N}\overset{+}{N}H_4 \xrightarrow{+NH(Si\equiv)_2} (\equiv Si)_3N + H_4\overset{+}{N}\overset{-}{O}Si\equiv \quad (5.17)$$

However, concentration of trisilylamino groups is apparently low, and the reflex of methyl protons, typical of these groups, was not observed in the NMRT-spectrum.

Transferred from model reactions to the organospirosilazasiloxane – diol system, it has been concluded that diol with lower acidity, for example, 1,2-bis-(hydroxymethyl)carborane gives a soluble product, whereas siloxane diols and diphenylolpropane form gels. In the case of carborane diol, the equilibrium reaction between it and ammonia is shifted to the left (see above) and the rate of siloxane ring disclosure is negligible due to low concentration of nucleophilic particles, whereas disclosure of the silazane ring and condensation of OH- and NH-groups proceeds more fully (ammonia conversion approaches 100%), which leads to formation of linear or low-branched polymer. In the system with organosilicon and cyclic diols, concentration of nucleophilic particles capable of disclosing the siloxane ring is higher. As a consequence, resulting comparatively rapid formation of branched polymers and polymer network and change in molecular mobility, the rates of polycondensation reactions and their completeness is lower than in the previous case.

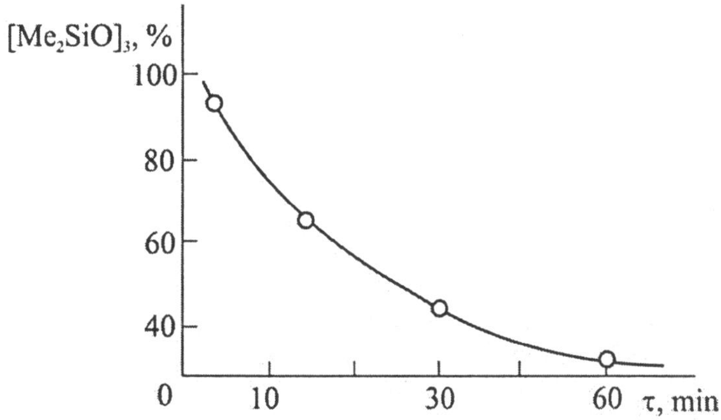

Figure 5.5. The exhaustion curve for $[Me_2SiO]_3$ in the reaction with HMCTS and diol ($n = 2$)

Of special attention is N,N'-bis(hydroxydimethylsilyl)tetramethylcyclodisilazane. The reaction is started from HMCTS disclosure and condensation of NH_2- and OH-groups. However, introduction of ammonia into the system initiate anionic regrouping, proceeding at an appreciable rate as follows:

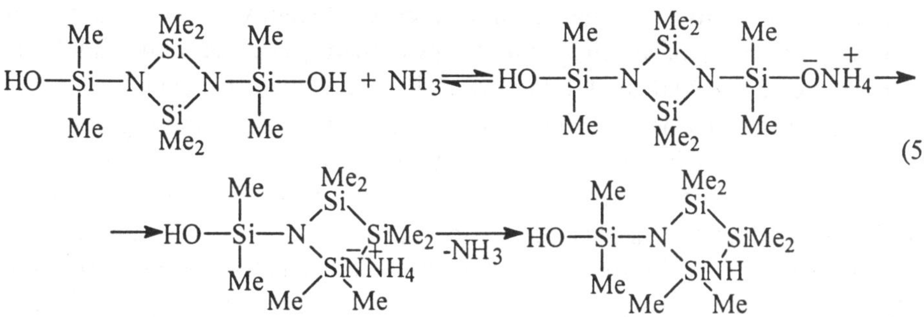

(5.18)

As a result, concentration of hydroxyl groups is decreased, polycondensation proceeds at lower rate and completeness, and low-molecular oligomers are synthesized. Copolymers obtained possess high resistance to thermal oxidation: weight loss is initiated at 350°C and the main degradation process proceeds at temperatures above 400°C.

Thus, application of polymerization of spirobicyclosiloxanes with cyclosiloxanes, as well as polymerizing polycondensation of organospirobicyclosilazasiloxanes with dihydroxy-derivatives may not always induce synthesis of organosiloxane card-type copolymers with the given structure of macromolecular backbone.

5.4. Organosilicon copolymers with card-type heterocyclic fragments in the linear chain

The card-type cyclolinear copolymers containing cyclocarbosiloxane and silaoxadihydrophenanthrene fragments in the chain were synthesized by HFC of dichlorocyclocarbosiloxane (structure IV, compound 1, Table 5.4) [16, 34, 35] and 1,1-dichloro-1-sila-2-oxadihydrophenanthrene (structure VI, compounds 1 and 2, Table 5.5) [36, 37] with α,ω-dihydroxydimethyl(diphenyl)siloxane with various length of the siloxane unit, at 1:1 ratio of initial compounds in the presence of pyridine as hydrogen chloride acceptor, was studied. As known from the literature, application of amine as the acceptor increases depth of HFC [38]. The reactions were studied at room temperature. At the final stage the reaction mixture was heated up to the boiling point of the solvent used. It is found that under these conditions secondary reactions of homofunctional condensation of α,ω–dihydroxydimethylsiloxanes do not proceed. Recently, it was shown [39] that in the presence of amines, more basic than pyridine, homofunctional condensation of 1,3-dihydroxytetramethyldisiloxane does not proceed.

It is shown that at short length of linear dimethylsiloxane segment ($n \leq 4$) the yield of copolymers is low, which may be explained by partial intramolecular cyclization proceeding with formation of cyclic compounds accompanying intermolecular condensation. The two-path proceeding of HFC during synthesis of cyclolinear copolymers were indicated before [40 - 44]. Copolymers with the card-type cyclic fragments in the chain are synthesized in accordance with the scheme as follows:

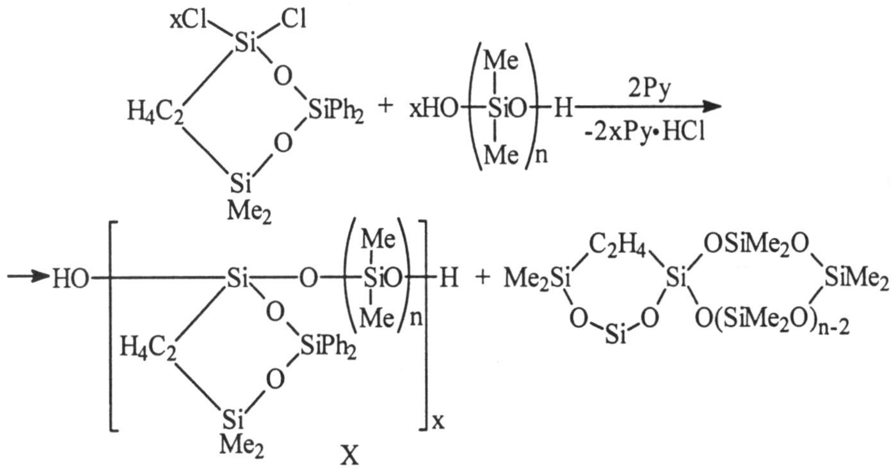

(5.19)

where $n = 2 - 36$.

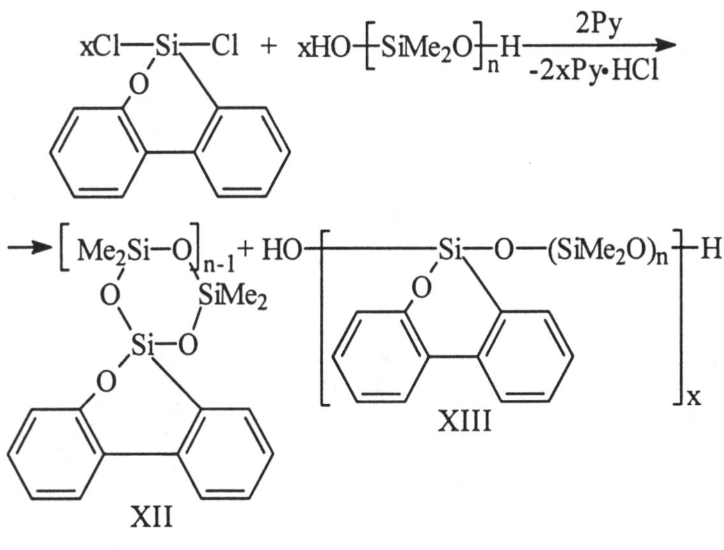

(5.20)

where $n = 2 - 27$.

Synthesized copolymers of structures X and XIII were reprecipitated from toluene solution by methyl alcohol. The yield of obtained copolymers is 73 - 95%. Depending on the length of linear dimethylsiloxane chain the copolymers are transparent or low opalescent substances, soluble in ordinary

Table 5.8

Some physical and chemical parameters of the card-type copolymers with cyclocarbosiloxane fragments in the chain

No.	Copolymer structure (X)	n_{SiO}	η_{spec}	T_g, °C	$\bar{M}_\omega \times 10^3$	d_1, Å	Residual mass, %	Yield, %
1	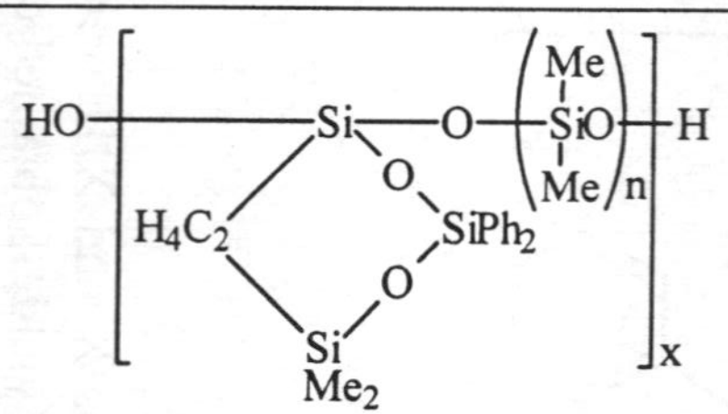	2	0.08	-51	78.4	9.50	46	74
2		4	0.12	-36	-	8.70	36	80
3		10	0.18	-85	-	-	-	89
4		21	0.20	-119	124	7.25	32	92
5		36	0.25	-123	-	7.25	-	95

Table 5.9

Some physical and chemical parameters of the card-type copolymers with silaoxadihydrophenanthrene fragments in the chain

No.	Copolymer structure (XIII)	n_{SiO}	$[\eta]$, dl/g	T_g, °C	5% mass losses	$\bar{M}_\omega \times 10^3$, (D)	d_1, Å	Yield, %
1	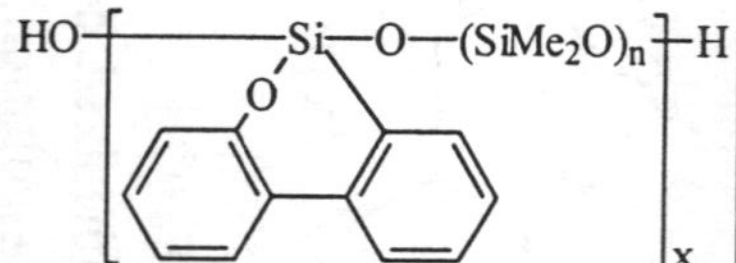 XIII	2	0.11	0	350	-	9.30	73
2		4	0.15	-32	300	150 (2.0)	8.23	86
3		8	0.28	-62	-	-	-	93
4		12	0.40	-85	-	-	-	94
5		19	1.82	-112	350	387 (2.6)	7.50	95
6		27	2.33	-123	300	648 (3.0)	7.50	94

organic solvents with η_{spec} = 0.11 – 2.33. Some physical and chemical parameters and yields of copolymers are shown in Tables 5.8 and Table 5.9.
As copolymer 1 (Table 5.8 and 5.9) is reprecipitated from the mother solution and solvents (toluene and methyl alcohol) are evacuated, low-molecular product were obtained with the yield of ~6%, corresponded to dicyclic structures XI and XII by composition and structure. These products were identified by determination of molecular masses and IR and NMR spectra. Based on these data, cyclic structure of these compounds of structures XI and XII was detected. Intramolecular cyclization product of structure XII (n = 2) was obtained by direct synthesis, e.g. HFC of 1,1-dichloro-1-sila-2-oxadihydrophe-nanthrene with 1,3-dihydroxytetramethyldisiloxane at 1:1 ratio of the initial reagents in the presence of the acceptor, aniline (5% anhydrous ether solution), at -5 ÷ -10°C. After fractionation of the reaction products, dicyclic compound of structure XII were obtained. The reaction proceeded in accordance with the following scheme:

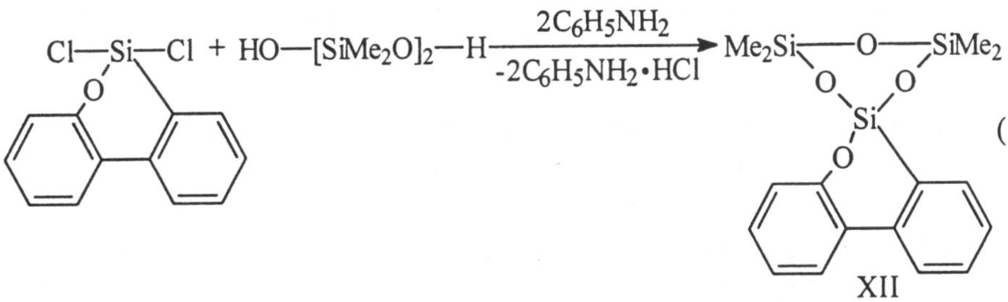

XII

(5.21)

Composition and structure of derived compound of structure XII were determined from data of ultimate analysis, determination of molecular mass and IR and ^{1}H NMR spectra.

Copolymers 3 (Table 5.8) and 4 (Table 5.9) were fractionated. It was shown that the ultimate analysis data on the fractions obtained are almost the same and similar to initial copolymers. This indicated that secondary processes are inhibited during HFC and the HFC is the main path of the reaction.

GPC studies of copolymers of structure XIII indicate that monomodal MMD of them. Molecular masses, $\bar{M}_\omega$, of copolymers fall within the range of 1.5×10^5 – 5.0×10^5 and polydispersity of copolymers varies from 2.0 to 3.0. Molecular masses, $\bar{M}_\omega$, and polydispersity are shown in Table 5.9.

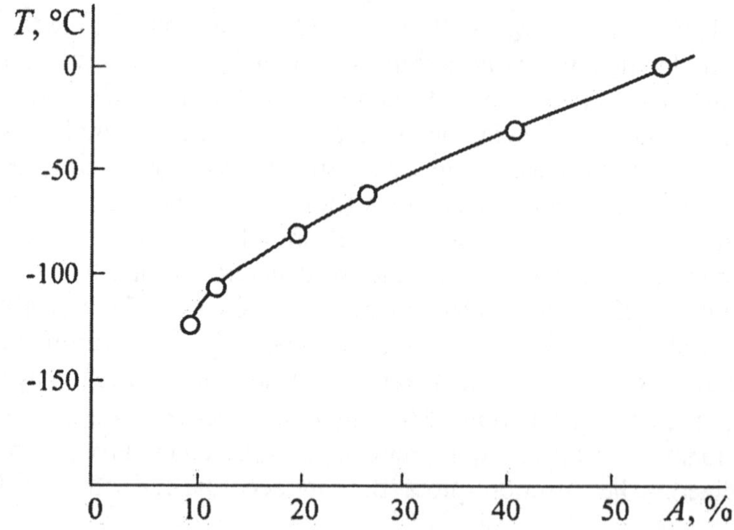

Figure 5.6. Dependence of T_g on molar concentration of silaoxadihydrophenanthrene (A) fragment

Thermomechanical and DSC studies of synthesized copolymers indicate (Tables 5.8 and 5.9) that the maximum of the glass transition temperature, T_g, of copolymers is reached at a short lengths of linear poly(dimethylsiloxane) chain. Figure 5.6 shows T_g dependence on the length of linear dimethylsiloxane unit for copolymers. This implies that T_g of copolymers (structure XIII) decreases with increase of dimethylsiloxane unit length down to T_g of linear poly(dimethylsiloxane).

The effect of cyclocarbosiloxane and 1-sila-2-oxadi-hydrophenanthrene fragments on dimethylsiloxane chain is sufficient up to $n = 19 - 21$, hence, T_g of copolymers 4 (Table 5.8) and 5 (Table 5.9) equals $-112 \div -119$°C.

The copolymers represent single-phase amorphous systems with maximal interchain distances, d_l, for short dimethylsiloxane unit. For copolymer 1, d_l = 9.30 - 9.50 Å (Table 5.8 and 5.9). Generally, d_l of synthesized copolymers is close to interchain distance typical of cyclolinear copolymers with organocyclopenta(hexa)siloxane and cyclocarbosiloxane fragments in the dimethylsiloxane backbone [42, 43]. Tables 5.8 and 5.9 show that the interchain distance is decreased with elongation of linear dimethylsiloxane unit (n) and for copolymer 5, d_l = 7.25 Å, which is also equal to the interchain distance of linear PDMS.

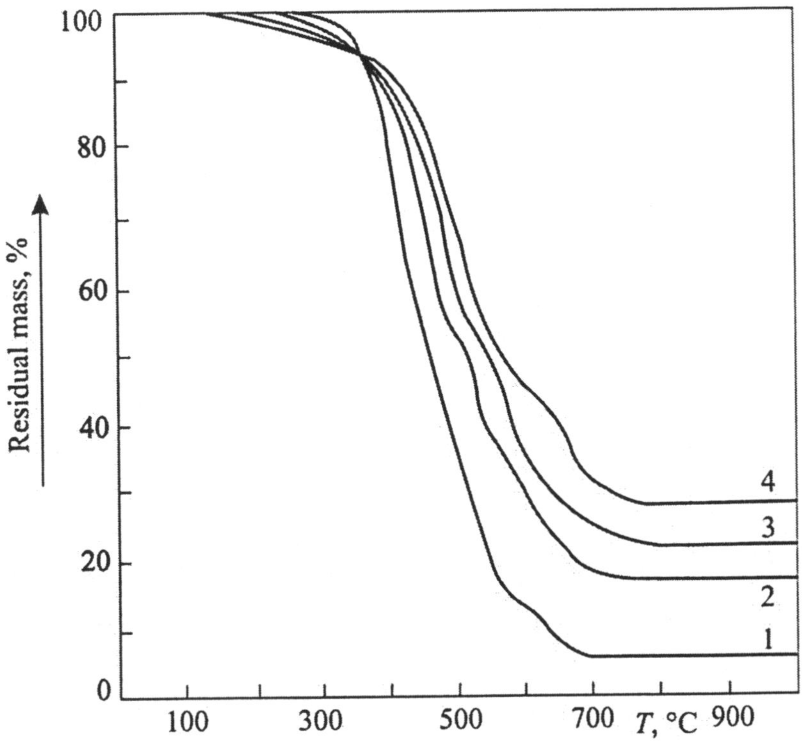

Figure 5.7. Thermogravimetric curves for copolymers with silaoxadihydrophenanthrene fragments in dimethylsiloxane chain: 1 - copolymer 1; 2 – copolymer 2; 3 – copolymer 5; 4 – copolymer 6 (Table 5.9)

Thermogravimetric studies for copolymers of structures X and XIII has shown that the card-type organosiloxane copolymers with cyclocarbosiloxane and silaoxadihydrophenanthrene fragments in the chain possess approximately the same thermal oxidative stability. Figure 5.7 shows thermogravimetric curves for copolymers of structure XIII. Copolymers with short dimethylsiloxane chain display high thermal oxidative stability (curves 1 and 2), which is decreased with dimethylsiloxane chain length increase. These copolymers do also display 5% mass losses in the temperature range of 300 - 350°C; main degradation proceeds at 450 - 700°C. Above 700°C no changes in the curves of mass losses are observed. Thus, introduction of the card-type heterocyclic fragments in the dimethylsiloxane chain, described in the literature [45, 46], disturbs the spiral structure of copolymers and changes their physical and chemical parameters.

Main thermal oxidative degradation is initiated at temperature by 80-100°C above that for unblocked linear PDMS [47].

5.5. Synthesis and studies of properties of card-type organosilicon oligomers with silaoxadihydrophenentrene fragments in the chain

For the purpose of synthesizing silaoxadihydrophenanthrene-diphenylsiloxane-containing oligomers with dihydroxyl end group, heterofunctional condensation of 1,1-dichloro-2-oxadihydrophenanthrene with dihydroxydiphenylsilane at 1:0.9 and 1:0.95 ratios of initial reagents in the presence of dry pyridine in 50 - 60% solution of anhydrous toluene was performed. It proceeded in accordance with the following scheme [48, 49]:

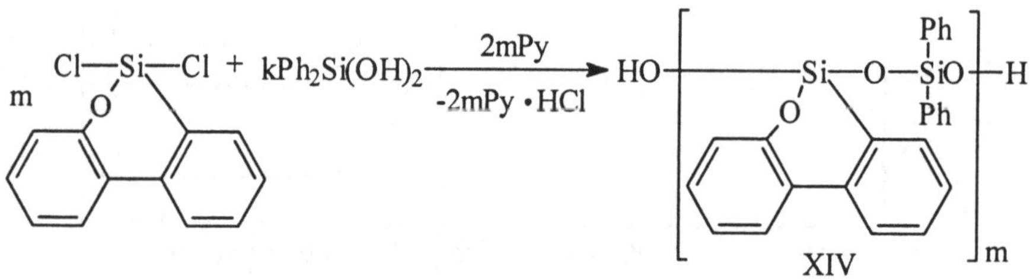

(5.22)

where m = 10 (I), 14 (II).

Table 5.10.

Some physical and chemical parameters of α,ω–dihydroxysilaoxadihydrophenanthrene-diphenylsiloxane oligomers (structure XIV)

No.	Structure	η_{spec}*	d_1, Å	T_{melt}, °C	OH, %	$\bar{M}$**	Yield, %
1	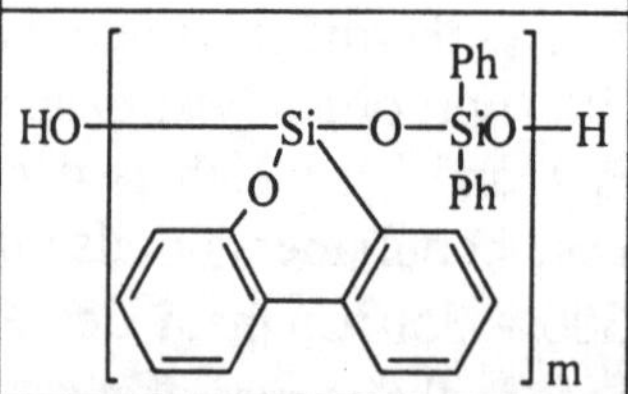	0.04	11.05	67÷72	0.83 / 0.81	4,111 / 4,200	92
2		0.05	11.05	74÷77	0.59 / 0.56	5,758 / 6,070	93

Notes: * In 1% toluene solution at 25°C.
**Numerator shows theoretical and denominator experimental values.

Reprecipitation of the reaction products from toluene by methanol has produced white α,ω–dihydroxysilaoxadihydrophenanthrene-diphenylsiloxane oligomeric powders, soluble in ordinary organic solvents.

The composition and structure of synthesized oligomers were determined by ultimate and functional analysis, determination of molecular masses and IR spectral data. Some physical and chemical parameters of synthesized oligomers are shown in Table 5.10.

Card-type oligomers with silaoxadihydrophenantrene fragments in silarylene chain were synthesized by dehydrocondensation of 1,1-dihydroxy-1-sila-2-oxadihyd-rophenantrene (compound 2, Table 5.5) with 1,4-bis(dimethylsilyl)benzene. The reaction proceeded at 1:1 ratio of initial reagents in the presence of anhydrous caustic potash as the catalyst (0.1 wt.%), at 40 - 60°C in anhydrous dimethylformamide (C = 0.3952 mol/l). Under current conditions, no polymerization of silaoxadihydrophenanthrene fragments was detected [50].

Catalytic dehydrocondensation of 1,1-dihydroxy-1-sila-2-oxadihydrophenantrene with 1,4-bis(dimethylsilyl)benzene proceeds in accordance with the following scheme [50, 51]:

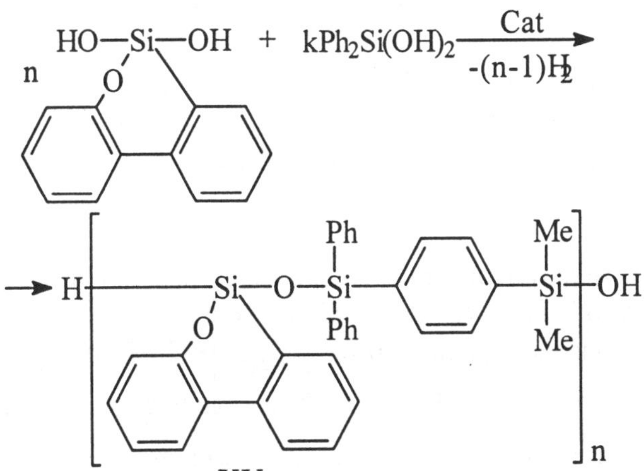

(5.23)

XV

where at 40°C (1^1), 50°C (1^2), 60°C (1^3).

Reprecipitation of synthesized oligomers from toluene by methanol gave yellow solids, well soluble in ordinary organic solvents with η = 0.09 – 0.11. It is found that with the reaction rate increases with temperature. Figure 5.8 shows hydrogen conversion in dehydrocondensation of dihydroxysilaoxadihydrophenanthrene with 1,4-bis(dimethylsilyl)benzene.

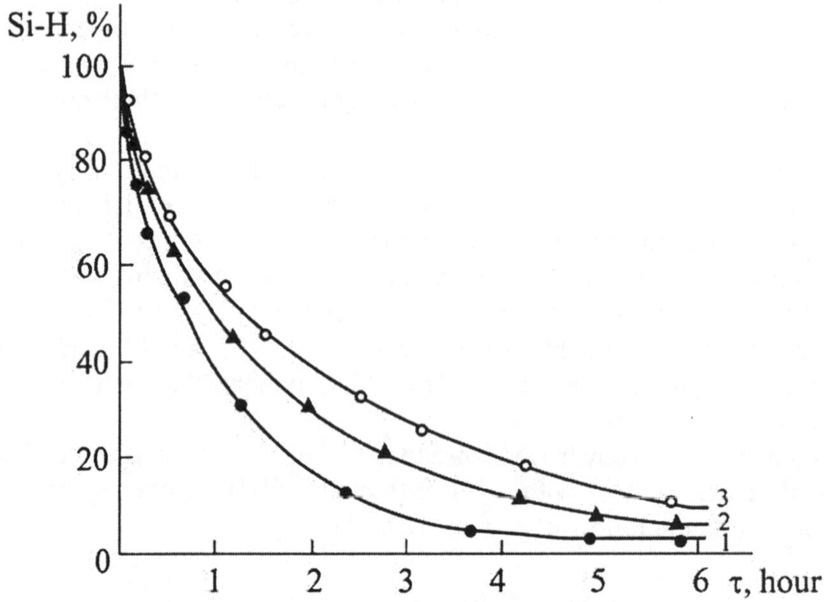

Figure 5.8. Time dependence of hydrogen conversion for catalytic dehydrocondensation reaction of dihydroxysilaoxadihydrophenanthrene with 1,4-bis(dimethylsilyl)benzene: 1 - at 60°C, 2 – at 50°C; 3 – at 60°C

It is found that dehydrocondensation is the second order reaction, the reaction rate constants of which for different temperatures were derived from time dependence of reverse concentration: $k_{40°C} = 3.3725\times10^{-2}$; $k_{50°C} = 4.6625\times10^{-2}$; $k_{60°C} = 6.3995\times10^{-2}$. The activation energy of dehydrocondensation equals $E_a = 29.44$ kJ/mol. Determined molecular mass of oligomer 1^3 indicated that dehydrocondensation proceeds deeper (for copolymer 1^3, $m = 17$) than heterofunctional condensation of 1,1-dichloro-1-sila-2-oxadihydrophenanthrene with 1,4-bis(hydroxydimethylsilyl)benzene (refer to reaction (5.23)). Some physical and chemical parameters of oligomers of structure XV (reaction (5.22)) are shown in Table 5.11.

Table 5.11

Some physical and chemical parameters of oligomers of structure XV (reaction (5.22))

No.	Oligomer structure	Reaction temperature, °C	η_{spec}	d_1, Å	H_2 conversion, %	*m*	T_{soft}, °C	Yield, %
1[1]	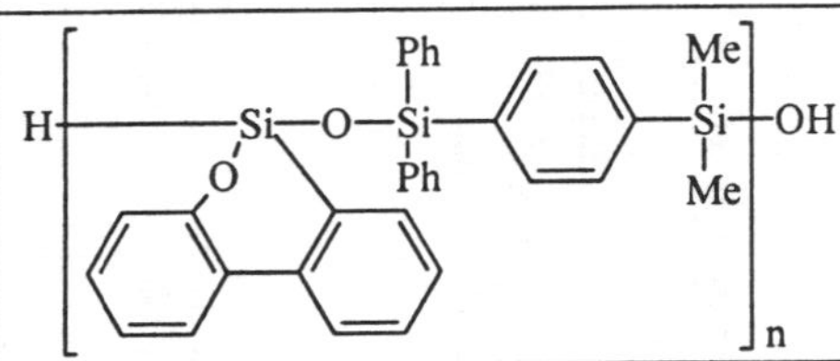	40	0.09	10.74	93	-	64-68	93
1[2]		50	0.10	-	95	-	65-70	95
1[3]		60	0.11	10.74	98	17	67-74	98

Silaoxadihydrophenanthrenesilarylene oligomers containing hydroxyl end groups were derived by HFC reaction of 1,1-dichloro-1-sila-2-oxadihydrophenanthrene (compound 1, Table 5.5) with 1,4-bis(hydroxydimethylsilyl)benzene at different ratios of initial reagents (1;0.95 and 1:1) in 50 - 60% solution of anhydrous toluene in the presence of pyridine. HFC proceeds in accordance with the following scheme [52, 53]:

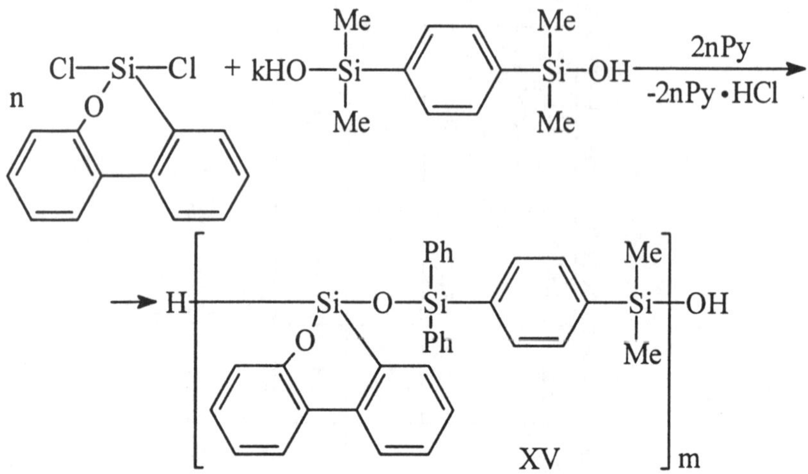

(5.24)

where $m = 6 - 12$.

Table 5.12

Some physical and chemical parameters of synthesized silarylenesilaoxa-dihydrophenanthrene oligomers (reaction (5.24))

No	Oligomer	η_{spec}	d_1, Å	OH, %	m	T_{soft}, °C	Yield, %
1	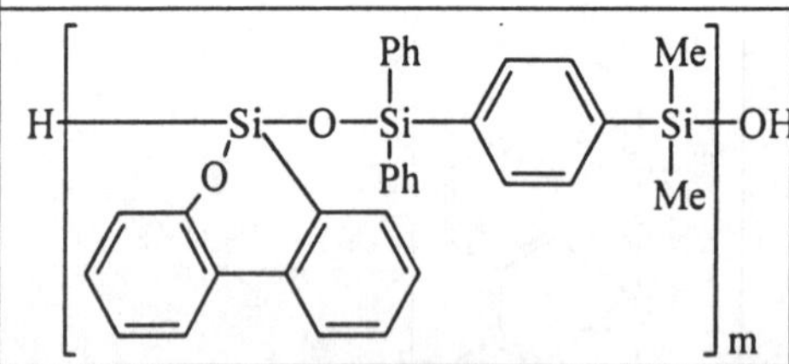	0.06	10.74	1.34 / 1.32	6	58-65	89
2		0.09	10.74	0.67 / 0.66	12	68-71	92

Note: * In numerator there are calculated values, in denominator – experimental values.

Synthesized oligomers, reprecipitated from toluene by methanol, represent white or yellow solids, well soluble in ordinary organic solvents with η = 0.06 – 0.09. Some physical and chemical parameters of synthesized silarylenesilaoxadihydrophenanthrene oligomers are shown in Table 5.12.

Thermogravimetric studies of synthesized silarylenesilaoxadihydrophenanthrene oligomers have displayed 10% mass losses at 340 - 360°C. The main destruction process proceeds in the temperature range of 450 - 680°C, and above 700°C no changes in the curves of mass losses are observed.

X-ray diffraction studies have shown that oligomers of structure XV represent single-phase amorphous systems with interchain distance d_1 = 10.74 Å. Synthesized oligomers are important substances for deriving block-copolymers of the $(AB)_nC_m$ type.

Thus, one may conclude that when introduced to macromolecular backbone, cardocyclic fragments behave themselves as cyclic ones in the backbone or in the side framing. Similar to cyclic fragments in the dimethylsiloxane backbone, they inhibit the chain transfer reaction during degradation, and the main degradation process is initiated by 60 - 80°C higher, than for linear poly(dimethylsiloxanes).

References

1. Wu T.C. and Hirt C.A., *J. Organomet. Chem.*, 1968, vol. **11**(1), p. 17.
2. Andrianov K.A. and Zachernyuk A.B., *Zh. Vses. Khim. Obshch. im. D.I. Mendeleeva*, 1974, vol. **19**(1), p. 107. (Rus)
3. Zachernyuk A.B., Burlova E.A., and Zhdanov A.A., *Zh. Obshch. Khim.*, 1985, vol. **55**(6), p. 1368. (Rus)
4. Meladze S.M., Mukbaniani O.V., Makarova N.N., and Khananashvili L.M., *Zh. Obshch. Khim.*, 1980, vol. **50**(11), p. 2493. (Rus)
5. Andrianov K.A., Zachernyuk A.B., and Danielov S.A., *KhGS*, 1972, No. 7, p. 893. (Rus)
6. Andrianov K.A. and Zachernyuk A.B., *KhGS*, 1972, No. 9, p. 1183. (Rus)
7. Zachernyuk A.B., *Candidate's Dissertation on Chemistry*, 1974, INEOS AS USSR, Moscow. (Rus)
8. Andrianov K.A., Zachernyuk A.B., and Lavrukhin B.D., *Doklady AN SSSR*, 1972, vol. **207**(3), p. 593. (Rus)
9. Andrianov K.A., Zachernyuk A.B., Lavrukhin B.D., and Vasilenko N.G., *Doklady AN SSSR*, 1972, vol. **204**(5), p. 1112. (Rus)
10. Hicton H.J. and Holt A., *J. Chem. Soc.*, 1966, vol. **??**, p.149.
11. Andrianov K.A., Zachernyuk A.B., and Zhdanova E.A., *Doklady AN SSSR*, 1974, vol. **214**(2), p. 325. (Rus)
12. Andrianov K.A., Zachernyuk A.B., and Solomatin G.V., *Doklady AN SSSR*, 1977, vol. **236**(3), p. 601. (Rus)
13. Zachernyuk A.B. and Solomatin G.V., *Thes. Rep. V All-Union Conference on Chemistry and Application of Organosilicon Compounds*, January 29 – February 1, 1980, Tbilisi, part 1, p. 281. (Rus)
14. Vdovin V.Sh. and Petrov A.D., *Zh. Obshch. Khim.*, 1960, vol. **30**, p. 838. (Rus)
15. Petrashko A.I., Andrianov K.A., Korneeva G.P., Kuntsova Z.M., Moiseenko A.P., and Asnovich A.Z., *Vysokomol. Soedin.*, 1967, vol. **9**, p. 2036. (Rus)
16. Inaridze I.A., *Candidate's Dissertation on Chemistry*, I. Javakhishvili Tbilisi State University, 1993, Tbilisi, Georgia.
17. Chernyishev E.A., Krasnova T.L., and Shchepinov S.A., *Zh. Obshch. Khim.*, 1970, vol. **40**, p. 1735. (Rus)
18. Chernyishev E.A., Krasnova T.L., and Shchepinov S.A., *Zh. Obshch. Khim.*, 1970, vol. **40**, p. 1058. (Rus)
19. Dewar M.J.S., *J. Amer. Chem. Soc.*, 1960, vol. **??**, p. 1344.
20. Dewar M.J.S. and Cubba P., *J. Amer. Chem. Soc.*, 1959, vol. **??**, p. 347.

21. Samsonia A.Sh., Mukbaniani O.V., and Karchkhadze M.G., *Georgian Eng. News*, 2000, No. 2, p. 144.
22. *A.C. No. 602,510* (USSR), 1976, Andrianov K.A., Zachernyuk A.B., and Burlova E.A.; *Bull. Izobr.*, 1978, No. 14, p. 85. (Rus)
23. Stabilization of Polymers and Stabilizing Processes. Advances in Chemistry, Series, *Amer. Chem. Soc.*, Washington, 1968.
24. Andrianov K.A., *High-molecular Compounds for Electrical Insulation*, Gosenergoizdat, 1961. (Rus)
25. Andrianov K.A. and Zachernyuk A.B., *Vysokomol. Soedin.*, 1974, vol. **16B**(6), p. 307. (Rus)
26. Andrianov K.A., Burlova E.A., and Zachernyuk A.B., *Vysokomol. Soedin.*, 1977, vol. **19B**(6), p. 211. (Rus)
27. Andrianov K.A. and Zachernyuk A.B., *Vysokomol. Soedin.*, 1974, vol. **16A**(6), p. 1435. (Rus)
28. Bunina L.O., Burlova E.A., and Zachernyuk A.B., *Thes. Rep. VII All-Union Conference on Chemistry, Production Technology and Application of Organosilicon Compounds*, 1990, Tbilisi, part 2, p. 405. (Rus)
29. *Patent No. 3,366,593* (USA), 1969, Breed L.W. and Elliott L.L.
30. Andrianov K.A., Lekishvili N.G., Nogaideli A.I., Kotrelev G.V., and Tkeshelashvili R.Sh., *Vysokomol. Soedin.*, 1974, vol. **16B**, p. 496. (Rus)
31. Andrianov K.A., Nogaideli A.I., Kotrelev G.V., Lekishvili N.G., Kamarnitski B.A., and Kryilov A.P., *Vysokomol. Soedin.*, 1975, vol. **17B**, p. 315. (Rus)
32. Andrianov K.A., Kotrelev G.V., Nogaideli A.I., Zhuravleva I.V., Lekishvili N.G., Tolchinski Yu.P., and Pushich V.I., *Vysokomol. Soedin.*, 1977, vol. **19A**, p. 451. (Rus)
33. Kotrelev G.V. and Zhdanova E.A., *Vysokomol. Soedin.*, 1981, vol. **23B**(4), p. 289. (Rus)
34. Inaridze I.A., Mukbaniani O.V., and Khananashvili L.M., *Bulletin of the Georgian Academy of Sciences*, 1994, vol. **149**(1), p. 84.
35. Mukbabiani O.V., *Doctor's Dissertation on Chemistry*, I. Javakhishvili Tbilisi State University, 1993, Tbilisi, Georgia. (Rus)
36. Khananashvili L.M., Mukbaniani O.V., Samsonia A.Sh., and Karchkhadze M.K., *XVI Mendeleev Meeting on General and Applied Chemistry*, Saint-Petersburg, May 25 – 26, 1998, p. 320. (Rus)
37. Mukbaniani O.V., Scherf U., Samsonia A.Sh., Karchkhadze M.G., and Khananashvili L.M., *Intern. J. Polym. Mater.*, 2001, vol. **48**(4), p. 381.

38. Andrianov K.A., Nogaideli A.I., Makarova N.N., and Mukbaniani O.V., *Izv. AN SSSR*, 1977, vol. ??, No. 6, p. 1388. (Rus)
39. Yuzhelevski Y.A., Kurlova T.V., Kagan E.A., and Suvorova M.V., *Zh. Obshch. Khim.*, 1972, No. 9, p. 2006.
40. Andrianov K.A., Nogaideli A.I., Slonimski G.L., Levin V.Yu., Kvachev Yu.P., Makarova N.N., and Mukbaniani O.V., *Vysokomol. Soedin.*, 1976, vol. **13B**(5), p. 359. (Rus)
41. Meladze S.M., Khananashvili L.M., and Mukbaniani O.V., *Vysokomol. Soedin.*, 1984, vol. **26B**(5), p. 250. (Rus)
42. Koyava N.A., Mukbaniani O.V., and Khananashvili L.M., *Vysokomol. Soedin.*, 1985, vol. **27A**(11), p. 2261. (Rus)
43. Mukbaniani O.V., Achelashvili V.A., Levin V.Yu., Meladze S.M., Inaridze I.A., and Khananashvili L.M., *Izv. AN GSSR*, 1990, vol. **16**(1), p. 20. (Rus)
44. Mukbaniani O.V., Achelashvili V.A., Khananashvili L.M., and Sturua G.I., *Izv. AN GSSR*, Coll., 1988, p. 53. (Rus)
45. Thomas T.H. and Kendrik T.C., *J. Polym. Sci.*, 1969, vol. **2A**, p. 537.
46. Alexander G. and Garzo G., *Chromatography*, 1974, vol. **7**, p. 225.
47. Andrianov K.A., Papkov V.S., Slonimski G.L., Zhdanov A.A., and Iakushina S.E., *Vysokomol. Soed.*, 1969, vol. **11A**, p. 2030. (Rus)
48. Mukbaniani O.V., Samsonia A.Sh, Karchkhadze M.G., and Khananashvili L.M., *J. Appl. Polym. Sci.*, 2002, vol. **84**, p. 9.
49. Samsonia A.Sh., Karchkhadze M.G., Mukbaniani O.V., and Khananashvili L.M., *Georg. Eng. News*, 2000, No. 2, p. 144.
50. Samsonia A.Sh., Mukbaniani O.V., and Karchkhadze M.G., *Georg. Eng. News*, 2000, No. 2, p. 142.
51. Samsonia A.Sh., *Candidate's Dissertation on Chemistry*, I. Javakhishvili Tbilisi State University, 2002, Tbilisi, Georgia. (Rus)
52. Mukbaniani O.V., Samsonia A.Sh., and Karchkhadze M.G., *Condensation Polymers: Synthesis, Structure, Properties*, (To the 90th anniversary of Academician V.V. Korshak), January 12 – 14, 1999, Moscow, pp. 1 – 15. (Rus)
53. Samsonia A.Sh., Karchkhadze M.G., and Mukbaniani O.V., *Georg. Eng. News*, 2000, No. 2, p. 142. (Rus)

Chapter 6. Organosilicon copolymers with bi-, tri- and tetracyclic structures in dimethylsiloxane backbone

For the purpose of tracing the influence of more rigid inclusions on properties of copolymers, synthesis of organobi-, organotri- and organotetracyclosiloxanes and bicyclocarbosiloxanes with the assigned disposition of functional groups at atoms of silicon and copolymers, derived from them, are discussed in this Chapter.

6.1. Synthesis of organobi-, organotri- and organotetracyclosiloxanes and bicyclocarbosiloxanes with functional groups

HFC reactions of cis-1,3,5,7-tetrahydroxy-1,3,5,7-tetraphenylcyclotetrasilocane (tetrole, T-1) and 1,1,5,5-tetrahydroxy-3,3,7,7-tetraphenylcyclotetrasiloxane (T-2) with various organochlorosiloxane and siloxanes for the purpose of obtaining polycyclic compounds with functional groups and some their transformations are discussed in this Chapter.

When polycondensation of tetrole with organodichlorosilanes is performed with 1:4.5 and 1:5.0 ratios of initial components in the presence of hydrogen chloride acceptor, spatial hindrances for interchain reaction due to *cis*-conformation of reactive hydroxyl groups may be expected. At the same time, in the interaction of tetrole with organochlorosilanes at 1:4.0 ratio of initial components or higher, the reaction may possibly proceed in two directions: formation of polycyclic compounds by intramolecular cyclization and synthesis of more complex structures at intermolecular condensation of intermediate products.

At the first stage model condensation reactions of tetrole with trimethylchlorosilane were studied [1]. Under the following reaction conditions: 1:5 ratio of initial components, temperature range 20 ÷ +25°C (under soft conditions), dilution C = 0.03 g/ml in ether solution, the presence of acceptor – pyridine, attempts to substitute all four hydroxyl groups to trimethylsiloxy-groups have failed. As a result, 1-hydroxy-3,5,7-tris(trimethylsiloxy)tetraphenylcyclotetrasiloxane (71%) and reaction products of intermolecular condensation of high transformation degree were obtained. The reaction proceeded as follows:

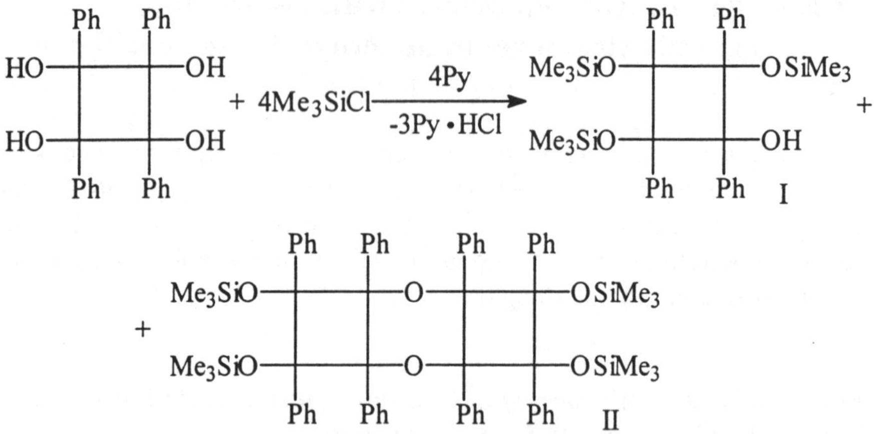

(6.1)

Thus, in spite of condensation proceeding with chlorosilane excess, attempts to substitute all four hydroxyl groups by trimethylsiloxy-groups failed under the above-mentioned conditions due to steric hindrances, and besides HFC product, homofunctional condensation product of the structure II is obtained.

^{1}H NMR spectrum of 1-hydroxy-3,5,7-tris(trimethylsiloxy)tetraphenyl-cyclotetrasiloxane displays a singlet reflex with chemical shift at $\delta = 0.20$ ppm. ^{1}H NMR spectrum of the product of the structure II displays expanded reflex for methyl protons, which is explained by formation of tricyclic compounds with different spatial structures:

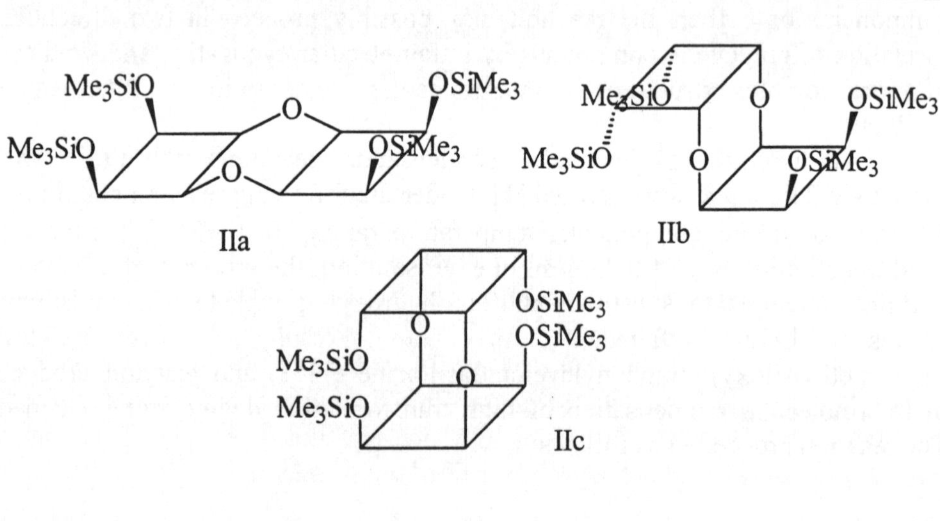

Compounds of the structure II can be synthesized by intermolecular homocondensation of products containing hydroxyl groups only. As a consequence, the authors suggest that mono- and dihydroxy-derivatives are formed as intermediate compounds:

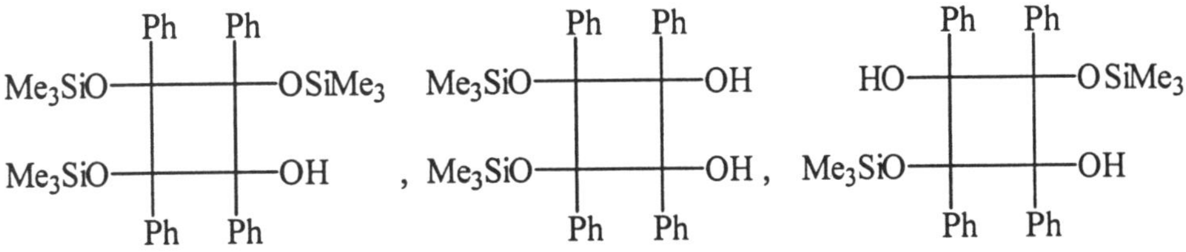

with future condensation of tetrole dihydroxy-derivatives, which leads to formation of tricycles of the structure II (a – c).

As the temperature mode is changed, i.e. when at the final stage the reaction proceeds at 80°C in benzene (under rigid conditions), it proceeds completely with complete substitution of all hydroxyl groups by trimethylsiloxy-groups. As a result, 1,3,5,7-tetrakis(trimethylsiloxy)-1,3,5,7-tetraphenylcyclotetrasiloxane with 78% yield was formed:

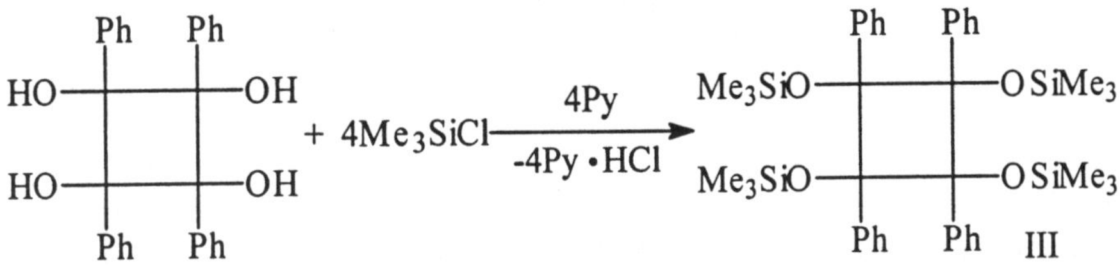

III (6.2)

It should be noted that intermolecular condensation products were not produced in this case. The compound of the structure III was produced in HFC reaction of the compound of structure I with trimethylchlorosilane in benzene at 1:1.5 ratio of initial components. ^1H NMR spectrum of the compound of the structure III displays a singlet with chemical shift $\delta = 0.19$ ppm for methyl protons due to their equivalence.

By analogy, HFC reaction of tetrole with dimethylchlorosilane proceeds under rigid conditions and at 1:4 ratio of initial components for soft conditions. In this case, three hydroxyl groups only are substituted, and the fourth one is preserved. However, contrary to reaction (6.1), intermolecular condensation product cannot be produced by fractionation of the reaction products. As a consequence, the reaction proceeds in accordance with the following scheme:

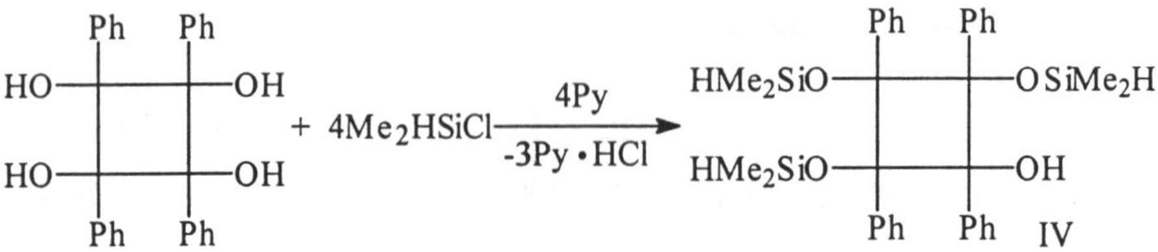

(6.3)

IV

Further HFC reaction of compound 1 (Table 6.1) of structure IV with dimethylchlorosilane gives an opportunity to substitute the fourth hydroxyl group [2] and synthesize tetrakis-substituted product with the following structure:

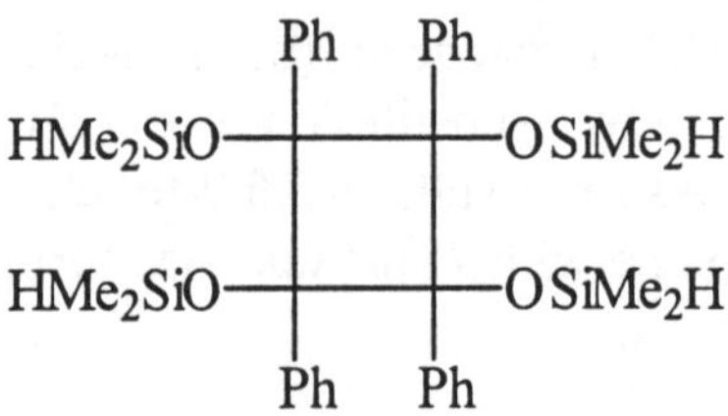

V

Compound (5) of structure V was also synthesized by a single-stage method by HFC reaction of tetrole with dimethylchlorosilane under rigid conditions with substitution of ether by toluene or benzene at the final stage.

To synthesize organocyclosiloxanes with chlorine functional groups at atoms of silicon, HFC of tetrole with dimethylchlorosilane proceeding under soft conditions at 1:4.0 and above 1:5.0 ratios of initial components was studied [3]. The reaction proceeded in the presence of acceptor – pyridine, in anhydrous ether in the temperature range of 20 - 30°C. At the final stage, the reaction mixture was boiled. Despite the excess of dimethylchlorosilane, substitution of all four hydroxyl groups by dimethylchlorosiloxy-groups under the mentioned conditions has failed. By analogy, the reaction with methylphenylchlorosilane proceeds. In the case of diphenyldichlorosilane, the reaction proceeds slowly and under more rigid conditions. Thus, HFC of dimethyldichlorosilane with methylphenyldichlorosilane produces bi- and tetracyclic compounds, proceeding in accordance with the following scheme [3, 4]:

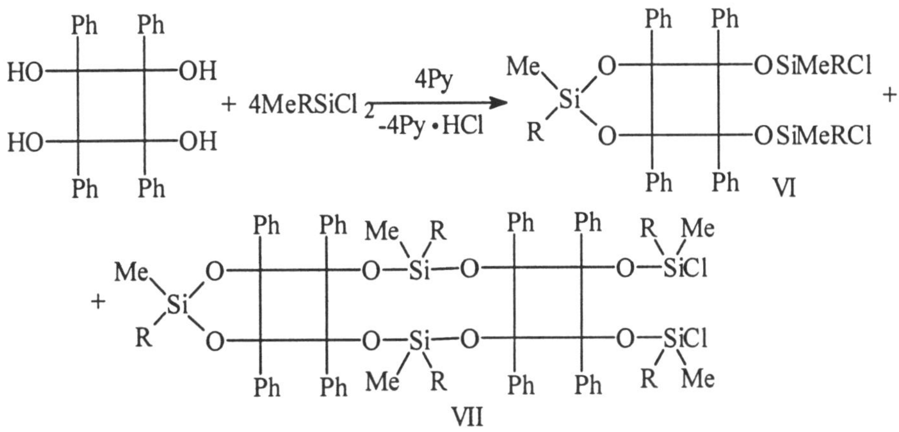

(6.4)

To prove the structure of bicyclic compounds, hydrolytic polycondensation of the compound of structure VI [R = Me (6), Ph (8), Table 6.1] in ether solution was performed [3, 4]. It is found that synthesis of organotricyclononasiloxane at the hydrolysis stage in diluted solution in the absence of hydrogen chloride acceptor does not proceed, because cyclization of this type requires more rigid conditions. Tricyclononasiloxanes of structure IX (compounds 10 and 11) were synthesized only after boiling of the products in 5% diluted benzene solution during several hours with the Dyne-Stark trap, at fractionation of the reaction products. The reaction proceeds as follows:

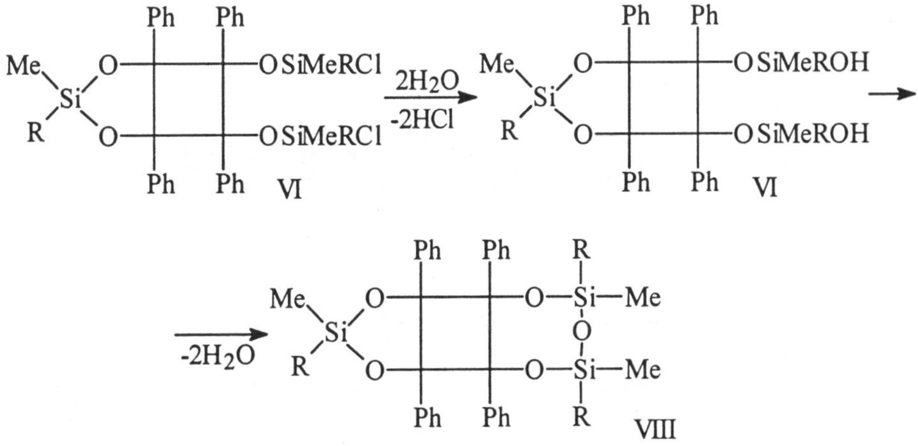

(6.5)

where R = Me (10), Ph (11).

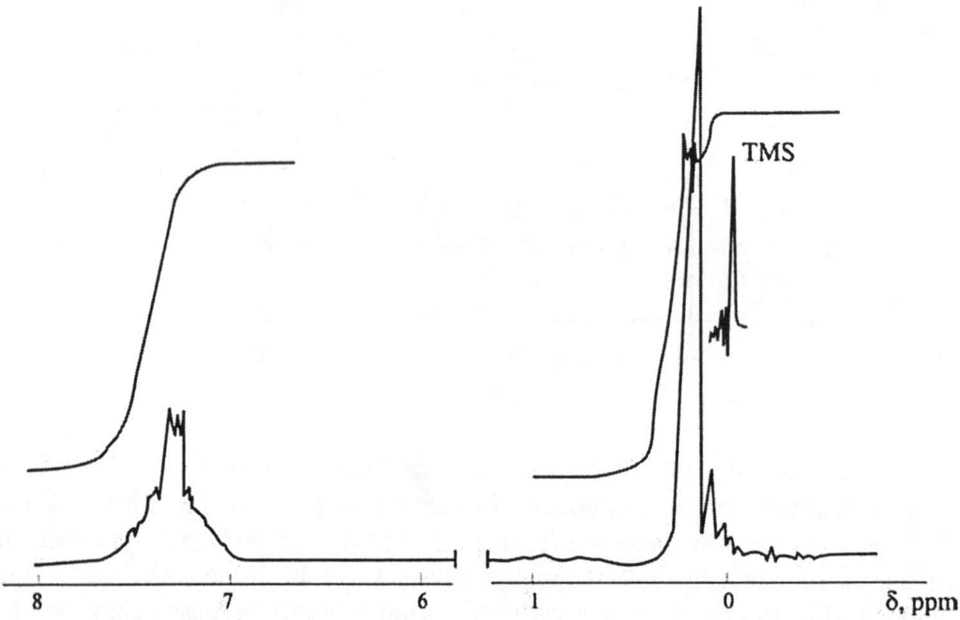

Figure 6.1. ^{1}H NMR spectrum of compound (10) (Table 6.1)

To obtain bis(diethylamino)bicyclosiloxanes, aminolytic splitting of *cis*-tetrole reaction products by diorganodichlorosilanes was performed [6]. Diethylamine was used as hydrogen chloride acceptor. The reaction proceeds as follows:

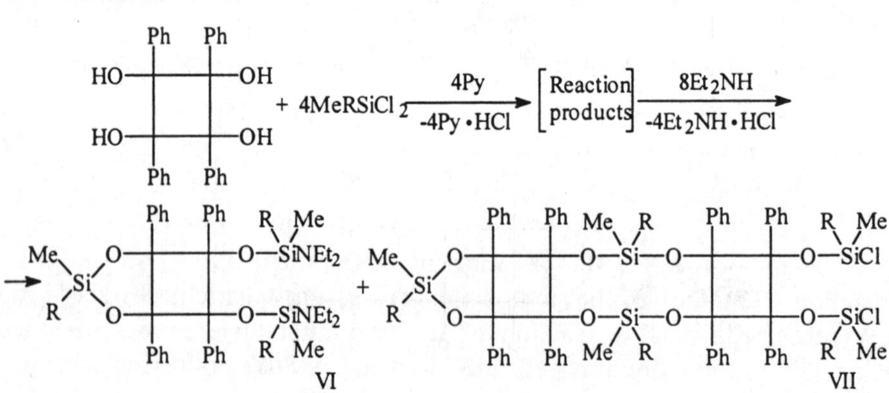

(6.6)

where R = Me, Ph.

It should be noted that in the case of phenyl framing, the compound of structure VII was not aminolytically split.

^{1}H NMR spectrum of compound (10) of structure IX (R = Me) contains three singlet reflexes with the intensity ratio 1:3:2 and chemical shifts 0.47, 0.31 and 0.12 ppm, respectively. The first of them is corresponded to protons of methyl groups in trisiloxane ring, the second one is composed of reflexes of methyl group protons in tetrasiloxane ring, and the third one is corresponded to protons of two methyl groups in tetrasiloxane ring (refer to Figure 6.1).

Under soft conditions, similar to scheme (6.4), HFC of tetrole with silicon tetrachloride at 1:4 ratio of initial components proceeds. However, the product of intermolecular condensation differs from the compound of structure VII. The following reaction scheme was suggested [5]:

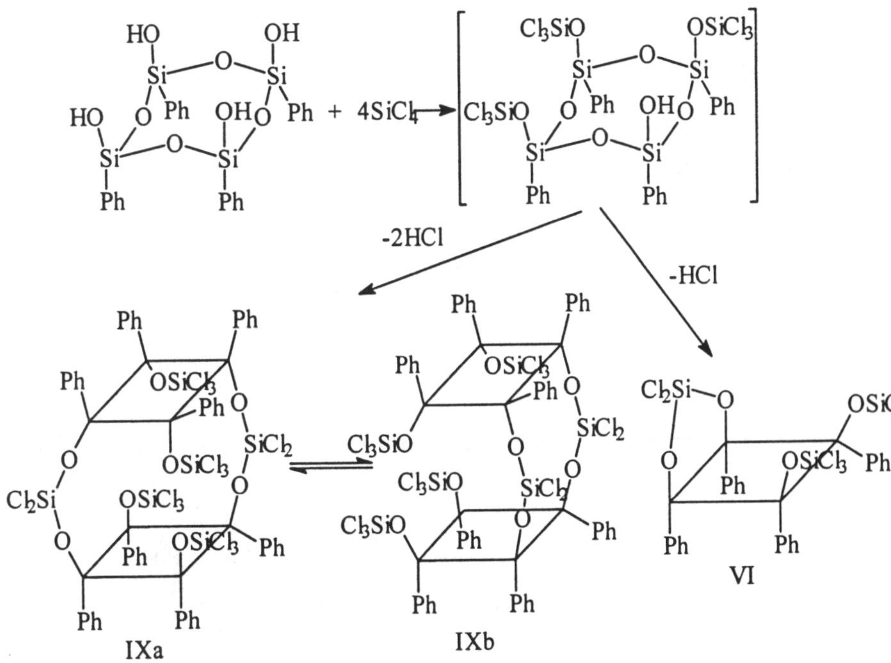

(6.7)

The authors suggest the presence of isomeric structures IXa and IXb in the reaction (6.7). Polycyclic compounds possessing no trichlorosiloxy-group were synthesized in HFC reaction of tetrole with double excess of silicon tetrachloride. The reaction proceeds in accordance with the following scheme:

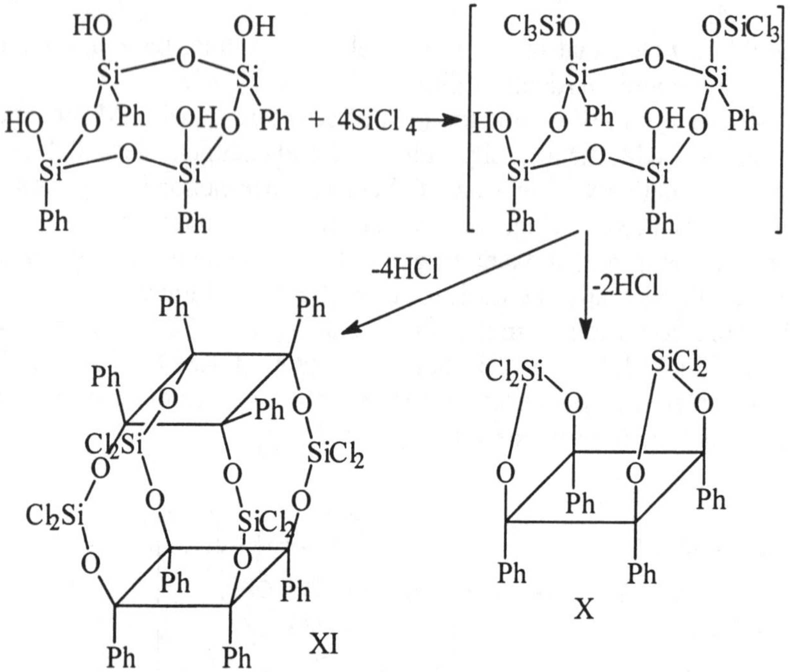

(6.8)

HFC reaction of tetrole with dimethyldichlorosilane was studied at the ratio of initial components 1:5 in diluted diethyl ether (solution concentration $C = 0.033$ g/ml) at 20 - 25°C, with future diluter substitution by benzene and temperature increase to 80°C. Fractionation of the reaction products gave tetrole derivatives, in which all four groups are substituted by chlorodimethylsiloxy-groups, and intermolecular condensation product. The reaction proceeds as follows:

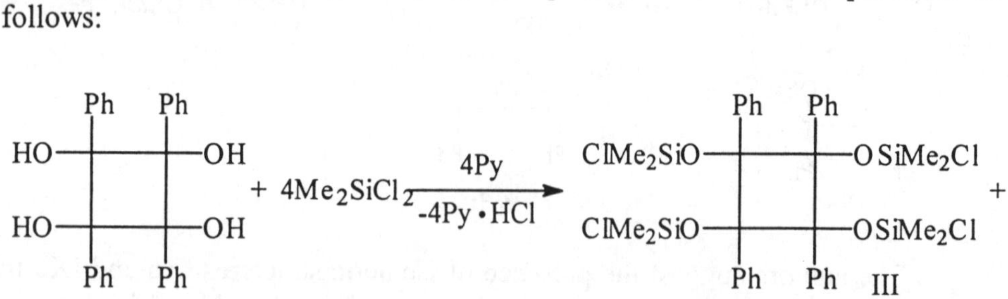

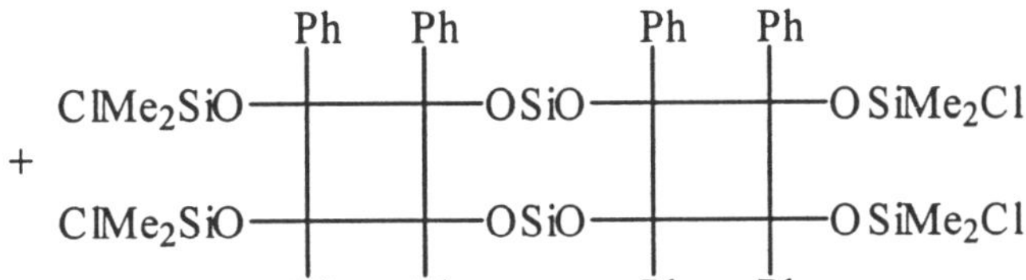

IXb (6.9)

Composition and structure of the compounds, synthesized by the scheme (6.9), are determined on the basis of ultimate and functional analysis, determination of the molecular mass, and from IR and NMR spectra. Figure 6.2 shows ^{1}H NMR spectrum of tetrakis(chlorodimethylsiloxy)tetraphenylcyclotetrasiloxane, which is compound (19) (Table 6.1). Due to equivalence of chlorodimethylsiloxy-groups, a singlet reflex with chemical shift $\delta = 0.56$ ppm is present in the spectrum.

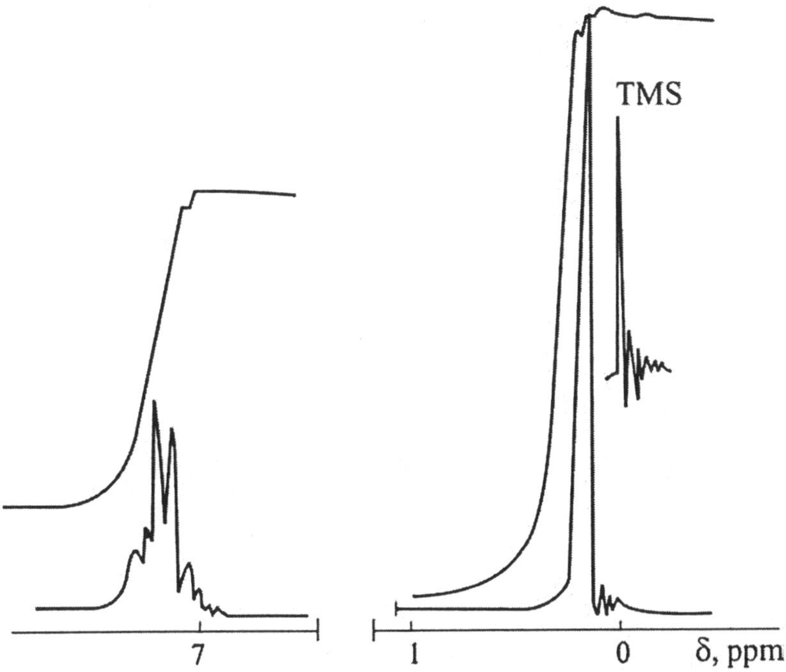

Figure 6.2. NMR spectrum of 1,3,5,7-tetrakis(chlorodimethylsiloxy)-1,3,5,7-tetraphenylcyclotetrasiloxane (19) (Table 6.1)

The influence of partial substitution of organochlorosilanes by 1,3-dichlorotetraorganodisiloxanes (at tetrole interaction with organochlorosilanes) on intra- and intermolecular cyclization, proceeding in the intermediate reaction products, is studied [5, 7].

Benched condensation of tetrole with 1,3-dichlorotetraorganodisiloxane and organochlorosilanes proceeding in diluted solutions in the presence of hydrogen chloride acceptor in the temperature range of 0 - 35°C may form organobicyclononasiloxanes of structure XII, organotricyclodecasiloxanes of structure XIII, organobicyclohexasiloxanes of structure VI and products of intermolecular condensation as the final products. The compounds of structures VI and XIII may be synthesized by tetrole interaction with the single reagent, i.e. by tetrole interaction with organochlorosilane or chlororganodisiloxane. General scheme of the reaction is the following:

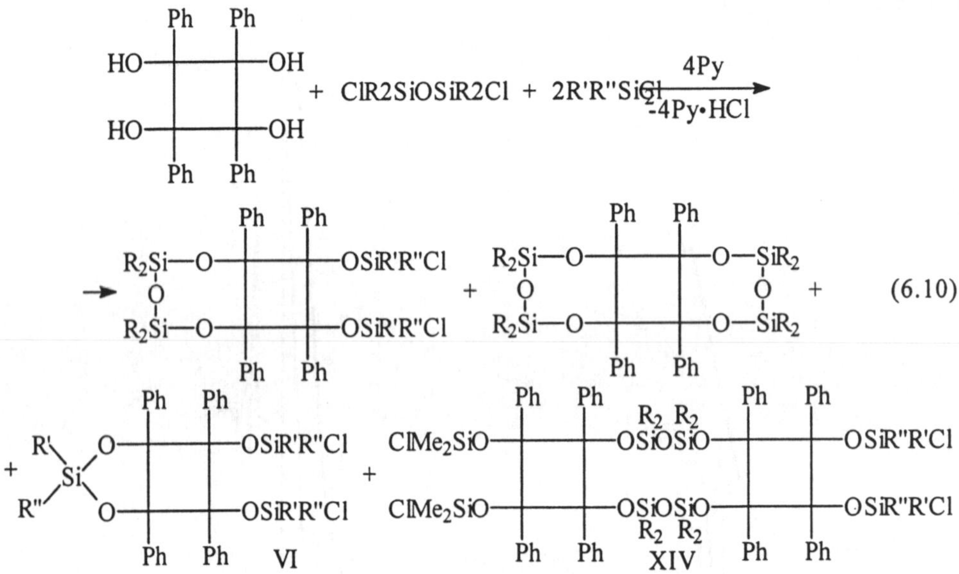

(6.10)

where R = Me, Ph; R' = R" = Me, Cl; R' = Me, R" = Ph, Vin, Cl.

Possible formation of polycyclic compounds of structures VI, VII and XIII is limited by introduction of 1,3-dichlorotetraorganodisiloxanes at the initial stage of the reaction. Negligible amounts of the compound of structure XIII can be formed at the interaction of tetrole with 1,3-dichlorotetraorganodisiloxane, but it is easily fractionated from the reaction products due to its solubility in used solvents at very high dilution degree only.

The reaction scheme (6.10) indicates that intermolecular condensation of tetrole with 1,3-dichlorotetraorganodisiloxane and organochlorosilanes proceeds through a tetraorganodisiloxane bridge. However, the authors do not exclude formation of isomeric structures XIV.

To prove the structure XIII, i.e. the fact that benched HFC reaction of tetrole with 1,3-dichlorotetraorganodisiloxanes generally proceeds by 1,3-positions of hydroxyl groups, the authors have performed full hydrolysis of dichlorobicycloheptasiloxane. As a result, corresponded dihydroxy-derivatives were synthesized, boiling of which in benzene solution induced further condensation proceeding with water extraction, collected by Dyne-Stark device. Fractionation of thermal condensation products gave compounds of structure XIII with both symmetrical and asymmetrical disposition of substituting agents. The reaction proceeds as follows:

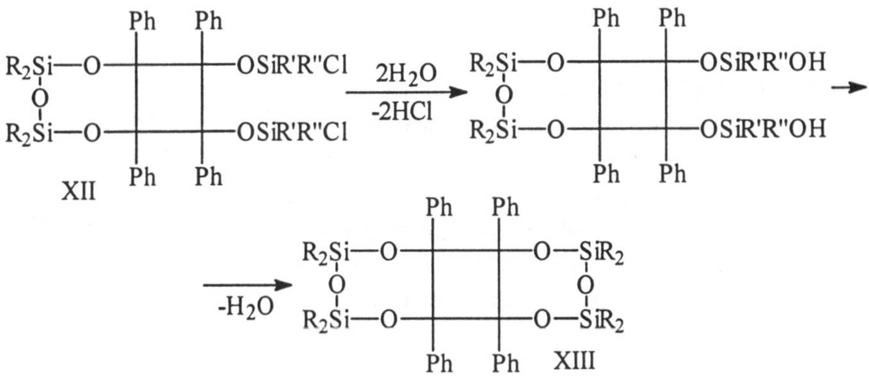

(6.11)

where R = R' = R" = Me; R = R' = Me, R" = Vin; R = Ph, R' = R" = Me.

By analogy, full hydrolysis of tetrachlorobicyclononasiloxane under soft conditions was performed, and intramolecular cyclization type of its proceeding with formation of dihydroxytricyclodecasiloxane by the following scheme was proved:

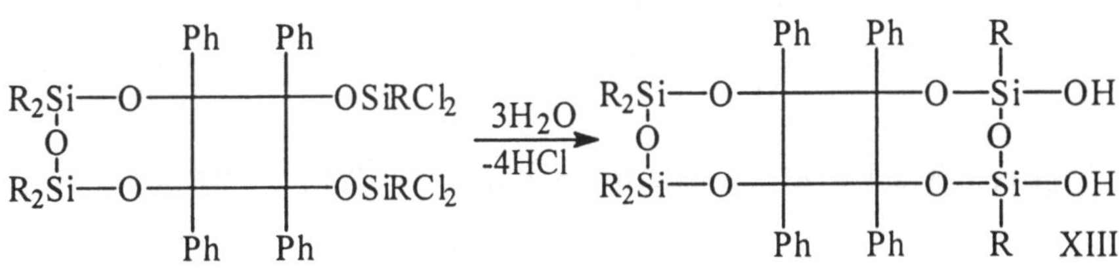

(6.12)

where R = Me, Ph.

By partial hydrolysis of tetrachlorobicyclononasiloxane of structure XII under soft conditions dichlorobicyclodecasiloxanes were synthesized by the scheme as follows:

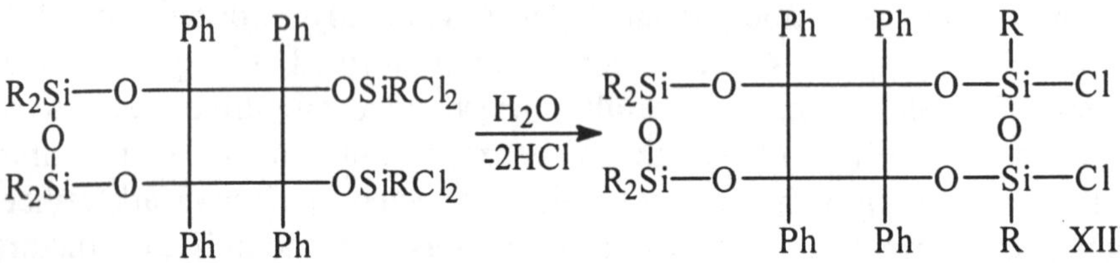

XIII (6.13)

where R = Me, Ph.

Composition and structure of the products, synthesized in the benched heterofunctional polycondensation, and the ones of their transformation were determined by functional and ultimate analysis, molecular mass measurement, and from IR and ^{1}H NMR spectral data.

^{1}H NMR spectrum of symmetrical tricyclodecasiloxane compound (29) (Table 6.1), synthesized by reaction (6.11) (Figure 6.3), displays two singlet reflexes of methyl group protons with chemical shifts $\delta = 0.57$ and $\delta = 0.77$ ppm and a complex multiplet of phenyl group protons with the chemical shift within the range of 7.5 - 8.2 ppm. Moreover, calculated ratios of methyl and phenyl groups correspond to the structural formula suggested.

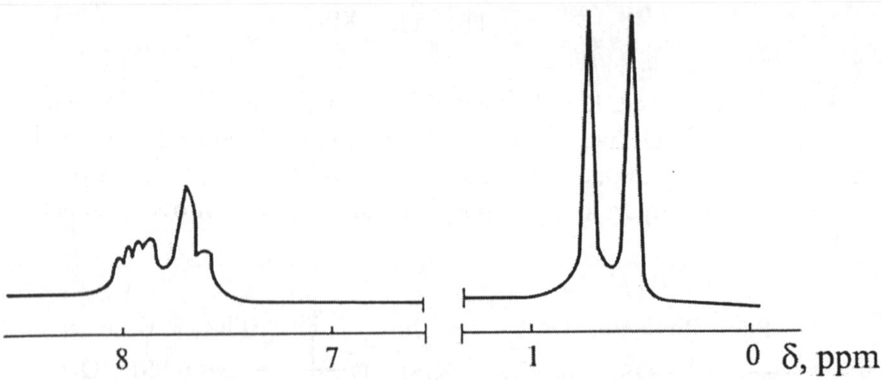

Figure 6.3. ^{1}H NMR spectrum of compound (29) (Table 6.1)

^{1}H NMR spectrum of asymmetrical tricyclodecasiloxane compound (29) (Table 6.1), synthesized by intramolecular cyclization of the hydrolysis product proceeding by the reaction (6.11), displays two singlet reflexes of methyl group protons at 0.55 and 0.75 ppm, which are most probable in formation of a *cis-syn-cis*-structure (Figure 6.4).

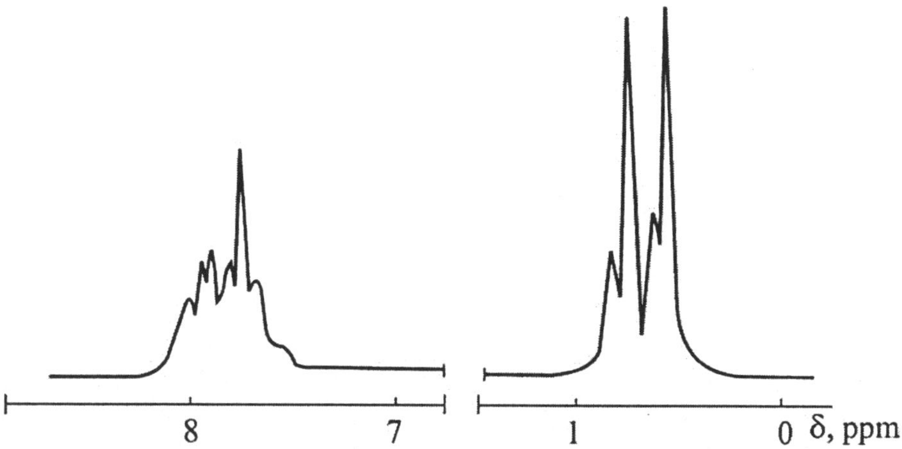

Figure 6.4. ^{1}H NMR spectrum of compound (30) (Table 6.1)

As far as the above-mentioned results indicate predominant proceeding of tetrole condensation with 1,3-dichlorotetraorganodiciloxane by 1,3-position of hydroxyl groups, further on, the benched HFC reaction of tetrole with 1,3-dichlorotetraphenyldisiloxane and 1,1,3,3-tetrachloro-1,3-diphenyldisiloxane was studied [5, 6]. The sequence of introduction of haloid-derivatives of phenylsiloxanes plays the important role, because fractionated condensation products form three-dimensional systems in the excess of 1,1,3,3-tetrachloro-1,3-diphenyldisiloxane. That is why, implementation of the reaction by suggested scheme (6.13) requires introduction of 0.05 – 0.10 mole excessive 1,3-dichlorotetraphenyldisiloxane, recalculated for tetrole, and a half of pyridine into the tetrole solution, first, and then 1,1,3,3-tetrachloro-1,3-diphenyldisiloxane and the rest of pyridine. The reaction proceeds by the scheme as follows:

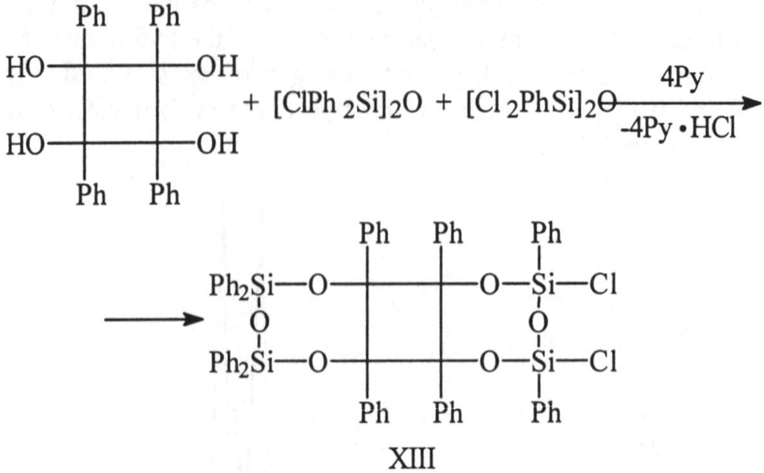

(6.14)

Synthesized compound (34) (structure XIII, Table 6.1) was identified on the basis of ultimate and functional analysis, molecular mass measurement and IR spectral data. Structure of the compound (34) was determined by hydrolytic condensation. Future processing of hydroxyl derivative by trimethylchlorosilane gave an opportunity to separate a compound with trimethylsiloxy-groups in accordance with the scheme as follows:

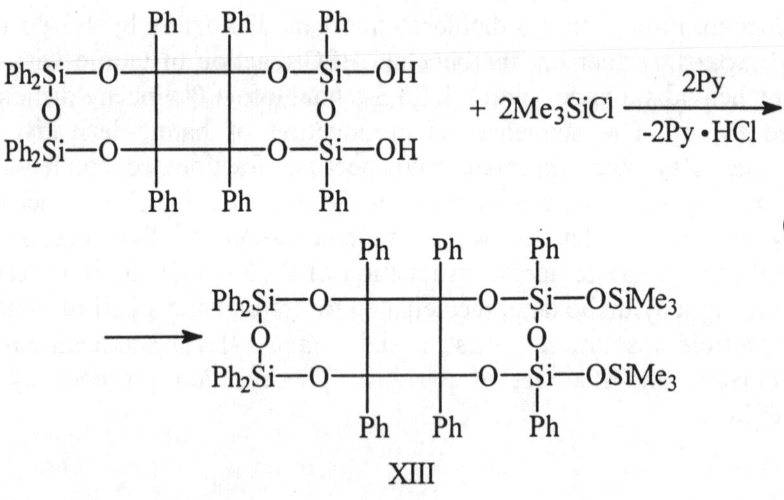

(6.15)

The final conclusion about the structure of decaphenyltricyclosiloxane was made on the basis of ^{1}H NMR spectrum [4, 6]. The influence of temperature of tetrole condensation with organochlorosilanes on the structure and yield of compounds formed were studied in detail in these works.

1,3-Divinylorganotricyclodecasiloxane was synthesized by benched HFC reaction of tetrole with 1,3-dichlorotetramethyldisiloxane and 1,3-dichloro-1,3-divinyl-1,3-dimethyldisiloxane at 1:1:1 ratio of initial components in the presence of pyridine by the scheme as follows:

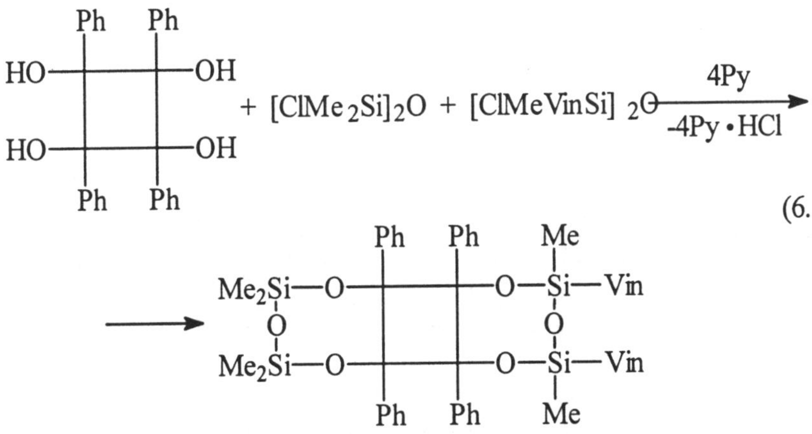

(6.16)

XIII

Physical and chemical parameters of the compound, synthesized by the reaction (6.16), coincide with these of the compound, synthesized by the reaction (6.11).

To obtain divinyltricyclohexasiloxane, the authors have studied HFC reaction of *cis*-tetrole with methylvinyldichlorosilane at 1:2 ratio of initial components in the presence of acceptor, triethylamine. Optimal conditions of obtaining 1,7-divinyl-1,7-dimethyl-3,5,9,11-tetraphenyltricyclohexasiloxane were selected, which proceeds in accordance with the following scheme [9]:

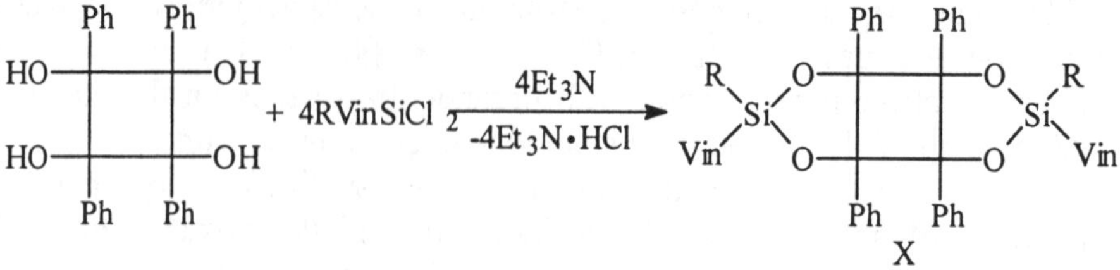

(6.17)

where R = Me, Ph.

^{1}H NMR spectroscopy method has indicated that the compound, synthesized by the reaction (6.17) (for R = Vin), displays three isomeric shapes: A, B and C, occurrence of which is associated with different dispositions of methyl and phenyl groups:

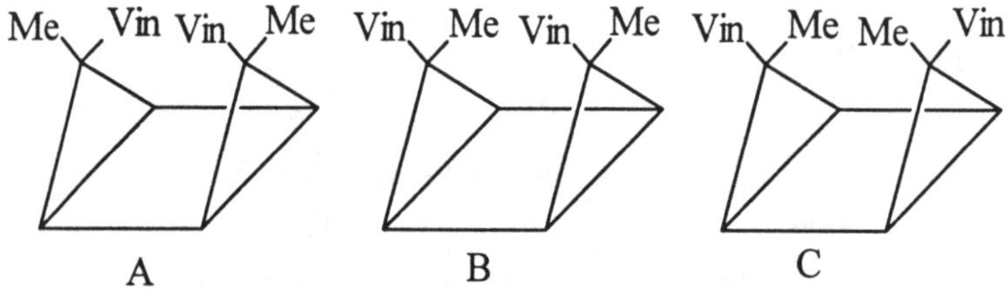

By positions in stereo-isomers, methyl groups are related to three nonequivalent groups that explains the presence of three reflexes from methyl protons in NMR spectra of divinyltricyclohexasiloxane. Analysis of reflex intensities in NMR spectra of separate fractions, obtained by fraction recrystallization of the reaction (6.17) product, by the method of selecting all possible alternatives allowed relation of each of three reflexes to a definite group of methyl protons. The only alternative suits experimental spectra: singlets with the shift at $\delta = 0.38$ ppm (A), $\delta = 0.41$ ppm (B) and $\delta = 0.51$ ppm (C). Based on this alternative, concentration of each of three stereo-isomers in the mixture is calculated from the reflex intensities: A – 18%, B – 57% and C – 25%.

Recently, it was shown that hydrolysis of 1,1,5,5-tetrachlorotetraphenylcyclotetrasiloxane produces 1,1,5,5-tetrahydroxytetra-phenylcyclotetrasiloxane (T-2) [10], which is *cis*-tetrole (T-1) isomer [11]. It was found that contrary to T-1, T-2 forms an aggregate with pyridine in accordance with the following scheme:

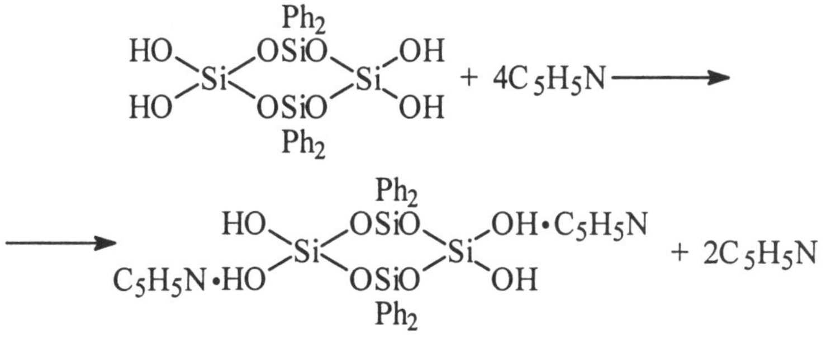

(6.18)

XV

White crystallites of T-2 complex with pyridine display low solubility in low-polar organic solvents and good solubility in polar ones. It is shown in the literature that tetraphenyldisiloxane diol and hexaphenyltrisiloxane diols form complexes with organic bases [12, 13]. Interaction of dipyridine-tetrahydroxytetraphenylcyclosiloxane [14] with excessive dimethyldichlorosilane in the presence of pyridine forms a difunctional bicyclic compound in accordance with the following scheme:

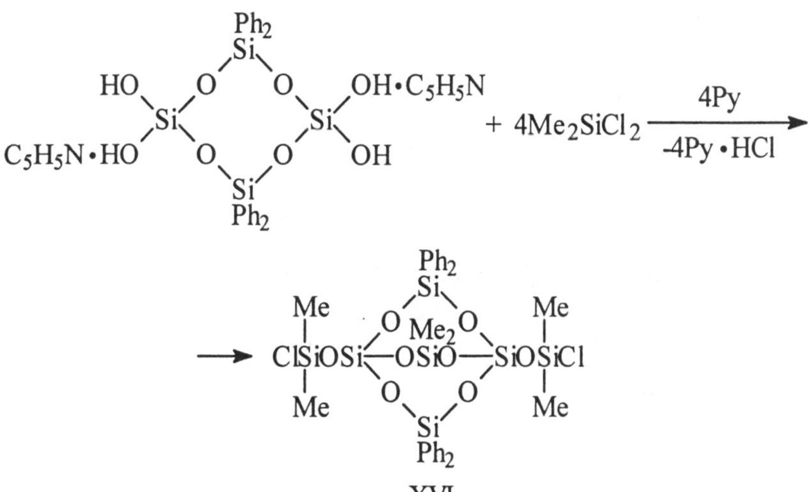

(6.19)

XVI

By analogy, condensation of the complex with excessive 1,3-dichlorotetramethyldisiloxane in the presence of pyridine proceeds:

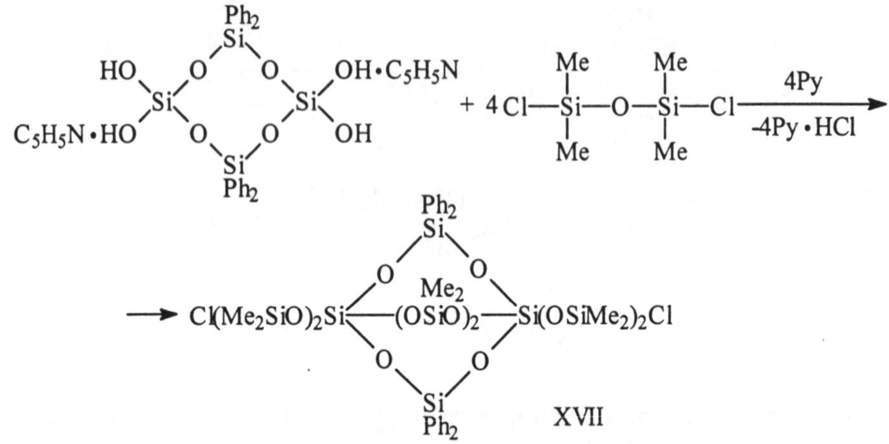

(6.20)

XVII

The interaction of T-2 with diorganodichlorosilanes in the presence of acceptor (pyridine) proceeds differently, giving tetrafunctional branched organocyclosiloxanes in accordance with the following scheme [15]:

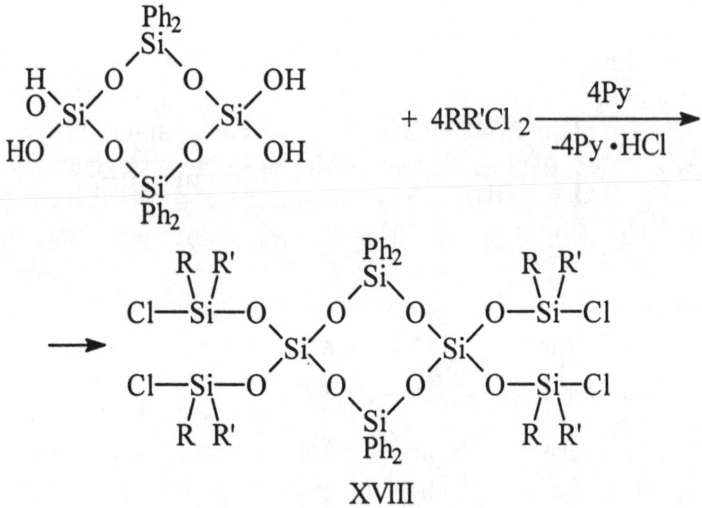

(6.21)

XVIII

where R = R' = Me, Ph, Et; R ≠ R'.

This gives grounds to assume that dipyridine-tetrahydroxytetraphenylcyclotetrasiloxane possesses the *cis*-structure. As suggested, formation of difunctional compounds proves [14] that free and linked hydroxyl groups in T-2 complex with pyridine differ by reactivity in relation to organochlorosilanes. Due to high basicity of oxygen in hydroxyl groups, HFC reaction rate of ≡Si-OH·NH_5C_5 group with organochlorosilanes is higher than that of ≡Si-OH group.

Hydrolysis of tetrakis(chlorodiorganosiloxy)tetraphenylcyclotetrasiloxane (structure XVIII) with water excess at 5 - 10°C has produced appropriate tetrahydroxy-derivatives of the following structure [14, 15]:

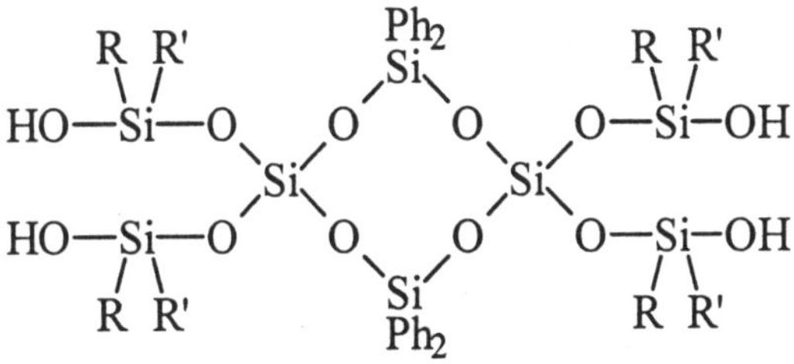

XVIII

where R = R' = Me, Ph; R ≠ R'.

Due to comparatively high reactivity of hydroxyl groups, which are easily condensed, the yield of hydroxyl derivative with methyl radicals (R = R' = Me) at the atom of silicon from hydrolysis proceeding in the presence of acceptor is negligible, and without acceptor was not obtained at all. On the contrary, stability of hydroxyl groups of the compound (R = R' = Ph) promoted synthesis of this product with high yield in hydrolysis of the compound both in the presence and in the absence of HCl acceptor.

HFC reaction of tetrakis(hydroxydiorganosiloxy)tetraphenylcyclotetrasiloxane (structure XVIII) with organotrichlorosilanes at 1:2 ratio of initial components was performed at 15 - 20°C in organic solvent, in the presence of HCl acceptor. Benzene, toluene, diethyl ether, tetrahydrofuran, methyl ethyl ketone or mixture of the solvents was used. Dichlorospirotricyclosiloxanes were synthesized in the presence of pyridine in accordance with the scheme as follows [14, 15]:

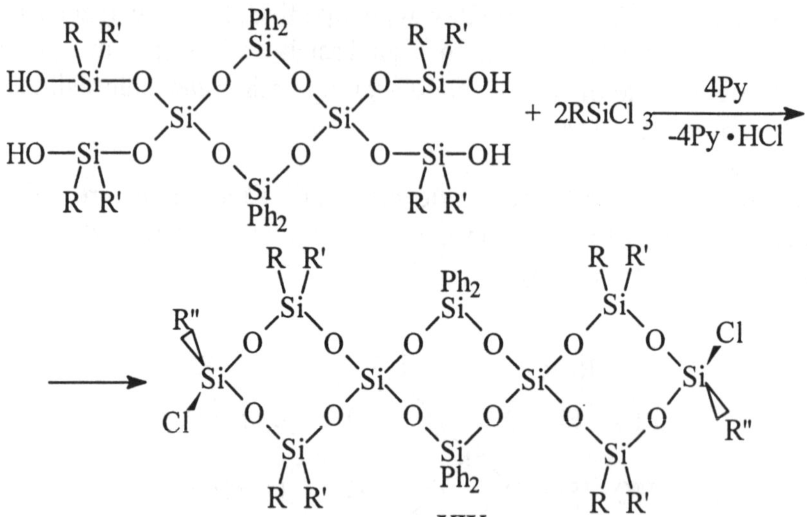

(6.22)

XIX

where R = R' = Ph; R" = Me, Ph.

By hydrolysis of dichlororganospirotricyclosiloxane, appropriate dihydroxy-derivatives were synthesized in accordance with the following scheme:

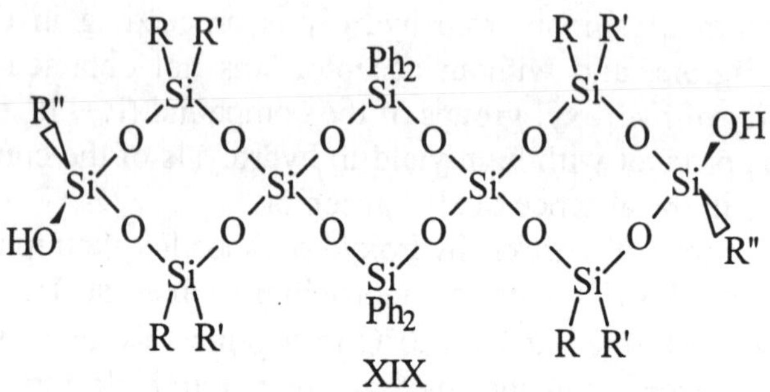

XIX

where R = R' = Me, Ph; R ≠ R'.

Composition and structure of synthesized difunctional spirotricyclosiloxanes were determined by functional and ultimate analysis, molecular mass measurement, and from IR and ^{1}H NMR spectral data. As indicated, found ratios of protons of methyl and phenyl groups correlate well with calculated values.

For the purpose of synthesizing arylenecyclosiloxane blocks with functional groups at silicon atom, the reaction was performed in two stages. At the first stage, HFC of 1,4-bis(dichlorophenylsilyl)benzene with dihydroxydiphenylsilane at 1:4 ratio of initial components in the presence of pyridine in the temperature range of -5 ÷ -10°C was performed. After recrystallization of the reaction product from toluene–heptane solution 52% yield of tetrakis-(hydroxydiphenylsiloxy)-1,4-bis(phenylsilyl)benzene was obtained. The reaction proceeds in accordance with the following scheme [16]:

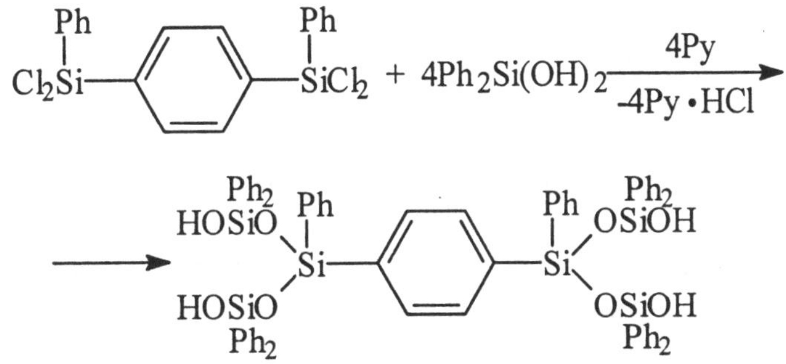

XX

(6.23)

At the second stage, HFC of the product of structure XX (compound 46, Table 6.1) with organotrichlorosilane at 1:2 ratio of initial components, in the presence of pyridine and temperature range of $-5 \div -10^0$C was carried out and dichloroarylenecyclosiloxanes were obtained. In this case, the reaction generally proceeds by intramolecular cyclization mechanism in accordance with the following scheme:

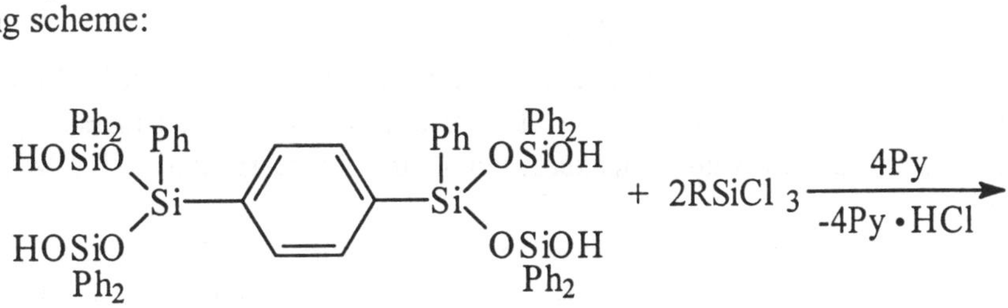

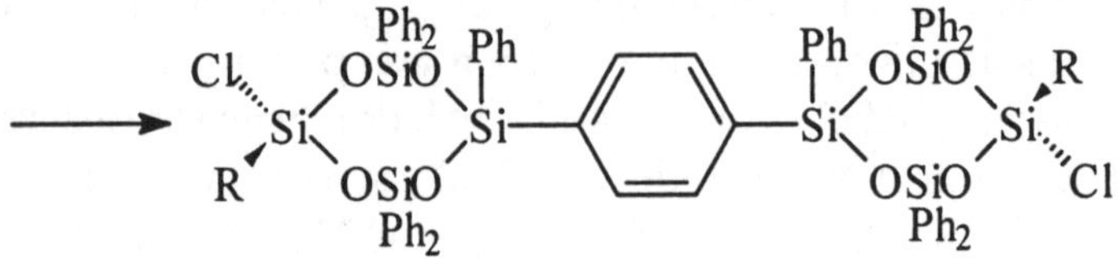

(6.24)

XXI

where R = Me, Ph.

Compounds of structure XXI (R = Me (47), Ph (48), Table 6.1) were also synthesized by a single stage HFC of 1,4-bis-(dichlorophenylsilyl)benzene with dihydroxydiphenylsilane and organotrichlorosilanes at 1:4:2 ratio of initial components in the presence of pyridine and temperature range of –5 ÷ -10°C, and the yield of compounds was below ~32 %.

Hydrolysis of dichloroarylenecyclosiloxanes of structure XXI, proceeding under soft conditions in the presence of sodium bicarbonate and further recrystallization from toluene-heptane solution produced appropriate dihydroxy-compounds as follows:

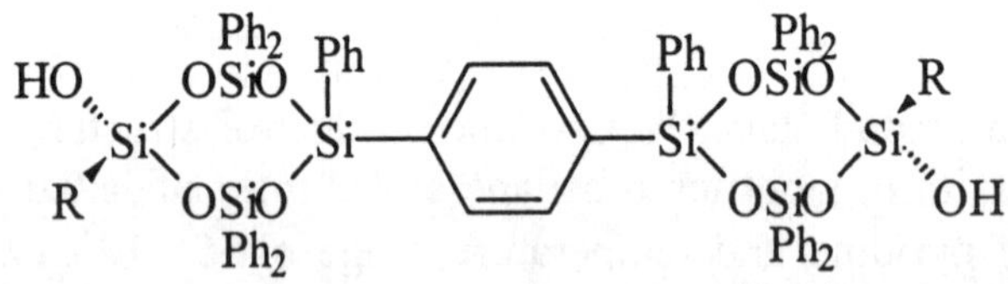

XXI

where R = Me (49), Ph (50), Table 6.1.

Composition and structure of synthesized dichloro- and dihydroxyarylene-cyclosiloxanes (46 - 50) of the structure XXI were determined by ultimate analysis, molecular mass measurements, and from IR and ^{1}H NMR spectral data.

^{1}H NMR spectrum of compound 47 (Table 6.1) displays two singlet reflexes with chemical shifts $\delta = 0.56$ and $\delta = 0.52$ ppm typical of protons of methyl groups of *cis*- and *trans*-configuration. The ratio of *cis*- and *trans*-configurations equals 60 and 40%, respectively.

Some physical and chemical parameters of all synthesized organobi-, organotri- and organotetracyclosiloxanes are shown in Table 6.1.

Table 6.1

Some physical and chemical parameters of organobi-, organotri- and organotetracyclosiloxanes

No.	Compound structure	Yield, %	T_{boil}, °C T_{melt}, °C	P, mmHg	References
1	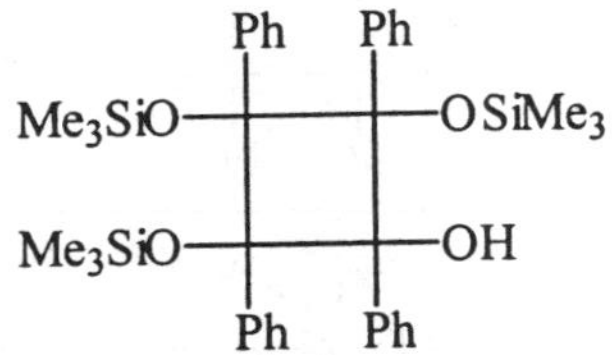	47	225-230	$1\text{-}2\times10^{-3}$	[1, 4]
2	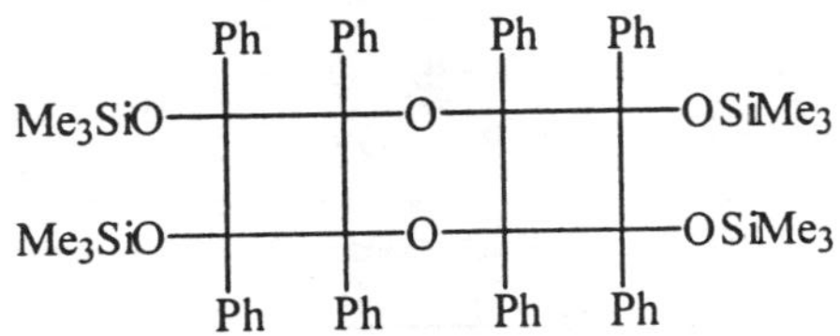	9	318-320	$1\text{-}2\times10^{-3}$	[1, 4]
3	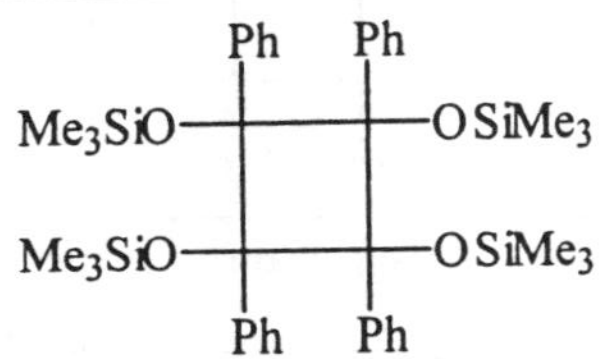	78	241-242	-	[1, 4]
4	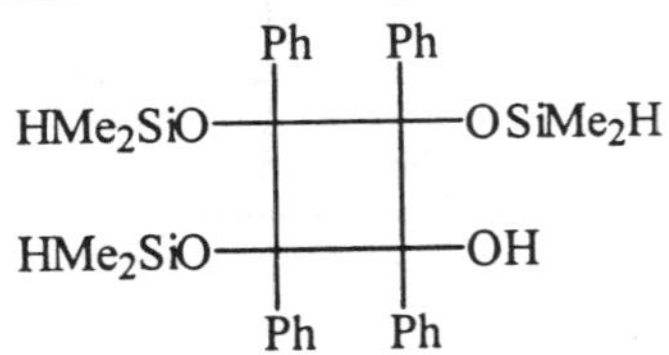	93	190-192	$1\text{-}2\times10^{-2}$	[2, 6]

5	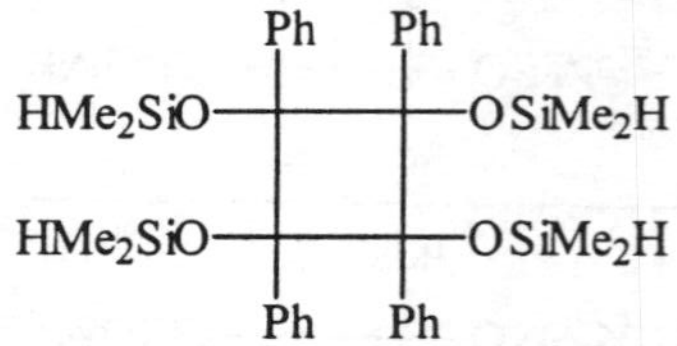	93	215-220	$1\text{-}2\times10^{-2}$	[2, 6]
6	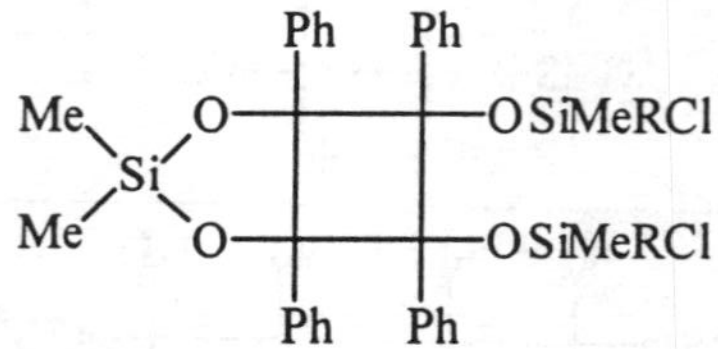	39	197-205	$1\text{-}2\times10^{-2}$	[3, 4]
7	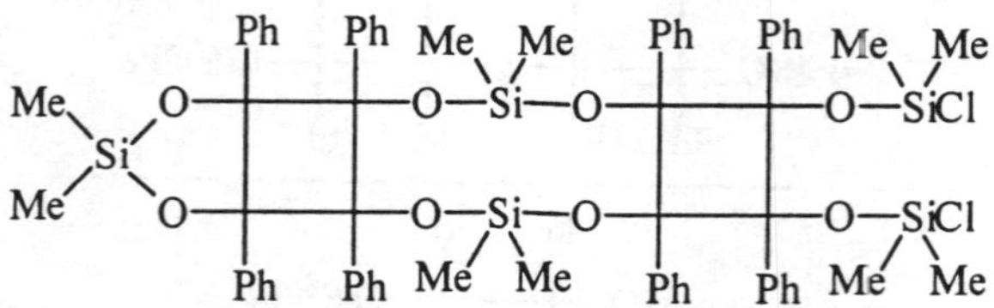	33	290-295	$1\text{-}2\times10^{-3}$	[3, 4]
8	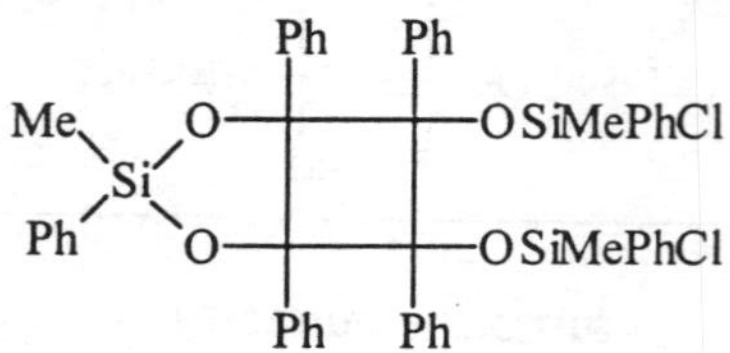	35	322-325	$1\text{-}2\times10^{-3}$	[3, 4]

9	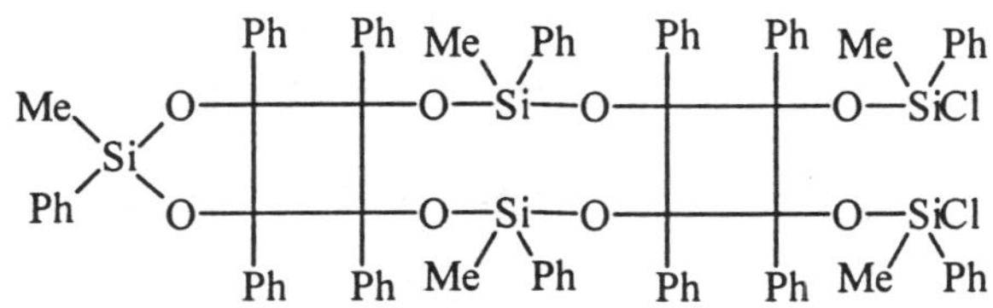	9	405-410	$1\text{-}2\times10^{-3}$	[3, 4]
10	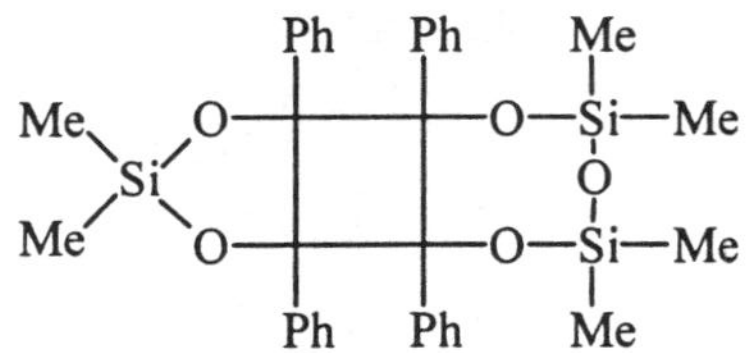	65	232-235	$6\text{-}8\times10^{-4}$	[3, 4]
11	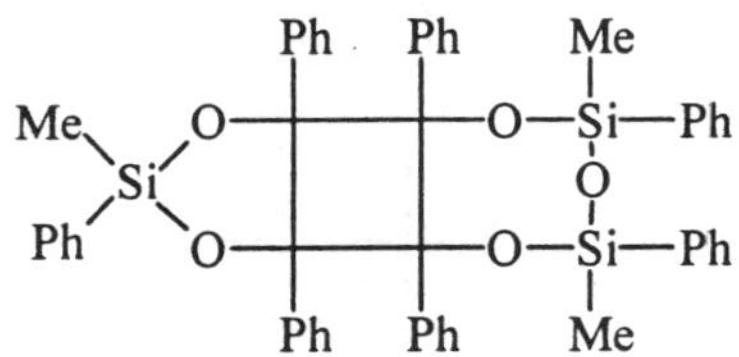	30	310-320	$1\text{-}2\times10^{-3}$	[3, 4]
12	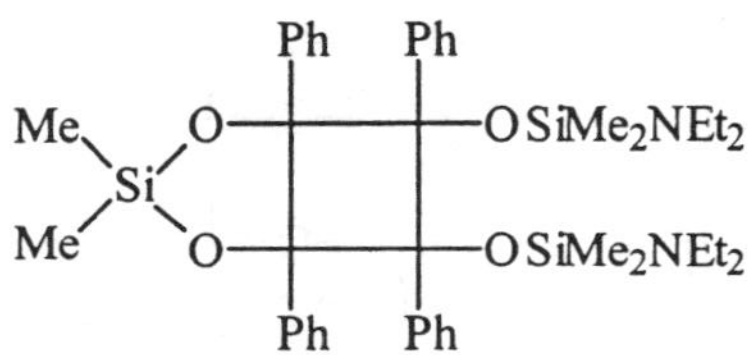	30	222-225	$1\text{-}2\times10^{-3}$	[4]

13	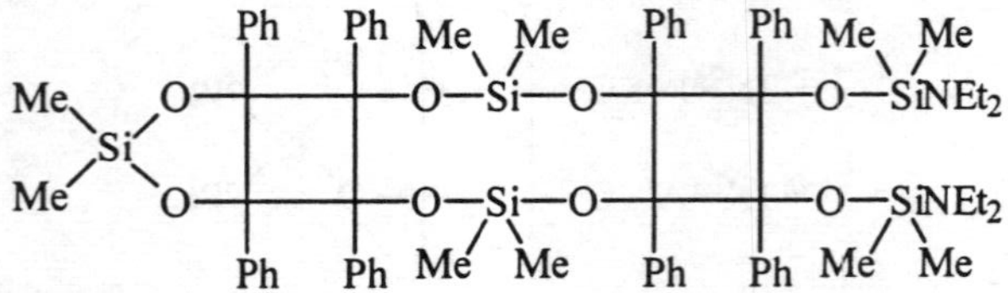	22	320-322	$1\text{-}2\times10^{-3}$	[4]
14	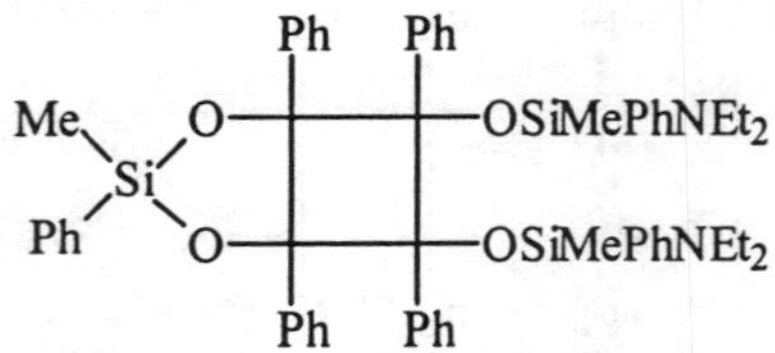	41.5	290-295	$1\text{-}2\times10^{-3}$	[4]
15	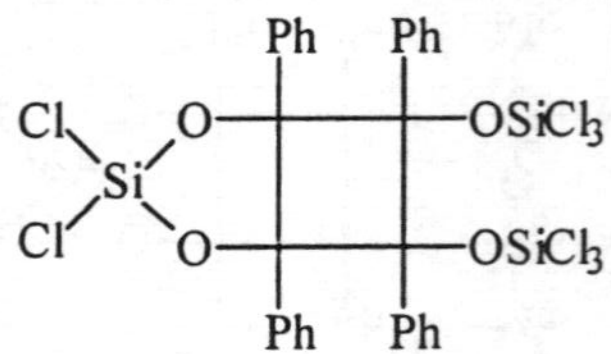				[5]
16	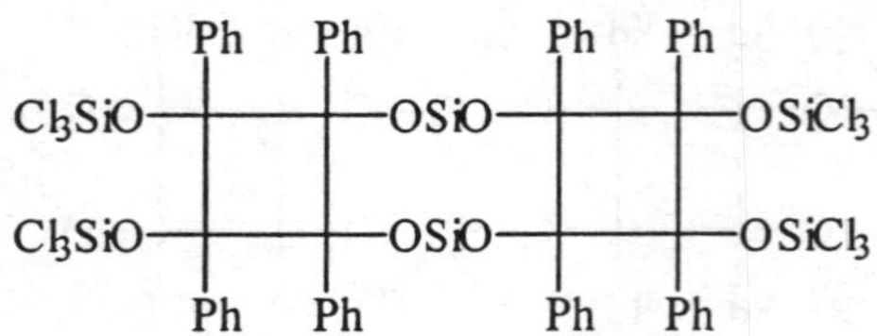				[5]

17	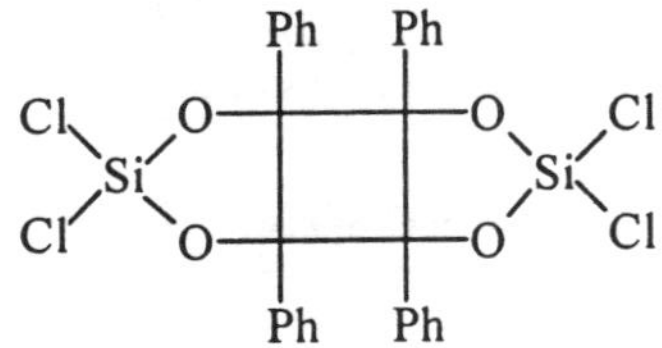				[5]
18	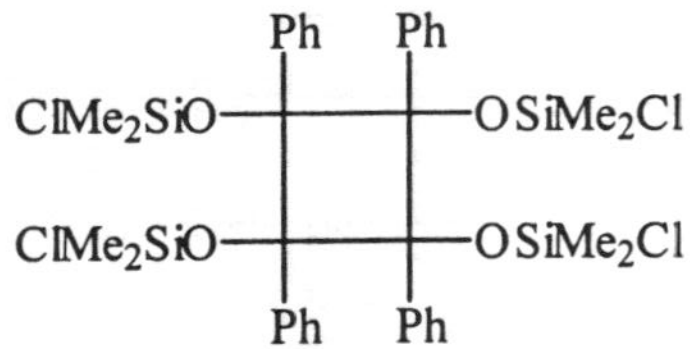	76	232-235	1-2×10^{-3}	[1, 4]
19	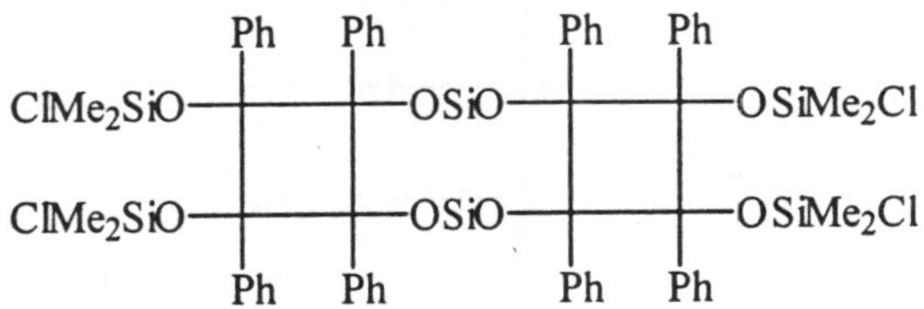	11	320-325	1-2×10^{-3}	[1, 4]
20	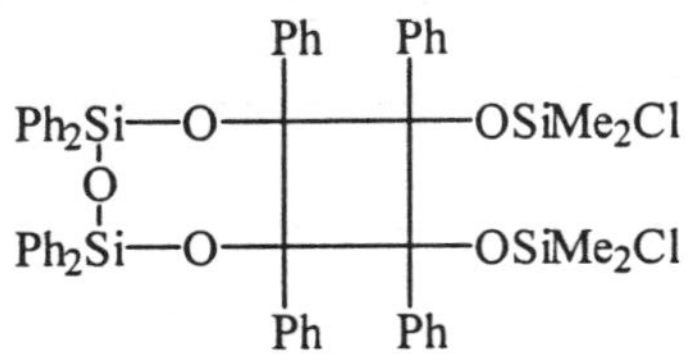	26	310-320	1-2×10^{-3}	[4, 7]

21	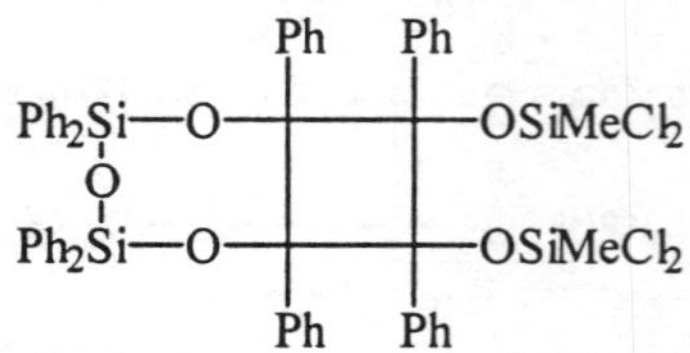	21	315-320	1-2×10^{-3}	[4, 7]
22	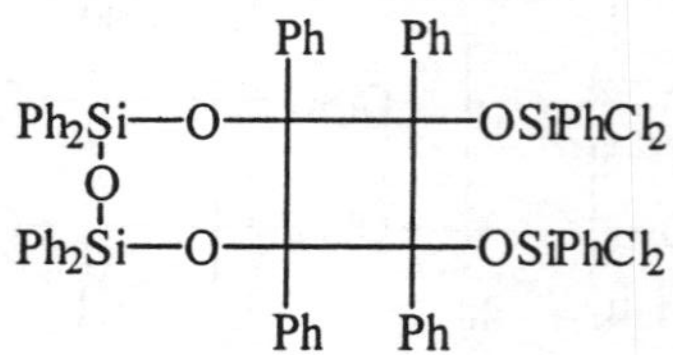	46	345-355	1-2×10^{-3}	[4, 7]
23	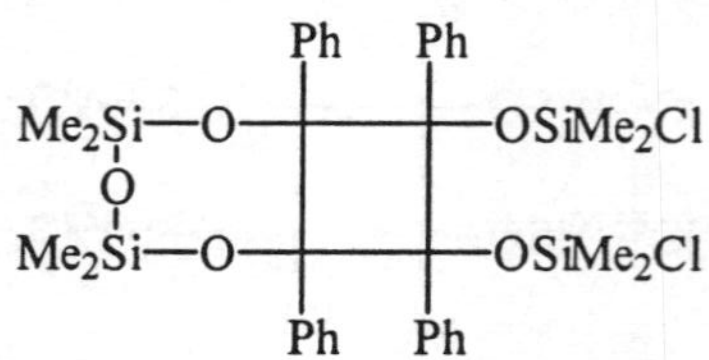	58	220-225	1-2×10^{-3}	[4, 7]
24	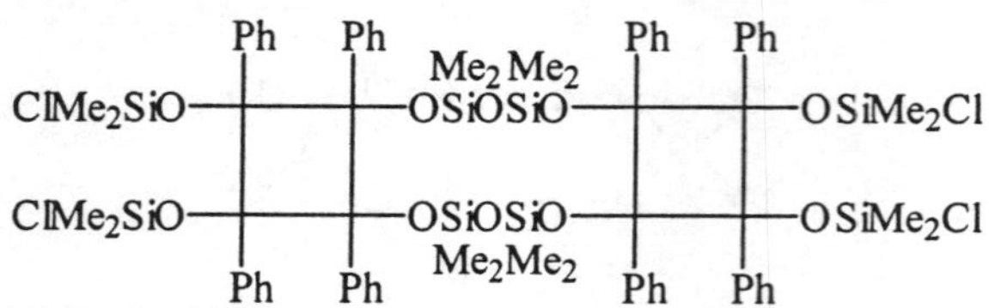	20	348-352	1-2×10^{-3}	[4, 7]

25	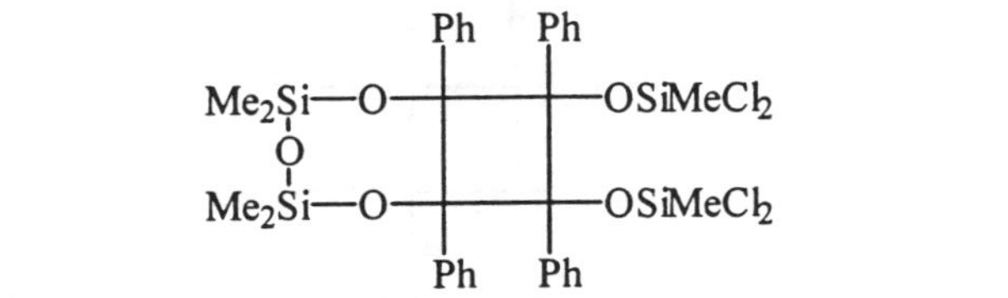	63	226-230	1-2×10^{-3}	[4, 7]
26	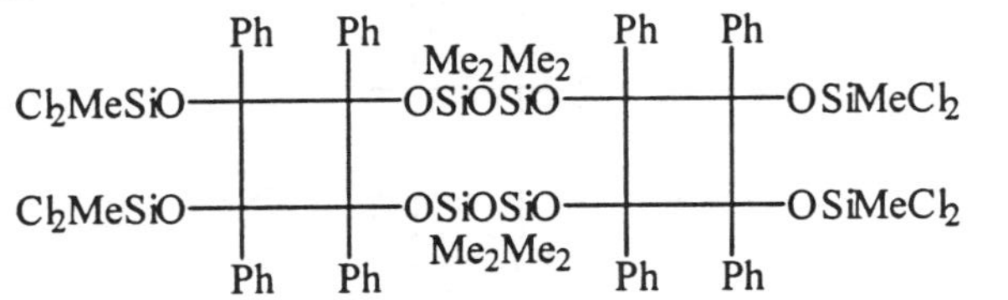	12	310-325	1-2×10^{-3}	[4, 7]
27	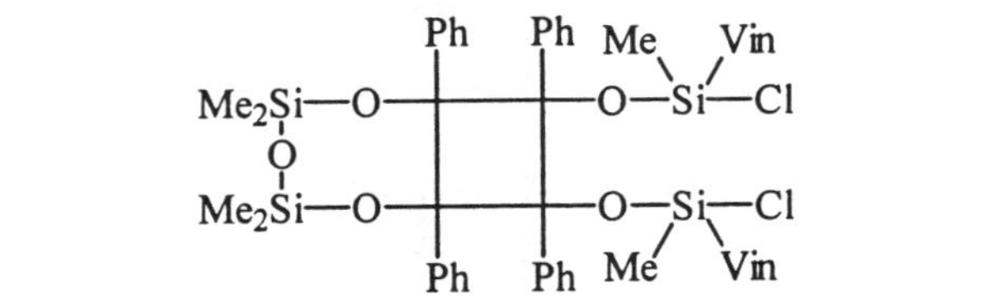	50	235-240	1-2×10^{-3}	[4, 7]
28	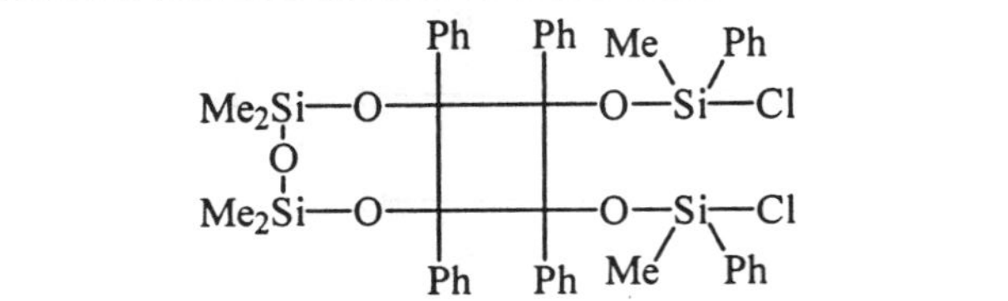	36	295-305	1-2×10^{-3}	[4, 7]

29	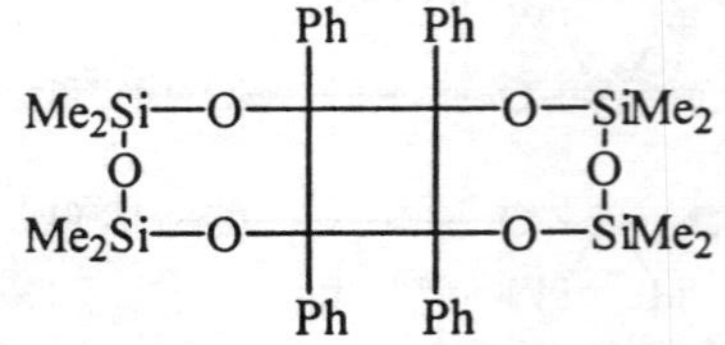	56	235-240	1-2×10^{-3}	[4, 7]
30	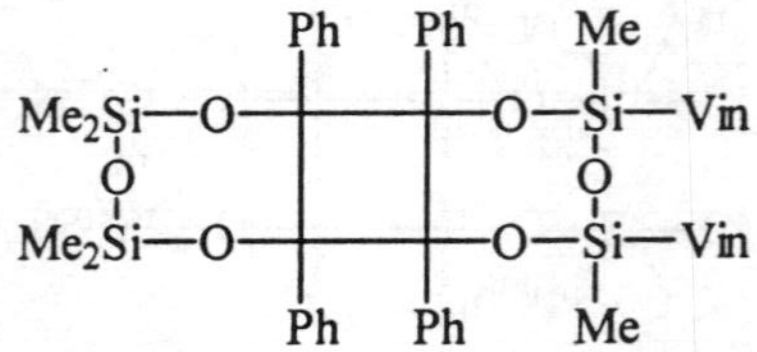	62	245-250	1-2×10^{-3}	[4, 7]
31	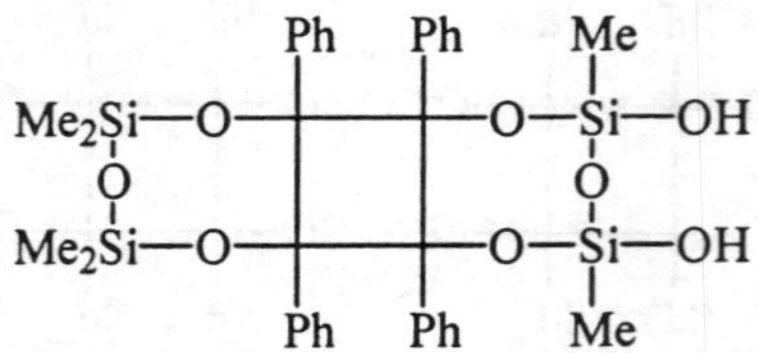	90	75-78	-	[4]
32	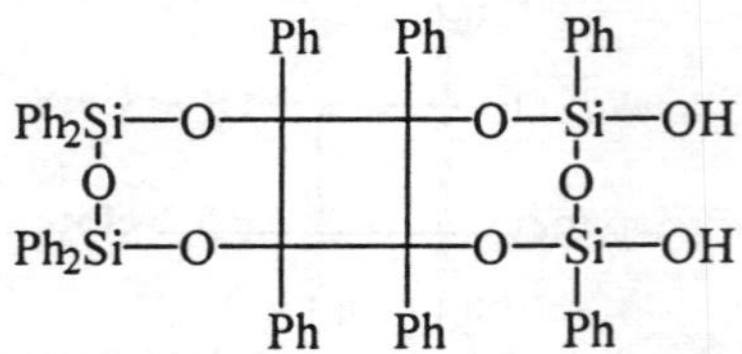	94	110-112	-	[4]

33	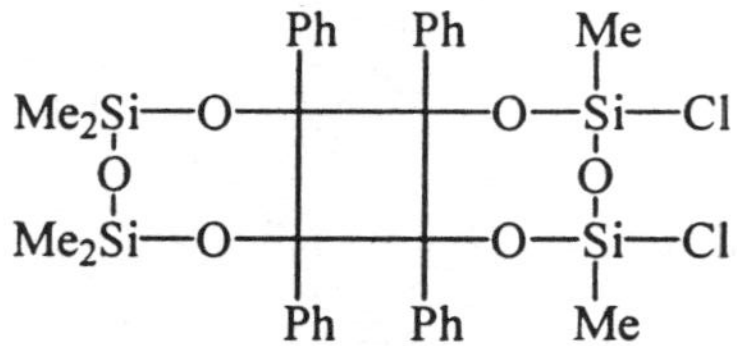	47	240-245	$1\text{-}2\times10^{-3}$	[1, 4]
34	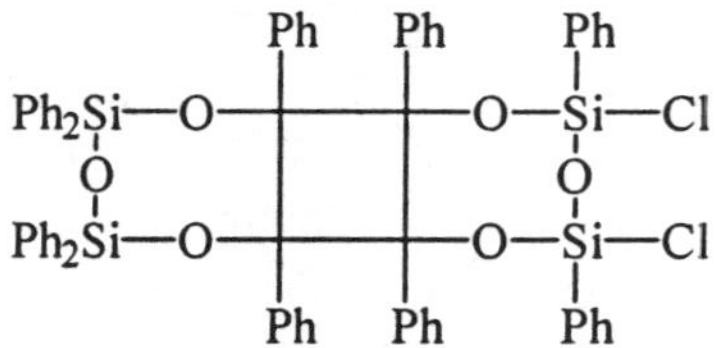	27	350-360	$1\text{-}2\times10^{-3}$	[4]
35	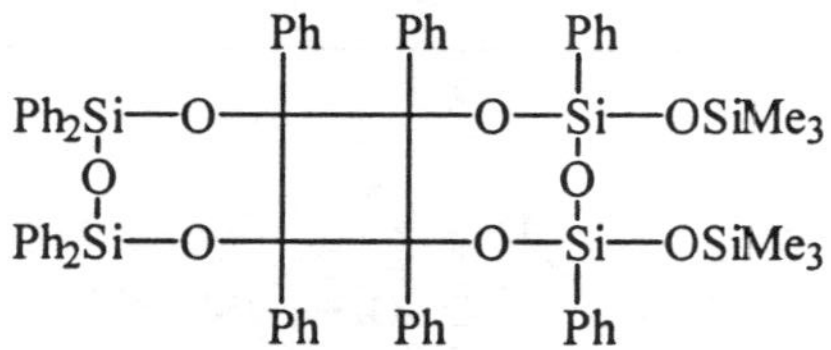	18	320-325	$1\text{-}2\times10^{-3}$	[13]
36	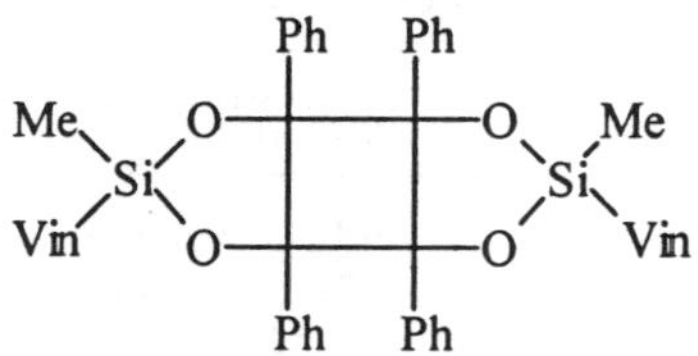	32	- 150-153	-	[9]
37	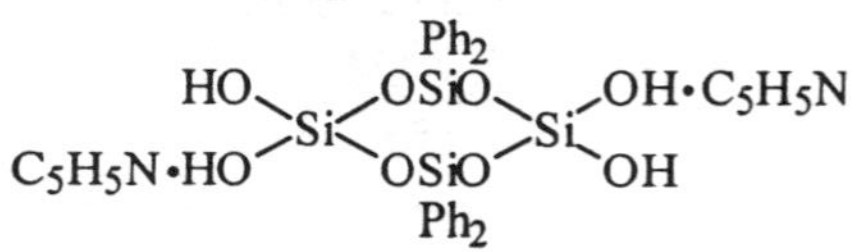	88	118-119	-	[10, 13]

38	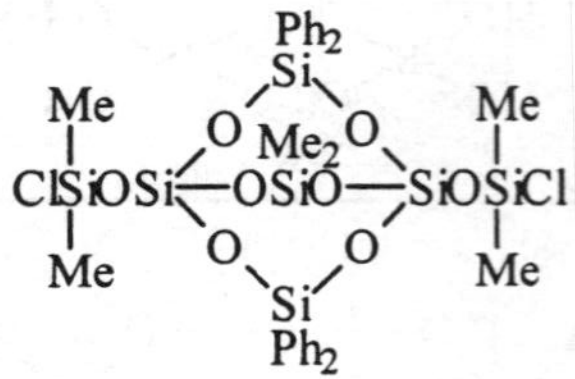	54	212-214	2×10^{-3}	[13]
39	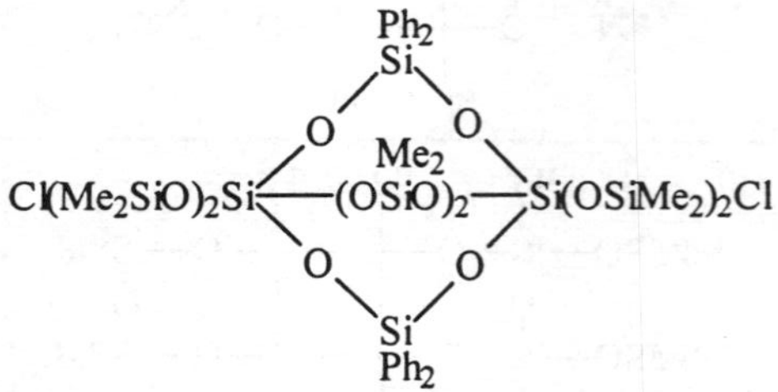	63	229-231	2×10^{-3}	[13]
40	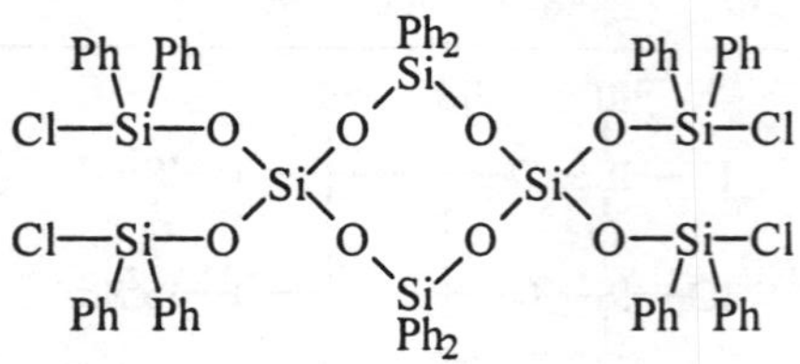	66	150-151	-	[15]
41	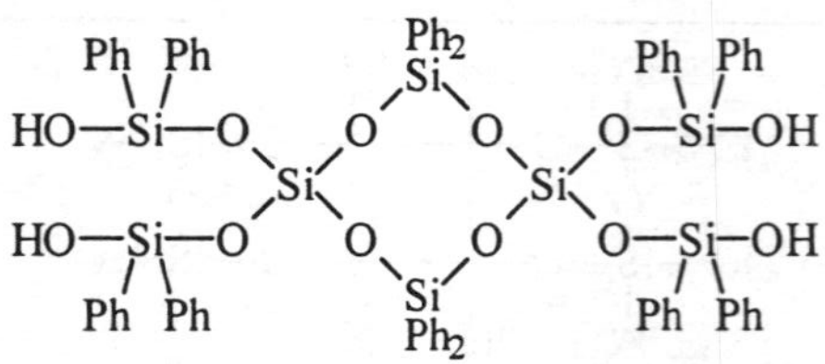	89	179-180	-	[15]

42	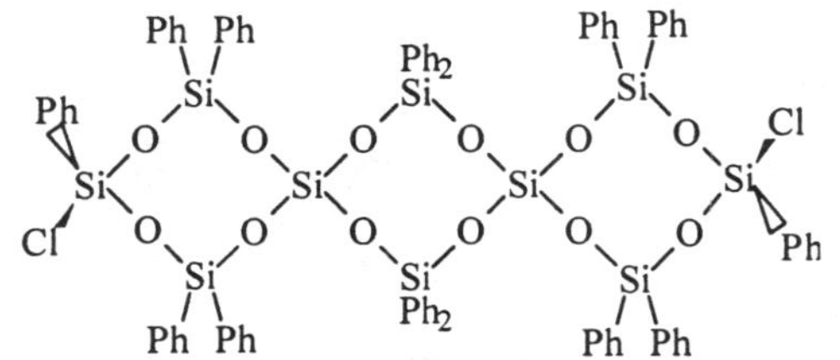	43	198-200	-	[15]
43	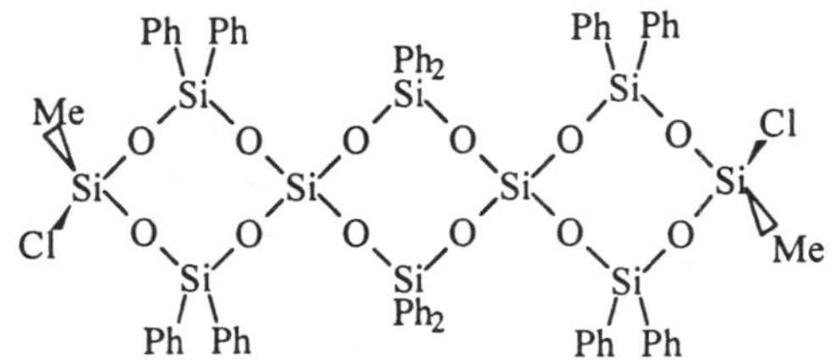	62	139-141	-	[15]
44	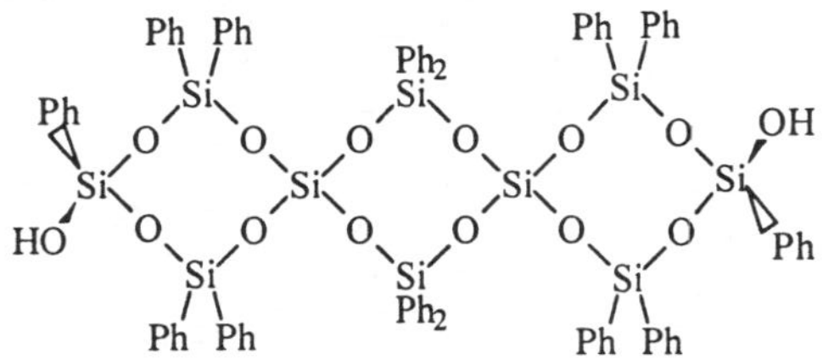	83	216-217	-	[15]
45	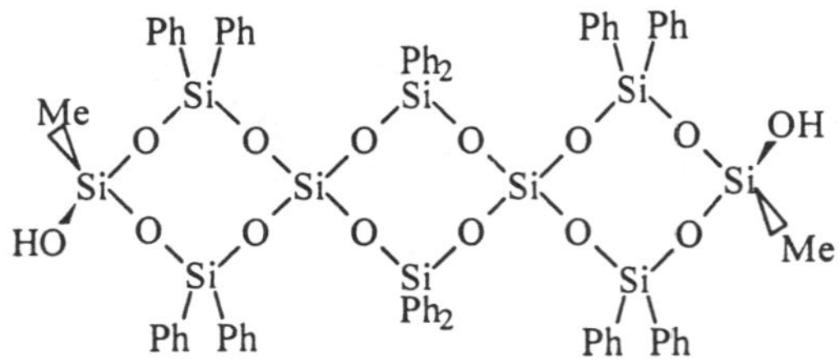	81	153-155	-	[15]

46	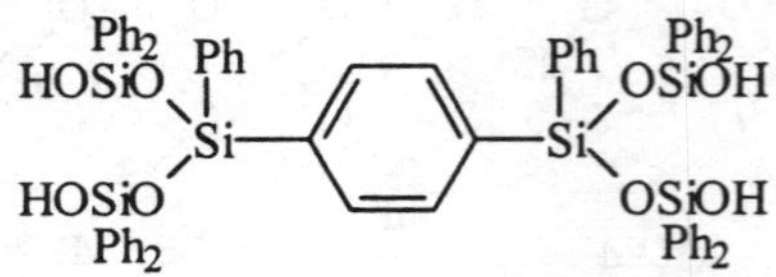	52	- 120-121	-	[16, 17]
47	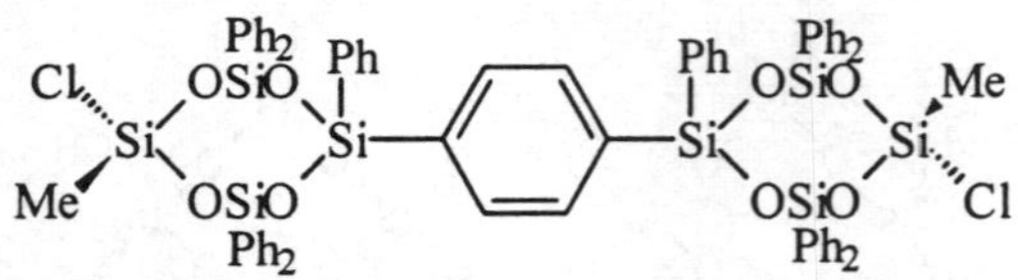	51 (31)	340-345 58-60	$1\text{-}2\times10^{-3}$	[16, 17]
48	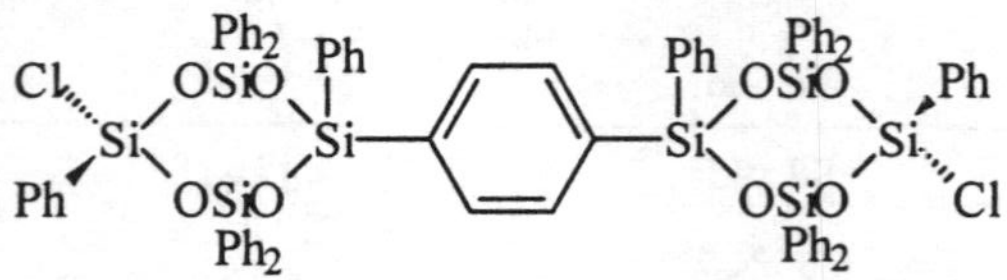	54 (33)	370-380 85-88	$1\text{-}2\times10^{-3}$	[16, 17]
49	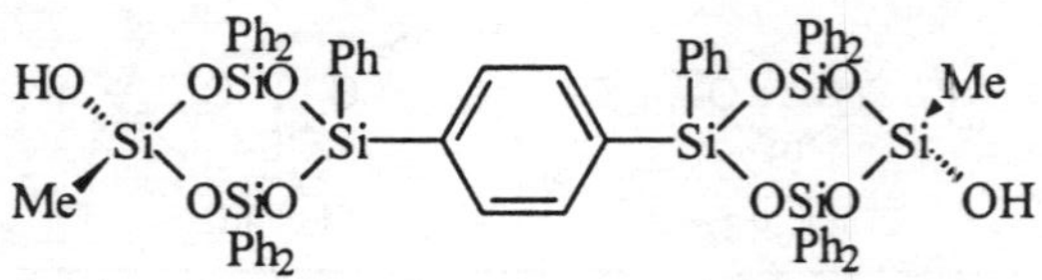	53	- 82-85	-	[16, 17]
50	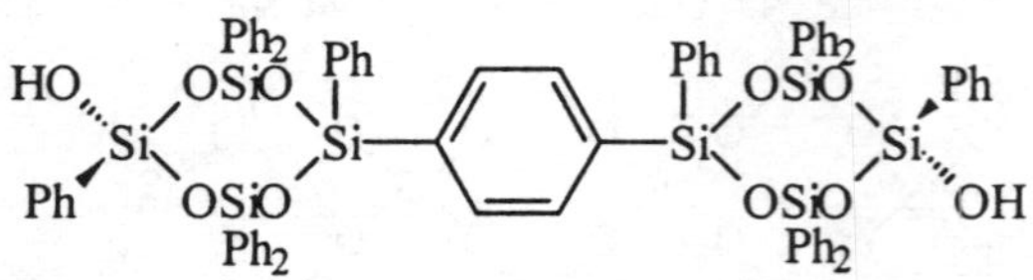	67	- 95-97	-	[16, 17]

6.2. Synthesis of difunctional bicyclocarbosiloxanes

For the purpose of synthesizing difunctional bicyclocarbosiloxanes, HFC reaction of 1,2-bis(trichlorosilyl)ethane with dihydroxydiphenylsilane was studied [18]. The reaction proceeded at 1:2 ratio of initial components in diluted solution of anhydrous toluene and in the presence of pyridine. In the temperature range of -5 ÷ -10°C dichlorobicyclocarbosiloxane with ~44% yield was synthesized. Higher yield of this compound (~69%) was obtained by the two-stage method: at the first stage, tetrachlorocyclocarbosiloxane was synthesized, and at the second stage, dichlorobicyclocarbosiloxane was produced by HFC reaction of tetrachlorocyclocarbosiloxane with dihydroxydiphenylsilane at 1:1 ratio of initial components, all other conditions being the same. In both cases, the reaction proceeds in accordance with the general scheme as follows:

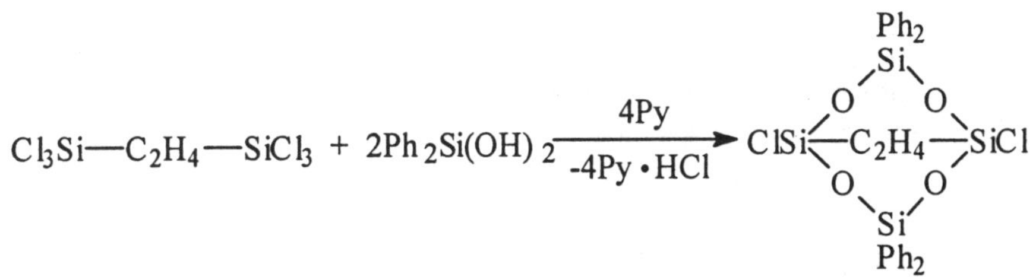

(6.25)

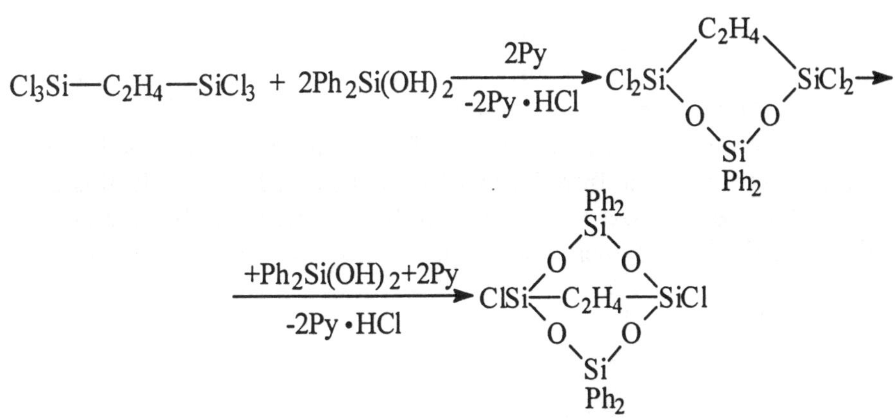

(6.26)

XXII

Composition and structure of synthesized dichlorobicyclocarbosiloxane (compound 1, Table 6.2) of structure XXII were determined from IR and ^1H NMR spectra that displayed a singlet reflex with chemical shift at $\delta = 1.078$ ppm, typical of methylene groups, due to the molecule symmetry.

Hydrolysis of dichlorobicyclocarbosiloxane in the neutral medium synthesized corresponded dihydroxybicyclocarbosiloxane by the scheme as follows:

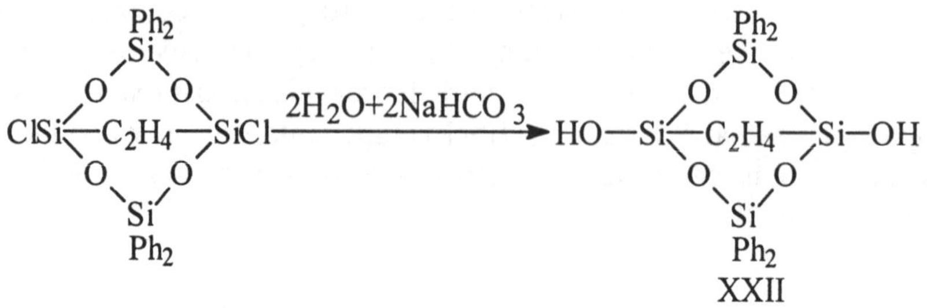

(6.27)

Some physical and chemical parameters of synthesized bicyclocarbosiloxanes are shown in Table 6.2.

6.3. Synthesis of organosilicon copolymers with organobi-, organotri- and organotetracyclosiloxane fragments in dimethylsiloxane backbone

For the purpose of synthesizing organosilicon copolymers with bicyclic fragments in the macromolecular chain, HFC reaction of dichlorobicycloorganosiloxanes of structure VI (compounds 6 and 8, Table 6.1) with α,ω–dihydroxydimethylsiloxanes in 70% anhydrous benzene solution at gradual temperature increase from 25 to 70°C was studied [6, 20]. In this case, it was found that conversion was at most 60% by hydrogen chloride (Figure 6.5), and the yield of copolymers after re-precipitation was below 40 – 45% [6, 20].

Table 6.2

Physical and chemical parameters of difunctional organobicyclosiloxanes

No.	Compound	Yield, %	T_{boil}, °C (P, mmHg)	T_{melt}, °C	[1]H chemical shift	References
1	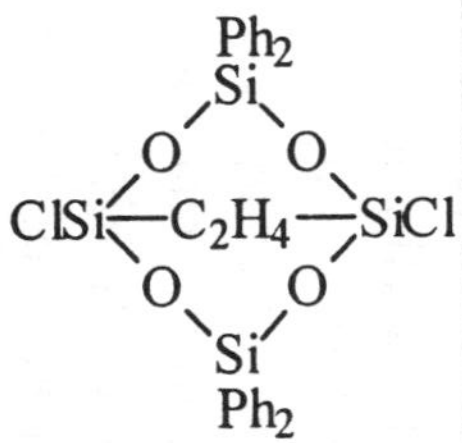	44 (69)	240-250 (0.02)	172-179	1.078	[18, 19]
2	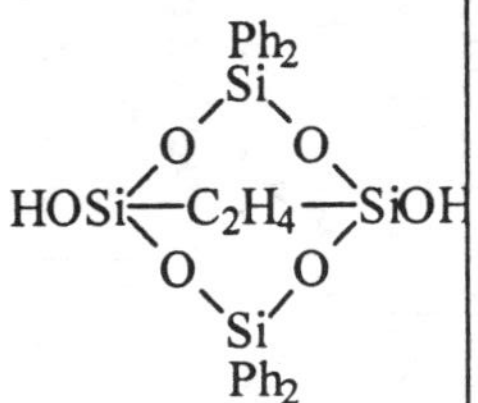	65	–	218-219	–	[18, 19]

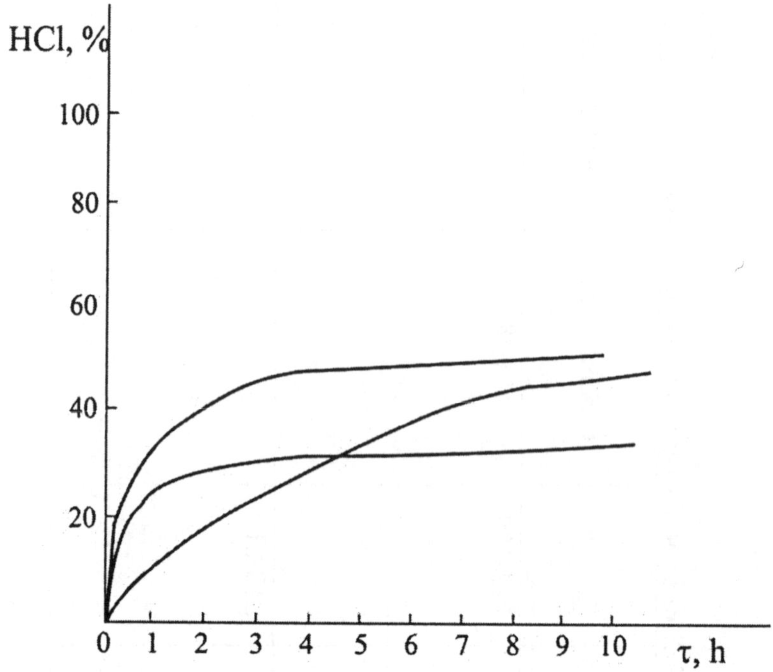

Figure 6.5. Dependence of hydrogen chloride liberation rate during compound 6 (table 6.1) condensation with α,ω–dihydroxydimethylsiloxanes: 1 - $n = 12$; 2 – $n = 4$; 3 – $n = 2$

Taking into account the fact that liberated hydrogen chloride cannot be immediately removed from the reaction mixture with argon flow, homocondensation of α,ω–dihydroxydimethylsiloxane may be excluded. As a result, besides formation of organocyclosiloxanes and increase of diols' length, the ratio of initial components is disturbed. Complete analysis of the reaction mixture, obtained at the interaction of dichlorobicyclosiloxane (compound 6, Table 6.1) with 1,3-dihydroxydimethylsiloxane, has indicated that 45% copolymer with $\eta_{spec} = 0.20$ is formed (after re-precipitation from benzene – methanol solution). Chromatographic analysis results have proved the absence of octamethylcyclotetrasiloxane in the reaction products. The compound with $T_{boil} = 220$ - 222°C at $P = 2\times10^{-3}$ mmHg was separated by fractionation of products, remained after elimination of benzene-methanol mixture from the mother solution of re-precipitated copolymer.

Based on data of the ultimate analysis, NMR spectrum and molecular weight, structure VIII was assigned to synthesized compound 10:

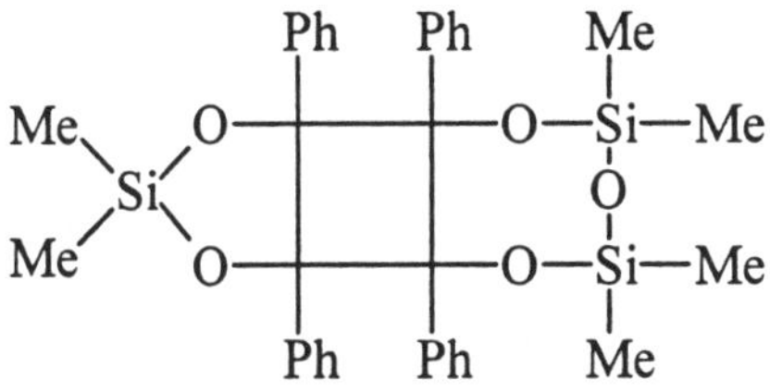

VIII

Moreover, about 13% of reaction products, in which the ratio of protons of methyl and phenyl groups equals 1.37:1.00, i.e. is close to that observed for the compound, synthesized by intramolecular cyclization of diol with bicyclic compound 6 (Table 6.1) (calculated 1.52:1.00), were obtained. However, an individual compound was not separated by fractionation of the reaction products [6], because a cross-linked polymer was synthesized. Counter synthesis of tricyclic compound of structure XXIII as follows was performed in diluted solution of diethyl ether:

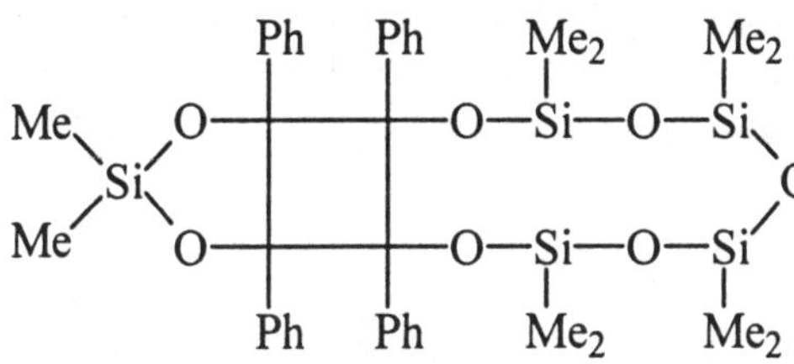

XXIII

However, fractionation of the reaction mixture also transferred it into cross-linked polymer.

Thus, based on the fractionated side products, the following reaction scheme may be suggested:

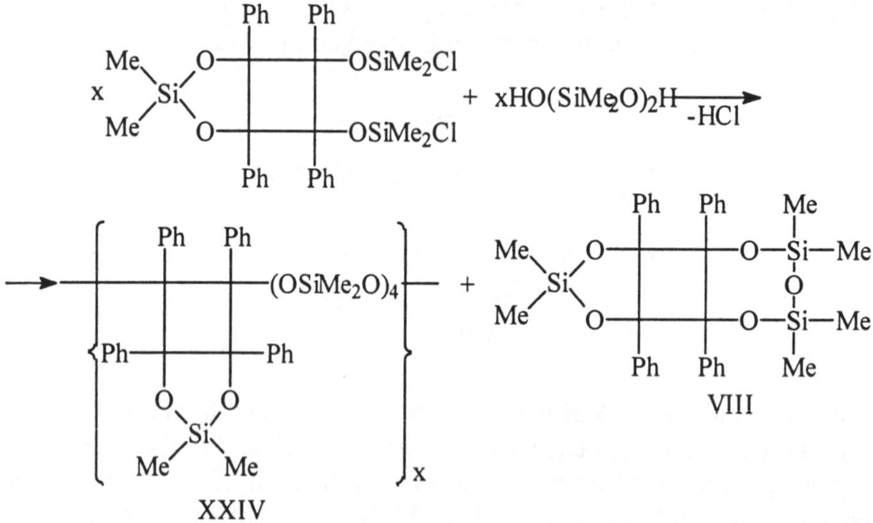

(6.28)

For the purpose of excluding homocondensation of diols, further study of structure VI compounds HFC with α,ω–dihydroxydimethylsiloxanes was implemented in anhydrous benzene solution in the presence of pyridine, which is hydrogen chloride acceptor. As the result, copolymers representing rubber-like substances or highly viscous liquids, completely soluble in different organic solvents, were synthesized with the yield equal 67 – 89%. Generally, the reaction proceeds as follows:

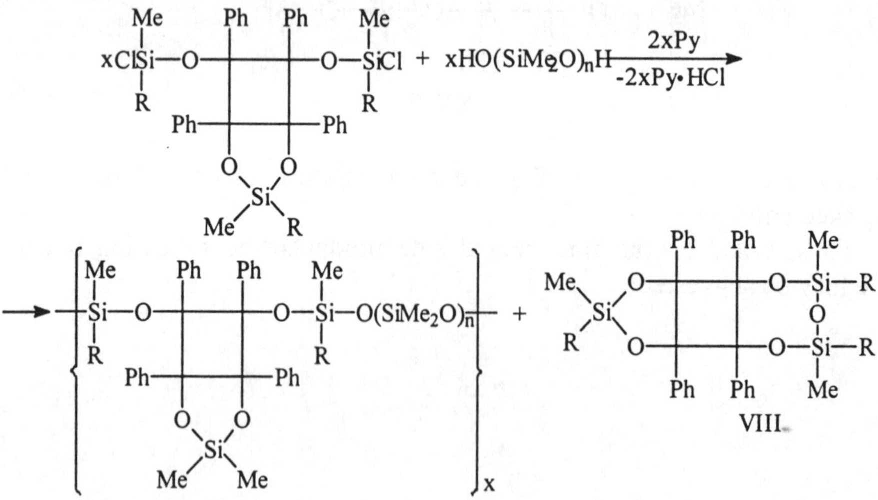

(6.29)

where R = Me, Ph; *n* = 2 - 70.

As noted, two suggested reaction schemes did not allow implementation of the study of polycondensation kinetics, in the first case, due to side homocondensation and intramolecular cyclization, and in the second case, due to poor replicability of obtained viscous parameters after elimination of hydrochloric pyridine on both washed and unwashed samples (apparently, as the samples approached constant weight, further polycondensation proceeded). That is why the authors have made an attempt to obtain kinetic parameters of bis-diethylaminobicyclosiloxane, compound 12 (Table 6.1), with 1,3-dihydroxytetramethyldisiloxane. It is found that polycondensation proceeds with dimethylamine liberation, which is identified by the boiling temperature and from chromatographic analysis. However, as conversion by diethylamine reaches 15 – 17%, a simple spatially cross-linked polymer is formed. A possibility of siloxane bond break by diethylamine in trimethyltriphenylcyclotrisiloxane with a complex formation with charge transfer was previously indicated in ref. [21].

Studying this reaction, one may hardly guarantee the absence of water traces, though the reactions were performed in absolutely anhydrous solvent and inert gas flow. As a consequence, hydrogen dissociation from bonded amine-water complex may cause occurrence of OH^-, H^+, H_3O^+, and $R_2H_2N^+$ ions in the solution. Further on, a nucleophilic particle attacks silicon atom in labile hexatomic ring in accordance with the scheme as follows:

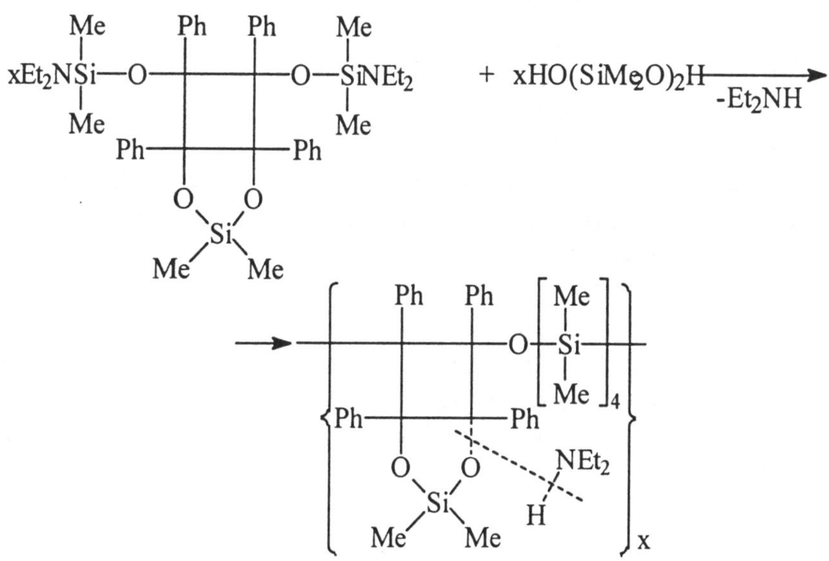

(6.30)

As a result, a 3D-structured copolymer is formed. Disclosure of the siloxane cyclic fragment by the attack of diethylamine on silicon atom is not excluded, which induces formation of bipolar zwitterion [22]. Finally, all attempts to study kinetics of polycondensation of polyorganobicyclohexasiloxanes with organosilicon diols have failed.

The results of studies of hydrodynamic properties (characteristic viscosity, translational diffusion and sedimentation coefficients) and birefringence in flow for the polymer, synthesized by interaction of compound 8 with dihydroxytetramethyldisiloxane, correspond to its linear copolymer structure:

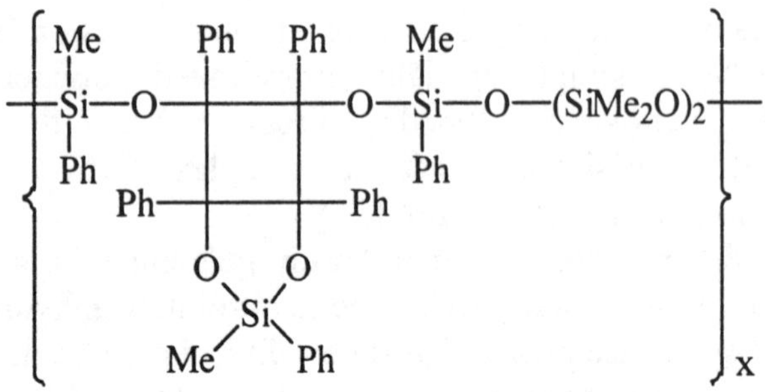

Actually, if the molecular weight of the sample equals M_{SD} = 57,000, characteristic viscosity is rather low, $[\eta]$ = 0.22 dl/g, which is untypical of rigid-chain polymers, for example, of the ladder-type copolymers [23].

For this polymer, birefringence is also low, $\Delta n/\Delta\tau = 7\times10^{-10}$, which coincides with that for linear poly(methylphenylsiloxane) [24].

HFC reaction of dichloroarylenecyclosiloxanes (compounds 47 and 48, Table 6.1) with α,ω-dihydroxydimethylsiloxanes (n = 2, 4), proceeded in diluted solution of anhydrous toluene, in the temperature range of 20 - 25°C, in the presence of pyridine, synthesized cyclolinear copolymers with the regular disposition of arylenecyclosiloxane fragments in linear dimethylsiloxane backbone. HFC of compounds 49 and 50 (Table 6.1) with α,ω–bis(dimethylamino)dimethylsiloxanes proceeding in melt in the temperature range of 100 - 120°C, and in vacuum at the final stage until constant values of viscosity are reached, synthesized copolymers with long linear dimethylsiloxane chain. Generally, HFC proceeds in accordance with the following scheme [16, 17]:

Table 6.3

Some physical and chemical parameters of organosiloxane copolymers (XXIV) with organobicyclohexasiloxane fragments in the backbone

No.	Copolymer structure	R	n_{SiO}	Yield, %	η_{spec}*	T_g, °C	d_1, Å
1	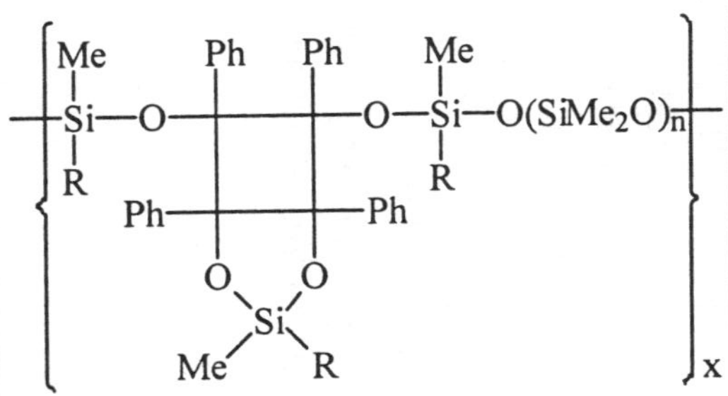	Ph	2	76	0.22	+6	10.40
2		Ph	4	76	0.13	-44	10.44
3		Ph	6	84	0.22	-57	9.70
4		Ph	12	61	0.15	-72	9.40
5		Ph	25	74	1.07	-96	8.20
6		Ph	35	70	0.15	-116	8.20
7		Ph	70	62	0.33	-123	7.70
8		Me	2	58	0.09	-26	-
9		Me	4	55	0.25	-69	-
10		Me	6	71	0.12	-	-
11		Me	12	62	0.16	-	-
12		Me	25	63	0.27	-114	-
13		Me	70	53	0.13	-123	-

Note: * 1% solution in benzene at 20°C.

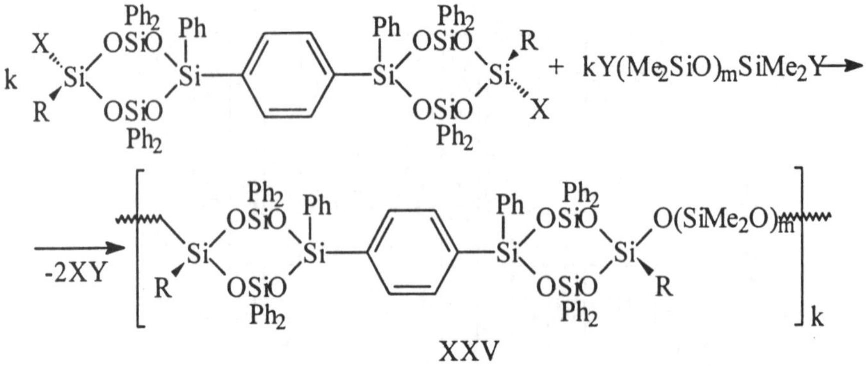

(6.31)

where X = Cl, OH; Y = OH, Me_2N; R = Me, Ph; *m* = 2 - 51.

After reprecipitation from toluene solution by methanol, synthesized copolymers (XXV) become light yellow or white viscous or solid substances (depending on *n*), well soluble in organic solvents with η_{sp} = 0.07 – 0.39. Molecular masses of copolymers fall within the range of $\bar{M}_\omega = 38\times10^3 - 13.5\times10^4$. Some physical and chemical parameters of copolymers with arylenecyclosiloxane fragments in the chain are shown in Table 6.4.

As initial difunctional arylenecyclosiloxanes represent a mixture of *cis*- and *trans*-isomers, synthesized copolymers possess atactic structure. Fractionation of copolymer 3 (Table 6.4) and ultimate analysis indicate that the reaction proceeds directly with formation of copolymers possessing regular disposition of arylenecyclosiloxane fragments in the chain. For copolymer 3, in the case of fraction 3^1 (19 mol%), characteristic viscosity equals $[\eta] = 0.20$ dl/g, which is untypical of rigid-chain polymers [23], $M_{SD} = 1.12\times10^5$. Bifurcation of this copolymers equals $\Delta h/\Delta\tau = 6.7\times10^{-10}$, which coincides with that for linear polymethyhylphenylsiloxane [24]. Some parameters of the substance are: diffusion coefficient, $D = 11.3\times10^{-7}$ cm^2/s; sedimentation index, $S = 4.4$; specific volume $\gamma = 0.93$ cm^3/g, refraction increment $dn/dc = 0.078$ cm^3/g.

Table 6.4

Some physical and chemical parameters of copolymers (XXV) with arylenecyclosiloxane fragments in the backbone

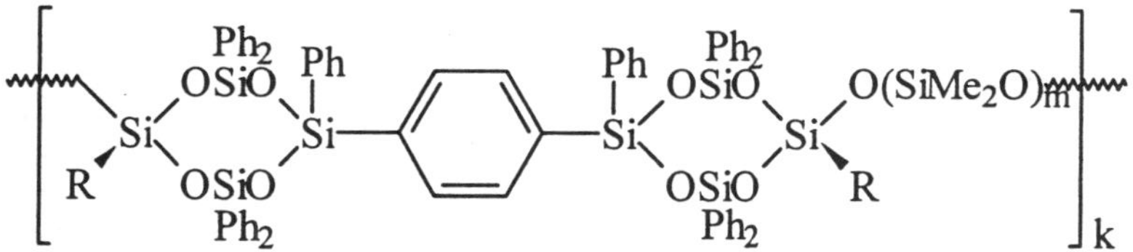

No.	R	m	Yield, %	η_{spec} at 25°C, 1% toluene solution	T_g, °C	d_1, Å	5% mass losses, °C	$M_w \times 10^{-3}$
1	Me	2	93	0.08	+41	9.54	420	38
2	Me	4	95	0.17	+30	9.41	400	-
3	Me	8	92	0.27	-10	-	-	87
4	Me	18	94	0.37	-51	-	-	-
5	Me	51	96	0.39	-123	7.21	350	135
6	Ph	2	93	0.07	+54	9.87	420	-
7	Ph	4	94	0.12	+38	-	-	-
8	Ph	8	94	0.16	-1	8.34	-	-
9	Ph	18	95	0.23	-42	-	330	68
10	Ph	51	93	0.29	-120	7.25	-	-

Thermomechanical studies of copolymers show that methyl group substitution by phenyl one at silsesquioxane atom of silicon in copolymers increases glass transition temperature (T_g) by ~10°C (Table 6.4), which coincides with previous data [20]. This is explained by a considerable effect of phenyl groups, the presence of which anywhere in the poly(dimethylsiloxane) chain increases T_g. Comparison of glass transition temperatures of arylenecyclosiloxane copolymers with and without bicyclic fragments in the backbone displays higher values for copolymers with arylenecyclosiloxane fragments in the backbone at the same length of flexible dimethylsiloxane unit. This indicates more rigid structure of the latter.

Figure 6.6 shows glass transition temperature dependence on the length of linear dimethylsiloxane chain. It is shown that T_g of copolymers decreases to -123°C as the length of linear dimethylsiloxane chain increases. The effect of arylenecyclosiloxane fragments on T_g of the copolymer is observed at linear dimethylsiloxane chain length $n = 18$.

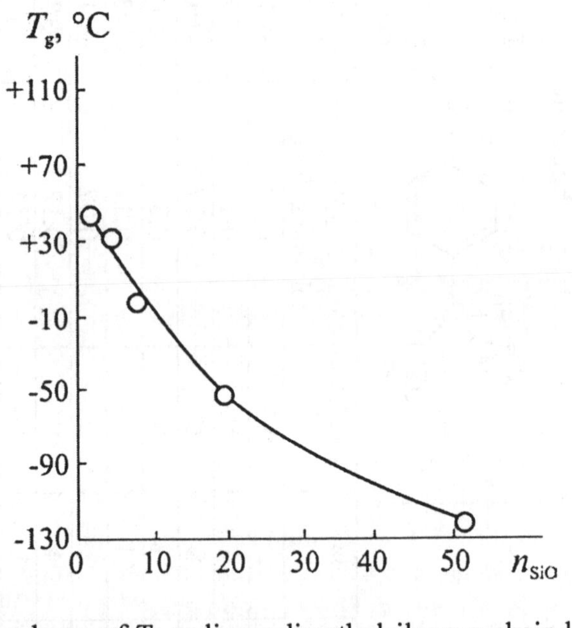

Figure 6.6. Dependence of T_g on linear dimethylsiloxane chain length for arylenecyclosiloxane copolymers

Thermogravimetric studies of synthesized copolymers indicate that the mass loss of dimethylsiloxane backbone with arylenecycloasiloxane fragments in it is initiated at 270 - 280°C, whereas unblocked poly(dimethylsiloxane) is completely degraded at this temperature. In the temperature range of 400 - 420°C 5% mass losses are observed. Substitution of a part of methyl groups at silsesquioxane atom of silicon by phenyl ones rises degradation initiation temperature to 420°C. This is because of high resistance to oxidation of phenyl groups and their inhibiting effect on oxidation of methyl groups [25]. The main degradation process proceeds in the temperature range of 450 - 700°C, above which no changes of mass losses are observed. Mass losses increase with the length of linear dimethylsiloxane backbone. As arylenecyclosiloxane fragments are included into linear dimethylsiloxane backbone, the main degradation process proceeds at temperatures by 80 - 100°C higher than for unblocked linear poly(dimethylsiloxane). Comparison of thermal oxidative stabilities of copolymers with arylenecyclosiloxane and bicyclohexasiloxane fragments in the backbone displayed higher values for the former compounds.

Wide angle X-ray studies show that synthesized copolymers are single-phase amorphous systems. Maximal values of interchain distances, d_1, are observed at short lengths of linear dimethylsiloxane backbone (d_1 = 9.54 – 9.87 Å, copolymers 1 and 6, Table 6.4). The interchain distance decreases to the value of poly(dimethylsiloxane) with the increase of dimethylsiloxane backbone length (refer to Table 6.4).

Heating of copolymer 3 at 100 - 120°C in the presence of 0.1 wt% of tertiary ammonium, the catalyst of anionic polymerization, an abrupt increase of viscosity and gel formation occur without gas products liberation. Varying lengths of linear dimethylsiloxane units between arylenecyclosiloxane fragments in the backbone, one can change the average distance between the crosslink sites.

For the purpose of clearing up the effect of cyclic structures' concentration in the macromolecular backbone on equilibrium rigidity, HFC reactions of difunctional organotricyclodecasiloxanes (compounds 31 – 34, Table 6.1) with α,ω–dichloro(dihydroxy)dimethylsiloxane in 60% anhydrous benzene solution at 20 - 25°C were studied [6, 20]. After 10 hours of interaction between 1,3-dichlorodecaphenyltricyclodecasiloxane (compound 24, Table 6.1) and 1,3-dihydroxytetramethyldisiloxane, conversion by hydrogen chloride did not exceed 7.0%. Contrary to these results, at the interaction of 1,3-dihydroxydecaphenyltricyclodecasiloxane with 1,3-dichlorotetramethyldisiloxane, 1,7-dichloroctamethyltetrasiloxane, 1,16-dichlorohexadecamethyloctasiloxane, conversion by hydrogen chloride was increased up to 80% and decreased from 80 to 25% with increase of α,ω-

dichlorodimethylsiloxane length, respectively. Figure 6.6 shows dependence of polycondensation degree. In the case of using dihydroxyorganotricyclodecasiloxane as the primary reagent in HFC reactions with α,ω-dichlorodimethylsiloxanes, the polycondensation degree is increased. HFC reaction proceeds in accordance with the following scheme:

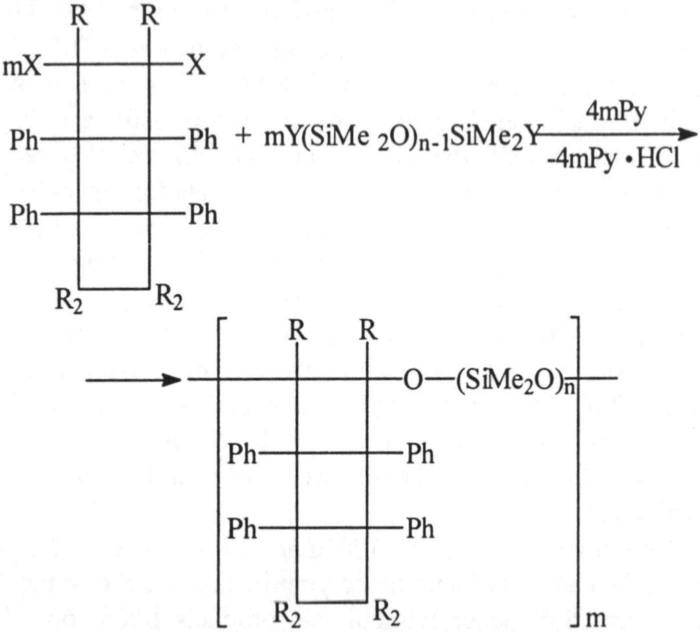

(6.32)

XXVI

where R = Me, Ph; X = Cl, OH; Y = OH, Cl; n = 2 - 70.

Copolymers with η_{spec} = 0.03 – 0.35 (Table 6.4), completely soluble in non-polar solvents, were synthesized by HFC reaction in the presence of pyridine.

Summing up data on the rate and quantity of hydrogen chloride, liberated during HFC of difunctional organocyclotetrasiloxanes (refer to Chapter 3) and organotricyclodecasiloxanes with α,ω-dichloro(dihydroxy)dimethylsiloxanes, one may recognize that reactivity of OH- and Cl-groups depends at least on two factors: induction effect of Me and Ph substituting groups and their steric effect [26, 27]. A conclusion was made that reactivity of hydroxyl groups present in cyclo- and tricyclodecasiloxanes considerably increases reactivity of chlorine-derived analogues.

Viscosity parameters indicate that low-molecular copolymers are formed at short lengths of dimethylsiloxane unit (n = 1, 2, 4). However, fractionation of copolymer 1 (structure XXVII, Table 6.5) with decaphenyltricyclodecasiloxane fragments in the backbone (n = 1) has produced a fraction with $[\eta]$ = 0.12 dl/g in benzene and $\bar{M}_{SD}$ = 74,000, close to molecular weight of polymer 1 (Table 6.3), which gives 28% of total weight of re-precipitated copolymer. Comparing structures of monomeric units of copolymer 1 (Table 6.3) and copolymer 1 (Table 6.5), one might expect that the equilibrium flexibility of copolymer 1 molecules (Table 6.5) is lower than that of copolymer 1 chains (Table 6.3), because the number of units in the backbone, around which internal rotation of a flexible fragment in copolymer 1 (Table 6.5), is lower than in copolymer 1 backbone (Table 6.3). In turn, this would increase the module of negative birefringence, $\Delta n/\Delta\tau$, if the ladder-type fragments defining (generally) optical anisotropy of the unit and molecular weights of these copolymers were identical. Experimentally determined anisotropy of copolymer 1 (Table 6.5) equals $\Delta n/\Delta\tau = 5\times10^{-10}$, which coincides with anisotropy of copolymer 1 (Table 6.3). Taking into consideration that chemical structures of cyclic fragments in these copolymers are different, the comparison of $\Delta n/\Delta\tau$ values gives no opportunity to make any conclusions about comparative rigidity of macromolecules of these two types of copolymers.

X-ray diffraction study of synthesized copolymers with bi- and tricyclic fragments in the linear chain was implemented. Independently of the single unit length, two amorphous halos are observed in diffraction patterns of all copolymers with both bi- and tricyclic fragments in the backbone. For copolymers with organobicyclohexasiloxane fragments in the backbone, d_1 varies from 7.70 to 10.40 Å and interplanar distance is increased with decrease of the single unit length.

In the case of tricyclic structure in the macrochain, the interplanar distance, d_1, varies from 10.75 Å for copolymer 7 (Table 6.5) to 11.20 Å for n = 1 (copolymer 1, Table 6.5). For copolymers with bi- and tricyclic structures, d_2 remains constant, 4.80 – 4.90 Å. Figure 6.6 showing variation of d_1 with the increase of flexible poly(dimethylsiloxane) unit length indicates that at n = 1, 2 the maximum of interchain distance is reached. For these polymers, d_1 approaches the interchain distance of poly(alkylphenylsilsesquioxane) polymer (d_1 = 11.50 Å) [28].

Figure 6.7 shows two most typical diffraction patterns for copolymers with low value of dimethylsiloxane unit (n = 1) and the average value of the flexible part equals n = 25. In the second case, the first halo displays two maximums of 10.80 Å and 7.80 Å. Occurrence of the second maximum in this area proves incompatibility of phases and formation of a two-phase structure.

Table 6.5

Some physical and chemical parameters of organosiloxane copolymers (structure XXVI) with organotricyclodecasiloxane fragments in the backbone

No.	Copolymer structure	n_{SiO}	Yield, %	η_{spec}*	T_g, °C	d_1, Å
1	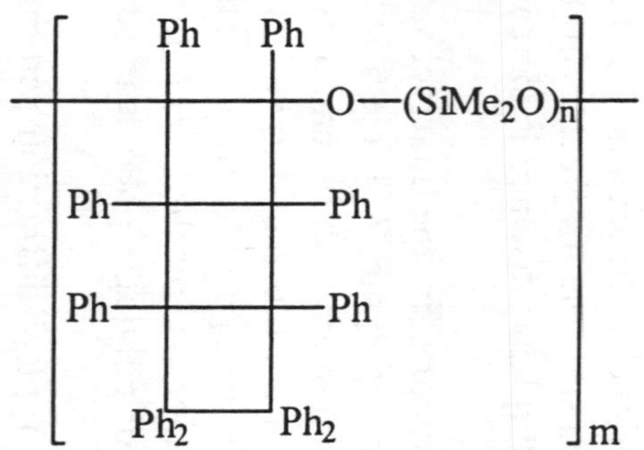	1	98	0.03	+68.5	11.20
2		2	67	0.03	+54.5	10.70
3		4	85	0.06	+30	10.8
4		6	90	0.08	+15.5	10.80
5		12	88	0.22	-10.5	9.80
6		25	79	0.15	-43	10.80 7.80
7		70	84	0.13	-	10.75 7.80
8	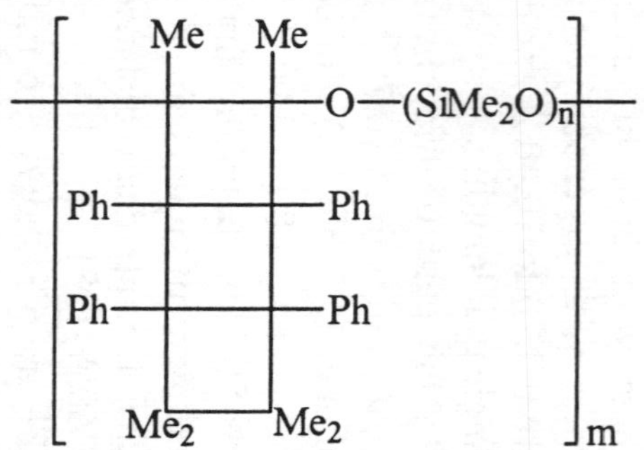	2	87	0.08	+2.5	-
9		4	86	0.11	-34	-
10		6	83	0.35	-65	-
11		8	81	0.14	-	-

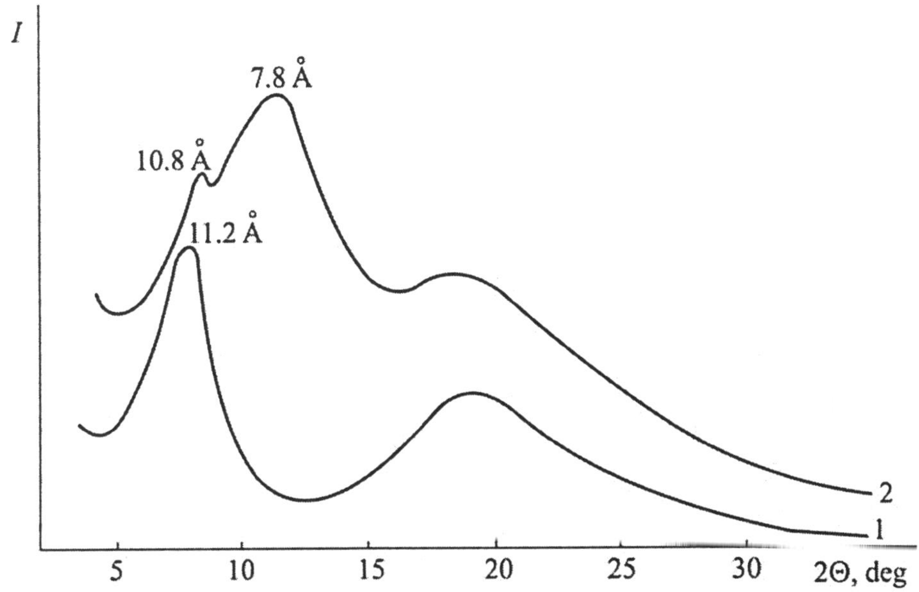

Figure 6.7. Diffraction patterns of copolymers with decaphenyltricyclodecasiloxane fragments in the backbone: 1 – copolymer 1; 2 – copolymer 6 (Table 6.5)

Hence, variations of interchain distances are associated with the length of poly(dimethylsiloxane) chain. At identical structure of rigid blocks, the maximal difference in interchain distances depends on the length (n) and equals 3.0 – 3.3 Å. The effect of structural variations of polycyclic compounds on interchain distance is much lower, and the difference in d_1 at identical actual length of the single unit equals 0.1 – 0.3 Å only. Nevertheless, a series of curves in Figure 6.8 indicates that increase of polycyclic fragment volume induces a negligible increase of the interchain distance in the whole range of single-unit block length.

For the purpose of synthesizing organosiloxane copolymers with tricyclohexasiloxane structure in the linear chain, HFC reaction of 1,1,7,7-tetrachloro-3,5,9,11-tetraphenyltricyclohexasiloxane (compound 17, Table 6.1) with α,ω–dihydroxydiorganosiloxanes at 1:2 ratio of initial compounds in inert gas (argon) and without acceptor was studied [29]. Hydrogen chloride, liberated during the reaction, was blown off with argon. HFC proceeding was judged by hydrogen chloride conversion level and viscosity increase. Polycondensation by *α,ω*-dihydroxydimethylsiloxane (n = 45) proceeded rather deep, and conversion by hydrogen chloride approached ~85% (Figure 6.9).

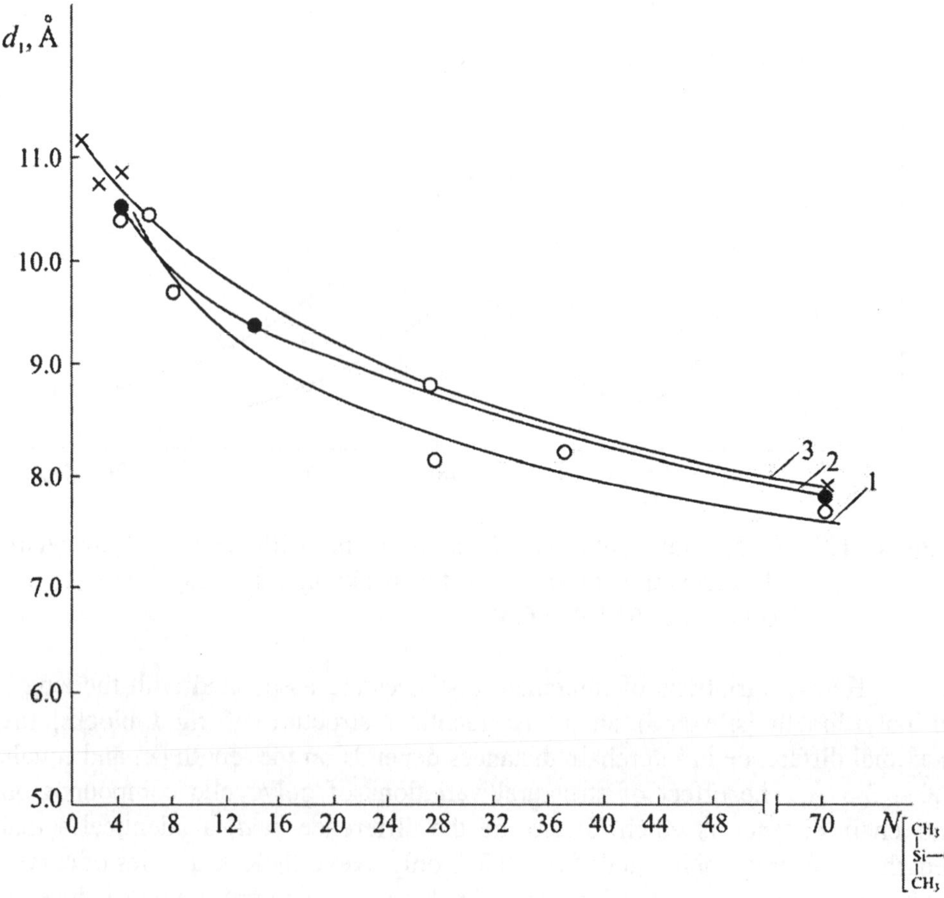

Figure 6.8. Dependence of interplanar distances on flexible part length in copolymers with bi- (1), tetra- (2) and tricyclic (3) structures

Figure 6.9 indicates that for 1,3-dihydroxytetramethyldisiloxane conversion equals 92%, and 87% for 1,3-dihydroxytetraphenyldisiloxane. The reaction proceeds as follows:

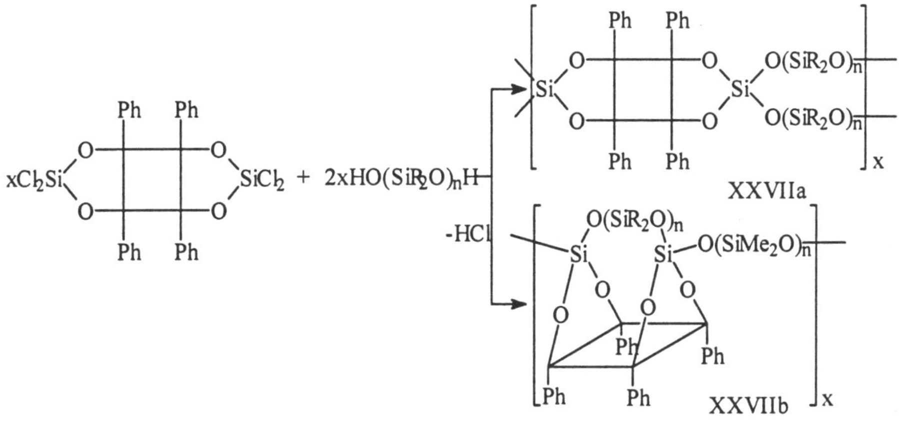

(6.33)

where R = Me, Ph; n = 2, 45.

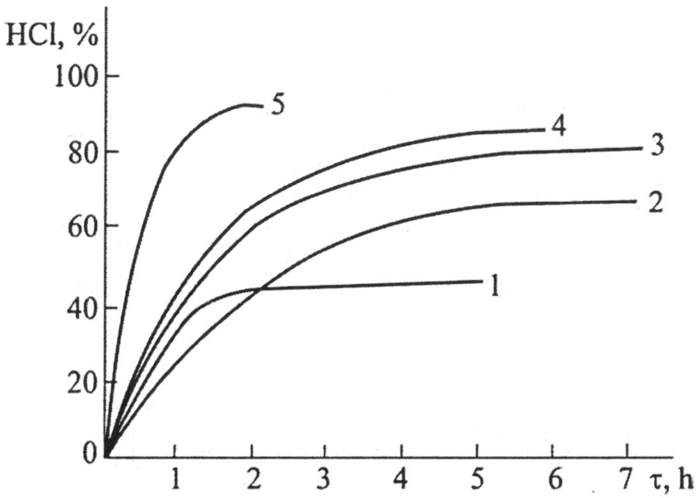

Figure 6.9. Conversion by hydrogen chloride in HFC reaction of 1,1,7,7-tetrachloro-3,5,9,11-tetraphenyltricyclohexasiloxane with α,ω–dihydroxydiorganosiloxanes: 1 – conversion at α,ω–dihydroxydiorganosiloxane chain length, n = 45; 2 – conversion with 1,3-dihydroxytetraphenyldisiloxane; 3 – conversion with 1,3-dihydroxytetramethyldisiloxane

Synthesis of completely soluble copolymers with η_{spec} = 0.05 – 0.29 in HFC reaction proves formation of polymers with linear chain containing polycyclic fragments of cyclolinear polymers with a spiroatom.

Based on a series of synthesized products (42 – 45) (Table 6.1), heterofunctional condensation of dihydroxyorganospirotricyclosiloxane with dichlororganospirotricyclosiloxane and dimethyldichlorosilane and HFC of dichlororganospirotricyclosiloxane with α,ω–dihydroxypolydimethylsiloxanes with n = 8, 70, 242 were studied. As the result, cyclolinear copolymers with organospirotricyclic fragments in the backbone, completely soluble in various organic solvents were synthesized. The mentioned reactions proceed by the following schemes [15]:

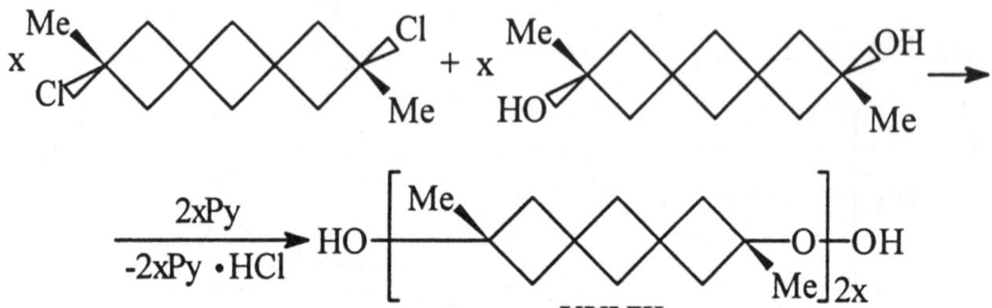

XXVIII

(6.34)

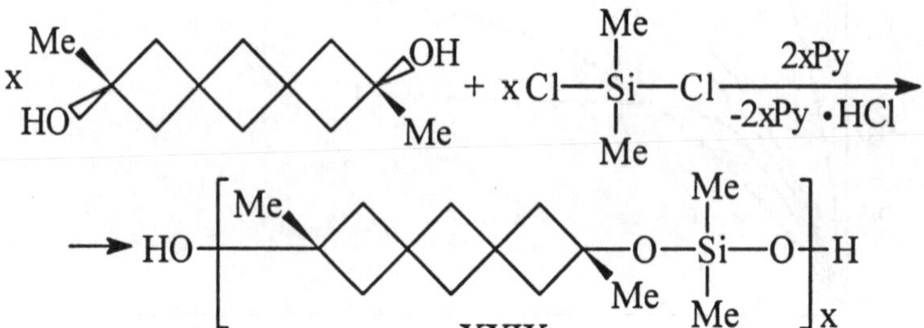

XXIX

(6.35)

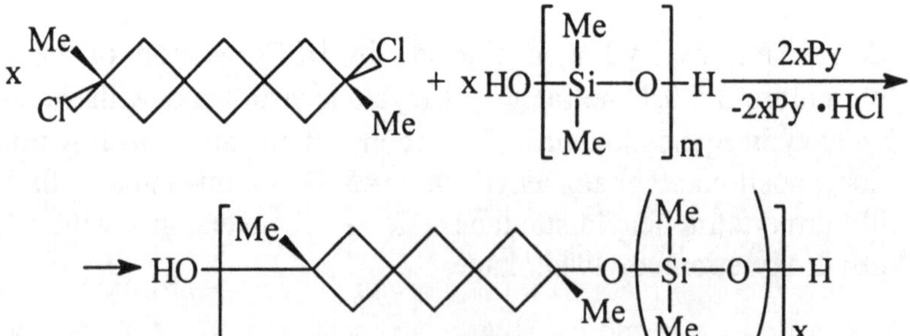

XXIX

(6.36)

where R = Me, $m = 8$; R = Ph, $m = 70, 242$; $n = 20 - 60$.

The reaction was performed in toluene, tetrahydrofurane or a mixture of solvents in the temperature range of 20 - 150°C in the presence of pyridine. Completeness of the reaction was determined by reaction mixture viscosity variation. Characteristic viscosity of cyclolinear polymers, measured at 25°C in toluene, equals $[\eta] = 0.10 - 0.40$ dl/g.

With regard to the initial reagents, polymers represent either a viscous substance, for example, at poly(dimethylsiloxane) fragment length $n = 242$, or solid friable substance (a cyclolinear polymer containing oxygen atom between spirocyclic structures).

Further on, to clear up the effect of polycyclic block volume on copolymer properties [6, 18], HFC reaction of dichlororganotetracyclosiloxane (compound 7, Table 6.1) with α,ω–dihydroxydimethylsiloxanes of different length proceeding at room temperature in the presence of pyridine (acceptor), in 60 – 70% anhydrous toluene solution was studied. At the final stage of the reaction, the mixture was heated up to the boiling point of the solvent used. The reaction proceeded by the following scheme:

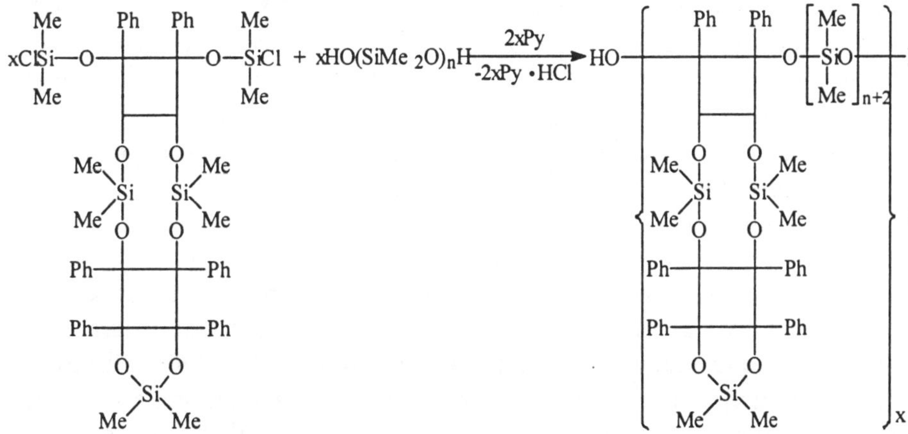

(6.37)

XXX

where $n = 2 - 70$.

Comparative X-ray diffraction study of copolymers with tetracyclic fragments in the backbone indicates their interchain distances, d_1, approaching the values for copolymers with organobicyclohexasiloxane and organotricyclodecasiloxane in the backbone. No increase of d_1 at short lengths was observed, because the minimal single-chain part was represented by octamethyltetrasiloxane.

As known from the literature [30], the interchain distance in linear or branched poly(dimethylsiloxane) varies from 7.70 Å to 8.80 Å, respectively. For the current case of synthesis of copolymers with rigid polycyclic blocks in the single-chain flexible macromolecular backbone, one may conditionally assume that these blocks form a branched structure. Then, apparently, at short lengths of linear dimethylsiloxane unit (n = 1, 2, 4), the first halo is corresponded to the interchain distance between polycyclic blocks. While the length increases (up to n = 35 – 70), d_1 approaches the interchain distance of poly(dimethylsiloxane), i.e. in block-copolymers with large flexible single-chain fragments the interchain distance is mainly defined by the flexible macromolecular chain. In the case, when two maximums can be detected in the first halo, Figure 6.6 indicates that 10.80 Å is the distance between polycyclic blocks, and 7.80 Å characterizes the packing of flexible segments. Table 6.6 shows several physical and chemical parameters of these copolymers.

As known from the literature, many works were devoted to the study of poly(dimethylsiloxanes) degradation. Obviously, the predominant direction of degradation is the process of depolymerization [31]. It is known that the type of degradation (depolymerization or ≡Si-C≡ bond break) depends on the nature of substituting groups at silicon atom [32, 33]. At the same time, it is known that introduction of groups different from ≡Si-O-bond into the poly(dimethylsiloxane) chain disturbs degradation by the depolymerization mechanism, because a transitional complex cannot be formed [34, 35].

The contribution of bi-, tri- and tetracyclic appendages to poly(dimethylsiloxane) chain and their effect on the degradation process was studied [36]. Thermogravimetric studies of the above-mentioned copolymers were performed on electronic B-60 thermal scales, Seteram Co., in argon at 5 deg/min heating rate. Gaseous reaction products were analyzed simultaneously. Low-molecular degradation products were analyzed by the ampoule technique. Figure 6.10 showing thermogravimetric curves indicates that the majority of studied polymers display 5%-degree mass losses in the temperature range of 175 - 200°C. Analysis of the products of copolymer thermal degradation with tricyclodecasiloxane fragments in the backbone (n = 2), proceeded at 280°C,

Table 6.6

Some physical and chemical parameters of organosiloxane copolymers (structure XXX) with organotetracyclosiloxane fragments in the backbone

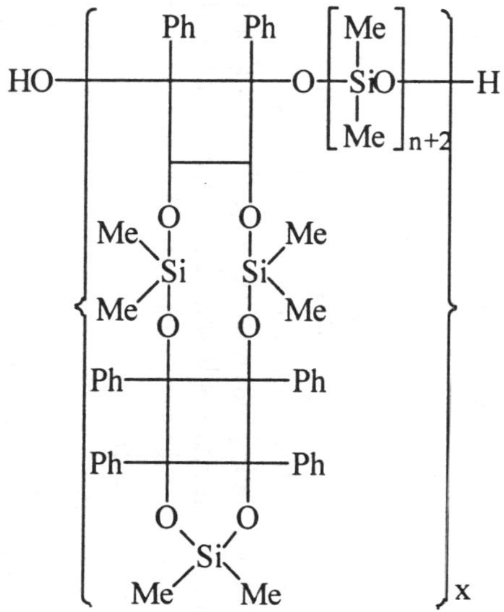

No.	Copolymer structure	n_{SiO}	Yield, %	η_{spec}*	T_g, °C	d_1, Å
1	(see structure above)	2	86	0.09	-26	10.50
2		4	87	0.15	-32	10.10
3		6	81	0.11	-	8.80
4		12	88	0.23	-72	9.40
5		25	82	0.69	114	8.80
6		70	89	0.22	-123	7.80

has displayed full absence of organosilicon and organic compounds in their composition. At 300 - 350°C, the rate of thermal decay of all studied copolymers is considerably increased reaching the maximum at 460 - 475°C, and at 560 - 600°C evanescent mass variation curves are observed. For all studied copolymers, it may also be observed that the mass losses are obligatorily increased with the number of dimethylsiloxane units in the chain. For example, copolymers with bicyclohexasiloxane fragments in the backbone (structure XXIV, R = Me) display typical mass change up to 31% for $n = 2$, 51.5% for $n = 4$, and 94% for $n = 25$ (Figure 6.10). Simultaneously, increase of the number of rings in the tab induces a noticeable decrease of the mass loss of copolymers containing equal amount of linear siloxane units in the backbone. In the case of decaphenyltricyclodecasiloxane fragments in the backbone, the mass loss for $n = 4$ equals ~40%.

To determine the composition of thermal degradation products, the ampoule-type analysis of copolymers containing decaphenyltricyclodecasiloxane tabs was performed. The results of the analysis indicate that degradation of the polymer with $n = 2$, proceeding in the temperature range of 280 - 580°C, did not produce any cyclosiloxane products of D_n composition. Thermal decay of the polymer with $n = 4$ and $n = 25$ is accompanied by liberation of negligible amounts of D_3 and D_4, whereas total mass losses of polymers significantly exceed these values. Such phenomena were observed during studies of thermal degradation of poly(organocarboranesiloxanes) [37, 38], the backbone of which is cyclolinear siloxane like composition. It has been shown that degradation of carborane-containing polymers proceeds by the radical mechanism, generally forming oligomeric products, the structure of which is similar to initial copolymers. Liberation of cyclic products does not also considerably contribute into total mass losses of copolymers. Such analogy allowed a supposition [36] that in the case of cyclolinear poly(organosiloxanes), thermal degradation proceeds by the same scheme, i.e. the main siloxane bond break by the law of randomness with formation of oligomeric products with the structure analogous to the initial polymers is predominant.

Simultaneously with the backbone degradation during thermal decomposition of all studied copolymers, liberation of hydrogen and methane was observed [39], associated with radical cross-linking of methyl and phenyl groups. Copolymers with bicyclohexasiloxane taps begin liberating hydrogen at 400°C, the process reaching maximal intensity at 560°C.

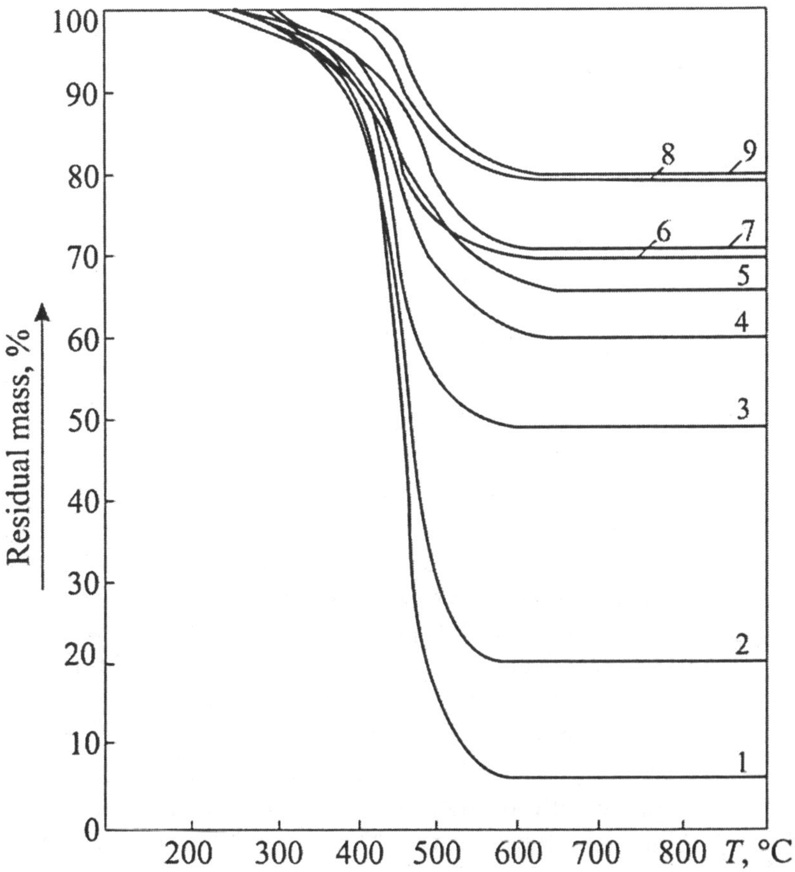

Figure 6.10. Thermal gravimetrical curves for linear organosiloxane copolymers: 1, 3, 6 – copolymer of structure XXIV (R = Me), n = 25, 4 and 2, respectively; 5, 4, 2 – copolymer of structure XXVI, n = 25, 4 and 2, respectively; 7, 8, 9 – copolymer of structure XXX, n = 25, 4 and 2, respectively

The second maximum is shifted by the sequence of copolymers from n = 2 to n = 25 and from 675°C to 625°C, respectively. Maximums of methane liberation from copolymer with bicyclohexasiloxane taps are also shifted from n = 2 to n = 25 and from 640°C to 600°C, respectively. As the number of rings in the tap is increased, maximums of hydrogen and methane liberation are shifted towards higher temperatures, preserving the shift regularity in every sequence of copolymers, simultaneously. It is found that the increase of dimethylsiloxy-groups' number in the backbone causes rise of gaseous degradation products' liberation, the methane quantity being increased first. Besides, hydrogen and methane, benzene liberation during degradation of cyclosiloxane copolymers is also observed.

Thus, based on the results obtained, the following scheme of thermal degradation of cyclolinear organosiloxane copolymers was suggested [36]: decomposition of these copolymers below 350°C is followed by condensation and structuring by means of residual hydroxyl groups, which is accompanied by liberation of gaseous products. As a result of introducing volumetric cyclic taps into the linear chain, the beginning of the basic degradation process is shifted by 50 - 100°C towards higher temperatures compared with linear poly(dimethylsiloxanes) [31, 40, 41].

Thermomechanical studies of copolymer with bi-, tri- and tetracyclic fragments in the poly(dimethylsiloxane) chain was performed. Copolymers with dominating methyl framing represent transparent colorless or light yellow viscous systems. Thermomechanical curves were plotted for continuous impact of constant compressing stress (0.5 kg/cm^3) by the technique, described in ref. [42]. The glass transition temperature (T_g) was determined by the initial deviation of thermomechanical curve in the area of positive deformations. Table 6.7 shows glass transition temperatures for copolymers with bi-, tri- and tetracyclosiloxane fragments in the backbone.

Comparison of T_g for copolymers of structures XXIV, XXVI and XXX, corresponded to equal lengths of linear siloxane component, *n*, clearly illustrates high efficiency of substitution of a part of methyl groups in the framing of the cyclic group by phenyl ones.

Substitution of a part of methyl groups in the side framing of siloxane fragments by phenyl ones is accompanied by T_g increase. In all cases, an abrupt fall of T_g and further gentle and slow decrease to the value, corresponded to poly(dimethylsiloxane) glass transition with *n* increase is observed.

Despite relatively high spatial complexity of tetracyclic grouping, glass transition temperature (T_g) of copolymer with tetracyclic fragments is close or even equal to analogous values of copolymers with bicyclohexasiloxane fragments in the backbone (with methyl framing). According to the authors' point of view, this effect is stipulated by presence of an original joint in a polycycle - two silicon atoms, not bonded by oxygen to one another. Note also that the same result is induced by predominance of methyl groups in the side framing with tricyclodecasiloxane fragments in the chain (copolymer 3, Table 6.6). Comparison of glass transition temperatures (T_g) of copolymers with decaphenyltricyclodecasiloxane fragments in the backbone with these of

Table 6.7

Glass transition temperatures for copolymers with bi-, tri- and tetracyclosiloxane fragments in the backbone

No.	Structural formula of repeating cyclic fragments in the dimethylsiloxane backbone	Number of dimethylsiloxane units, *n*					
		2	4	6	12	25	70
1	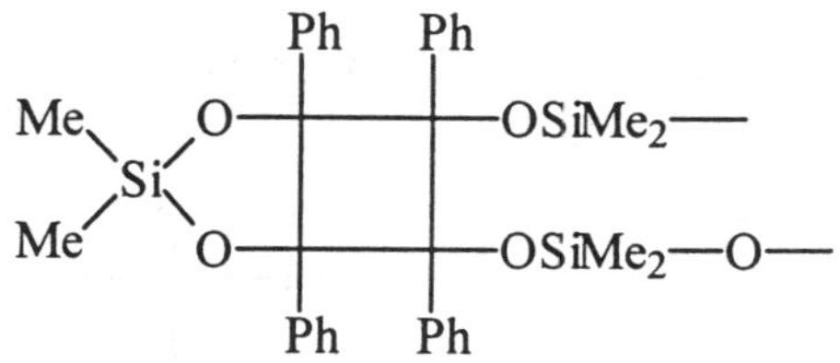 XXIV	-26	-69	-	-	-114	-123
2	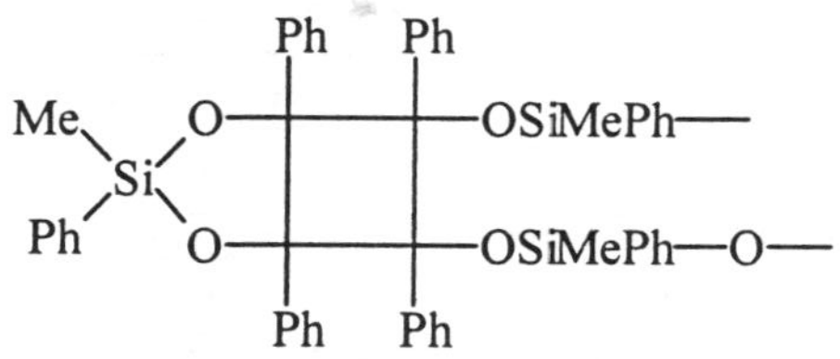 XXIV	+6	-44	-57	-72	-96	-123
3	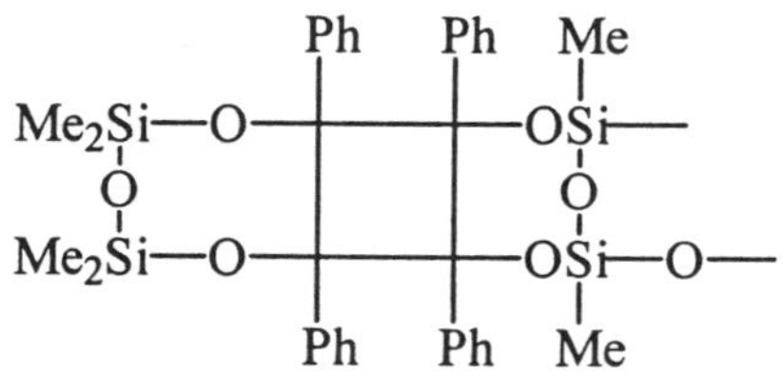 XXIV	+2	-34	-65	-	-	-123

4	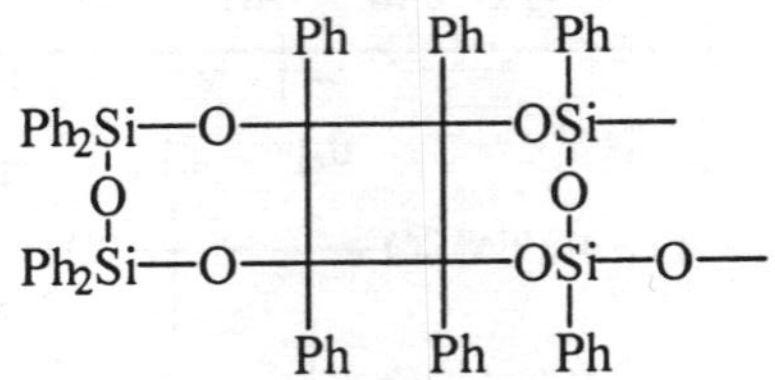 XXIV	+54	+30	+16	-11	-43	-123
5	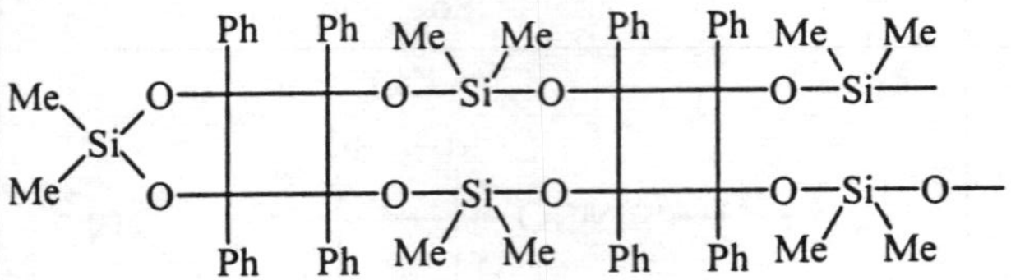 XXX	-26	-31	-	-72	-114	-123

Note: T_g of copolymer with structure 4 ($n = 1$) equals +68°C.

copolymers with 1,3-disposition of hexaphenylcyclotetrasiloxane fragment in the backbone (Chapter 3.1) gives a possibility to determine T_g with a single "ladder" unit, introduced into the ring. On the average, it equals 70°C.

6.4. Organosilicon copolymers with bicyclocarbosiloxane fragments in the backbone

Copolymers with bicyclocarbosiloxane fragments in the backbone were synthesized by HFC reaction of dichloro(dihydroxy)-containing organobicyclocarbosiloxanes of structure XXII (compounds 1 and 2, Table 6.2) with α,ω–dihydroxy(dichloro)dimethylsiloxanes. Low-temperature HFC of dichlorobicyclocarbosiloxanes with α,ω–dihydroxydimethylsiloxanes was implemented in 70% anhydrous toluene solution in the presence of pyridine. At the final stage, the reaction mixture was heated up to the boiling point of the solvent used. HFC of dihydroxybicyclocarbosiloxane with α,ω–bis(dimethylamino)dimethylsiloxanes was implemented in dry nitrogen flow both in block and in solution. As for conditions in block, the reaction mixture was heated up to 50 - 60°C till a homogeneous mixture is obtained. After that the reaction was continued under vacuum (P = 3 – 5 mmHg), in the temperature range of 100 - 120°C up to constant viscosity. In both cases, the reaction proceeds by the scheme as follows [43]:

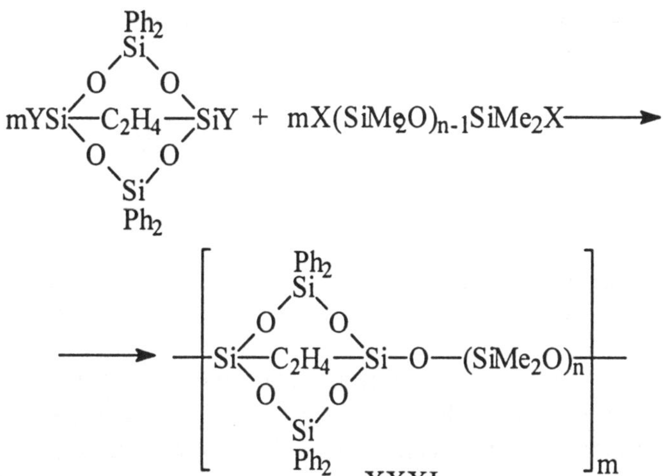

XXXI

(6.38)

where Y = Cl, OH; X = OH, Me_2N; n = 2 - 37.

After re-precipitation by methyl alcohol from toluene solution, synthesized copolymers represent solid or viscous products (depending on *n* value), well-soluble in usual organic solvents with η_{spec} = 0.10 - 0.28.

Table 6.8 shows some physical and chemical parameters of cyclolinear copolymers of the structure XXXI.

Comparative assessment of thermomechanical properties of bicyclocarbosiloxane copolymers with the ones possessing cyclocarbosiloxane fragments in the backbone (refer to Section 3.2, structures VIII and IX, Table 3.14, copolymers 1 – 10) has displayed their high glass transition temperatures. Figure 6.11 shows T_g dependence for cyclolinear organosilicon copolymers with bicyclocarbosiloxane and cyclocarbosiloxane fragments in the backbone.

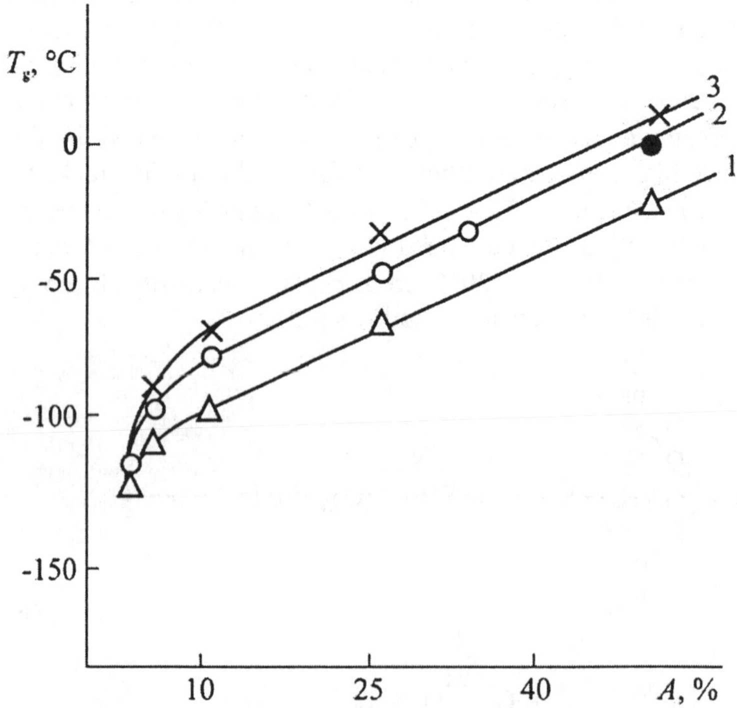

Figure 6.11. T_g dependence for cyclolinear copolymers on content (%) of cyclocarbo- and bicyclocarbosiloxane fragments in the backbone: 1 – copolymers (1 – 5) of structure VIII (refer to Section 3.2); 2 – copolymers (6 – 10) of structure IX (Section 3.2, Table 3.13); 3 – copolymers of structure XXXI (Table 6.7)

Table 6.8

Some physical and chemical parameters and yields of copolymers (XXXI) with bicyclocarbosiloxane fragments in the backbone

No.	Copolymer	n_{SiO}	Yield, %	η_{spec} at 25°C	T_g, °C	d_1, Å	5% mass losses
1	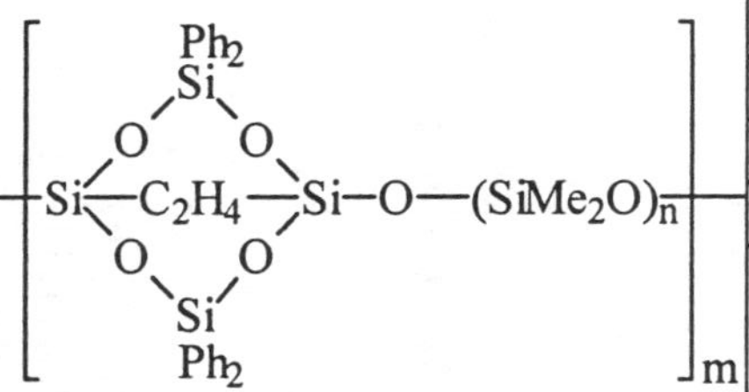	2	81	0.10	-10 ÷ -12	9.51	420
2		4	84	0.12	-30	-	-
3		10	84	0.18	-66	-	-
4		20	86	0.23	-93	7.32	360
5		37	88	0.28	-123	7.31	-

Wide-angle X-ray diffraction analysis of copolymers has indicated their amorphous single-phase state. They possess two diffraction maximums, typical of amorphous copolymers. The above-shown values of the interchain distance, d_1 (Table 6.7), display its decrease with dimethylsiloxane unit increase. As n = 37, d_1 becomes equal to the interchain distance of PDMS. At the same time, d_1 for copolymers with bicyclocarbosiloxane fragments in the backbone exceeds the interchain distance of copolymers (structures VIII and IX) with cyclocarbosiloxane fragments in the backbone and succumbs to the ones with bicyclohexasiloxane fragments.

Similar to copolymers of the poly(dimethylphenylsiloxane) sequence [44], the effect of units different from dimethylsiloxane ones on T_g of copolymers with cyclocarbosiloxane fragments in the backbone is primarily observed at 3 mol% concentration of the latter. Moreover, the Figure indicates also that the effect of extended cyclocarbosiloxane fragment (structure IX, m = 2) on T_g of studied copolymers is greater than for copolymers of structure VIII, m = 1. Bicyclocarbosiloxane structures, introduced into dimethylsiloxane backbone, induce higher influence on copolymers than cyclocarbosiloxane fragments. The general feature of cyclolinear copolymers is that dimethylsiloxane chain length increase causes decrease of the glass transition temperature of copolymers.

Table 6.7 shows that copolymers with bicyclocarbosiloxane fragments in the backbone are characterized by high thermal oxidative stability. 5% Mass losses are observed at 420°C (n = 2) and 360°C (n = 20), whereas for copolymers with cyclocarbosiloxane fragments - at 300°C. Thermal oxidative stability of copolymers with bicyclocarbosiloxane fragments is similar to that of copolymers with monocyclic or bicyclohexasiloxane fragments in the dimethylsiloxane backbone.

6.5. Carbosiloxane copolymers with tricyclic fragments in the backbone

The present Chapter discusses synthesis and studies of carbosiloxane copolymers containing flexible dimethylsiloxane and decaorganotricyclodecasiloxane fragments in the backbone [8, 45]. For the purpose of synthesizing carbosiloxane copolymers, hydride addition of α,ω-dihydridedimethylsiloxane to 1,3-divinyl-1,3,9,9,11,11-hexamethyl-5,7,13,15-tetraphenyltricyclodecasiloxane (compound 29, Table 6.1) was performed at temperature below 90°C.

Therefore, cyclosiloxane ring disclosure did not take place under conditions of hydride polyadditions.

Preliminary heating of initial divinylorganotricyclodecasiloxane (compound 30, Table 6.1) during 10 hours by temperature of 70 - 90°C in the presence of rhodium acetylacetonatedicarbonyl or platinum-hydrochloric acid as a catalyst did not initiate polymerization of the primary divinyltricyclodecasiloxane. Thorough analysis of the reaction mixture by gas-liquid chromatography method has detected the presence of initial organosiloxanes. Besides, there are no changes in the NMR and IR spectra of divinyl-containing compounds and dihydridedimethylsiloxanes.

Hydride polyaddition of divinyl-containing compounds was carried out for various lengths of α,ω-dihydridedimethylsiloxanes. The reaction run was searched by a decrease of active ≡Si-H groups' concentration. It was found that for rhodium acetylacetonatedicarbonyl as a catalyst, copolymers soluble in organic solvents were obtained, which were structured after some time. This may be explained by the fact that in spite of polymers reprecipitated from toluene solution by methyl alcohol, rhodium catalyst remains in polymeric systems, which decompose and induce structuring (cross-linking) of copolymers.

Therefore, copolymers were synthesized in the presence of platinum-hydrochloric acid as the catalyst. The rate and depth of polyaddition are decreased with the increase of α,ω-dihydridedimethylsiloxane chain length. Figure 6.12 shows that conversion of ≡Si-H bond is incompletely and decreases from 95% ($n = 4$) to 83% ($n = 12$).

Hydride polyaddition of α,ω-dihydridedimethylsiloxane to divinylorganotricyclodecasiloxane (compound 30, Table 1) proceeds in accordance with the general scheme as follows [8, 45]:

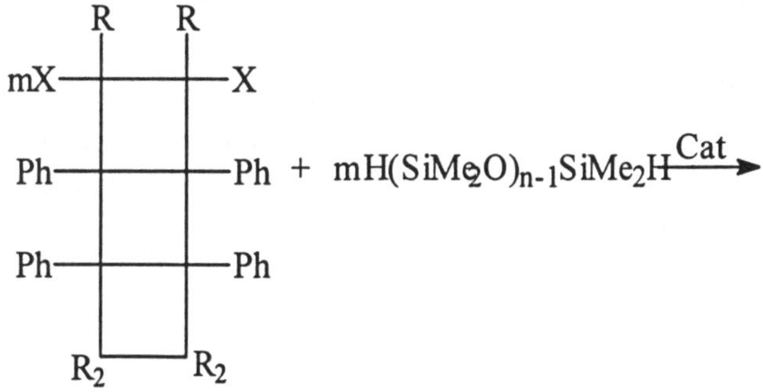

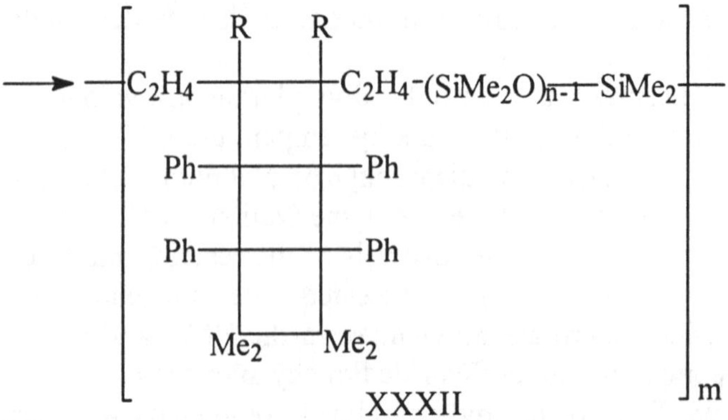

XXXII (6.39)

where $n = 2 - 21$; Cat is H_2PtCl_6.

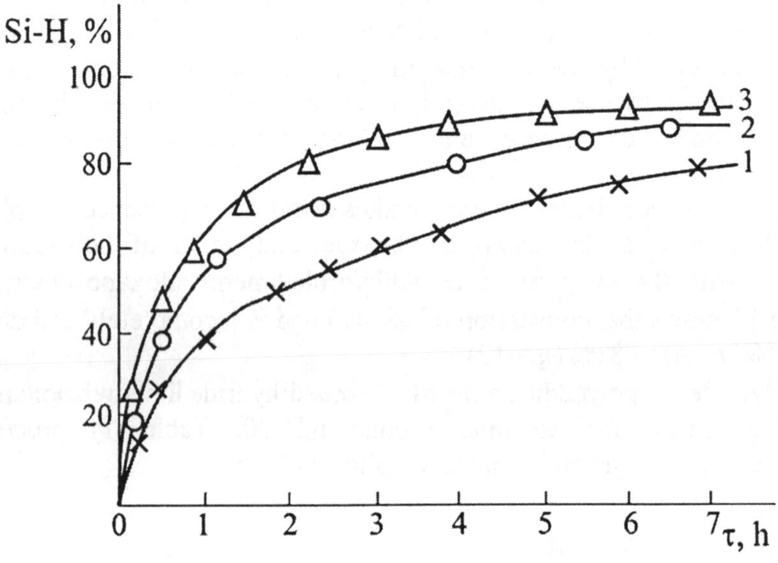

Figure 6.12. Time dependence of ≡Si-H group concentration (%) for polyaddition of α,ω-dihydridedimethylsiloxane to divinylorganotricyclodecasiloxane at 90°C for dimethylsiloxane backbone lengths: 1) $n = 12$; 2) $n = 6$; 3) $n = 4$

Table 6.9

Some physical and chemical parameters of carbosiloxane cyclolinear copolymers with tricyclodecasiloxane fragments in the backbone

No.	Copolymer structure	n_{SiO}	Yield, %	T_{react}, °C	η_{spec}*	T_g, °C	d_1, Å	$\bar{M}_\omega \times 10^{-3}$
1	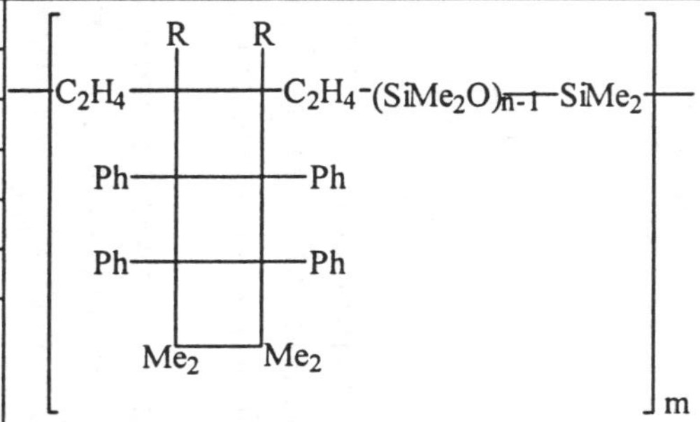	2	80	90	0.08	-12	10.20	72
2		4	83	70	0.09	-	-	-
3		4	88	80	0.11	-	-	-
4		4	91	90	0.11	-50	-	85
5		6	92	90	0.14	-96	8.68	-
6		12	93	90	0.20	-123	-	-
7		21	94	90	0.26	-123	7.54	110

Note: * In toluene at 25°C; molecular masses were determined by the gel chromatography method.

As a result of the reaction, synthesized copolymers possess η_{spec} = 0.08 - 0.26 and represent liquid or glassy-like light yellow transparent products, soluble in ordinary organic solvents. Some physical and chemical parameters, molecular weights and yields of synthesized copolymers are shown in Table 6.8.

The reaction proceeding was also monitored by viscosity increase of synthesized copolymers. It was found that viscosity of copolymers and the hydride polyaddition degree increase with temperature rise to 70 - 90°C.

In hydride polyaddition of α,ω-dihydridedimethylsiloxane to divinylorganotricyclodecasiloxane, conversion of ≡Si-H bond increases with temperature as follows: from 85% (70°C) to 95% (90°C). Figure 6.13 shows time dependence of ≡Si-H concentration (%) decrease for various temperatures.

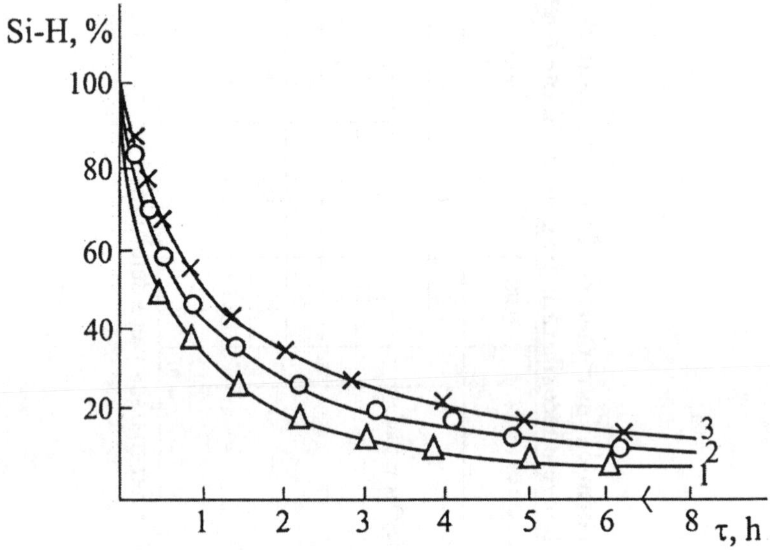

Figure 6.13. Time dependence of ≡Si-H group concentration (%) in polyaddition of α,ω-dihydridedimethylsiloxane (n = 4) to divinylorganotricyclodecasiloxane: 1 - 90°C; 2 - 80°C; 3 - 70°C

Time dependence of reverse reagent concentration displays the second order of hydride polyaddition. Further on, reaction rate constants for various temperatures were calculated: $k_{70°C} = 1.1086\times10^{-2}$, $k_{80°C} = 1.6196\times10^{-2}$, $k_{90°C}$

= 2.3834×10^{-2}. It was shown that the reaction rate constants increase by 1.5 times, approximately, for every 10°C of temperature rise.

Activation energy of hydride polyaddition was derived from dependence of the reaction rate constant logarithm on reverse temperature: E_a = 64.4 kJ/mol.

Study of ^{1}H NMR spectrum for copolymer 2 (Table 6.8) displays that catalytic hydride polyaddition mainly proceeds by the Farmer rule with formation of dimethylene bridges. NMR spectrum also displays a reflex of -CH_2-CH_2- group with chemical shift of δ = 0.35 ppm. A duplet reflex centered at the chemical shift of δ = 1.12 ppm, corresponded to methyl protons in =CH-CH_3 groups with 5 – 6% concentration was also observed. Integral ratios of methyl and phenyl protons correspond to the formula of copolymer 2 (Table 6.8).

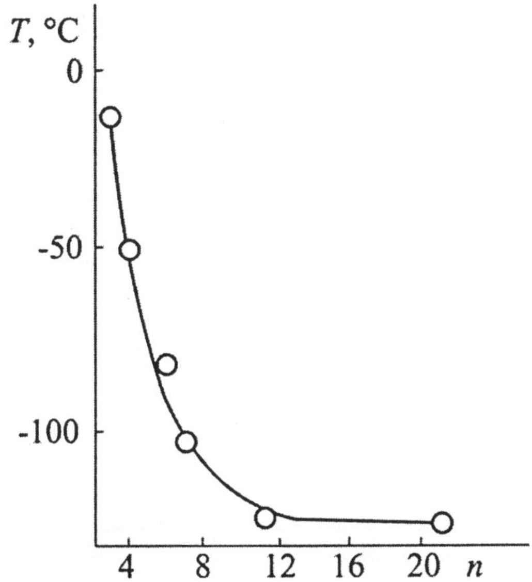

Figure 6.14. T_g dependence of cyclolinear carbosiloxane copolymers (XXXII) on length of linear poly(dimethylsiloxane), *n*

Thermomechanical studies of synthesized copolymers indicate that the glass transition temperature of copolymers is decreased with an increase linear dimethylsiloxane backbone length, *n* (Figure 6.14). Since *n* = 12, carbotricyclodecasiloxane fragments in copolymers cause no effect on the dimethylsiloxane backbone and T_g of copolymer 6 (Table 6.8) remains equal -123°C.

Contrary to previous considerations [18], dimethylsiloxane backbone length increase (n = 21) does not cause formation of two-phase systems in cyclolinear copolymers with rigid decaphenyltricyclodecasiloxane fragments and flexible dimethylsiloxane units (n = 25).

Comparison of copolymers containing carbotricyclodecasiloxane and tricyclodecasiloxane fragments [18] displayed lower glass transition temperatures of the former, which may be explained by excessive concentration of flexible -CH_2-CH_2- groups in its backbone.

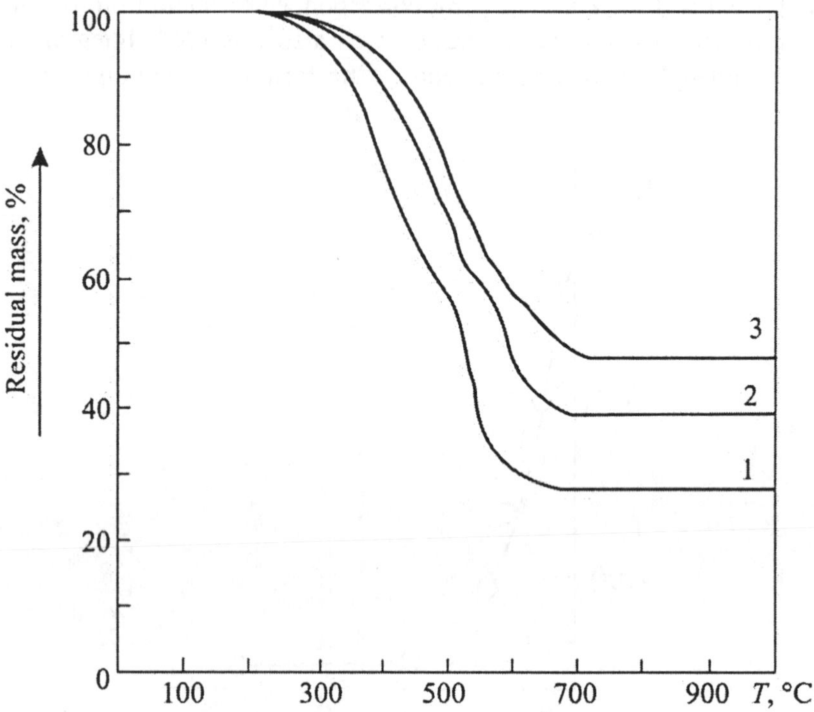

Figure 6.15. Thermogravimetric curves of cyclolinear carbosiloxane copolymers: 1 - copolymer 7; 2 - copolymer 5; 3 - copolymer 1 (Table 6.8, in air, at 5 deg/min heating rate)

Thermogravimetric studies of copolymers show (Figure 6.14) their higher thermal oxidative stability for short length of the dimethylsiloxane backbone, n. As the length of dimethylsiloxane backbone increases, thermal oxidative stability of copolymers decreases. Compared with siloxane analogies, thermal oxidative stability of carbotricyclodecasiloxane-containing copolymers is lower [18]. In the temperature range of 300 - 350°C mass losses of the polymer are below 3 - 7%, and the main degradation process proceeds at 400-650°C. Above 650°C, the curves of mass losses are preserved unchanged (Figure 6.15).

As carbotricyclodecasiloxane fragments are introduced into the dimethylsiloxane backbone, the main degradation process proceeds at temperature by 80 - 100°C higher than for unblocked poly(dimethylsiloxane).

Synthesized copolymers were studied by the X-ray diffraction method. Diffraction patterns of amorphous polymers (Figure 6.16) show that the interchain distance reaches its maximum (d_1 = 10.24 Å) at short lengths of flexible dimethylsiloxane backbone, *n*. As the length of dimethylsiloxane backbone increases (*n* = 21), the interchain distance decreases and for copolymer 5 reaches 7.54 Å (Table 6.8).

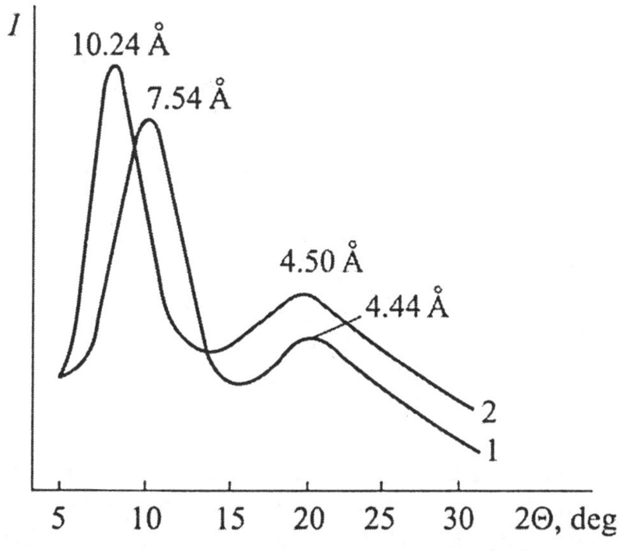

Figure 6.16. Diffraction patterns of copolymers: 1 - copolymer 5; 2 - copolymer 1 (Table 6.8)

Thus copolymer 7 does not form a two-phase system, as observed for copolymers with decaphenyltricyclodecasiloxane fragments in the dimethylsiloxane backbone. This may be explained by the presence of combination of rigid carbotricyclosiloxane and flexible dimethylsiloxane fragments in it. Therefore, copolymers represent single-phase systems.

For the purpose of synthesizing carbosiloxane copolymers with tricyclohexasiloxane fragments in the backbone, hydride polyaddition of α,ω–dihydridepoly(dimethylsiloxanes) to 1,7-divinyl-1,7-dimethyl-3,5,9,11-tetraphenyltricyclohexasiloxane was studied [9]. The reaction was implemented in anhydrous toluene in the presence of platinum-hydrochloric acid in tetrahydrofurane and in the temperature range of 70 – 170°C according to the following scheme:

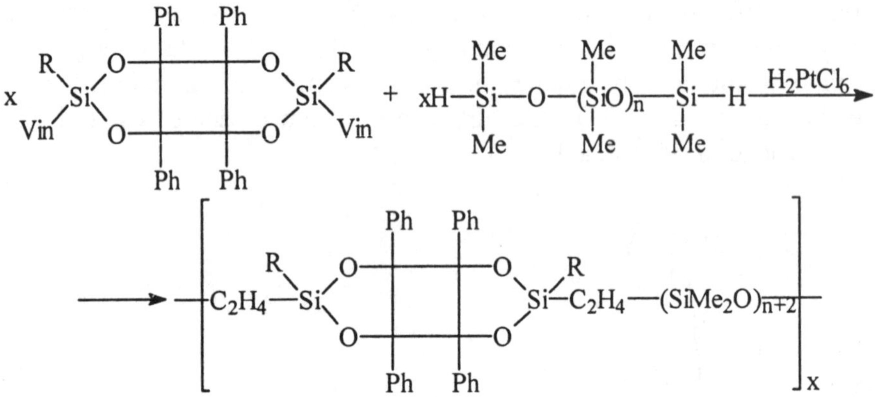

(6.40)

XXXIII

where R = Me, Ph; n = 0 - 86.

New cyclolinear carbosiloxane copolymers containing organotricyclohexasiloxane fragments in backbones were synthesized in the reaction. They represent viscous liquids with molecular mass varying in the range of 35,000 - 45,000.

The influence of organotricyclohexasiloxane fragments in carbosiloxane copolymer on thermal oxidative degradation proceeding was studied. It is found that the increase of concentration of rigid organotricyclohexasiloxane fragments in linear chains of polymers rise their resistance to thermal oxidative degradation. For example, 15% mass loss of structure XXXIII copolymers (R = Ph, n = 74 and R = Me, n = 86) is observed at 380 and 420°C, respectively. For copolymer with n = 0, it is observed at 540°C.

Thus, the increase of cyclic fragments' concentration in the linear chain induces rise of macromolecular chain rigidity and in some cases leads to formation of a two-phase system. The increase of cyclic fragments' concentration in the macromolecular chain raises thermal oxidative stability of the copolymers.

References

1. Meladze S.M., Mukbaniani O.V., Makarova N.N., and Khananashvili L.M., *Zh. Obshch. Khim.*, 1980, vol. **50**(11), p. 2493. (Rus)
2. Mukbaniani O.V., Khananashvili L.M., and Inaridze I.A., *Int. J. Polym. Mater.* 1994, vol. **24**, p. 211.
3. Andrianov K.A., Nogaideli A.I., Makarova N.N., Raspopova K.N., and Mukbaniani K.A., *Doklady AN SSSR*, 1975, vol. **224**(4), p. 825. (Rus)
4. Mukbaniani O.V., *Candidate's Dissertation on Chemistry*, Tbilisi State University, 1976, Tbilisi, GSSR. (Rus)
5. Andrianov K.A., Makarova N.N., and Chizhova N.V., *Izv. AN SSSR, Ser. Khim.*, 1975, p. 2554. (Rus)
6. Mukbaniani O.V., *Doctor's Dissertation on Chemistry*, Tbilisi State University, 1993, Tbilisi, Georgia.
7. Andrianov K.A., Nogaideli A.I., Makarova N.N., and Mukbaniani O.V., *Soobshch. AN GSSR*, 1976, vol. **81**(2), p. 349. (Rus)
8. Mukbaniani O.V., Khananashvili L.M., Karchkhadze M.G., Tkeshelashvili R.Sh., and Mukbaniani N.O. In: *Synthesis and Properties of Polymers*, Nova Science Publishers, Inc., Commak., 1996, p. 89.
9. Klementiev I.Yu., *Candidate's Dissertation on Chemistry*, M.V. Lomonosov Institute of Fine Chemical Technology, 1979, Moscow, USSR. (Rus)
10. Andrianov K.A., Zhdanov A.A., Kurasheva N.A., and Khyinku E.S., *Doklady AN SSSR*, 1977, vol. **236**(2), p. 354. (Rus)
11. *Patent No. 3,378,575* (USA).
12. *Patent No. 3,222,369* (USA).
13. *Patent No. 3,317,578* (USA).
14. Andrianov K.A., Zhdanov A.A., Kurasheva N.A., and Khyinku E.S., *Doklady AN SSSR*, 1978, vol. **240**(3), p. 591. (Rus)
15. Zhdanov A.A., Khyinku E.S., Kurasheva N.A., and Shkvoler V.E., *Doklady AN SSSR*, 1980, vol. **250**(6), p. 1392. (Rus)
16. Mukbaniani O.V., Koyava N.A., Karchkhadze M.G., Shengelia M.G., and Khananashvili L.M., *Int. J. Appl. Polym. Mater.*, 2001, vol. **82**, p. 3142.
17. Mukbaniani O.V., Koyava N.A., Karchkhadze M.G., Shengelia M.G, and Khananashvili L.M., In: *Radical and Ion Reactions. Problems and Ways of Their Solution*, Ed. by G.E. Zaikov, 2001, Nova Science Publishers, Inc, p. 99.
18. Inaridze I.A., Doctor's Dissertation on Chemistry, Tbilisi State University, 1993, Tbilisi, Georgia.

19. Mukbaniani O.V., Inaridze I.A., and Meladze S.M., *Thes. Rep. IV Republ. Research Conference "Chemistry and Scientific Progress"*, 1988, Kutaisi, p. 64. (Rus)
20. Andrianov K.A., Slonimski G.L., Zhdanov A.A., Levin V.Yu., Tsvankin D.Ya., Kvachev Yu.P., Obolenkova E.S., Makarova N.N., Belavtseva E.M., and Mukbaniani O.V., *Vysokomol. Soedin.*, 1977, vol. **19A**(7), p. 1507. (Rus)
21. Andrianov K.A., Zhdanov A.A., Zavin B.G., and Evdokimov A.M., *Doklady AN SSSR*, 1971, vol. **199**, p. 597. (Rus)
22. Andrianov K.A., Zhdanov A.A., Zavin B.G., Evdokimov A.M., Biryukova T.V., and Lavrukhin B.D., *Vysokomol. Soedin.*, 1972, vol. **14B**, p. 337. (Rus)
23. Tsvetkov V.N., Andrianov K.A., Vitovskaya M.G., Makarova N.N., Zakharova E.N., Bushin S.B., and Lavrenko N.P., *Vysokomol. Soedin.*, 1972, vol. **14A**, p. 2369. (Rus)
24. Tsvetkov V.N., Andrianov K.A., Vinogradov E.L., Shtennikova I.N., Iakushkina S.E., and Pakhomov V.I., *J. Polym. Sci.*, 1968, vol. **C23**, p. 385.
25. Bulkina A.F., Papkova V.S., Zhdanov A.A., and Andrianov K.A., *Vysokomol. Soedin.*, 1978, vol. **20A**, p. 70. (Rus)
26. Makarova N.N., Nogaideli A.I., and Mukbaniani O.V., *Thes. Rep. I All-Union Symposium "Structure and Reactivity of Organosilicon Compounds"*, Irkutsk, 1977, p. 273. (Rus)
27. Andrianova K.A., Nogaideli A.I., Makarova N.N., and Mukbaniani O.V., *Izv. AN SSSR*, 1977, No. 6, p. 1388. (Rus)
28. Andrianov K.A., Slonimski G.L., Tsvankin D.Ya., Papkov V.S., Levin V.Yu., Kvachev Yu.P., Il'ina M.N., and Makarova N.N., *Vysokomol. Soedin.*, 1974, vol. **16B**, p. 208. (Rus)
29. Andrianova K.A., Makarova N.N., Paspopova K.N., and Mukbaniani O.V., *Doklady AN SSSR*, 1975, vol. **223**(4), p. 861. (Rus)
30. Andrianov K.A., Kononov A.M., and Tsvankin D.Ya., *Vysokomol. Soedin.*, 1968, vol. **10B**, p.320. (Rus)
31. Patnode W. and Wilcock D.F., *J. Amer. Chem. Soc.*, 1946, vol. **68**, p. 358.
32. Luyot A., Cuidard R., and Bartolin M., *J. Polym. Sci.*, 1969, vol. **C22**, p. 785.
33. Tessler M.M., *Amer. Chem. Soc., Polym. Prepr.*, 1967, vol. **8**(1), p. 152.
34. Thomas T.H. and Kendrick T.C., *J. Polym. Sci.*, 1969, vol. **2A**(7), p. 537.
35. Blazho M.B., Garzo G.I., and Zekeliy T.S., *Chromatographia*, 1972, No. 5, p. 485.

36. Andrianov K.A., Pavlova S.S., Zhuravleva I.V., Tolchinski Yu.P., Makarova N.N., and Mukbaniani O.V., *Vysokomol. Soedin.*, 1977, vol. **19A**(6), p. 1387. (Rus)
37. Toplchinski Yu.P. and Astapov B.A., *Thes. Rep. I Conference on Ageing and Stabilization of Polymers*, Tashkent, 1976, p. 86. (Rus)
38. Andrianov K.A., Pavlova S.S., Zhuravleva I.V., Tolchinski Yu.P., and Astapov B.A., *Vysokomol. Soedin.*, 1977, vol. **19A**(4), p. 896. (Rus)
39. Sobolevski M.V., Skorokhodov I.I., Ditsent V.I., Sobolevskaya A.V., and Efimov V.M., *Vysokomol. Soedin.*, 1969, vol. **11A**, p. 109. (Rus)
40. Andrianov K.A., Papkov V.S., Slonimski G.L., Zhdanov A.A., and Yakushina E.S., *Vysokomol. Soedin.*, 1969, vol. **11A**, p. 2030. (Rus)
41. Hunter M.J., Hyde J.F., Warrick E.L., and Fletcher H.J., *J. Amer. Chem. Soc.*, 1946, vol. **68**, p. 667.
42. Tsetlin B.L., Gavrilov V.I., Velikovskaya N.A., and Kochkin V.V., *Zavodskaya Laboratoria*, 1956, vol. **22**(3), p. 352. (Rus)
43. Mukbaniani O.V., Inaridze I.A., Khananashvili L.M., and Esartiya I.G., *Izv. AN GSSR*, 1993, vol. **19**(1), pp. 21 – 27. (Rus)
44. Andrianov K.A., Slonimski G.L., Levin V.Yu., Godovski Yu.K., Kuznetsova N.K., Tsvankin D.Ya, Moskalenko V.A., and Kuteinikova L.I., *Vysokomol. Soedin.*, 1970, vol. **12A**(6), p. 1268. (Rus)
45. Mukbaniani O.V., Khananashvili L.M., Karchkhadze M.G., Tkeshelashvili R.Sh., and Mukbaniani N.O., *Intern. J. Polym. Mater.*, 1996, vol. **33**, p. 47.

Chapter 7. Synthesis of organosilicon cyclolinear block-copolymers

The possibility of realizing novel technical solutions in many branches of industry is always defined by thermal and frost resistance of modern elastic materials. That is why development of methods for obtaining new polymers for composites with improved resistance to destabilizing influence of high and low temperatures is always relevant.

One of the most interesting ways in this direction is synthesis of block-copolymers, which macromolecules represent "hybrids" of blocks different by chemical structure or composition. In the majority of cases, thermodynamic incompatibility of blocks induces stable microphase splitting, which, finally, allows an original combination of properties of heterogeneous fragments of block-copolymer macromolecules (BCP). Depending on difference in the chemical origin of blocks, their length, number and alternation sequence, as well as their ability to crystallize, materials with the structure and properties, significantly different from the properties of initial components, may be synthesized.

A combination of flexible poly(dimethylsiloxane) blocks with more rigid high-melting siloxane (for example, diphenylsiloxane) or rigid non-siloxane blocks in the polymer molecule imparts high strength and other valuable properties to non-vulcanized copolymer. Of special interest are block-copolymers (BCP), composed of blocks with different physical parameters: glass transition temperature, ability to crystallize, solubility parameters, etc.

Poly(dimethylsiloxane) blocks possessing glass transition temperature equal -123°C and dielectric constant equal 28 impart high elasticity at low temperatures, good dielectric properties, high resistance to ozone-induced aging, durability and good gas permeability to BCP. The requirements to the rigid block are dictated by the necessity of increasing mechanical strength (in some cases, thermal stability also) to BCP. Mechanical properties of BCP generally depend on both length of flexible and rigid blocks and their ratio.

The recent main attention of investigators in the branch of physical chemistry and technology of high-molecular compounds is attracted by BCP of A_mB_n and $(AB)_mC_n$ types. Besides BCP containing cyclic and dimethylsiloxane fragments, the ones with other different inclusions are considered in this Section.

7.1. Block-copolymers with poly(phenylsilsesquioxane) fragments in the dimethylsiloxane backbone

Representatives of rigid-chain polymers are poly(phenylsilsesquioxanes) (PPSSO). The unique properties of these polymers (high heat and thermal resistance, photo- and thermal oxidative resistance) have made them one of the perspective polymeric materials.

It is known that properties of polymers are generally dependent on the structure of macromolecules and molecular parameters, which, in their turn, depend on the synthesis reaction conditions. At present, there is a high scope of material, accumulated in the literature, indicating that PPSSO properties are induced by the synthesis conditions [1, 2].

That is why the study of reasons affecting formation of the PPSSO macromolecule structure during their synthesis and BCP derived from them is quite up-to-date.

7.1.1. Synthesis of poly(phenylsilsesquioxanes) with functional groups

High interest to synthesis and study of poly(organosilsesquioxanes) is caused by good thermal properties of synthesized polymers and specific features of organosilsesquioxane polymers to combine regular linear structures using tri- and tetrafunctional compounds as monomers.

The interest to polycondensation of *cis*-1,3,5,7-tetrahydroxy-1,3,5,7-tetraphenylcyclotetrasiloxane (tetrole T_4) is cause by the modern multistage proceeding of the synthesis of organosilsesquioxane polymers, based on phenylchlorosilane: hydrolysis, polycondensation, and polymerization. Such technique for obtaining poly(organosilsesquioxanes) with functional groups concludes in restructuring of siloxane bonds in the macromolecule and, accordingly, obtaining of the most profitable conformation by them.

Hydrolytic condensation of phenyltrichlorosilane gives polycyclic phenylsilsesquioxanes of comparatively low molecular mass, which contain from 3 to 10% of residual OH-groups. Anionic polymerization of hydrolysis products proceeds at injection of an alkaline initiator into the system, and condensation by residual silanemonol groups with water extraction and partial restructuring of the siloxane skeleton of primarily formed oligomers proceed simultaneously with polymerization.

To clear up the reasons for PPSSO macromolecules formation, a stage-by-stage study of PPSSO chain formation was executed [3].

The full hydrolysis product of 1,3-diphenyltetrachlorosiloxane (PSSO-H) was chosen as the study object. This monomer is selected, because raw stocks from disiloxane should induce formation of regulated ladder structures [4]. The results of hydrodynamic and conformation studies were compared with similar data for PPSSO, obtained during 6 minutes of the reaction (PPSSO-C), as well as for high-molecular PPSSO, synthesized during several hours [5].

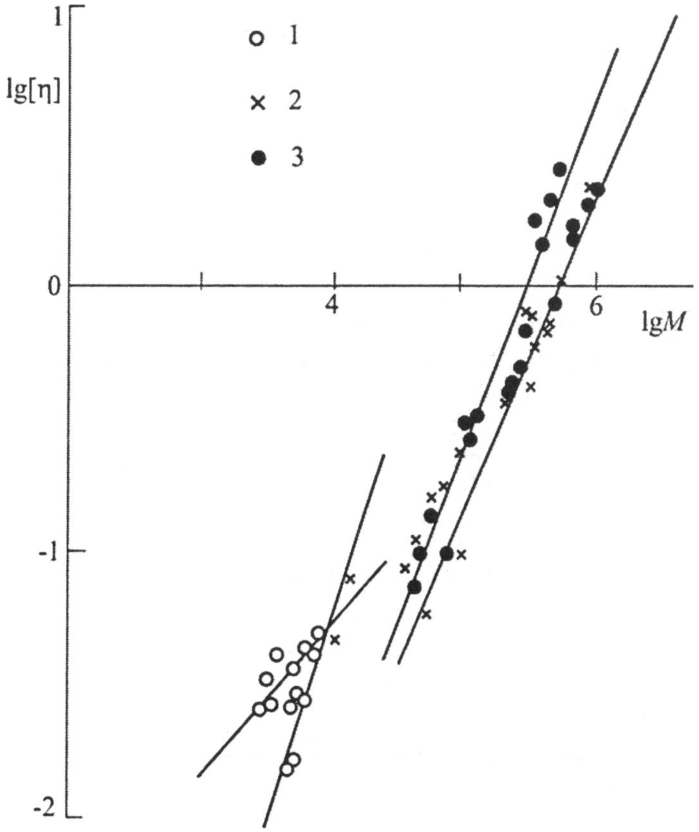

Figure 7.1. Dependence of lg[η] on lg$M$ for PSSO-H (1), PSSO-C (2) and PPSSO (3) [6]

The studies have shown that experimental data on fractions of every sample of PSSO-H and PPSSO-C (Figure 7.1) are subdivided into two groups. Appropriate points fit two straight lines with different α и K_υ parameters. This testifies about the presence of, at least, two types of macromolecule structures in the samples. The following Mark-Kuhn-Hauvink equations correspond to these polymers:

PSSO-H	$[\eta] = 2.89\times10^{-5}\, M^{0.58}$	(7.1)
	$[\eta] = 3.13\times10^{-9}\, M^{1.85}$	(7.2)
PSSO-C	$[\eta] = 1.09\times10^{-7}\cdot M^{1.48}$	(7.3)
	$[\eta] = 2.24\times10^{-7}\cdot M^{1.16}$	(7.4)

Probably, the exponent value $\alpha > 1.7$ indicates that the structure of macromolecules in the current sequence of fractions varies with increase of molecular mass. The plot in Figure 7.1 shows that experimental data on lg$[\eta]$ and lg$M$ for the fractions of PPSSO-C and final products, i.e. the product of anionic polymerization, fall on the same straight lines.

For PSSO-H and PPSSO-C, high values of α parameters allowed assumption of high molecular rigidity of these polymers. Based on experimental hydrodynamic parameters, equilibrium rigidity of polymeric chain was quantitatively estimated in accordance with the simulation of a persistent chain as the most universal one in a wide range of molecular masses. For the persistent chain model various theoretical equations, which allow obtaining of quantitative characteristics for rigidity of macromolecular chain, have been suggested. Because the exponent is $\alpha > 1$ in equations (2) – (4), the known relations [7] for models of persistent chains and Gaussian coils cannot be applied to estimation of the rigidity parameter. This may be explained by the fact that the range of molecular mass variation of the studied samples is, probably, related to the area of low-bended rod conformations, which is known to be wider for rigid-chain polymers than for flexible-chain ones and shifted towards high-molecular area.

The Yamakawa-Fuji equation, modified by Bogdanetskii [8], gave satisfactory results:

$$(M^{1/2}[\eta])^{1/2} = [\Phi_0(<h^2>/M^{3/2})]^{1/2}[1 + b(<h^2>/M)^{1/2}M_L/M^{1/2}].$$

In future studies, the linear shape of the Bogdanetskii dependence in the range of low molecular masses, which is especially difficultly extrapolated, allowed successful application of this equation to comparison of PPSSO chain rigidity in a broad range of molecular masses (Figure 7.2). Experimental data [4] were also treated with the help of the Bogdanetskii equation.

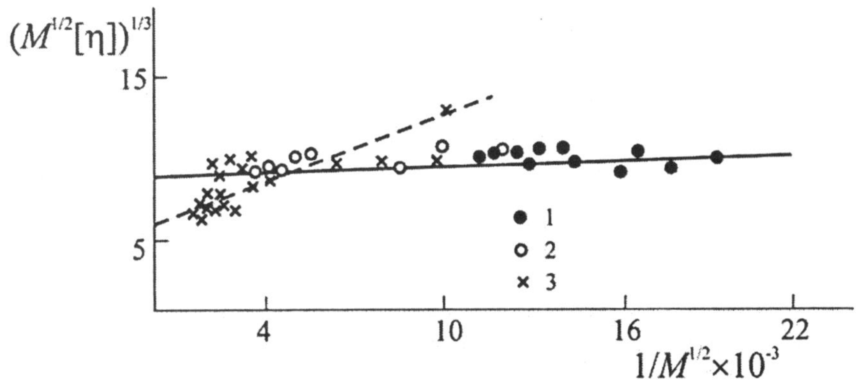

Figure 7.2. Extrapolation by the Bogdanetskii equation for PSSO-H (1), PSSO-C (2) and PPSSO (3)

Experiments performed and calculations of hydrodynamic parameters executed indicate two types of molecules with considerably different rigidity parameters (Table 7.1), formed during hydrolytic condensation of 1,3-phenyltetrachlorodisiloxane and polycondensation and anionic polymerization of its hydrolysis products.

Table 7.1

PPSSO rigidity parameters for persistent simulation of chain molecule, calculated by the Bogdanetskii equation

No.	Polymer	$(1/k_\theta)^{1/3}$	$\langle h^2\rangle/M\times10^{-16}$	*A*, Å
1	PSSO-H	10.3 2.6	0.5 7.7	50 800
2	PSSO-C	11.5 4.7	0.4 2.4	40 250
3	PPSSO	11.4 4.7	0.4 2.3	40 240
4	PPSSO-L	10.5	0.5	50

The data from the Table show that besides hydrolytic condensation, macromolecules with the Kuhn statistical segment equal 50 or even 250 Å are formed at all stages of PPSSO synthesis. It is probable that their ratio is the reason stipulating differences in properties of synthesized PPSSO. It is improbable that the defects, caused by incomplete condensation by OH-groups are responsible for the differences in rigidity parameters, because the synthesis proceeds under homogeneous conditions, and such "defect" would be statistically distributed.

Poly(phenylsilsesquioxane) (PPSSO-L), obtained by *cis*-1,3,5,7-tetrahydroxy-1,3,5,7-tetraphenylcyclotetrasiloxane (*cis*-tetrole, T_4) polycondensation in the absence of alkaline initiator (i.e. without ring disclosure), was applied as the standard compound to estimation of PPSSO molecule structure. The studies have shown that PPSSO-L contains molecules of a single type (Figure 7.3), the equilibrium rigidity of which, calculated by the Bogdanetskii equation, is comparable with that of one of two types of PPSSO molecules, synthesized by anionic polymerization of phenyltrichlorosilane hydrolysis products in the presence of the base-type initiator (Table 7.2).

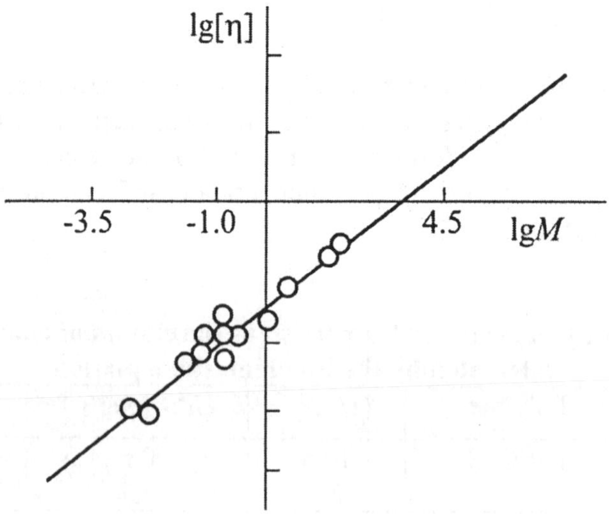

Figure 7.3. Dependence of lg[η] on lg$M$ for PPSSO-L [9]

Analysis of a possibility of forming cyclolinear PPSSO and application of Steward-Briegleb simulations indicate that PPSSO synthesis may give the following types of regular cyclolinear structures of macromolecules: *cis*-syndiotactic, *cis*-isotactic, *cis*-anti-tactic and a mixed one, characterized by alternation of the former conformations [10, 11].

These conformations may be composed using a fragment of T_4 chain as the repeating unit.

Table 7.2

PPSSO rigidity parameters, calculated for persistent simulation of the chain molecule by the Bogdanetskii equation

No.	Polymer	$(1/K_\theta)^{1/2}$	$<h^2>/M \times 10^{-16}$	A, Å
1	PSSO-H	10.3 2.6	0.5 7.7	50 800
2	PPSSO-C	11.5 4.7	0.4 2.4	40 250
3	PPSSO	11.4 4.7	0.4 2.3	40 240
4	PPSSO-L	10.5	0.5	50

A combination of T_4 fragments with *trans*-disposition of functional groups may give formation of *cis*-syndiotactic chain structure as follows:

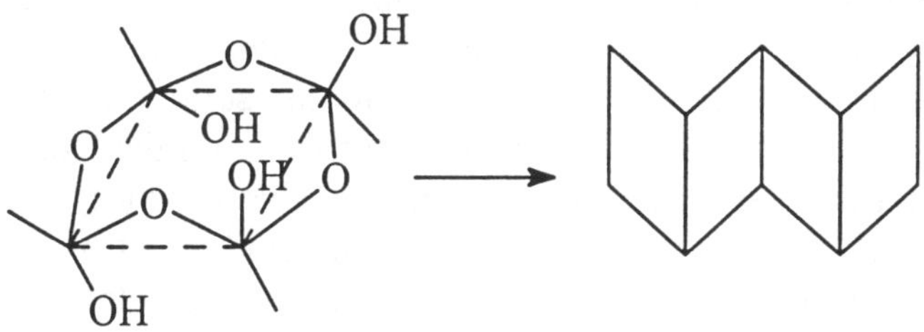

trans-T_4-coach

cis-syndiotactic configuration

Such composition of the chain is sterically more profitable and is characterized by increased rigidity. For defect-free *cis*-syndiotactic structure the theoretical Kuhn segment equals ~1,000 Å. In the present case of hydrolytic condensation, one of the structures possesses the Kuhn segment value ~800 Å. Thus the *cis*-syndiotactic chain structure may be assigned to PPSSO macromolecules possessing such equilibrium rigidity.

As T_4 fragments with *cis*-disposition of functional groups interact, PPSSO with *cis*-isotactic conformation can be formed, if atoms of silicon in tetrole molecule are located in the same plane, and with *cis*-anti-*cis*-tactic conformation, if T_4 cycles are located in parallel planes.

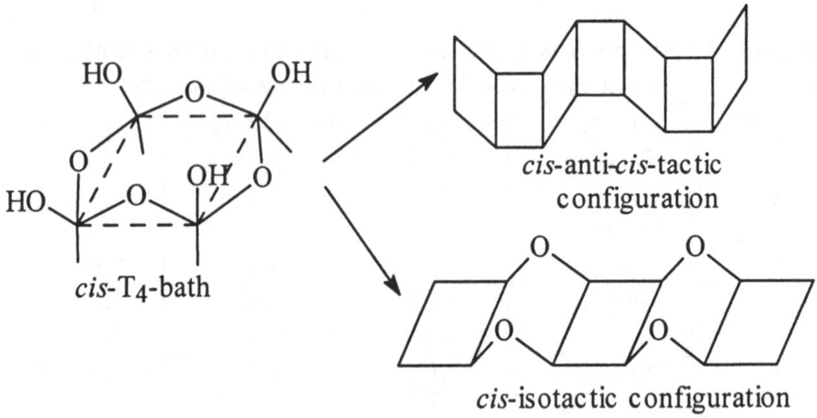

However, realization of the *cis*-isotactic structure is associated with several steric hindrances and characterized by a short length of molecules. During tetrole polycondensation in the absence of initiator of the main type, which allowed obtaining of a polymer without siloxane bond break in organocyclosiloxane, the synthesis conditions somewhat design a perfect cyclolinear ladder structure, and macromolecules with the Kuhn segment equal ~50 Å are formed. Spectroscopic studies [9] give grounds to suggest that the structure of PPSSO-L macromolecules, synthesized by tetrole T_4 condensation, differs from that of PPSSO, synthesized from phenyltrichlorosilane products by anionic polymerization in the presence of initiator of the main type. Experimental studies allowed an assumption that macromolecules with the Kuhn segment equal ~50 Å possess the structure close to the *cis*-anti-*cis*-tactic one. Statistical Kuhn segment value equal ~250 Å should correspond to macromolecules of the mixed structure with predominance of *cis*-syndiotactic conformation fragments in the macromolecule backbone.

Thus, ambiguous structure of PPSSO molecules is designed at the stage of hydrolytic condensation of phenyltrichlorosilane and products of its partial hydrolysis. Correlation between experimental data and parameters of high-molecular PPSSO [5], treated by the Bogdanetskii equation (Table 7.1), allowed a conclusion that formation of silsesquioxane-type macromolecule structure with high equilibrium rigidity is mainly affected by macromolecule formation processes, namely, condensation or anionic polymerization. The studies indicate that polycondensation of tetrole T_4 in the absence of an initiator of the alkaline type produces comparatively low-molecular compounds with probable *cis*-anti-*cis*-tactic molecular structure and the Kuhn segment equal ~50 Å. In the case of anionic polymerization, besides molecules with low equilibrium rigidity, thermodynamically profitable macromolecules with the equilibrium structure and high rigidity parameters are produced. Such macromolecules are formed during restructuring of molecules, caused by initiator of the main type.

For the purpose of synthesis of oligotetroles with *cis*-anti-*cis*-tactic structure of molecules, homofunctional polycondensation of *cis*-1,3,5,7-tetrahydroxy-1,3,5,7-tetraphenylcyclotetrasiloxane (tetrole) was performed both in melt and in solution and without a catalyst or in the presence of various quantities of activated coal [12 – 14]. Depending on the condensation time, phenylsilsesquioxane oligomers of the ladder structure (oligotetroles) with different condensation degree were synthesized:

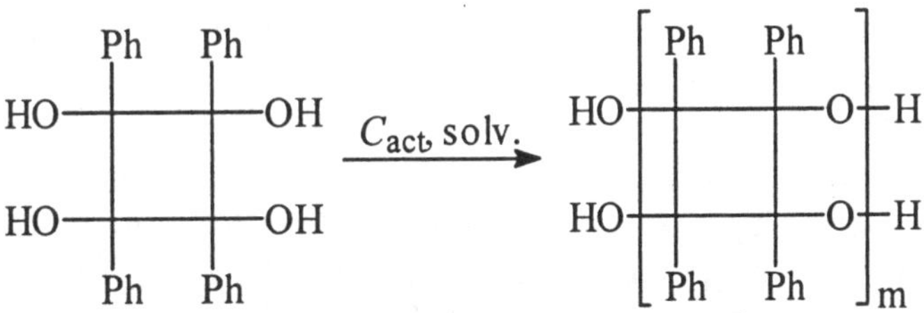

I

(7.1)

where m = 3 - 46. Hereinafter, *cis*-1,3,5,7-tetrahydroxy-1,3,5,7-tetraphenylcyclotetrasiloxane will be denoted as follows:

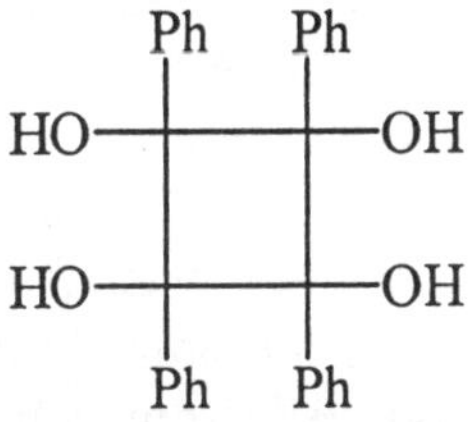

It is found that homofunctional condensation of tetrole (T_4) with *cis*-disposition of hydroxyl groups in toluene, both in the presence of "non-equilibrating" catalyst, activated coal, and without it produces poly(phenylsilsesquioxanes) (PPSSO) of the *cis*-isotactic conformation, i.e. when silicon atoms in tetrole are located in the same plane, and *cis*-anti-*cis*-tactic structure, i.e. when T_4 cycles are disposed in parallel planes. The presence of incomplete cyclization structures in obtained products is not excluded [12 –

14]. In the case of reaction (7.1) proceeding in solution, quantity of these structures is reduced to the minimum.

However, it is indicated [3] that realization of *cis*-isotactic structure is associated with several steric hindrances and characterized by short chain length, and such macromolecules are produced in negligible amounts.

In the range of asymmetrical valence oscillations at 1,000-1,200 cm^{-1}, IR-spectra of synthesized oligotetroles, produced both in the presence and in the absence of activated coal, display a broad absorption band with a single clear maximum at 1,140-1,145 cm^{-1}, whereas IR-spectra of PPSSO, produced by anionic polymerization of phenyltrichlorosilane hydrolysis products, always displays bifurcation with the maximums at ν_{as} – 1,045 cm^{-1} and ν_{as} – 1,145 cm^{-1}. As a consequence, it has been shown that homofunctional condensation of *cis*-tetrole in the presence of activated coal proceeds without changes in the siloxane skeleton.

For the purpose of studying homofunctional condensation of tetrole with higher boiling and polar medium than toluene, dimethylformamide was used as a solvent. The use of polar solvent leads to PPSSO formation with *cis*-syndiotactic conformation (compound 7, Table 7.3), which is proved by IR-spectrum of produced PPSSO. For asymmetrical and symmetrical valence oscillations of ≡Si-O-Si≡ bond in oligotetrole 7, bifurcation of bands with the maximums at ν_{as} – 1,045 and 1,145 cm^{-1} and ν_{as} – 485 and 505 cm^{-1} was observed, which corresponds to the literary data [15].

As a consequence, one may assume that oligotetrole 8 (Table 7.3) possesses the structure close to PPSSO produced by anionic polymerization. In turn, this allows a conclusion that the presence of nucleophilic catalysts not only promotes formation of the active center zone of propagating polymer, but are themselves equilibrating catalysts and promote a change in configuration of cyclotetrasiloxane skeleton. Similar structure is displayed by low polymerized oligotetrole 9 (Table 7.3), produced from tetrole mother solution. Tetrole mother solution in benzene was washed out, dried above anhydrous sodium sulfate, after which benzene was removed and the product vacuumed.

As a result, oligotetrole 9 with the transformation degree m = 4 (Table 7.3) and *cis*-syndiotactic molecular structure was obtained.

All synthesized oligotetroles represent glassy-like transparent solids, well soluble in ordinary organic solvents. Data from Table 7.3 show that the polycondensation degree, m, and oligotetrole melting temperature increase with percent concentration of activated coal.

Table 7.3

Physical and chemical parameters of tetrahydroxyoligophenylsilsesquioxanes I

Oligomer No.	m	C_{act}, %	Solvent	Time, min	η_{spec}*	OH, %		M		T_{melt}, °C
						Exper.	Calc.	Exper.	Calc.	
1	3	-	toluene	20	0.03	4.5	4.3	1,500	1,576	128-135
2	5	-	toluene	120	0.03	2.62	2.59	2,700	2,616	152-158
3	10	-	toluene	240	0.03	1.24	1.25	5,440	5,203	210-215
4	15	-	toluene	30	0.04	0.79	0.87	7,880	7,776	220-225
5	20	7	toluene	270	0.04	0.63	0.66	10,200	10,371	226-230
6	37	15	toluene	270	0.04	0.36	0.35	19,000	19,156	250-255
7	46	20	toluene	270	0.07	0.32	0.28	23,700	23,772	-
8	42	-	DMF	270	0.06	0.29	0.31	21,500	21,708	232-237
9**	4	-	-	-	0.03	3.33	3.23	2,025	2,100	148-154

Note: * Hereinafter, specific viscosity of 1% solution in toluene was determined at 20°C.
** Oligotetrole is produced from tetrole mother solution.

Difunctional phenylsiloxanes of the ladder structure with different disposition of functional groups in polycyclic fragment were derived from produced oligotetroles.

Difunctional PPSSO were synthesized by partial interlocking of hydroxyl groups in oligotetroles, produced in reactions (7.1). Partial interlocking was performed by two reactions: 1) oligotetrole HFC with different transformation degrees, *m*, with 1,3-dichlorotetraorganodisiloxanes in 5% anhydrous benzene or toluene solution at 0 ÷ -5°C and 1:1 ratio of initial components, in the presence of pyridine in accordance with the following scheme [13, 16, 17]:

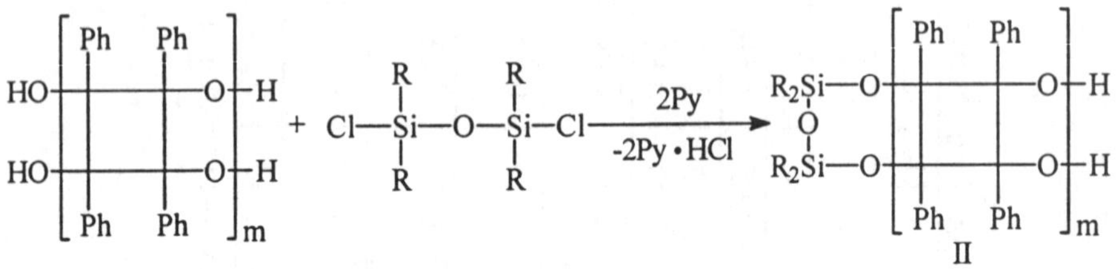

(7.2)

where R = Me, Ph; *m* = 3 - 37.

The reaction completeness was judged by a change of hydroxyl groups and the ratio of methyl and phenyl groups' protons in NMR spectra. In the cases of *m* = 3 and 10, ratios of methyl and phenyl protons found coincide with the calculation.

Table 7.4

Physical and chemical parameters of structure II dihydroxy-poly(phenylsilsesquioxanes)

Oligomer No.	R	*m*	η_{spec}	T_{melt}, °C	OH, %		*M*	
					Exper.	Calc.	Exper.	Calc.
1	Me	3	0.03	210-215	2.01	1.98	1,680	1,714
2	Me	10	0.03	220-225	0.64	0.63	5,412	5,333
3	Me	37	0.04	314-322	0.16	0.17	19,460	10,286
4	Ph	10	0.03	225-230	0.56	0.60	5,623	5,582
5	Ph	37	0.04	320-325	0.17	0.17	19,300	19,534

If the ratio of initial components is disturbed, compounds completely interlocked by functional groups may be produced at partial interlocking [18]. However, the presence of these compounds does not prohibit application of the

basic product with structure II to further polycondensation. Physical and chemical parameters of synthesized dihydroxy-PPSSO are shown in Table 7.4.
2) HFC of oligotetrole with trimethylchlorosilane in anhydrous diluted toluene solution (5 – 7%) at 0 ÷ -5°C and 1:2 ratio of initial reagents, in the presence of pyridine. The reaction apparently proceeds in two directions by the following scheme [19, 20]:

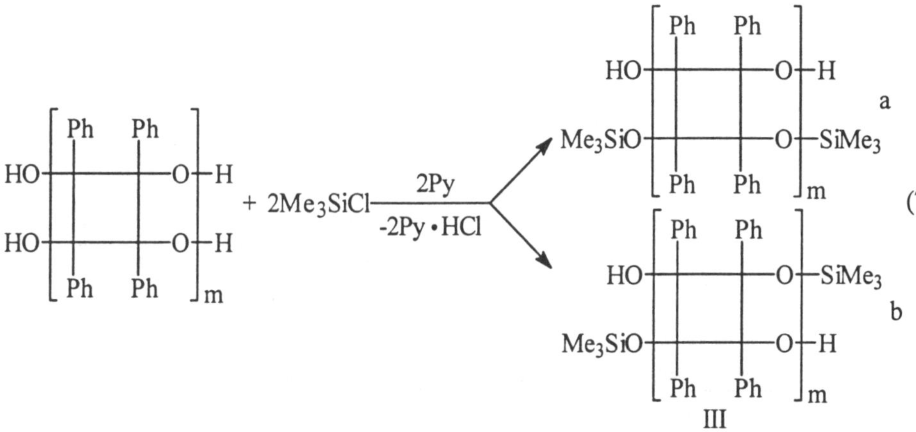

(7.3)

where for structure III (a and b) $m = 3 - 20$.

Formation of a mixture of products IIIa and IIIb in the reaction is also probable. At the same time, HFC proceeding by the following scheme:

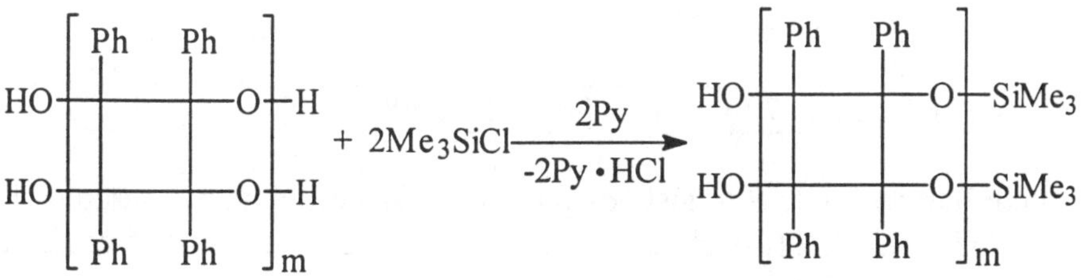

(7.4)

is unlikely due to the screening effect of neighboring trimethylsiloxy-groups, which correlate well with the published data [21]. Physical and chemical parameters of PPSSO, interlocked by trimethylsiloxy-groups, are shown in Table 7.5. ^{1}H NMR spectra of structure III oligomers display coincidence of experimental and theoretical values for the ratios of methyl and phenyl protons.

Synthesized di- and tetrafunctional phenylsilsesquioxane oligomers are valuable products for synthesis of block-copolymers of the A_mB_n type.

Oligotetrole 4 (Table 7.3) with *cis*-anti-*cis*-tactic molecular structure and oligomer 2 (Table 7.5), partly interlocked by trimethylsiloxy-groups, were tested as binders of fiberglass plastics. The mentioned oligomers were applied to creation of fiberglass laminates using fabric of TS-8/3 VM-TO trademark. Tests were executed on a dynostate. Sample dimensions were 55×15×2.2 mm. Table 7.6 shows the tests results obtained.

Table 7.5

Physical and chemical parameters of structure III dihydroxy-poly(phenylsilsesquioxanes)

No.	m	η_{spec}	T_{melt}, °C	OH, %		M	
				Exper.	Calc.	Exper.	Calc.
1	3	0.03	215-220	2.05	1.94	1,700	1,752
2	10	0.04	210-220	0.76	0.78	4,400	4,314
3	20	0.04	230-235	0.55	0.63	5,200	5,340

Table 7.6

Strength parameters of fiberglass laminates

No.	Binder type	Binder content, %	Initial bending σ, kgf/cm^2	Bending σ after aging at 350°C during 1,000 h, kgf/cm^2
1	Poly(phenylsiloxane)	30	1,910	1,280
2	Poly(phenylvinylsiloxane)	30	2,900	630
3	Oligotetrole-4 (Table 7.3)	30	2,000	900-1,000
4	Oligomer-2 (Table 7.5)	30	2,500-2,900	1,100-1,300

According to Table 7.6, tests of poly(phenylsilsesquioxane) samples in the material indicated that fiberglass laminates, derived from them, possess high strength and are of real interest for creation of new prospective materials.

7.1.2. Phenylsilsesquioxane-dimethylsiloxane copolymers. Block copolymers of A_mB_n type

For a long time, the basic directions of activities aimed at improvement of thermal resistance of siloxane elastomers were associated with the introduction of regularly distributed groups in a varying degree inhibiting processes, which induce operational properties' loss by materials, i.e. creation of statistical copolymers, into the backbone. In this direction, considerable successful results were obtained, which made the modern stock of thermally stable rubbers mainly consisting of statistical polymers. However, a significant disadvantage is typical of statistical polymers restricting the range of heat and frost resistance, which is deterioration of low-temperature properties with the increase of concentration of groups (mostly of the cyclic type) in macromolecules, which increase heat resistance of the rubber.

On the other hand, recently, from positions of heat and frost resistance the prospectivity of block-copolymers of microheterophase structure containing rigid blocks with high "glass transition temperature" and "flexible" low glass transiting diorganosiloxane ones was indicated. However, processing from solution is the unique method of producing articles from the known block-copolymers of this type because of high temperature of rigid block softening.

That is why plasticization of ladder polymers with linear dimethylsiloxane units is a quite relevant process.

Discussed in this Section is HFC of difunctional PPSSO with α,ω–dichloro[(bisdiethylamino)]dimethylsiloxanes in order to modify ladder polymers and obtain block-copolymers of the A_mB_n type.

For the purpose of synthesizing block-copolymers with PPSSO fragments in the framing and along the backbone, HFC of dihydroxy-containing oligotetroles of structures II and III with α,ω-dichlorodimethylsiloxanes and dimethyldichlorosilane in 60 - 70% anhydrous toluene solution in the presence of pyridine and at 1:1 ratio of initial reagents and 20 - 25°C was studied. At the final stage, the reaction mixture was boiled during 4 – 5 hours. HFC reaction proceeds by the following scheme [19 – 22]:

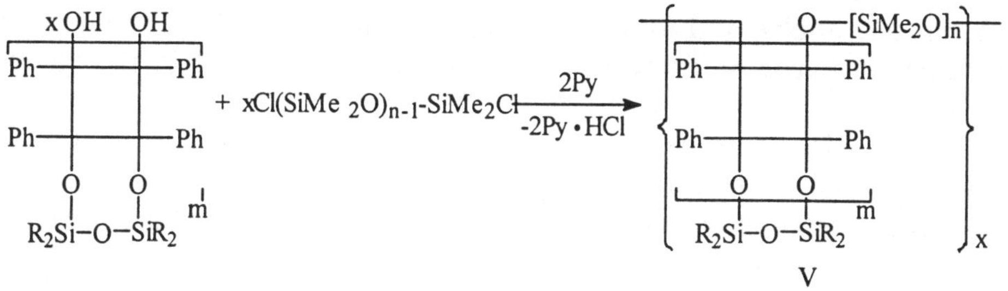

(7.5)

where R = Me, Ph; m = 3 - 37; n = 2 - 108.

Similarly, the HFC reaction of dihydroxy-containing oligotetroles of structure III with α,ω-dichlorodimethylsiloxanes proceeds (refer to the scheme below):

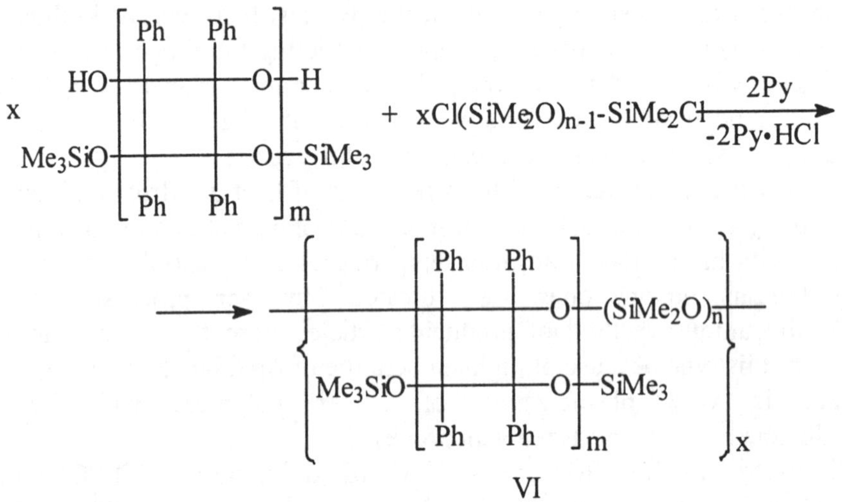

(7.6)

where R = Me, Ph; m = 3 - 37; n = 2 - 35.

Re-precipitation of copolymers from toluene solution with methanol has produced cyclolinear block-copolymers representing white powder-like products, well soluble in various organic solvents, with η_{spec} = 0.05 – 0.27. Some physical and chemical parameters and yields of synthesized block-copolymers are shown in Tables 7.7 and 7.8.

In the range of 1,000 – 1,200 cm^{-1} typical of valence asymmetrical oscillations of ≡Si-O-Si≡ groups, IR spectra of synthesized block-copolymers display a broad absorption band with clear single maximum at 1,145 cm^{-1}, analogous to that observed for oligotetroles (refer to Section 7.2). This proves that *cis*-anti-*cis*-tactic structure of PPSSO blocks is preserved in the block-copolymers.

For purified non-fractionated copolymer (structure V, copolymer 12, Table 7.7) hydrodynamic parameters and sedimentation index are determined as follows: $[\eta]$ = 0.10 dl/g, V_p = 0.764 cm^3/g, $\bar{M}_{SD}$ = 43,000, D = 11 ± 1, and S = 6.7×10^{-7} g, respectively.

Comparative study of thermal oxidative stability of synthesized block-copolymers indicates that the effect of disposition of PPSSO fragments of the same length in the dimethylsiloxane backbone on thermal stability is negligible.

Figure 7.4 shows that block-copolymers (1 and 3, Table 7.7) containing PPSSO fragments in the framing and along the dimethylsiloxane backbone (1 and 4, Table 7.8) display similar thermal oxidative stability. Moreover, it is found that the effect of PPSSO fragment length, *m*, increase in the dimethylsiloxane backbone on thermal oxidative stability of block-copolymers is also negligible (Figure 7.5).

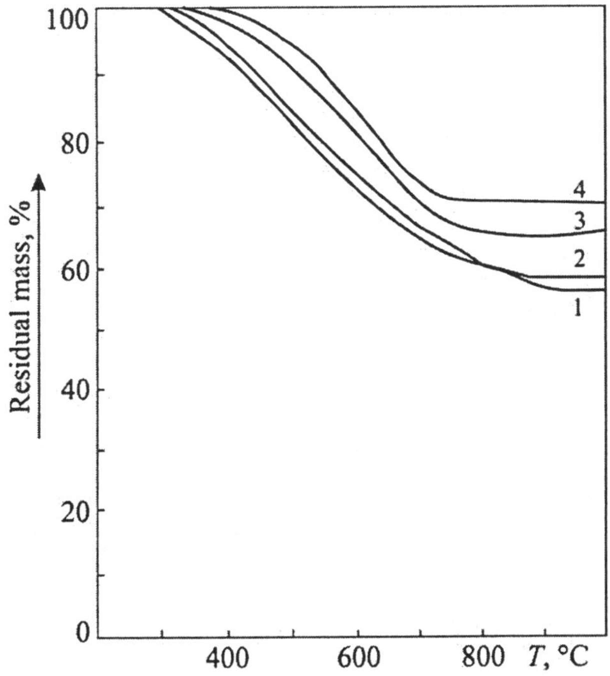

Figure 7.4. Thermogravimetric curves for block-copolymers: 1 – copolymer 1 (Table 7.8); 2 – copolymer 1 (Table 7.7); 3 – copolymer 4 (Table 7.8); 4 – copolymer 4 (Table 7.7). (Tests were performed in air, at the heating rate $v = 5$ deg/min)

Thermogravimetric studies of structure V copolymers were carried out on electron thermal weights B-60, Seteram Co., in argon, at 5 deg/min heating rate, with simultaneous sampling and analysis of gaseous degradation products [25]. Figure 7.5 shows that mass losses regularly increase with the length of dimethylsiloxane unit in the chain. Mass losses of copolymers are decreased with transition from tricyclic to polycyclic fragments (structure VI, $m = 10$, R = Me). However, it is apparent that after definite values of polycyclic fragments (*m*), the length increase of a rigid polycyclic fragment (*m*) is a negligible factor affecting thermal oxidative stability of the copolymers (Figure 7.5).

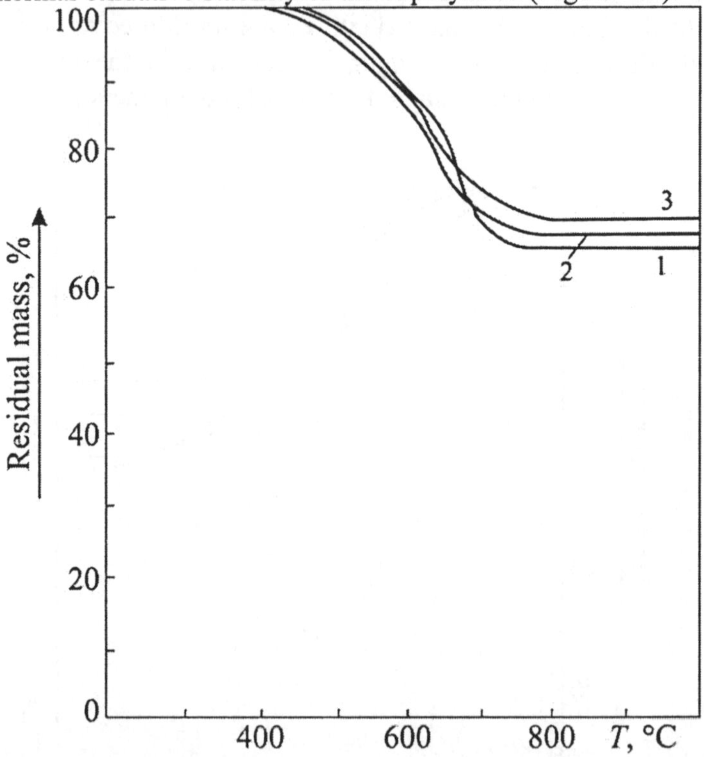

Figure 7.5. Thermogravimetric curves for block-copolymers: 1 – copolymer 1 (Table 7.8); 2 – copolymer 6 (Table 7.8); 3 – copolymer 12 (Table 7.8). Tests proceeded in air, at the heating rate $v = 5$ deg/min

Data on analysis of structure V block-copolymer (m = 10, R = Me; Figure 7.5) degradation products show that no compounds of D_n composition are formed during copolymer 6 (Table 7.7) degradation. Liberation of gaseous degradation products (H_2 and CH_4) is increased with the number of dimethylsiloxane units in the chain, the amount of methane being increased first. According to ref. [26], liberation of hydrogen and methane is associated with radical cross-linking of methyl and phenyl groups. Hydrogen is liberated above 400°C. Besides hydrogen and methane, benzene is also liberated during degradation.

Table 7.7

Some physical and chemical parameters of cyclolinear organosiloxane block-copolymers of structure V

No.	Polymer unit	n	m	R	η_{spec}	d_1, Å (d_1^1)	T_{soft}, °C	Yield, %
1	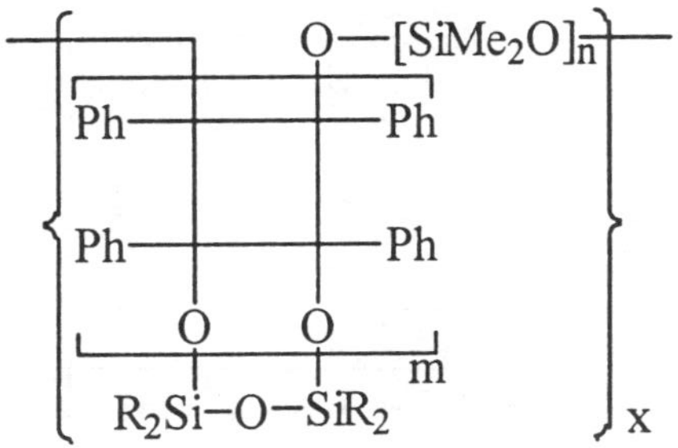	2	3	Me	0.07	-	+160	93
2		4	3	Me	0.07	12.89	+129	91
3		8	3	Me	0.13	-	-	91
4		11	3	Me	0.14	-	-	94
5		35	3	Me	0.24	12.30 (7.71)	-	96
6		1	10	Me	0.07	13.50	+170	92
7		2	10	Me	0.08	-	+173	90
8		4	10	Me	0.08	13.50	+146	96
9		11	10	Me	0.12	-	-	95
10		55	10	Me	0.12	12.54 (7.49)	+132	92
11		108	10	Me	0.13	-	-	94
12		1	10	Ph	0.06	-	212	96
13		2	10	Ph	0.05	13.51	+150	94
14		4	10	Ph	0.08	-	+130	93
15		11	10	Ph	0.25	-	-	94
16		55	10	Ph	0.15	12.58 (7.45)	-	96

17	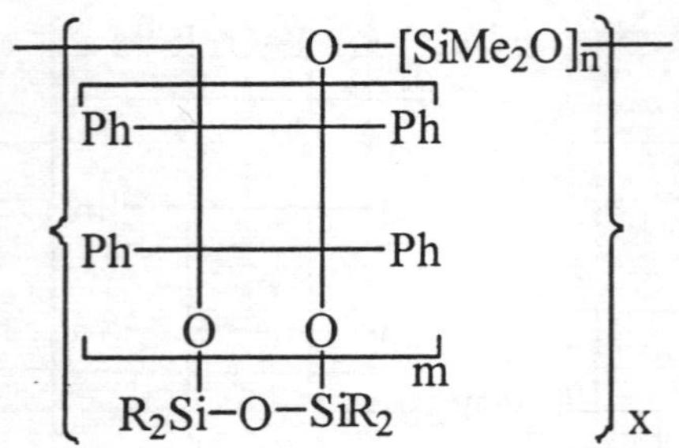	108	10	Ph	0.18	-	-	93
18		2	37	Me	0.06	13.52	+180	93
19		4	37	Me	0.06	-	+180	96
20		34	37	Me	0.08	-	+190	93
21		55	37	Me	0.09	12.78 (7.60)	+137	93
22		108	37	Me	0.12	-	-	94
23		2	37	Ph	0.05	13.54	+150	93
24		4	37	Ph	0.06	-	+212	95
25		34	37	Ph	0.07	-	-	91
26		55	37	Ph	0.09	12.81 (7.70)	+135	94
27		108	37	Ph	0.11	-	+190	95

Table 7.8

Some physical and chemical parameters of cyclolinear organosiloxane block-copolymers of structure VI

No.	Polymer unit	n	m	η_{spec}	d_1, Å (d_1^1)	T_{soft}, °C	Yield, %
1	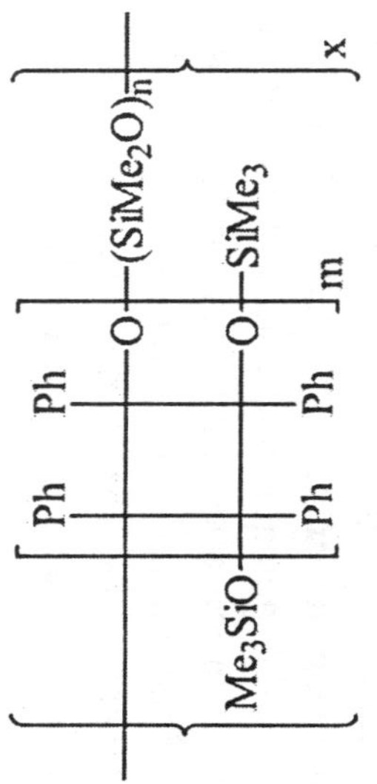	2	3	0.05	12.56	+163	95
2		4	3	0.07	-	+158	94
3		8	3	0.10	12.50	+156	92
4		11	3	0.11	-	-	94
5		35	3	7.27	11.90 (7.50)	-	96
6		2	10	0.06	13.10	+173	93
7		4	10	0.06	-	-	94
8		8	10	0.07	-	+160	95
9		11	10	0.11	-	-	95
10		35	10	0.20	12.40 (8.0)	-	95
11		1	20	0.06	-	-	95
12		2	20	0.06	13.20	+188	93
13		4	20	0.06	-	+180	93
14		8	20	0.07	-	+175	95
15		11	20	0.10	-	-	93
16		35	20	0.19	12.60 (8.10)	-	94

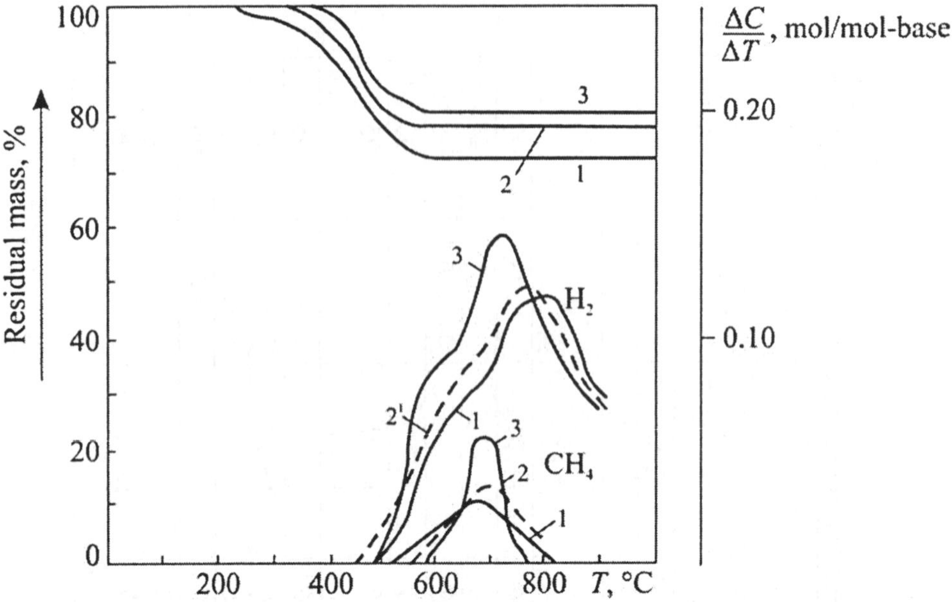

Figure 7.6. Thermogravimetric curves for block-copolymers: 1 – copolymer 7; 2 – copolymer 8; 3 – copolymer 9 (Table 7.7). Tests were performed in argon, at the heating rate v = 5 deg/min

Thus, based on the results obtained, the following scheme of structure V block-copolymer thermal degradation was suggested: condensation and structuring by residual hydroxyl groups proceed at copolymer heating up to 350°C, which are accompanied by liberation of gaseous products. Due to introduction of bulky ladder tabs into the linear siloxane chain, the main degradation process is shifted by 100 - 120°C towards higher temperatures compared with linear non-interlocked PDMS [27 – 29]. Thermal decomposition itself proceeds by the radical mechanism with formation of oligomeric products, but not by the chain ion one with formation of cyclosiloxanes. Simultaneously with the backbone degradation, structuring by methyl and phenyl groups proceeds, accompanied by liberation of hydrogen, methane and benzene. These processes decelerate backbone degradation, and at definite temperatures terminate it completely.

Existence of defect structures in obtained block-copolymers is also assumed possible [20, 22, 24]. Such structural defects are caused by incomplete cyclization of initial polycyclic block in cases of $m \geq 10$ and, to the authors' point of view, they should not deteriorate thermal characteristics of studied block-copolymers.

Thermomechanical studies of synthesized block-copolymers were performed. The glass transition temperature (T_g) was determined by initial deviation of thermomechanical curves in the area of positive deformations. Data from Tables 7.7 and 7.8 show high softening temperatures (T_{soft}) of block-copolymers. At the same time, the plasticizing effect is obvious, because T_{soft} of copolymers with polycyclic fragments in the backbone is lower than that of PPSSO ($T_{soft} = 410°C$).

It is found that the optimal softening temperature is attributed to copolymers with the number of units in the polycycle below 3, but not 10, as it is indicated in ref. [30]. It is shown that further increase of the ladder fragment length causes no considerable changes of T_{soft}.

Detailed thermomechanical studies of properties of synthesized block-copolymers indicate that the effect of the disposition type of polycyclic ladder fragments in the dimethylsiloxane backbone is negligible, because at the same lengths of ladder and dimethylsiloxane fragments in appropriate block-copolymers glass transition temperatures are equal (Figure 7.7).

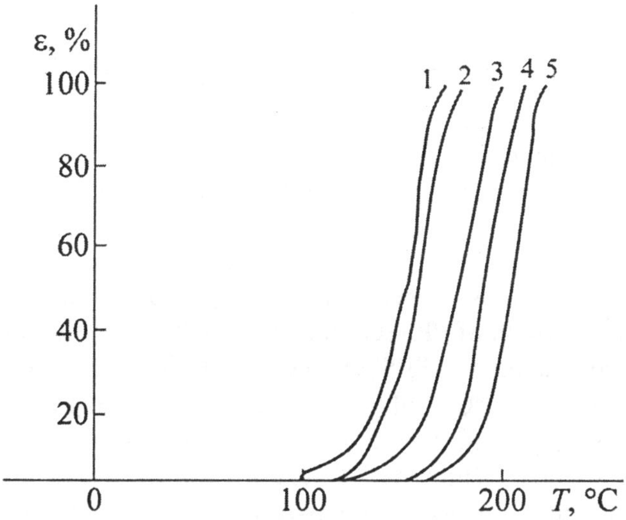

Figure 7.7. Thermomechanical curves for cyclolinear block-copolymers: 1, 2, 4 – copolymers 4, 2 and 1, respectively (Table 7.7); 3, 5 – copolymers 2 and 1, respectively (Table 7.8). Load is $\sigma = 0.1$ kg/cm^2

These data coincide with the previous assumption that 1,3- and 1,5-positions of cyclotetrasiloxane fragments in dimethylsiloxane backbone cause similar effects on physical and chemical properties of synthesized polymers [31].

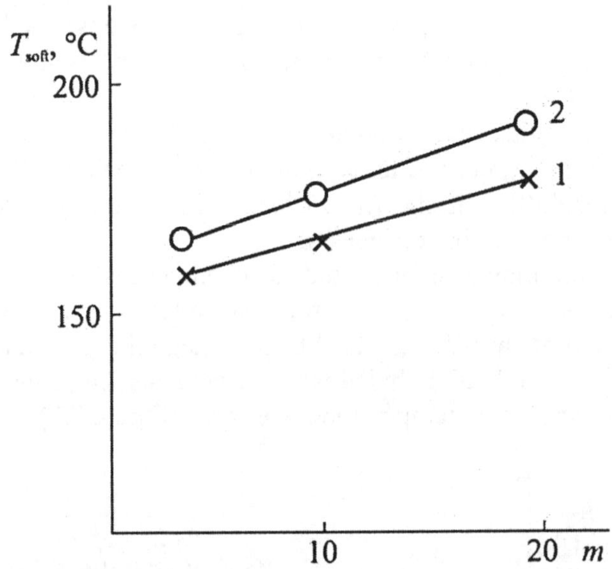

Figure 7.8. Dependence of T_{soft} on polycyclic block length for structure VI block-copolymers at different lengths of dimethylsiloxane unit (n): 1 – $n = 2$; 2 – $n = 8$

Figure 7.8 shows dependence of T_{soft} of organosiloxane block-copolymers on the length of PPSSO block at dimethylsiloxane fragment value $n = 2, 8$. Clearly the increase of PPSSO block length from $m = 3$ to $m = 20$ causes T_{soft} rise by 15 - 20°C only, whereas at the length of plasticizing dimethylsiloxane fragment, $n = 2 – 11$, copolymers represent a one-phase system with a single T_g (Figures 7.5 and 7.8), and increase of the fragment length ($m \geq 35$) induces formation of a two-phase system (Figure 7.9, curves 5 – 8). In this case, the effects caused by devitrification and melting of the microphase, formed by flexible blocks, and devitrification of the microphase, formed by rigid blocks, are clearly observed.

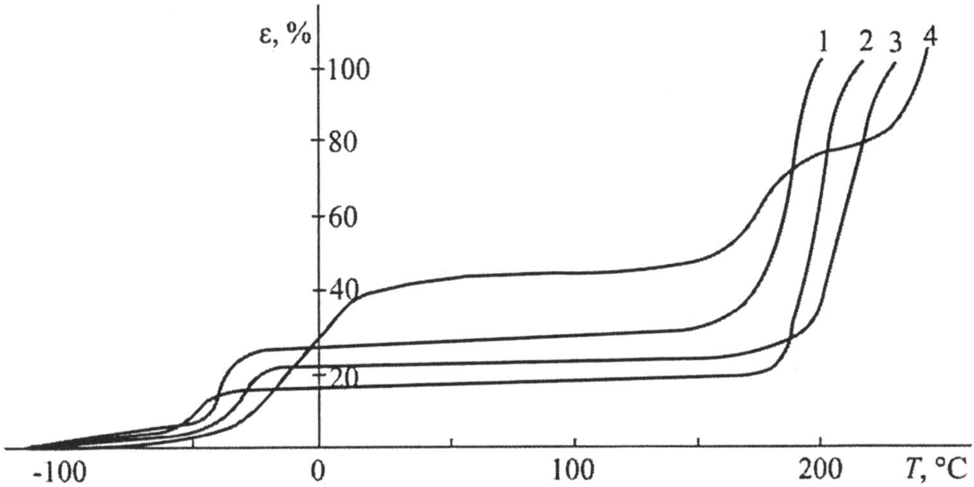

Figure 7.9. Thermomechanical curves of block-copolymers: 1 -6 - copolymers 8, 6, 14, 11, 10 and 16, respectively (Table 7.8); 7, 8 - copolymer 22 and 27, respectively (Table 7.7). Load equals $\sigma = 0.1$ kg/cm^2

For such block-copolymers, formation of a two-phase system is explained by high incompatibility between dimethylsiloxane fragments and ladder blocks.

The two-phase system indicates the presence of microdomain structure in these block-copolymers. If the concentration of ladder blocks in copolymer is high, they represent a continuous phase with distributed discrete aggregates of dimethylsiloxane fragments. Microdomain structure is formed at definite values of rigid and flexible fragments ($m \geq 3$, $n \geq 20$) only.

Synthesized structure V and VI block-copolymers were studied by the X-ray diffraction method. Data in Tables 7.7 and 7.8 indicate that the presence of two or three amorphous halos is general for block-copolymers, and clear dependence of diffraction patterns, i.e. the number of diffraction maximums, on the ratio of linear and polycyclic fragments is indicated (refer to Figures 7.10 and 7.11 and Tables 7.7 and 7.8).

For all copolymers with short dimethylsiloxane units (n_{SiO} = 1 - 11), independently of PPSSO block length, two diffraction maximums in the ranges of 11.0 - 13.4 Å and 4.0 - 5.0 Å are observed. As the length of flexible dimethylsiloxane fragment is increased, $n \geq 20$, the third diffraction maximum in the range of 7.8 - 8.1 Å occurs. It should be noted that for structure V block-copolymers, maximal d_1 value exceeds the interchain distance, d_1, of structure

VI ones by 0.5 unit, which matches with the fact that poly(organosilsesquioxane) macromolecules form aggregates, composed of closely packed parallel macromolecules, interchain distances of which vary due to framing and strictly match with the macromolecule diameter [32].

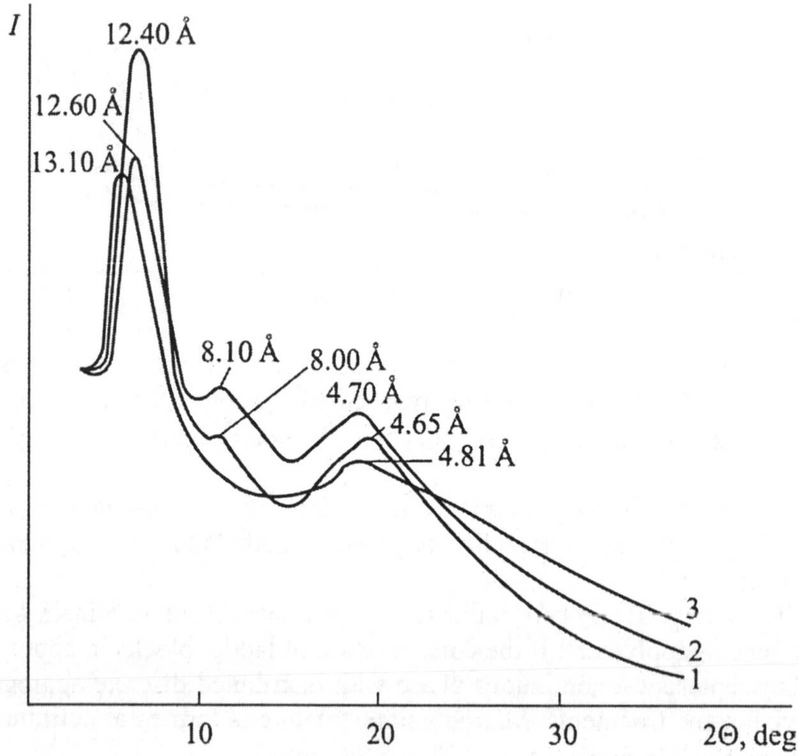

Figure 7.10. X-ray diffraction curves for structure VI block-copolymers: 1 – copolymer 16; 2 – copolymer 10; 3 – copolymer 6 (Table 7.8)

The presence of an intense peak on diffraction patterns of studied block-copolymers, which location corresponds to the maximum of poly(organosilsesquioxane) polymers, testifies about the presence of structures, identical to these detected in poly(organosilsesquioxanes), in these polymers.

If supermolecular structure of poly(organosilsesquioxanes) is formed by means of packing of macromolecular backbones, then studied block-copolymers display the packing of both backbones (structure VI) and side branches from the backbone (structure V).

As at low values of flexible fragments, $n \leq 11$, no maximum in the range of 7 - 8 Å on diffraction patterns is observed, such structures occupy the main volume in the polymer.

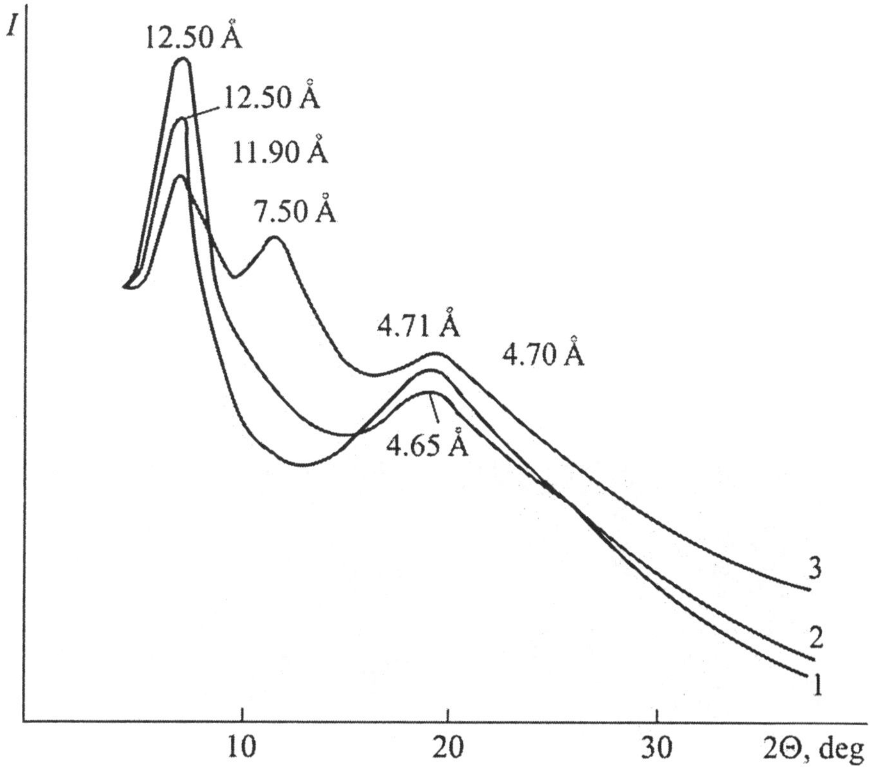

Figure 7.11. X-ray diffraction curves for block-copolymers of structure VI: 1 – copolymer 1; 2 – copolymer 3; 3 – copolymer 5 (Table 7.8)

Besides the presence of diffraction maximum corresponded to the structures, composed of PPSSO fragments (11.0 - 13.5 Å), at definite values of *n* clear maximum at 7.0 - 8.1 Å is observed for PDMS. It proves that after reaching a sufficient distance between polycyclic blocks ($n \geq 20$) a two phase structure is formed. One of the phases is presented by structures, composed of closely packed parallel polycyclic fragments (both in the framing and along the macromolecular backbone) of the ladder type, and displays a mesomorphous attitude. The second phase is presented by an amorphous structure, similar to that observed in PDMS.

For block-copolymers of structure V with $n = 1, 2$, parallel packing of polycyclic blocks of the same chain cannot be realized, because based on the analysis of molecular models, side polycyclic tabs are disposed at different angles between them. Such disposition of side branches should be accompanied by smearing of the diffraction maximum in the area of 11.0 - 13.5 Å. However, this was not proved by the study.

This gives grounds to assert that parallel packing of polycyclic groups, belonged to neighboring macromolecules, is valid at $n = 1, 2$. When such possibility is realized, at large n tabs belonged both to the same and different macromolecules are packed in parallel.

Electron microscopic studies prove the supposition that the attitudes of supermolecular structure and the one, observed in PPSSO studies, are similar. As an example, Figure 7.12 shows a micrograph of structure V block-copolymer, which indicates that supermolecular structure of block-copolymers 6 and 7 (Table 7.7) display clear fibril type ($n = 1, 4$; $m = 10$) (Figure 7.12, a and b).

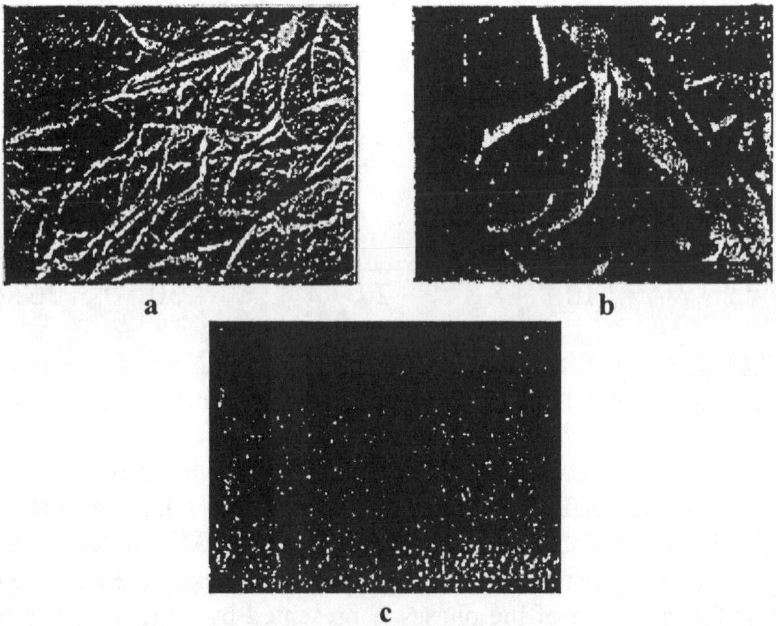

Figure 7.12. Electron micrographs of structure V block-copolymers: a – copolymer 6; b – copolymer 7; c – copolymer 10 (Table 7.7)

At large n = 55 and m = 10 (copolymer 10, Table 7.6) a globular supermolecular structure is observed (Figure 7.12, b). The increase of rigid polycyclic block length, m, to 37 (structure V, copolymer 21, Table 7.7) and n = 55 is accompanied by occurrence of the fibril structure, i.e. at the same lengths of linear fragments, at which block-copolymers with shorter polycyclic branches display the globular structure.

The integrated approach to the study of block-copolymers consisting of PPSSO and flexible dimethylsiloxane units helps in determining specific features of these polymers. For example, X-ray diffraction studies have detected the interchain distance, typical of ladder and linear polymers. However, spectroscopy data display the structure of polycyclic blocks different from *cis*-syndiotactic ladder ones, which is also proved by the data of thermomechanical studies. This gives an opportunity to state that this class of copolymers possesses clear differences from ladder or linear polymers.

Thus, an optimal alternative of ladder PPSSO plasticization was found, and cyclolinear block-copolymers with ladder fragments (of *cis*-anti-*cis*-tactic structure) in the macromolecular chain were synthesized. These block-copolymers possess increased thermal oxidative stability and may be used as binders for solid lubricants. Antifriction material, derived from copolymer 1 (Table 7.8), was tested by the front scheme in air, at room and increased temperature, at the load P = 20 kgf/cm^2 and sliding rate v = 0.5 m/s. Test duration was 3 hours. The assessment criterion for workability was the friction index values. The friction index of the material, derived from copolymer 1 (Table 7.8), equals f = 0.08 – 0.15 at 300°C. This makes it perspective as the material for friction type bearings.

Block-copolymers containing poly(phenylsilsesquioxane) fragments in diorganosiloxane chain of structure V were synthesized by exchange polymerization of polycyclic diols (structure II) with organocyclosiloxanes [33]. The reaction proceeds as follows:

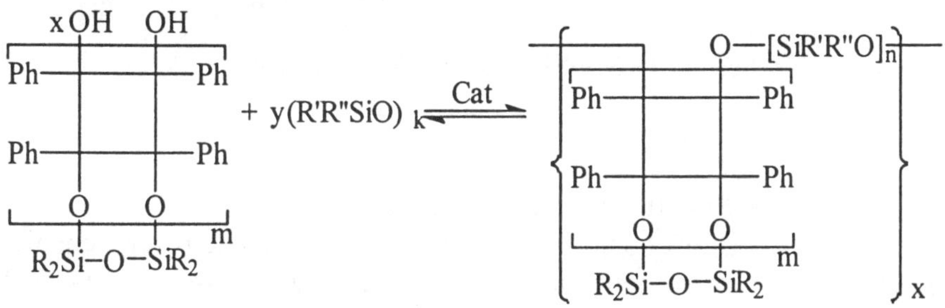

V

(7.7)

where R' = R" = Me, Ph; R' ≠ R"; R = Me, Ph; k = 3, 4; m = 10, 20.

The reaction proceeded with different ratios of diorganosiloxane units, in the temperature range of 60 – 100°C in block and in different solvents in the presence of nucleophilic type catalyst, for example, $Me_4NO(Me_2SiO)_nNMe_4$.

It is shown that polycyclic diol participate in the macromolecule composition. The size of polycyclic diol has a significant effect on the polymerization proceeding. The ratio of dimethyl and diphenylsiloxane units is the dominant factor for compatibility of copolymers and polymerization type. As the fracture of phenyl groups decreases from 75 to 0%, flexibility of the chain formed is increased. In turn, this induces occurrence of branchings and gel formation. Note also that this method may help in synthesizing block-copolymers of irregular structure. Moreover, apparently, this method of synthesizing cyclolinear copolymers was rejected, because no more data on it can be found in the literature.

Contrary to the above-considered methods, in which blocks of different origins are previously formed and block-copolymers are synthesized by heterofunctional polycondensation of oligomeric blocks, another principle of block-copolymers synthesis can was used [34]. In this method, rigid blocks are formed directly during copolymer synthesis. Using polyfunctional oligomers - α,ω–bis(organochlorosiloxy)dimethylsiloxanes (D-units, Me_2SiO), produced under controlled conditions, and phenyltrichlorosilane, rigid phenylsilsesquioxane blocks (T-units, $PhSiO_{1.5}$) may be expanded in block-copolymers, which are sharply different by type of macromolecular chains formed (for poly(dimethylsiloxane), persistent length equals 10 Å [35] and 150 Å for PPSSO [36]), by hydrolytic co-condensation proceeding by the scheme as follows:

$$yCl_xR_{3-x}SiO[Me_2O]_nSiR_{3-x}Cl_x + zPhSiCl_3 \xrightarrow[-HCl]{H_2O} \qquad (7.8)$$

$$\longrightarrow \left\{[O_{x/2}R_{3-x}SiO(Me_2SiO)_nSiR_{3-x}O_{x/2}]_y(PhSiO_{1.5})_z\right\}_k$$

VII

where R = Me, Ph; n = 20 - 80; z = 5 - 20; x = 2, 3.

In the mentioned reactions, the minimal size of flexible blocks equaled 10 – 12 D-units, which corresponded to the mean molecular mass of 750 – 900. The size of rigid blocks varied from 2 to 200 T-units per dimethylsiloxane block.

With regard to existing data on high sensitivity of structural parameters of ladder poly(phenylsilsesquioxane) to reaction conditions [37, 38], block-copolymers were synthesized under strictly identical conditions [34]. Thus, at the given concentration of trifunctional monomer the relative content of T-units in block-copolymers is the function of the length of used oligodimethylsoloxane. To the authors' point of view, three parameters defining both composition and structural features of formed macromolecules, consecutively associated with mechanical, hydrodynamic and other properties of the mentioned block-copolymers, can be realized in the approach suggested. These parameters are: the value of rigid block at fixed length of flexible one, the value of flexible block at fixed value of rigid one, and proportional variation of sizes of both blocks at constant D:T ratio.

Using the mentioned approach, block-copolymers differing by the D:T unit ratio were synthesized [34]. Parameters of these copolymers are shown in Table 7.9.

Table 7.9

Some physical and chemical parameters of structure VII poly(phenylsilsesquioxane)-dimethylsiloxane block-copolymers

No.	Flexible block $\{O_{x/2}R_{3-x}SiO[Me_2SiO]_nSiR_{3-x}O_{x/2}\}$		Rigid block $(PhSiO_{1,5})_m$	D/T	[η], dl/g at 25°C, in benzene	$\bar{M}_n$
1	2	20	5	4	0.18	4,600
2	2	20	10	2	0.11	3,260
3	2	20	20	1	0.13	8,530
4	2	40	10	4	0.19	4,500
5	2	80	20	4	0.21	9,100
6	3	20	5	4	0.16	520
7	3	80	10	8	0.14	6,900

X-ray diffraction analysis has indicated heterogeneity, i.e. incompatibility of the phases formed for copolymer 3 ($D_{20}T_{20}$, Table 7.9). Diffraction patterns display three diffraction maximums. The first maximum of synthesized block-copolymers displays scattering by the phase, formed by rigid fragments possessing d_1 = 12.3 Å. The second maximum is corresponded to the phase from flexible dimethylsiloxane fragments. The third maximum represents the sum of two maximums, observed in diffraction patterns of poly(dimethylsiloxane) and poly(phenylsilsesquioxane). Intensities of the first

Table 7.10

Parameters of studied block-copolymers*, structure VII

Group	Synthesis conditions				$[\eta]$, dl/g at 20°C, benzene	$\bar{M}\times10^{-3}$ paraphase osmometry	Type of GPC curves**	[OH]-groups, ^{29}Si NMR, wt.%	Notes
	Nota-tion	*n*:*m*	[HCl], mol/l [39]	Solvent					
I	BS-1	20:10	9.0	Benzene	0.06	2.2	M	5.4	50% of fraction is insoluble
	BS-2	20:10	4.5		-	3.6	U, T	7.3	50% of fractions are insoluble
	BS-3	20:10	Neutral		-	5.8	U	-	50% of fractions are insoluble
II	BS-4	20:5	9.0	Benzene	0.18	2.0	U	-	Insoluble
	BS-5	20:20	9.0		0.06	3.0	B	-	-
	BS-6	80:10	9.0		-	3.0	B	-	-
III	BS-8	20:10	9.0	Toluene	0.19	24	M	1.1	All fractions are soluble
	BS-9	20:10	9.0	Benzene: Ethyl-acetate (1:1)	0.05	2.2	U, T	3.9	All fractions are soluble

Notes: * $\bar{M}_n$ of $(Me_2SiO)_n$ block equals 1,360 ($n = 20$).

** M is multimodality; U is unimodality; T is threshold; B is bimodality.

and the second maximums in the block-copolymers are proportional to concentrations of flexible and rigid fragments, respectively.

For the purpose of detailed study of the reaction conditions effect on block-copolymer structure formation, three groups of polymers, synthesized under different conditions, were studied by GPC and molecular hydrodynamics methods [38]. The reaction conditions differed as follows: I – different acidity of the medium; II - different ratio of block contents; III – different solvents (Table 7.10).

Analysis of MMD curves obtained by GPC method and deduced from fractionation results shows that their type is significantly dependent on the synthesis conditions: medium acidity and the ratio of block contents, in particular (Table 7.10, Figure 7.13). In the majority of cases, block-copolymers display either bimodal or multimodal type of chromatograms. Unimodal chromatograms are displayed by polymers only, synthesized in neutral or low acidic medium, as well as in the case of the shortest silsesquioxane block, BS-4.

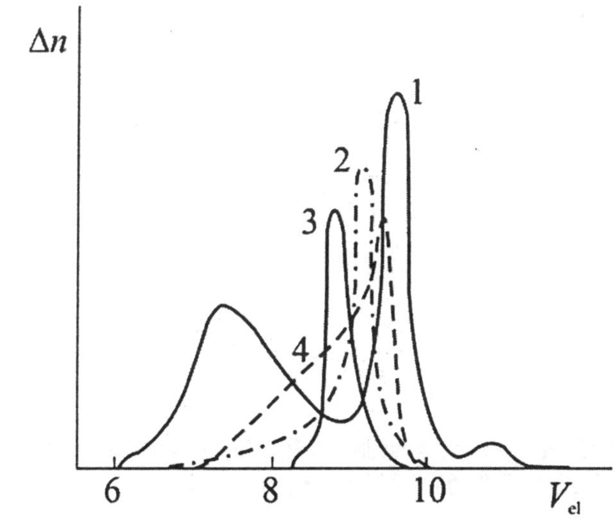

Figure 7.13. GPC curves for BS-1 (1), BS-2 (2), BS-3 (3) and BS-4 (4)

The possible reasons for bimodal distribution are as follows: "separate" hydrolysis due to difference in reactivity of reagents used; "separate" condensation due to micro-fractionation, caused by thermodynamic incompatibility of blocks; limitation of chain propagation due to interlocking of functional groups, cyclization, etc. These processes induce polydispersity of formed macromolecules by molecular mass and composition. In particular, PPSSO homopolymer may be produced in acidic medium. This homopolymer is the product of phenylsilanetriol [40]. Presumably, limited compatibility of dimethylsiloxane (DMS) and phenylsilsesquioxane (PSSO) blocks is more clearly displayed during synthesis under these conditions (i.e. in acidic medium), which results in the change of conditions of macromolecule structure formation up to synthesis of PPSSO homopolymer.

This suggestion is also proved by the results of analysis of BS-7 sample fractions. NMR and GPC studies (Table 7.11) indicate that the ratio of dimethylsiloxane (D) and phenylsilsesquioxane (T) units (i.e. the ratio of blocks, *n*) in the first fraction is highly shifted towards T groups compared with the given ratio of 20:10. Because this is the most low-molecular fraction, formation of low-molecular homopolymer is quite strong argument for such shift. Bimodal distribution in the gel-chromatogram of the fraction also confirms this supposition (Figure 7.14, curve 2).

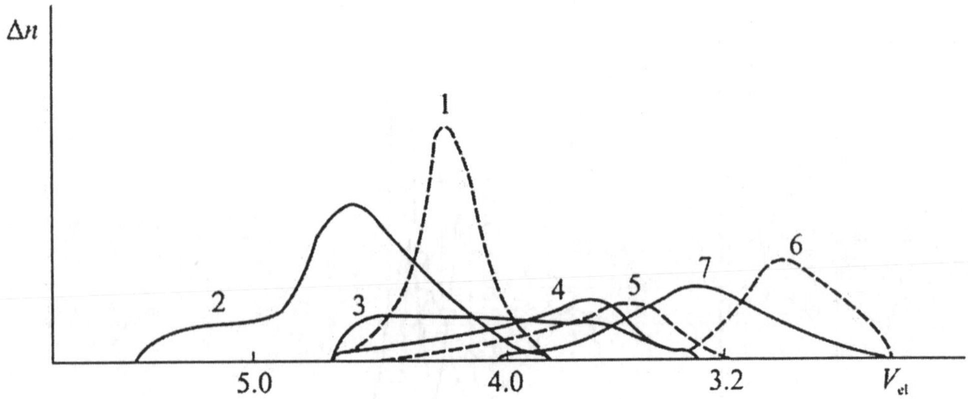

Figure 7.14. GPC curves for standard PPSSO (1) and BS-7 fractions: 1 (2); 4 (3); 7 (4); 8 (5); 12 (6); 13 (7)

Inhomogeneity of block-copolymer composition, first, due to variation of the ratio of the lengths of blocks due to polydispersity of initial ODMS may be another possible reason for multimodal distribution; secondly, as the result of forming three-block macromolecules of the ABA type instead of $(AB)_n$. This is also promoted by micro-fractionation (formation of microphases) during synthesis, stipulated by limited thermodynamic compatibility of PDMS and PPSSO blocks.

Table 7.11

Parameters of BS-7 and BS-8 sample fractions

Sample	Fraction No.	$\bar{M}_n \times 10^{-3}$	[η], dl/g at 20°C, in benzene	[Me:Ph] ^{1}H NMR	[Me:Ph] GPC	[OH-groups] (^{29}Si NMR), wt.%
BS-7	Initial	14.56	0.19	20:17	-	1.1
	1	2.70	0.03	20:45	20:50	0.66
	2	5.67	0.05	-	-	-
	3	7.30	0.056	20:18	20:18	0.89(0.81)**
	4	10.90	0.09	20:17	20:19	-
	5	14.46	0.13	-	-	1.4
	6	20.40	0.136	20:17	-	-
	7	21.80	0.15	-	-	-
	8	19.85	0.165	20:12	20:12	-
	9	-	0.19	-	-	-
	10	31.20	0.20**	-	-	
	11	43.20	0.25	20:10	-	1.2
	12	48.50	0.30	20:12	-	-
	13	54.60	0.30	20:13	20:10	-
	14	111.1**	0.56	20:10	20:16	1.8 (1.3*)
BS-8	Initial	2.70	0.05	20:8	-	3.9
	1	1.10	0.018	20:30	-	-
	2	2.90	0.029	20:32	-	-
	3	3.60	0.036	20:30	-	-
	4	5.30	0.050	20:14	-	-
	5	8.50	0.072	20:6	-	-
	6	11.80	0.10	20:3	-	-

Notes: * Chemically determined (the method of active hydrogen determination with the help of lithium aluminum hydride).

** Calculated from Mark-Kuhn-Hauvink equation.

Inhomogeneity of composition is proved by the results of ^{1}H NMR and GPC analysis of produced fractions composition, shown in Table 7.11. The ratio of blocks was calculated from under-eluent curve areas of fractions containing D and T units and the "standard", for which PPSSO containing 100% of phenylsiloxane units was taken [40]. The analysis was performed for equal volumes with equal concentration. Quite good coincidence of the results of both methods was obtained (Table 7.11, Figure 7.14). Figure 7.14 shows that some fractions display bimodal chromatograms. These results prove the supposition about composition inhomogeneity of studied samples.

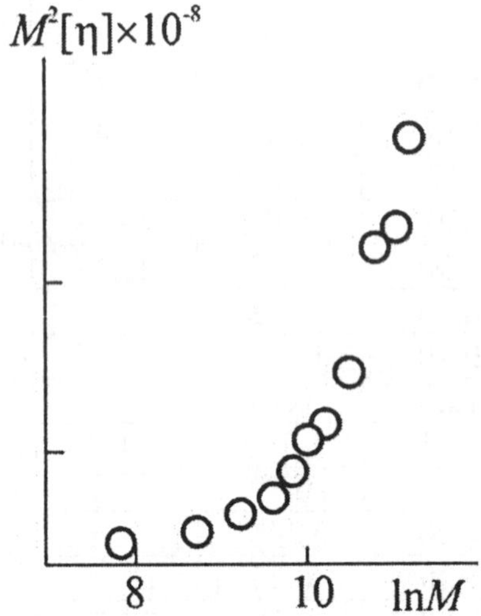

Figure 7.15. Dependence $M^2/[\eta]$ - ln*M* [42] for BS-7 sample

The last column in Table 7.11 shows that the majority of fractions become insoluble after drying by vacuum treatment, Except for BS-7 and BS-8 polymers containing the lowest amount of uncondensed OH-groups. BS-7 polymer was produced via additional condensation stage, and thus OH-group concentration is low in it (Table 7.11). That is why intermolecular condensation by OH-groups is the probable reason for solubility loss by a part of fractions during drying by vacuum treatment. Therefore, BS-7 and BS-8 samples displaying the general dependence $[\eta] = 2.37\cdot 10^{-5}\cdot M^{\alpha}$ were chosen for more detailed study of hydrodynamic and conformation properties. Exponent α in this equation equals 0.88. Under the same conditions, for PDMS it equals 0.79 [41], and for PPSSO falls within the range of 0.9 – 1.46 [38]. The value of parameter α and significant quantity of uncondensed OH-groups testify about high defectiveness of polycyclic block in block-copolymers.

Among the equations used for assessment of mean-square sizes [42 – 44] (Figures 7.15 – 7.17), the one deduced in ref. [44] and the Yamakawa – Fuji equation, modified by Bogdanetskii [45], are the most suitable in this case.

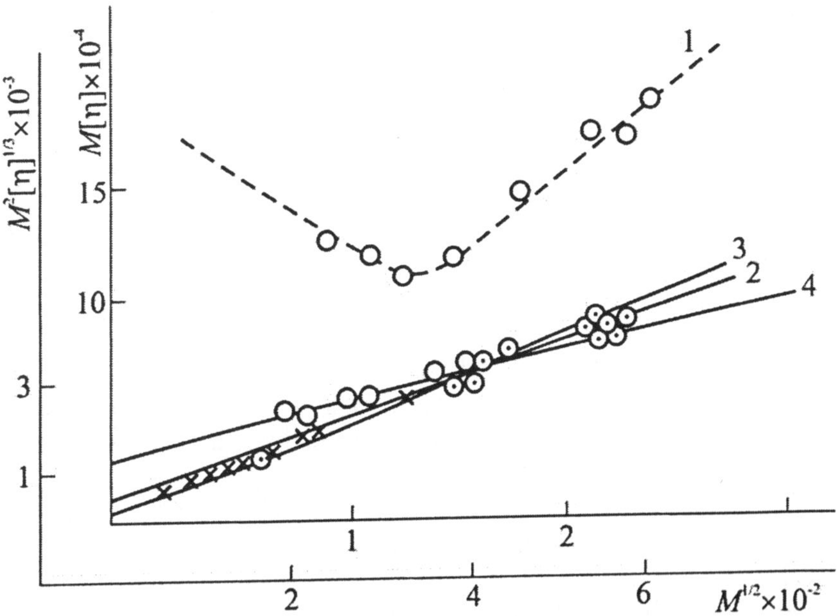

Figure 7.16. Dependences $M/[\eta]$ - $M^{1/2}$ for BS-7 sample (1) and $(M^2/[\eta])^{1/2}$ for BS-7 (2), PDMS (3) and PPSSO (4), composed by data of refs. [43], [44], [41] and [37], respectively

Figure 7.17. Dependence $(M^{1/2}/[\eta])^{1/3}$ - $1/M^{1/2}$ for BS-7 sample, composed by data of ref. [45]

Even in the area of quite low molecular masses nonlinear dependence $M^2/[\eta]$ - $\ln M$ (Figure 7.15) indicates flexibility of the chains of studied polymers

[42]. This is also proved by the curve of $M/[\eta]$ - $M^{1/2}$ (Figure 7.1, curve 1) [9]. Rigidity parameters, $\bar{h}_0^2/M$, calculated for BS-7, PDMS and PPSSO from tangents of dependences, composed by the technique [44], equal 0.46, 0.85 and 1.87, respectively. It is impossible to calculate the Kuhn segment and chain diameter for studied block-copolymers, because data on macromolecule structure are required. Comparison of conformation parameter, $\bar{h}_0^2/M$, values (Figure 7.15) and the totality of all obtained data agrees with the assumption that PPSSO blocks in the macromolecule represent incomplete polycyclic fragments with different defectiveness. The quantity of these defective structures in synthesized block-copolymers is apparently greater than in structure V and VI block-copolymers, which induces side cross-linking reactions during drying of some fractions.

Supermolecular structure of poly(dimethylphenylsilsesquioxane) block-copolymers with different contents of dimethylsiloxane and phenylsilsesquioxane blocks was studied [46].

Morphology of poly(dimethylphenylsilsesquioxane) block-copolymers, cross-linked at the interaction with γ-aminopropyltrietoxysilane, is described in refs. [47, 48]. It was concluded that supermolecular structure of block-copolymers, cross-linked in this manner, is realized in fibrillar formations (domains), located against the background of a structure-free matrix. Simultaneously, no phase fractionation of non-cross-linked block-copolymers was detected by electron microscopy method. Recently [49, 50], a microphase fractionation of di- and tri-block copolymers with large molecular masses was observed, the domains being spheroid [49] and pocket [50] shaped.

Siloxane block-copolymers containing heterogeneous oligomeric siloxane blocks were studied [46]. Macromolecule chains of such block-copolymers are composed of alternating flexible and rigid fragments, which are linear dimethylsiloxane and polycyclic oligophenylsilsesquioxane blocks, respectively. Studied block-copolymers were of the following general formula: $\{Me_2SiO\}_n([PhSiO_{1.5}]_a[PhSi(O)OH]_{1-a}\}_m$.

Table 7.12 shows parameters of these copolymers. Note also that the quantity of dimethylsiloxanes, identified as admixtures to initial oligomers, is decreased with increase of phenyl blocks' concentration in block-copolymers with equal content of DMS-blocks. Presumably, this is stipulated by the features of synthesis of these compounds.

Magnification ×50,000 makes observable spheroids in block-copolymers 1 – 3 (Table 7.12), uniformly and densely distributed in the field of view (Figure 7.18). For block-copolymers 4 and 5 (Table 7.12) large domains were observed, which were spheroid-shaped, located against the background of a continuous matrix. The average diameters of spheroids in block-copolymers 1 – 5 equaled ~0.02, 0.04, 0.06, 0.08, and 0.15 μm, respectively (Figure 7.18). It is found that the average size of spheroids in block-copolymer 6 reaches several μm and this copolymer represents a mixture of block-copolymers and homopolymers. No effect of block-copolymer $\bar{M}_n$ on the spheroid diameter was observed. The effect of possible admixtures (a mixture of dimethylsiloxanes), unavoidable during synthesis, on supermolecular structure of block-copolymers was not observed, too.

Table 7.12

Parameters of studied siloxane block-copolymers

Block-copolymer No.	Number of units per block		$\bar{M}_n$x10^3	OH-group content, %	Relative density, d (g/cm^3)	Dimethycyclo-siloxane content, D_n (mg/g)
	DMS (n)	PSSO (m)				
1	20	5	21	1.10	1.12	77.0
2	20	10	17.9	2.30	1.09	21.8
3	20	20	12.5	4.10	1.07	Traces
4	40	10	4.4	0.95	1.04	Traces
5	40	20	5.9	2.34	1.06	Traces
6	80	40	10.3	1.80	1.05	2,562

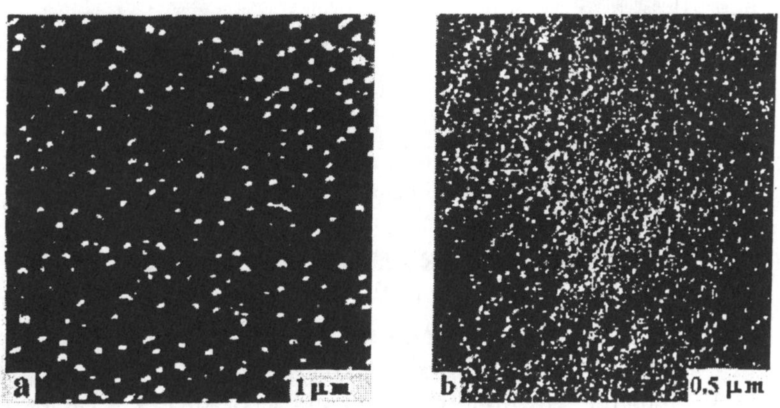

Figure 7.18. Block-copolymers 1 (a) and 4 (b) chip surface photographs

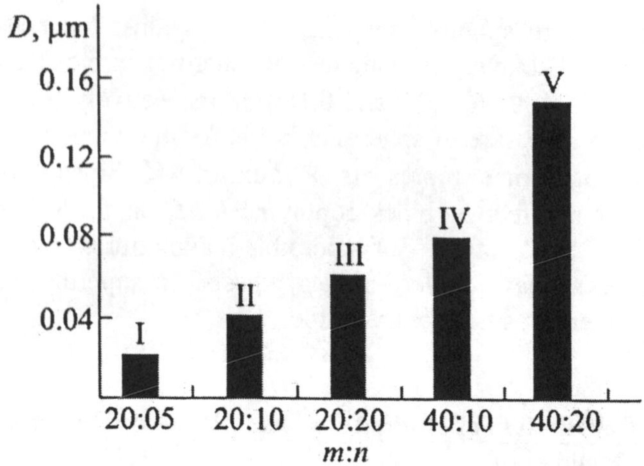

Figure 7.19. Dependence of spheroid diameter, D, on the number of units in DMS (n) and PSSO (m) blocks for block-copolymers 1 – 5 (Table 7.12)

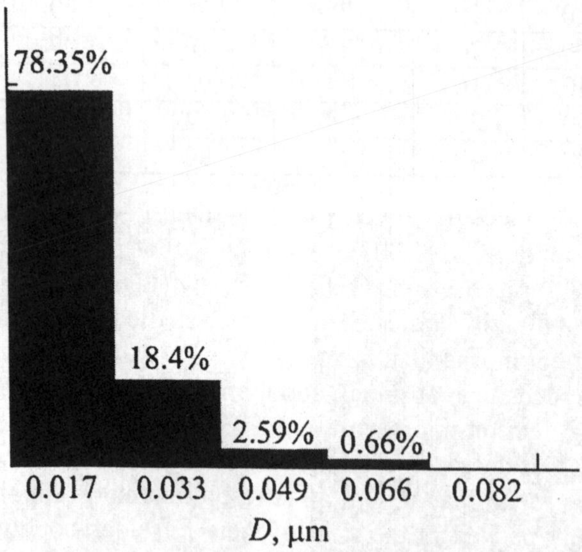

Figure 7.20. Spheroid distribution in block-copolymer 1 (Table 7.12) by equivalent diameters

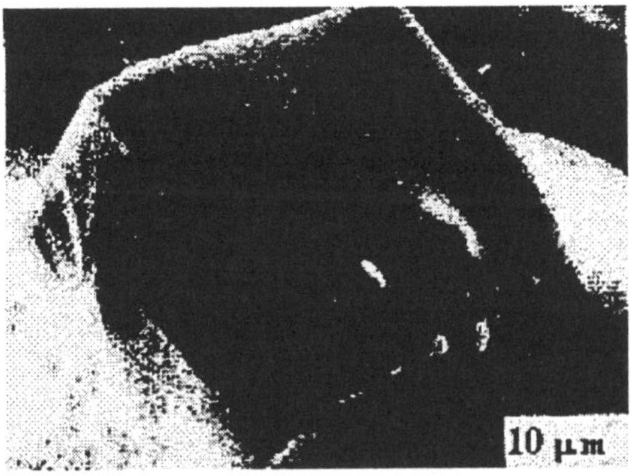

Figure 7.21. A photo image of PDMS and PPSSO mechanical mixture (20:5) surface

Calculated typical curve of spheroid distribution by sizes is shown in Figure 7.19. Of interest is that average sizes of spheroids, determined for block-copolymers 1 – 3 (Table 7.12; $\bar{M}_n$ = 21,000, 17,900 and 12,500, respectively), are comparable with macromolecule size, if the following parameters are taken into calculation: ≡Si-O- bond length equals 1.68 Å, ≡Si-O-Si≡ and -O-Si-O bond angles equal 150° and 110°, respectively.

Surfaces of poly(dimethylsiloxane) and poly(phenylsilsesquioxane) mechanical mixtures of the same ratios, as in block-copolymers, were studied on an electron microscope, which indicated a sharp difference between block-copolymers and mechanical mixtures. Against the "continuous" phase background, formations of different sizes (1 – 100 μm) and shapes (spheroid, cubes), chaotically disposed in the field of view (Figure 7.21), are observed in the mixtures. All this indicates that the studied substances are block-copolymers, but not mechanical mixtures.

Relative densities of block-copolymers 1 – 4 were determined. The above results of electron microscopy study correlate with their relative density, i.e. relative densities decrease with increase of the block size.

Thus it is found that at 20 – 40 Å resolution block-copolymers represent spheroids: closely packed in the field of view for block-copolymers 1 – 3 and observed against the background of continuous matrix for 4 – 6 ones. Spheroid diameters in studied block-copolymers are independent of their molecular masses.

The totality of obtained and cited data indicates that despite incompatibility of composing blocks, siloxane block-copolymers, studied in the current monograph, are able to form clusters due to affinity to coiling inherent to all organosiloxanes.

Block-copolymers containing ladder fragments in macromolecular chain were synthesized by homo- and heterofunctional condensation by hydroxyl-containing alkyl(aryl)silsesquioxanes with silanediols and organochlorosilanes in accordance with the following scheme [51, 52]:

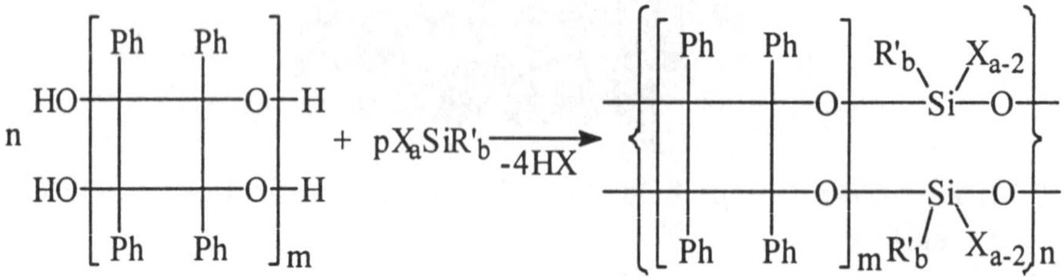

(7.9)

As during homofunctional condensation the probability of reaction proceeding between siloxanediols of the same structure is high, the accurate number of linear siloxane units in blocks can be hardly regulated. In the case of heterofunctional polycondensation without proceeding of homofunctional polycondensation, block-copolymers with the given structure without admixtures of homopolymers may be synthesized. However, different spatial disposition of hydroxyl groups in PPSSO tetroles of *cis*-syndiotactic structure will promote reaction proceeding with a network structure formation.

Many works [53 – 56] are devoted to the study of heterofunctional polycondensation of ladder-type oligotetroles of *cis*-syndiotactic structure and with $f_{av} = 4$:

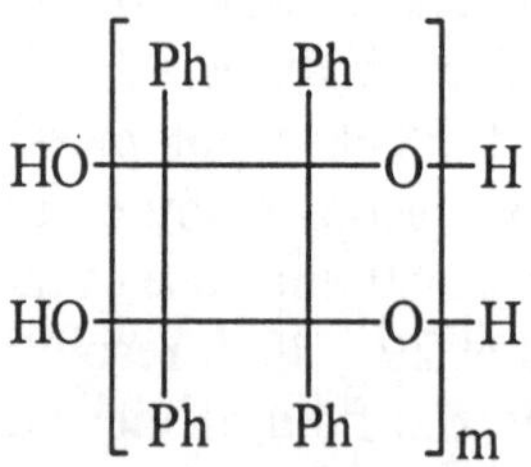

where $m = 5 – 50$;

with *α,ω*–bis(diethylamino)diorganosiloxanes of the general structure as follows:

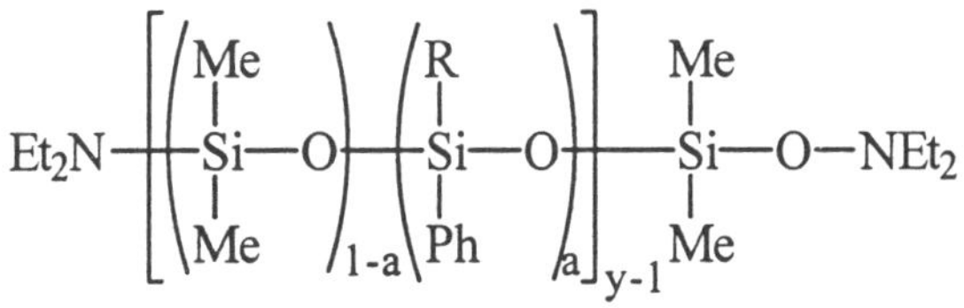

where $a = 0 – 0.5$; $y = 20 - 250$; R = Me, Ph.

Polycondensation was performed at different molar ratios of α,ω-bis(diethylamino)diorganosiloxane and initial oligotetrole. At 1:2 molar ratio of initial components or more, at definite transformation degrees gel formation was observed. That is why gel formation conditions in the current system were studied [56].

Basic regularities of gel formation depending on reagent functionality are known. They were formulated by Caroserse and Flory for the ideal case under assumption of low probability of the processes producing cyclic structures, equal and invariable activity of functional groups during the whole process, as well as independence of critical conversion in the gel point of concentration, temperature, catalyst and other factors. To ascertain if the process abides "classical" Flory's laws, the completeness degree of the reaction near the gel point was determined and compared with the critical one, calculated by the Flory equation. In particular, according to Flory dependence of the critical completeness degree of the process (conversion) in the gel point (P_c) on functionality of branching structural unit (f) and stiochiometrical balance (r) is expressed by the following equation:

$$P_c = 1/[r + r(f - 2)]^{1/2}.$$

As specially shown in the present work by comparing the mean molecular masses of PPSSO samples containing functional hydroxyl groups, for the case of PPSSO interaction with silamines, the average functionality of the branching structural unit equals four.

In this case, stoichiometrical imbalance equals the ratio of molar concentrations of diethylamino groups in silamine and hydroxyl groups in PPSSO: $r = [NEt_2]/[OH]$. At $r = 1$, calculated value of P_c equals 0.58.

The study of the PPSSO – silamine system shows that usually, real conversions near the gel point exceed calculated P_c value. For example, for reaction between PPSSO (m = 6) and silamine (y = 6) at r = 1 and total reagents' concentration in chloroform C = 0.02 mol/l, the reaction completeness by OH-groups equaled P = 0.83 after two hours of the reaction. Therefore, no gel formation was observed. Thus, the current system is much different from the ideal one.

For the purpose of studying features of nonlinear condensation between PPSSO and silamines, a series of particular dependencies was studied. As the feature of the critical system conditions, time of gel formation was recorded by a sharp increase of relative viscosity of the reaction mixture. The effect of reaction mixture concentration, lengths of oligomers and imbalance on time of gel formation was also studied.

Dilution of the reaction mixture increases time of the gel formation. Hence, it becomes longer than it may be expected, based on supposed second order of the reaction. This phenomenon is associated with the increase of the part of intramolecular cyclization reactions proceeding with the loss of functionality.

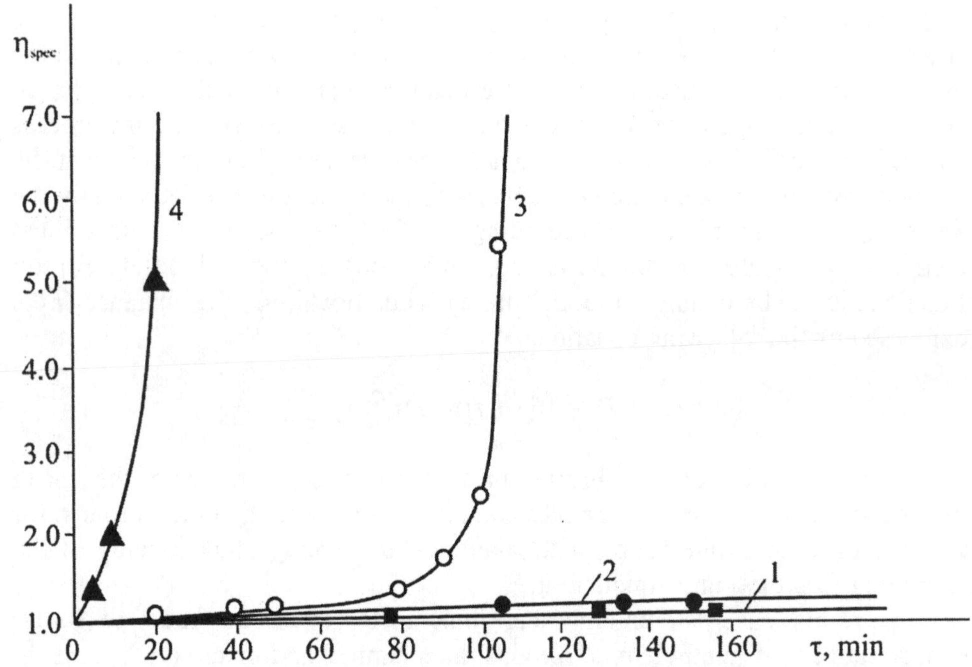

Figure 7.22. Dependence of gel formation time on the length of dimethylsiloxane oligomer with y = 2, 6, 25 and 130 at m = 6: curves 1 - 4, respectively. The process proceeds at C = 0.1 mol/l, r = 1, in chloroform, at room temperature

It has been found that time of gel formation in polycondensation of α,ω–bis(diethylamino)dimethylsiloxanes ($y = 2 - 130$) with PPSSO ($m = 6$) is decreased with the increase of dimethylsiloxane oligomer length (Figure 7.22).

Figure 7.23 shows variation of the transformation degree with time during PPSSO reaction with silamines, recorded by high resolution NMR spectroscopy method alternatively by intensity increase of signal (quartet) from methyl protons of liberated diethylamine, $(CH_3CH_2)_2NH$ (2.30 ppm), or by decrease of similar signal from protons of diethylamino groups of silamine, $(CH_3CH_2)_2NSi\equiv$ (2.86 ppm). The Figure shows that the rate of PPSSO ($m = 6$) reaction with silamine ($y = 130$) is, at least, not higher than that with silamines ($y = 6$ and 25). As a consequence, the observed dependence of gel formation time on the length of siloxane oligomer can be explained by greater part of intramolecular cyclization with decrease of flexible oligomer length.

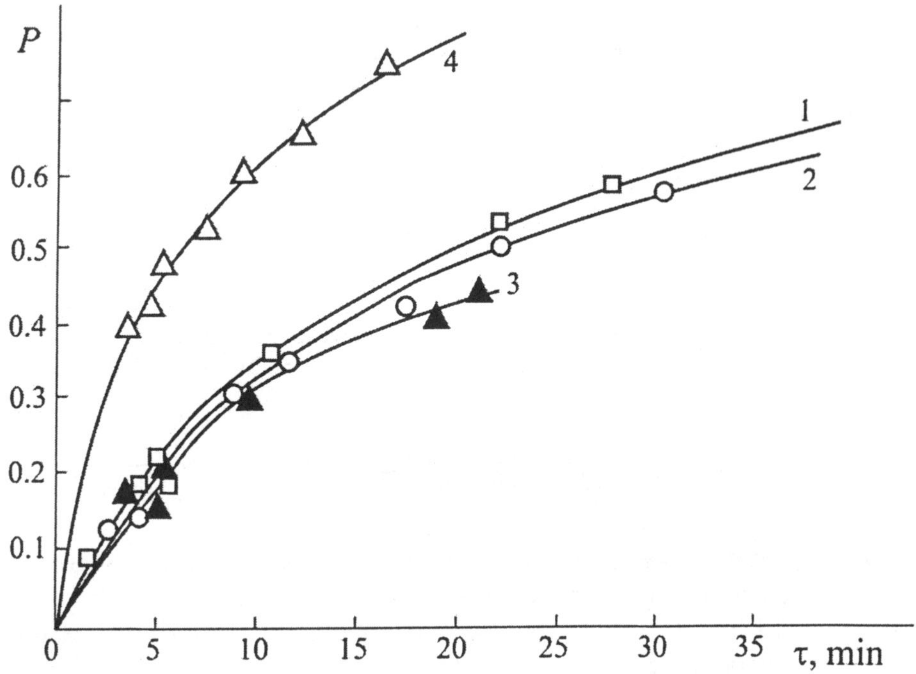

Figure 7.23. Variation of transformation degree with time during PPSSO ($m = 6$) reaction with silamines: 1 – $y = 6$, 2 – $y = 25$, 3 – $y = 130$ at $r = 1$; 4 - $y = 25$ at $r = 0.5$

However, it should be noted that recorded conversion is not the result of HFC reaction only. It is found that diethylamine liberated in the main process induces proceeding of PPSSO side homofunctional condensation and, as a consequence, probable hydrolysis of ≡Si-N= bond, caused by water formed. Hence, disclosure of cyclosiloxane ring by diethylamine may not be excluded, as it was indicated in Chapter by reaction (6.30), which gives a bipolar zwitterion [57]. In some cases, relative part of these reactions was estimated as ~15 – 20% of the main one. It is found that time of gel formation in this system is independent of length of the rigid block.

It is also found that dependence of gel formation time on imbalance of functional groups is of the extreme type with no regard to the length of silamine and concentration of the reaction solution (Figure 7.24).

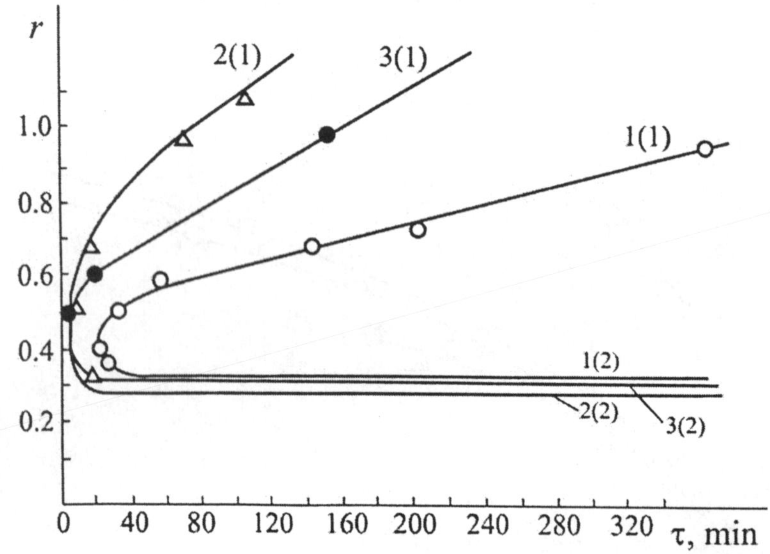

Figure 7.24. Dependence of gel formation time on imbalance in PPSSO ($m = 6$) reaction with silamines: 1 – $y = 25$, $C = 0.1$ mol/l; 2 – $y = 25$, $C = 0.1$ mol/l; 3 – $y = 6$, $C = 0.4$ mol/l; in chloroform, at room temperature

The observed dependence may be hardly unambiguously explained. That is why, some possible factors defining regularities of gel formation in the studied system were put under consideration [56]: statistical, kinetic, and some other. Statistical analysis of data by Caroserse determines intervals of gel formation, which fall within the imbalance range r = 0.5 - 1.6. Therefore, maximal affinity to gel formation is observed at r = 1. Presumably, the observed shape of the curves in the area (1) of the dependence of gel formation time on imbalance (Figure 7.24) may be explained by the totality of the above considered factors. The shape of curves in the area (2) is probably explained by location of it outside the gel formation range due to material chain termination at the stage of ultra low-molecular products formation.

Studies of regularities of nonlinear polycondensation determined the most favorable conditions of the process proceeding without gel formation, which were taken into account in the synthesis of various modifications of poly(phenylsilsesquioxane)-poly(diorganosiloxane) block-copolymers.

Preliminary experiments on PPSSO (m = 5 – 50) condensation with α,ω–bis(diethylamino)diorganosiloxane (y = 20 – 250) taken in different ratios indicate that both soluble and insoluble rubber-like and resin-like polymers, capable of forming film materials during processing from solution, can be the reaction products.

It is shown that despite similar chemical origin of blocks, a low-temperature relaxation transfer at -120°C is displayed, corresponded to the transfer in homopoly(dimethylsiloxane), and at sufficient length (over 1,000 units) dimethylsiloxane fragment displays the ability to crystallize in the temperature range of -50 ÷ -60°C. It should be noted that crystallization of dimethylsiloxane phase proceeds at the length of dimethylsiloxane linear fragments equal 20 units [58].

Poly(phenylsilsesquioxane)-dimethylsiloxane block-copolymers are used for preparing a rubber mixture. Materials derived from them possess high physical and chemical parameters: non-filled composites display rupture strength up to 100 kgf/cm^2, relative elongation of 285 – 433% (rupture strength of linear diorganosiloxanes equals 2 – 4 kgf/cm^2), and much higher heat resistance. Vulcanizates derived from block-copolymer display increased thermal and frost resistance. Frost resistant sealers of cold hardening are also derived from block-copolymers. The frost resistance index of composites, derived from the above-mentioned block-copolymers, equals 0.26 at -70°C.

There are data in the literature about synthesis of ladder copolymers, synthesized by catalytic dehydrocondensation. An attempt was made to use tetrole dehydrocondensation with tetramethylcyclotetrasiloxane in diethyl alcohol with caustic potash or caustic soda methylate as a catalyst.

It is shown that soluble polymers are obtained at low transformation degrees, and insoluble polymer yield is increased with conversion. The yield of insoluble part is similarly increased with temperature and concentration of the solution. Presumably, this may be explained by different positions of hydrogen relative to the plane of tetramethylcyclotetrasiloxane, which may promote formation of branched system. Besides, cyclosiloxane structure disclosure is also possible.

Synthesis and properties of phenylsilsesquioxane polymers, modified by linear dimethylsiloxane units with *cis*-anti-*cis*-tactic structure of macromolecules by catalytic dehydrocondensation of *cis*-1,3,5,7-tetrakis-(hydridedimethylsiloxy)-1,3,5,7-tetraphenylcyclotetrasiloxane (refer to Table 6.1) with tetrole and olirotetroles [60], will be discussed in the present Section.

At the first stage, model reaction of catalytic dehydrocondensation of 1,3,5-tris(hydridedimethylsiloxy)-7-hydroxytetraphenylcyclotetrasiloxane (Chapter 6, structure IV, compound 4, Table 6.1) was studied.

Catalytic dehydrocondensation of compound 4 (Table 6.1) was performed in 5% solution of absolute toluene in the presence of catalyst (0.1 M platinum-hydrochloric acid solution in tetrahydrofurane). The reaction proceeds directionally, mostly with formation of a compound of bicyclic structure IX with the yield equal ~83% and a compound of structure X with 13% yield. The reaction scheme is shown below [60]:

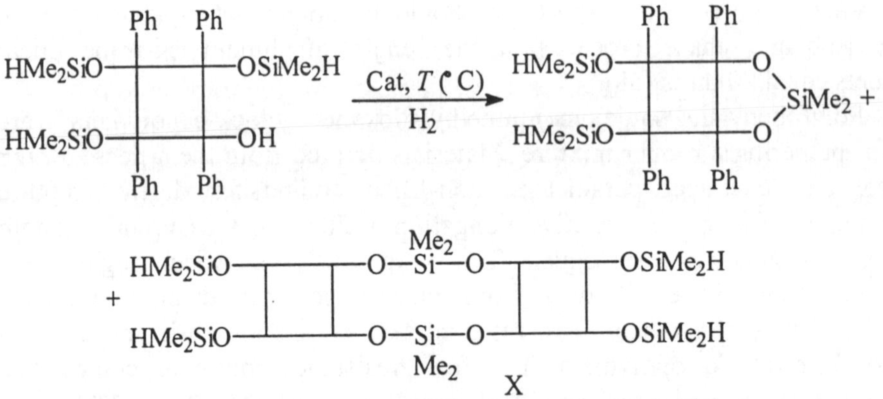

(7.10)

X

It is found that at low temperatures dehydrocondensation proceeds slowly, and conversion reaches ~58% at 40°C. While temperature is increased to 60°C, the reaction proceeding becomes faster, and conversion by hydrogen is almost 100%. In the case of reaction performance in melt, intermolecular proceeding of dehydrocondensation with formation of a tricyclic compound of structure X and 85% yield dominates. Compounds of structures IX and X were identified by the ultimate analysis, determination of molecular masses, IR and NMR spectra.

^{1}H NMR spectrum of structure IX compound displays a triplet for methyl protons with the following chemical shifts: $\delta = 0.22$ ppm (3H), $\delta = 0{,}46$ ppm (12H), broadened reflex with the center at $\delta = 4.3$ ppm for ≡Si-H bond, and broadened multiplet in the range of 6.8 – 7.6 ppm for phenyl protons. It is found that catalytic dehydrocondensation proceeds without break of labile cyclosiloxane fragment.

Based on the results obtained, for the purpose of synthesizing ladder copolymers linked by linear dimethylsiloxane units, catalytic dehydrocondensation of *cis*-1,3,5,7-tetrakis(hydridedimethylsiloxy)-1,3,5,7-tetraphenylcyclotetrasiloxane (compound 5, Table 6.1) with tetrole and oligotetrole (*cis*-anti-*cis*-tactic molecular structure) at 1:1 ratio of initial components was studied. Both 0.1 M solution of platinum-hydrochloric acid in tetrahydrofuran and caustic potash powder were used as catalysts in the reaction [60].

Preliminary heating of initial reagents ($C = 0.03639$ mol/l) at 40 - 50°C during 6 hours in the presence of platinum-hydrochloric acid ($8 - 10\times10^{-5}$ g per 1 g of the substance) or anhydrous caustic potash powder (0.1% per 1 g of the substance) did not initiated either polymerization or polycondensation of the components.

Dehydrocondensation of the compound (compound 5, Table 6.1) with tetrole at 20 and 30°C in the presence of platinum-hydrochloric acid proceeded very slowly. That is why the reaction was studied at higher temperatures (40 - 60°C). The increase of dehydrocondensation temperature from 40 to 60°C induces consecutive rise of hydrogen conversion from 83 to 97%. In the case of KOH presence and 50°C reaction temperature, hydrogen conversion is more intensive. Dehydrocondensation proceeds by the scheme as follows [60]:

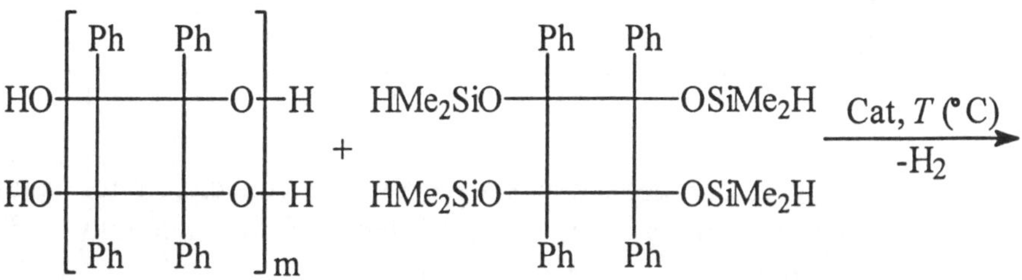

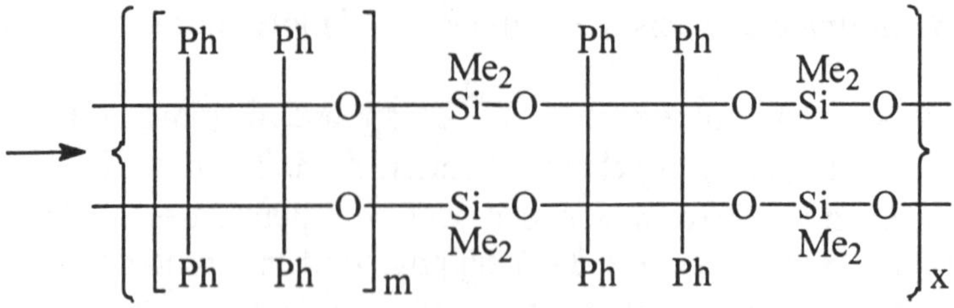

(7.11)

XI

where: $m = 1 - 10$; $T_{react} = 40 - 60°C$; catalysts: H_2PtCl_6 and KOH.

As a result, solid transparent products, well soluble in ordinary organic solvents, were obtained. The above-mentioned reaction is the second order process. Figure 7.25 shows time dependence of hydrogen conversion for catalytic dehydrocondensation of compound 5 (Table 6.1) with oligotetroles (m = 5 and 10) at 60°C. The Figure indicates that the rate and depth of the reaction decrease with oligotetrole length increase.

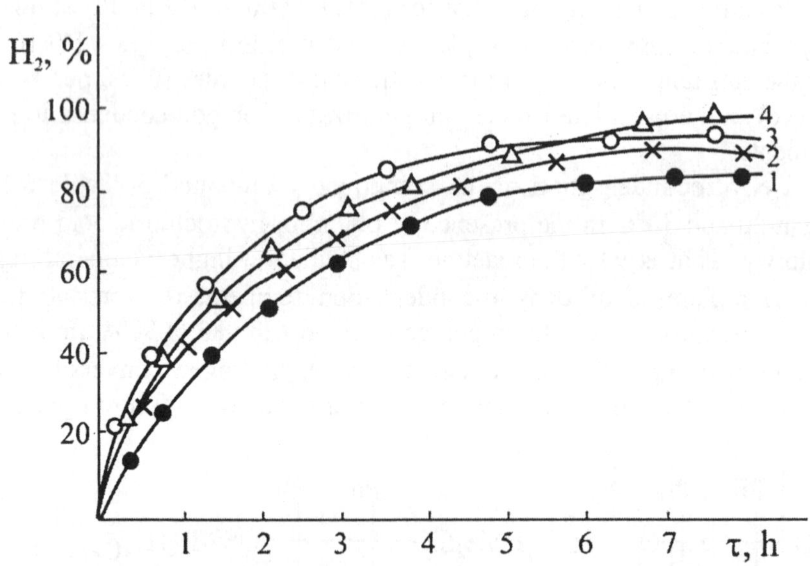

Figure 7.25. Temperature dependence of hydrogen conversion for catalytic dehydrocondensation of 1,3,5,7-tetrakis(dimethylsiloxy)-1,3,5,7-tetraphenylcyclotetrasiloxane with tetrole: 1 - 40°C; 2, 3 - 50°C; 4 - 60°C. 1, 2, 4 – in the presence of H_2PtCl_2; 3 – in the presence of KOH

The following rate constants of the reaction were calculated: $K_{40°} = 2.22\times10^{-2}$, $K_{50°} = 3.15\times10^{-2}$ and $K_{60°} = 4.14\times10^{-2}$. As a consequence, the rate constant increases by 1.4 times every 10°C of dehydrocondensation, and the activation energy equals 29.4 kJ/mol.

It is shown [60] that double-stranded polymeric chain may possess the following structures:

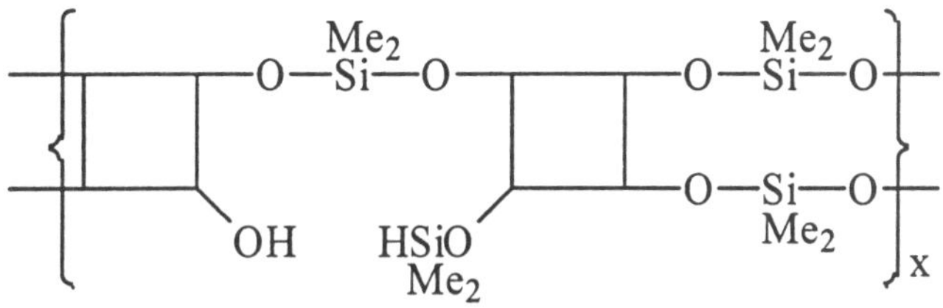

,

which may be formed at incomplete catalytic dehydrocondensation only. The presence of faulted structures increases flexibility of the polymeric segment and decreases softening temperature. Defectiveness of the structure decreases with temperature increase, which causes increase of the softening temperature and molecular mass. Some parameters of synthesized copolymers of structure XI are shown in Table 7.13.

IR spectra of copolymers for asymmetrical valence oscillations of ≡Si-O-Si≡ bonds in the range of 1,000 – 1,200 cm^{-1}, recorded in the presence of platinum-hydrochloric acid, in the range of 1,000 – 1,200 cm^{-1} display absorption bands at 1,040, 1,070, 1,110 and 1,130 cm^{-1}, whereas in the presence of KOH, bands at 1,040 and 1,100 cm^{-1} are absent. This indicates that in the presence of caustic potash as the catalyst, catalytic regrouping of cyclosiloxane skeleton with formation of a different *cis*-syndiotactic structure of macromolecular chain takes place.

Thus, for the first time the possibility of synthesizing ladder poly(organosiloxanes) by catalytic dehydrocondensation of 1,3,5,7-tetrakis(dimethylsiloxy)tetraphenylcyclotetrasiloxane with tetrole and oligotetroles was proved [60].

For the purpose of modification of PPSSO with *cis*-anti-*cis*-tactic structure by dimethylsiloxane units, its heterofunctional polycondensation with α,ω-dichloro- or α,ω-bis(dimethylamino)dimethylsiloxanes with different transformation degree, at 1:2 and 1:1 molar ratios on initial components was studied. The reaction scheme is shown below [61]:

Table 7.13

Some physical and chemical parameters of structure XI copolymers

No.	Copolymer structure	*m*	Yield, %	Catalyst	T_{react}, °C	η_{spec}, 25°C	T_{soft}, °C	d_1, Å
1	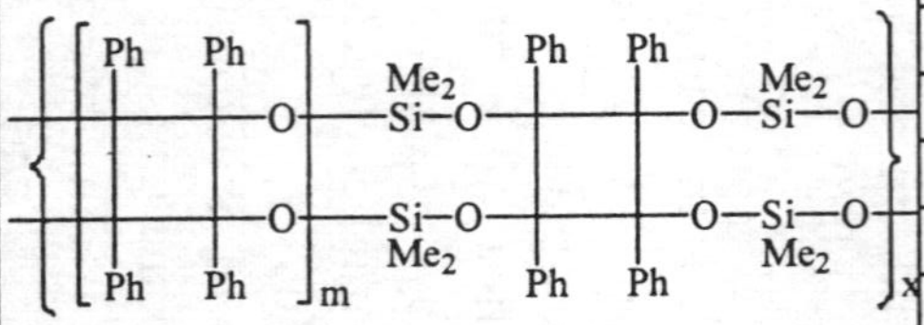	1	80	H_2PtCl_2	40	0.07	123-130	-
2		1	86	H_2PtCl_2	50	0.09	127-135	-
3		1	94	H_2PtCl_2	60	0.12	140-145	11.65
4		1	91	KOH	50	0.10	126-132	-
5		5	82	H_2PtCl_2	60	0.10	175-180	12.03
6		10	80	H_2PtCl_2	60	0.08	161-170	-

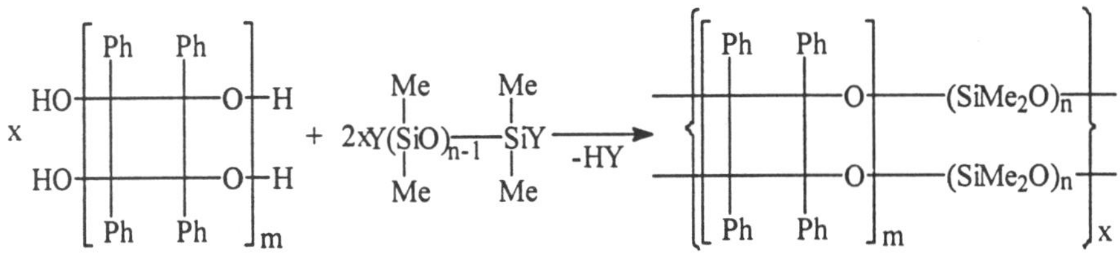

(7.12)

XII

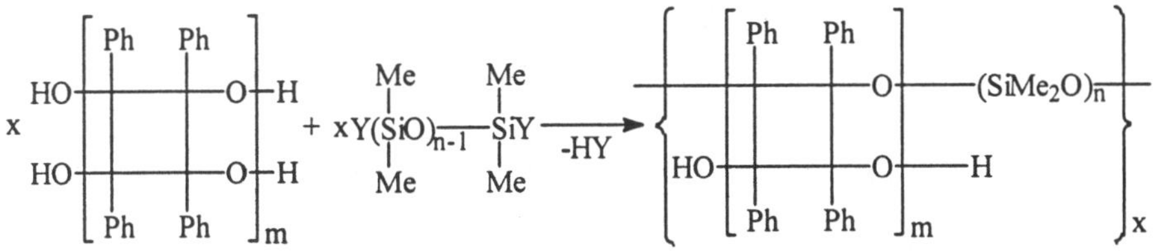

(7.13)

XIII

where: Y = Cl, Me_2N for structures XII and XIII; $m = 5 - 46$; $n = 1 - 66$.

In the case of Y = Cl, HFC was performed in 60% solution of anhydrous toluene and in the presence of HCl acceptor, pyridine. At the final stage, the reaction mixture was heated up to boiling of the diluter used. In the case of Y = Me_2N, HFC was performed in block. It is found that HFC of oligotetroles of *cis*-anti-*cis*-tactic structure with α,ω-bis(dimethylamino)dimethylsiloxanes even at 1:2 ratio of initial components ($f_{av} = 4$) produce completely soluble polymers. This testifies that despite observations made in works [53 – 56], dimethylamine, liberated in the reaction, does not cause disclosure of the cyclosiloxane ring. Moreover, despite reactions shown in refs. [53 – 56], *cis*-anti-*cis*-tactic configuration of initial PPSSO is, apparently, the reason for reactions (7.12) and (7.13) proceeding without gel formation. In accordance with the data obtained [53 – 56], obviously, copolymers derived from initial PPSSO with dominating *cis*-syndiotactic configuration possess faulted structures containing non-reacted groups in the chain. Presumably, disposition of these and end hydroxyl groups promotes gel formation proceeding, which is testified by HFC reaction of PPSSO (compound 9, Table 7.3), produced from the mother tetrole solution, and α,ω-dichloro- or α,ω-bis(diethylamino)dimethylsiloxanes. It is found that immediately when equimolar ratio of the initial components is shifted towards increase of dimethylsiloxanes concentration, the gel formation proceeds. Moreover, diethylamine liberated in the reaction displays higher basic properties than dimethylamine, which, in turn, likely effects the gel formation.

Thus, solid or viscous (with regard to *n* and *m* values) copolymers, completely soluble in organic solvents, possessing $\eta_{sp} = 0.11 - 0.32$ were synthesized in reactions (7.12) and (7.13).

Besides the above-mentioned structure XIII, other alternatives of macromolecule structuring by diagonal XIII[1] and edge XIII[2] are possible in HFC, preformed in accordance with the scheme (7.13):

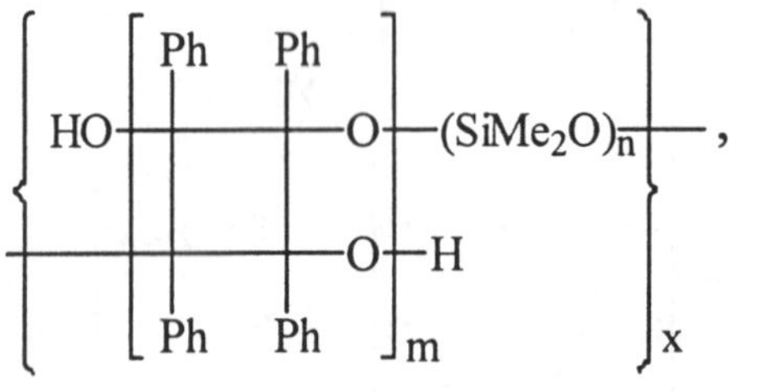

XIII[1]

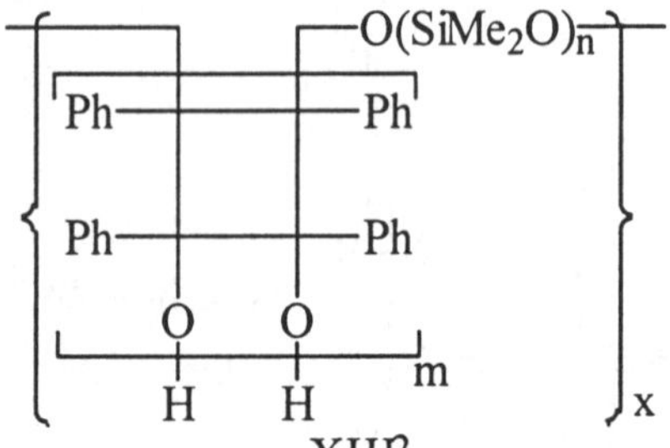

XIII[2]

However, the effect of disposition of polycyclic fragments on thermomechanical properties and thermal oxidative stability of copolymer is negligible [62 – 65].

Moreover, the authors suppose formation of different-chain structures due to incomplete cyclization in copolymers with double-stranded dimethylsiloxane units in reaction (7.12).

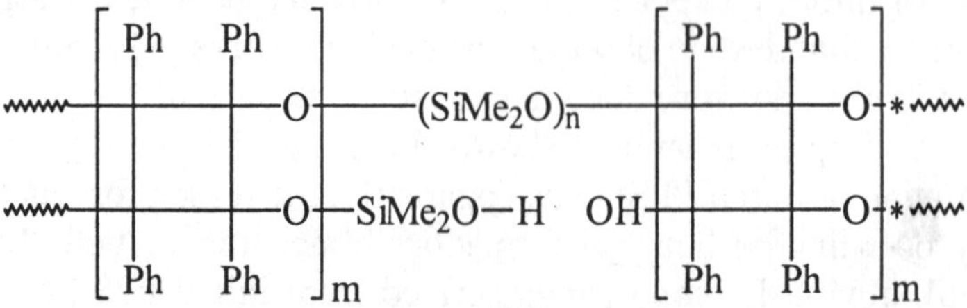

XII

Tables 7.14 and 7.15 show yields, ultimate analysis results and properties of synthesized block-copolymers of structures XII and XIII.

A single broad absorption band at 1,145 $см^{-1}$ for asymmetrical valence oscillations of ≡Si-O-Si≡ bonds is preserved in IR spectra of both synthesized block-copolymers and initial PPSSO. This proves the preservation of *cis*-anti-*cis*-tactic structure, i.e. the given structure of polycyclic blocks in synthesized block-copolymers. As a consequence, dimethylamine liberated in the reaction causes no regrouping of separate areas of the macromolecular chain.

Table 7.14

Some physical and chemical parameters of block-copolymers with PPSSO and double-stranded dimethylsiloxane units (structure XII)

No.	Copolymer structure	*m*	n_{SiO}	Yield, %	η	d_1, (d_1'), Å	T_g, °C
1	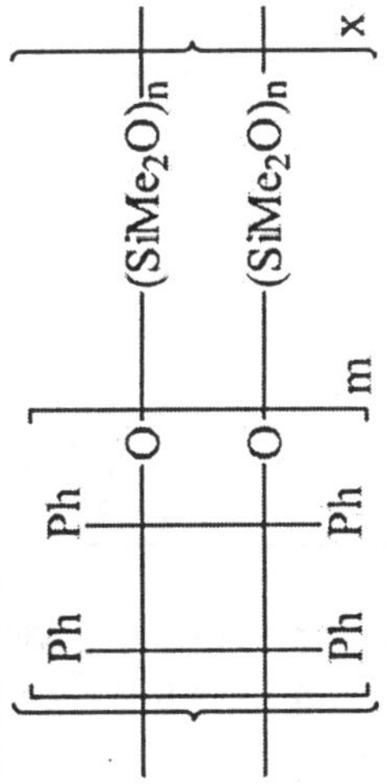	5	1	85	0.12	12.63	+210
2		5	2	86	0.13	12.60	+195
3		5	9	86	0.19	12.45	+156
4		5	21	87	0.24	12.38 7.80	-123 +146
5		5	32	88	0.26	12.03 7.30	-123 +140
6		5	38	89	0.27	12.54 7.29	-123 +137
7		5	66	90	0.28	12.52 7.18	-123 +129
8		10	1	85	0.11	12.46	+231
9		15	1	86	0.13	12.63	+237
10		15	2	86	0.14	-	+207
11		15	9	88	0.20	12.45	+172
12		15	21	89	0.23	12.39 7.30	-119 +157
13		15	32	89	0.27	12.11 7.30	-123 +148
14		15	38	89	0.27	12.20 7.36	-
15		15	66	90	0.29	-	-123 +140
16		20	1	87	0.13	12.57	+242
17		46	1	89	0.15	12.68	+241

Table 7.15

Some physical and chemical parameters of block-copolymers with PPSSO and double-stranded dimethylsiloxane units (structure XIII)

No.	Copolymer structure	m	n_{SiO}	Yield, %	η	d_1, (d_1'), Å	T_g, °C T_{soft}, °C
1	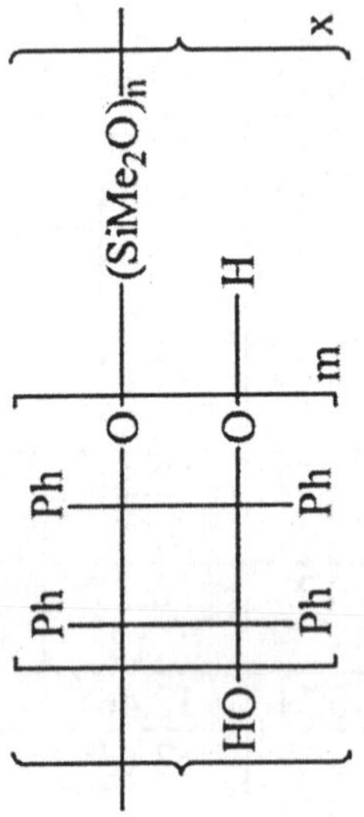	5	1	86	0.12	-	+185
2		5	2	86	0.12	12.60	+182
3		5	9	87	0.13	-	+170
4		5	21	88	0.13	12.40 (7.14)	-118 +158
5		5	38	89	0.14	12.34 (7.37)	-123 +146
6		5	66	90	0.15	12.10 (7.37)	-123 +140
7		15	1	86	0.16	-	+190
8		15	2	87	0.16	-	+180
9		15	9	87	0.21	-	+170
10		15	21	88	0.22	12.20 (7.30)	-116 +145
11		15	38	90	0.28	12.10 (7.37)	-123 +149
12		15	66	91	0.30	12.10 (7.37)	-123 +138

Thermogravimetric analysis of synthesized block-copolymers was performed. Figures 7.26 and 7.27 show TGA curves of structure XII double-stranded copolymers. It is indicated that at the same value (n) of dimethylsiloxane unit, thermal oxidative stability of block-copolymers is not significantly increased with polycyclic block length (m). Similar situation is observed for block-copolymers of structure XII at different values of n and m (Figure 7.26).

Comparison of thermal oxidative stability of block-copolymers of structures XII and XIII indicates that these with double-stranded dimethylsiloxane units are more stable to thermal oxidation than the ones with single-stranded dimethylsiloxane units (Figure 7.27). On the other hand, block-copolymers of structures V and VI are more stable to thermal oxidative degradation than the ones, synthesized in reactions (7.12) and (7.13).

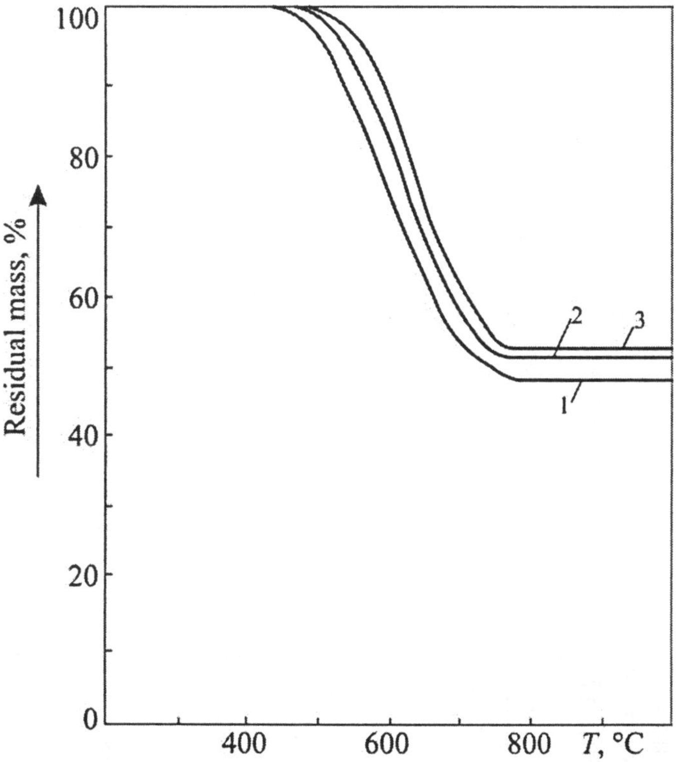

Figure 7.26. Thermogravimetric curves, measured for block-copolymers of structure XII: 1 – copolymer 17; 2 – copolymer 16; 3 – copolymer 8 (Table 7.14). The analysis was performed in air, at the heating rate of 5 deg/min

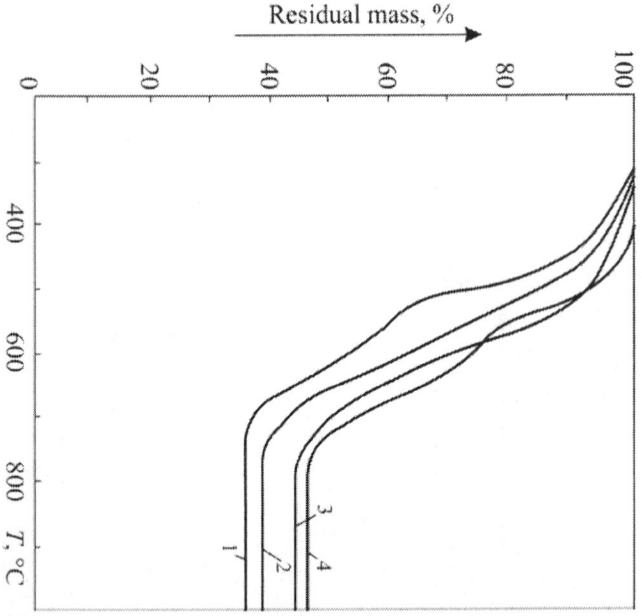

Figure 7.27. Thermogravimetric curves for block-copolymers of structure XII: 1 – copolymer 10; 2 – copolymer 2; 3 – copolymer 11; 4 – copolymer 3 (Table 7.14). The analysis was performed in air at 5 deg/min heating rate

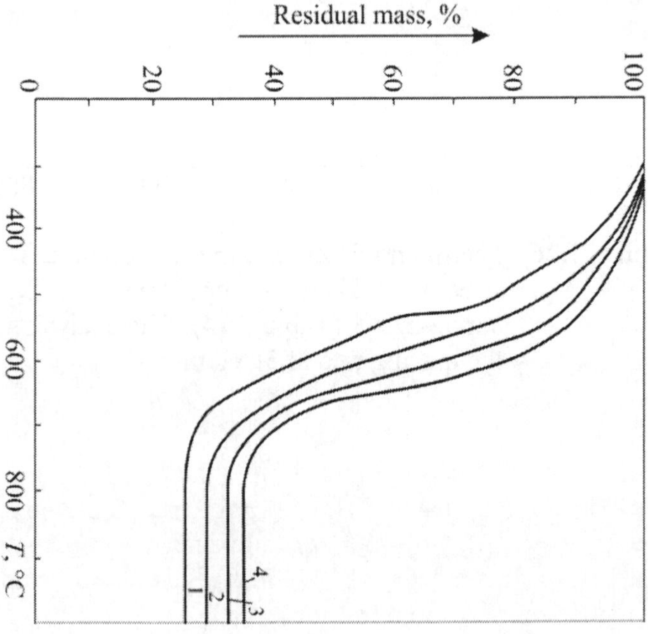

Figure 7.28. Thermogravimetric curves, measured for block-copolymers of structure XIII: 1 – copolymer 8; 2 – copolymer 2; 3 – copolymer 9; 4 – copolymer 3 (Table 7.15). The analysis was performed in air at 5 deg/min heating rate

Figure 7.29 shows thermogravimetric curves for block-copolymers indicating that the two-phase state of copolymers is displayed at large lengths of dimethylsiloxane unit ($n \geq 20$). The effects of devitrification and melting of dimethylsiloxane units and softening of polycyclic ladder block are clearly observed in the Figure. Mutual influence of linear dimethylsiloxane units and ladder phenylsilsesquioxane blocks in block-copolymers with both single- and double-stranded dimethylsiloxane units is indicated by T_g increase of dimethylsiloxane parts of macromolecular chain and decrease of polycyclic block softening point, T_{soft}.

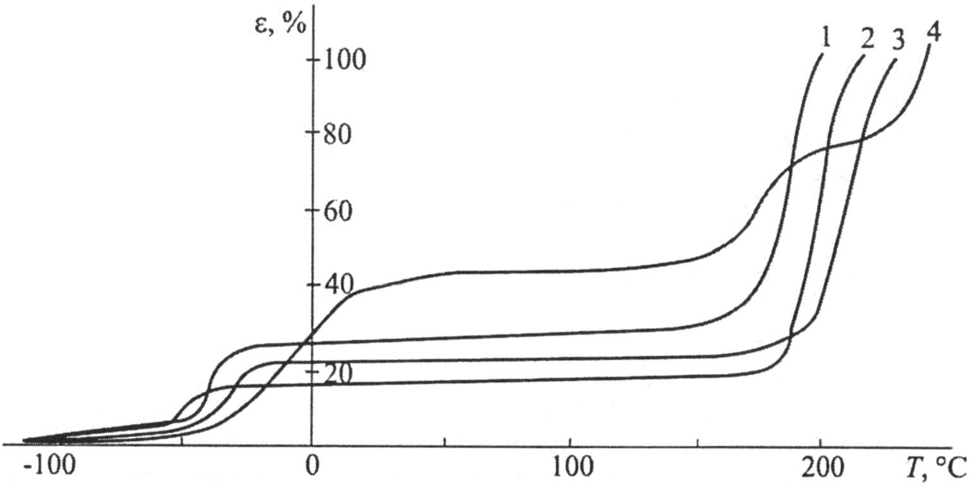

Figure 7.29. Thermomechanical curves for block-copolymers: 1 – copolymer 13 (Table 7.14); 2 – copolymer 10; 3 – copolymer 11; 4 – copolymer 5 (Table 7.15, $\sigma = 0.1$ kg/cm^2)

At short lengths of flexible dimethylsiloxane fragments ($n = 1 - 9$) block-copolymers are characterized by a single T_{soft}. The increase of dimethylsiloxane unit length, n, decreases T_{soft} of copolymers (refer to Tables 7.14 and 7.15).

X-ray diffraction studies of block-copolymers indicate clearly the dependence of number of diffraction maximums on the ratio of lengths of rigid PPSSO (m) and flexible dimethylsiloxane (n) fragments. At short lengths of dimethylsiloxane units ($n = 1 - 9$), independently of the ladder block length (m), two diffraction maximums are observed in the areas $d_1 = 12.0 - 12.63$ Å and $d_2 = 4.40 - 5.0$ Å. The absence of a diffraction maximum, corresponded to linear siloxane component, also proves the absence of phase stratification at short lengths of dimethylsiloxane unit (*n*). Maximal interchain distance, d_1 = 12.68 Å, for both types of block-copolymers (refer to Tables 7.14 and 7.15) is reached at dimethylsiloxane unit length *n* = 1. As the unit length increases, the first diffraction maximum, d_1, decreases.

Occurrence of the third diffraction maximum, d_1^{1}, in the range d_1 = 7.20 - 7.40 Å, corresponded to the interchain distance of dimethylsiloxane unit, proves that at definite lengths of the rigid block (*m* = 5 – 15) and flexible siloxane fragment (*n* = 21 -66) a two-phase system with both single- and double-stranded dimethylsiloxane units in macromolecular chain is formed in block-copolymers. Similar to block-copolymers of structures V and VI, one of the phases is represented by parallel, closely packed ladder fragments, and another is formed by dimethylsiloxane linear fragments. The ratio of intensities of the diffraction maximums, d_1 and d_1^{1}, is proportional to the ratio of rigid and flexible fragments.

Thus, block-copolymers containing phenylsilsesquioxane blocks of *cis*-anti-*cis*-tactic structures V, VI, XII and XIII and single- and double-stranded dimethylsiloxane units in macromolecular chain are ahead test industrially produced *Lestosyl SF* copolymer [54] containing phenylsilsesquioxane and double-stranded diorganosiloxane fragments in the chain. The advantage of the block-copolymers is that they do not form cross-linked systems during synthesis. They are of interest for deriving cold hardening heat and frost resistant rubbers and vulcanizates from them.

Organosiloxane block-copolymers with oligocyclotetra(hexa)siloxane fragments in linear diorganosiloxane chain were synthesized by HFC of produced dihydroxy-containing oligomers of the bead-like type of structures II and II (Section 2.1) with α,ω–diamino(bisdimethylamino)dimethylsiloxanes at 1:1 ratio of initial components. The reaction proceeded in solution or in block, in the temperature range of 20 - 100°C, at *m* < *n*. For higher reaction degree at *m* > *n*, it was performed at 100 - 150°C. Stripping of ammonia or dimethylamine, liberated during synthesis, was performed in vacuum (*P* = 3 – 5 mmHg). General scheme of the reaction is shown below [13, 19, 66 – 68]:

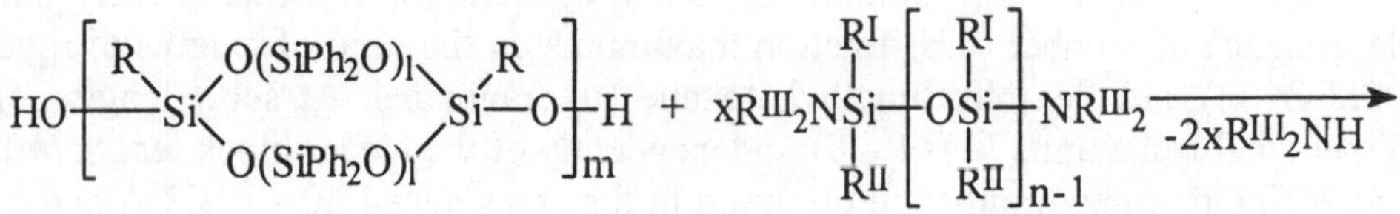

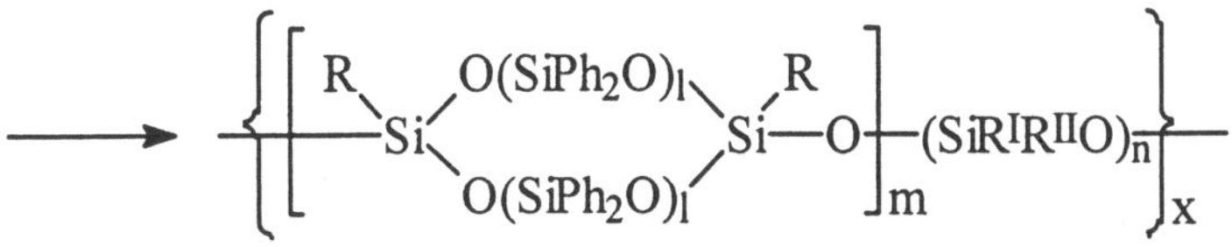

(7.14)

XIV, XV

where R^{III} = Me, H; R = Me, Ph; $R^{I} = R^{II}$ = Me, $R^{I} \neq R^{II}$; l =1, 2; m = 7 - 10, n = 2 - 66.

Similarly, HFC reaction of dihydroxy-containing α,ω-dihydroxy-1.3-diphenyl-2-oxaindane and α,ω-dihydroxybicyclocarbosiloxanes, produced by homofunctional condensation of corresponded dihydroxy-derivatives (refer to Chapters 1 and 2) with α,ω-dichloro- and α,ω-bis(dimethylamino)dimethylsiloxanes. The reaction conditions are the same as for reaction (7.14). The reaction scheme is shown below [69 – 72]:

$$xHO\text{-}[A\text{-}O]_mH + xY(SiMe_2O)_{n-1}SiMe_2Y \xrightarrow[-2xHY]{} \quad (7.15)$$

$$\longrightarrow HO\{[A-O]_{\overline{m}}-(Me_2SiO)_{\overline{n}}\}H$$

where A-O is

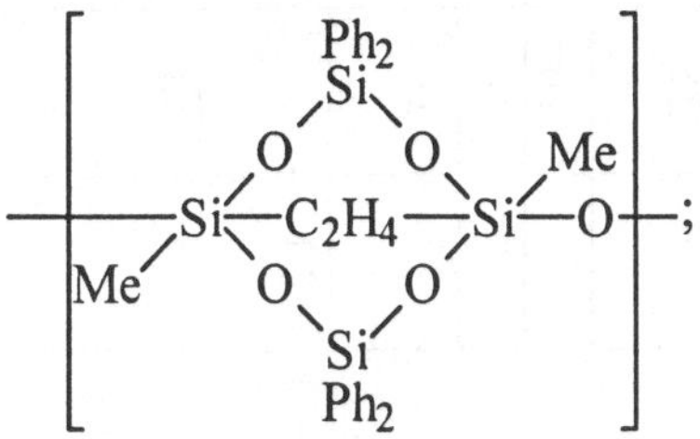

XVI

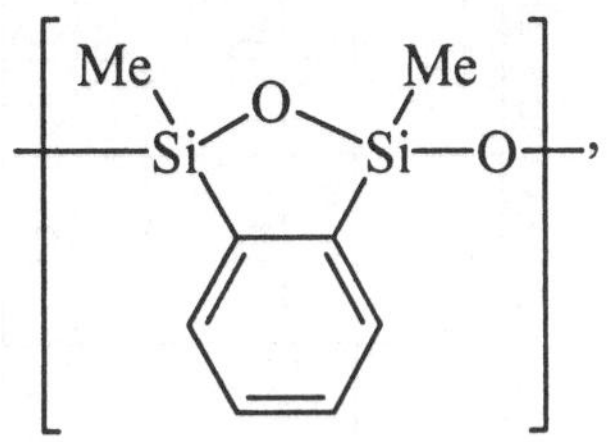

XVII

Y = Cl, Me_2N; m = 6 – 11; n = 2 – 66.

Table 7.16

Some physical and chemical parameters of structure XIV block-copolymers

No.	Polymer unit	n	m	R	η_{spec}	d_1, Å d^1_1, Å	T_g, °C T_{soft}, °C	Yield, %
1	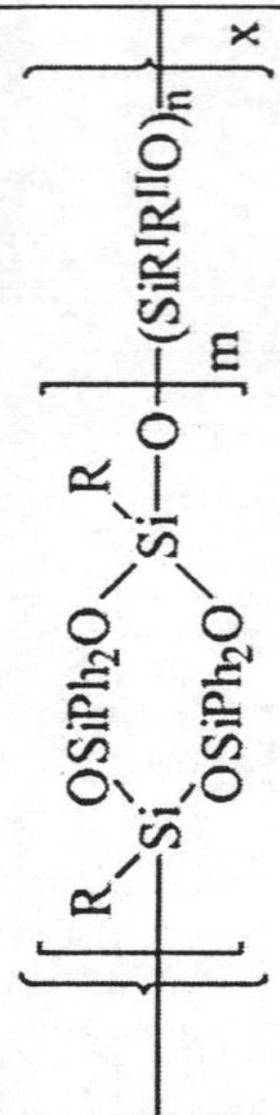	7	10	Me	0.21	-	+61	94
2		34	10	Me	0.24	-	-123 +47	95
3		7	10	Me	0.18	11.83	-	95
4		53	20	Me	0.26	11.80	-	97
5		53	28	Me	0.28	11.47 7.44	-123 +43	96
6		3	10	Ph	0.13	-	+72	98
7		7	10	Ph	0.16	11.92	+67	96
8		25	10	Ph	0.22	11.80 7.81	-120 +51	96
9		53	10	Ph	0.25	10.8 7.63	-123 +46	97
10		53	14	Ph	0.21	11.05 7.44	-	96

Note: x = R^I = Me, R^{II} = Ph; for the rest cases R^I = R^{II} = Me.

Table 7.17

Some physical and chemical parameters of structure XV block-copolymers

No.	Polymer unit	n	m	R	η_{spec}	d_1, Å d^1_1, Å	T_g, °C T_{soft}, °C	Yield, %
1	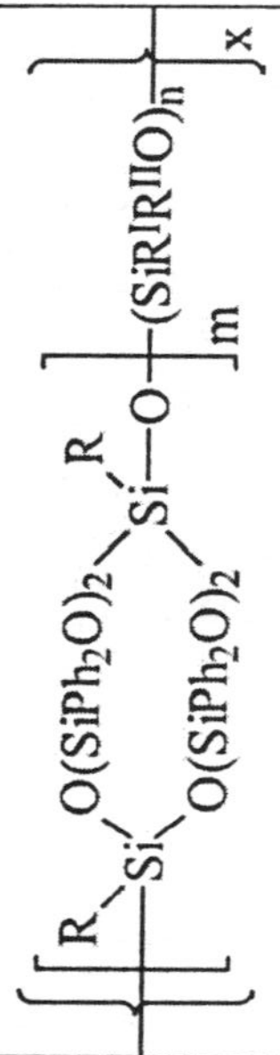	32	7	Me	0.24	10.15 7.31	-123 +37	87
2		66	7	Me	0.31	10.15 7.29	-123 +32	89
3		2	10	Me	0.09	10.78	+73	79
4		12	10	Me	0.15	10.56	+48	83
5		24	10	Me	0.13	10.31 8.29	-	85
6		66	10	Me	0.25	10.27 8.23	-123 +34	90
7		2	10	Ph	0.10	10.86	+72	79
8		32	10	Ph	0.24	10.65 7.51	-119 +46	87
9		66	10	Ph	0.30	10.45 7.50	-123 +43	90

Note: x = R^I = Me, R^{II} = Ph; for the rest of cases, R^I = R^{II} = Me.

Table 7.18

Some physical and chemical parameters of structure XVI block-copolymers

No.	Polymer unit	n	m	η_{spec}	d_1, Å	d_1^I, Å	T_g, °C T_{soft}, °C	Yield, %
1	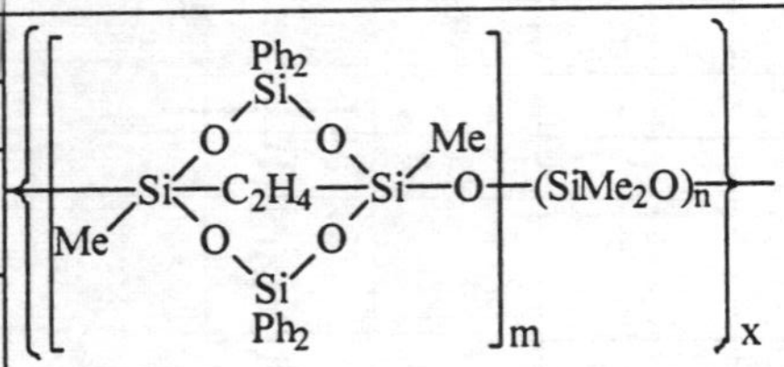	2	6	0.12	10.86	-	+74÷+78	91
2		10	6	0.15	10.81	-	+47÷+52	92
3		36	6	0.25	10.80	7.62	-123 +41	94
4		66	6	0.30	10.41	7.42	-123 +37	95

Table 7.19

Some physical and chemical parameter of block-copolymers of structure XVII:

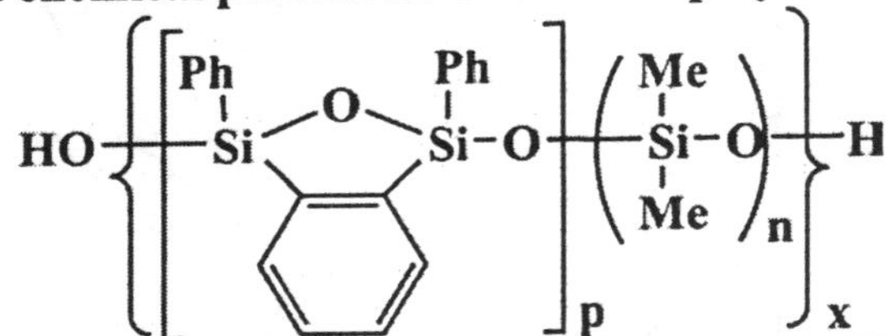

No.	n	m	Yield, %	$[\eta]$ dl/g	d_1, Å d_1^1, Å	T_g, °C	T_{melt}, °C T_{soft}, °C	5% mass loss, °C	Residual mass, %
1	2	3	4	5	6	7	8	9	10
1	2	17	93	0.12	11.05	-	+88÷+91	420	51
2	4	17	94	0.14	10.70	-	+75÷+78	-	-
3	12	17	95	0.21	10.35	-	+84÷+87	390	46
4	32	17	96	0.26	-	-123	-48; +73;	-	-
5	66	17	96	0.31	9.41 7.63	-	-44; +68;	-	-
6	2	29	92	0.12	11.11	-	+87÷+89	-	-
7	4	29	93	0.21	-	-	+79÷+82	-	
8	12	29	93	0.23	-	-	+77÷+79	380	36
9	32	29	94	0.34	10.79 7.68	-123	-45; +78;	-	-
10	66	29	96	0.37	9.56 7.70	-123	-43; +78;	330	32

Block-copolymers, synthesized by schemes (7.14) and (7.15), represent transparent or light-yellow viscous or solid products (η_{spec} = 0.09 – 0.32), well soluble in usual solvents. Tables 7.17 and 7.18 show physical and chemical parameters, ultimate analysis results and yields of synthesized block-copolymers.

In the case of α,ω–bis(dimethylamino)dimethylsiloxane, the method of block-copolymer synthesis by schemes (7.14) and (7.15) is quite simple and does not require additional elimination of side reaction products.

GPC studies of some block-copolymers of structures XIV - XVII were performed. Figure 7.30 shows molecular mass distribution (MMD) of structure XVII block-copolymer, which indicates its monomodality and molecular masses, $\bar{M}_{\omega}$, falling within the range of $1.37\times10^5 - 1.52\times10^5$ and polydispersity varying from 1.32 to 1.93.

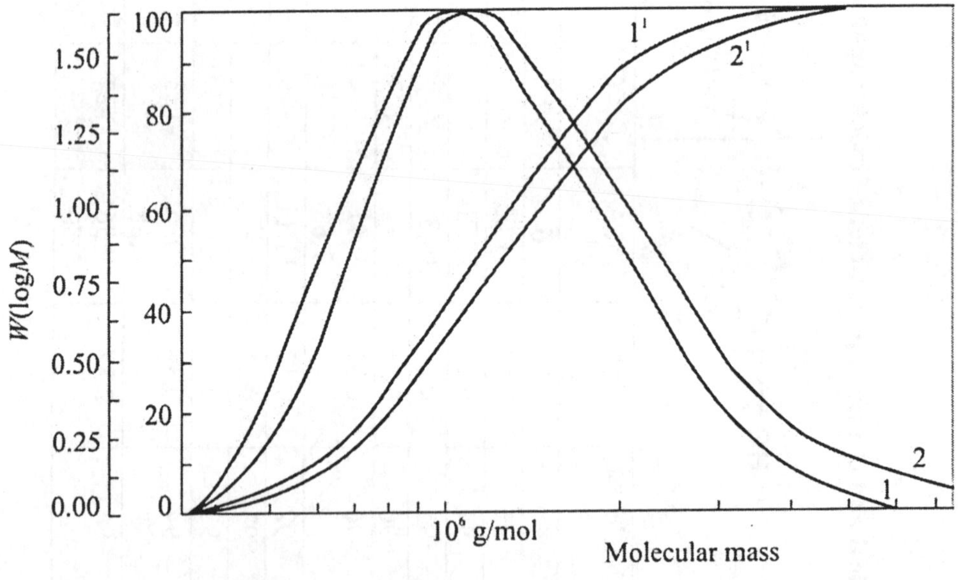

Figure 7.30. Gel permeation chromatography curves of poly(disilaoxaindane)-dimethylsiloxane block-copolymer: 1 and 1' – block-copolymer 10; 2 and 2' – block-copolymer 8 (Table 7.18)

X-ray diffraction analysis of synthesized block-copolymers shows (Figures 7.31 and 7.32) that for short lengths of dimethylsiloxane chain (n = 2 – 12), copolymers represent single-phase amorphous systems. Diffraction patterns display two maximums: 2θ = 8.0 – 8.5° and 2θ = 19.25 – 19.50°. The main maximum $2\theta = 8.0 - 8.5°$ corresponds to the maximum value of the interchain distance, $d_1 = 11.05 - 10.40$ Å, and the second maximum $2\theta = 19.25 - 19.50°$ corresponds to $d_2 = 4.49 - 4.55$ Å, which characterizes both intramolecular and interatomic interactions [73]. As the length of dimethylsiloxane chain increases, interchain distances become shorter and for copolymer 3 ($n = 12$) (Table. 7.19) equals $d_1 = 10.35$ Å. For higher values of dimethylsiloxane unit length ($n \geq 32$), two-phase systems are formed. For such block-copolymers, two-phase system formation can be explained by the high incompatibility of flexible dimethylsiloxane fragments with rigid oligomeric cyclosiloxane blocks.

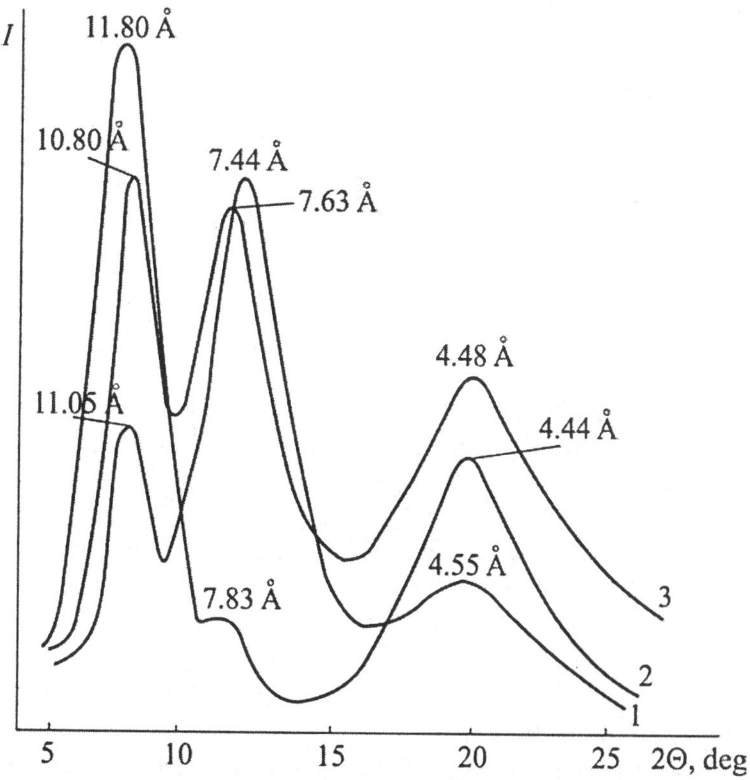

Figure 7.31. Wide angle X-ray diffraction patterns for structure XIV block-copolymers: 1 – block-copolymer 9; 2 – block-copolymer 7; 3 – block-copolymer 8 (Table 7.16)

Figure 7.31 shows three diffraction maximums at $2\theta = 9.25 - 9.40°$, $2\theta = 11.50 - 11.60°$ and $2\theta = 19.25 - 20.40°$. The maximum at $2\theta = 9.25 - 9.40°$

corresponds to highest interchain distances between rigid oligomeric cyclosiloxane blocks, d_1 = 9.41 – 11.80 Å. Occurrence of the third diffraction maximum at 2θ= 11.50 – 11.60° corresponds to the interchain distance between linear flexible dimethylsiloxane blocks, d_1 = 7.63 – 7.70 Å. The maximum of 2θ = 19.25 – 20.40° corresponds to d_2 = 4.23 – 4.61 Å (refer to Figure 7.31).

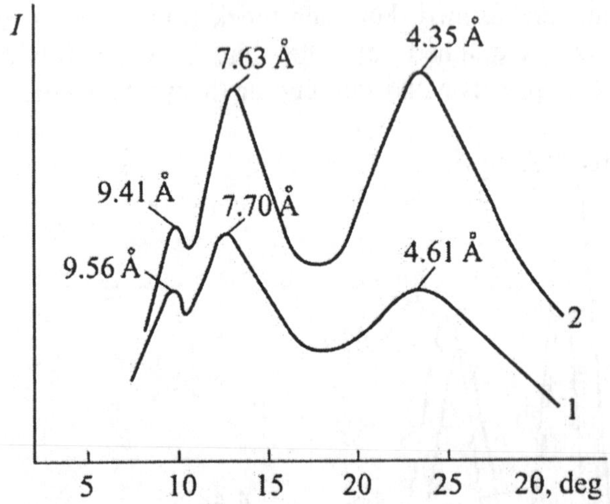

Figure 7.32. Wide angle X-ray diffraction patterns for structure XVII block-copolymers: 1 – block-copolymer 10; 2 – block-copolymer 5 (Table 7.19)

Thermomechanical and differential scanning calorimetry studies of synthesized BC were performed. It is found that at short lengths of linear dimethylsiloxane unit ($n \leq 12$), these copolymers represent single-phase systems with one transition temperature, T_{melt} = +61 - +91°C. Tables 3.16 - 3.19 show that at the same length of flexible dimethylsiloxane unit the of rigid fragments' length increase on the melting point of copolymers is negligible. Figures 7.33a and 7.33b show DSC thermograms of structure XVII block-copolymers. Curves 1" and 2" show temperature transitions for scanning at cooling (Figure 7.33a). One temperature transition of copolymers was also displayed by thermomechanical studies, and the values obtained are close to DSC results.

As the length of dimethylsiloxane unit increases ($n \geq 32$) (Figure 7.33b), DSC curves display glass transition temperature in the range of -120 ÷ -123°C and melting temperature at -45°C. This T_g value is typical of linear dimethylsiloxane fragments and agrees with data from the literature [5]; T_{melt} = +75°C is typical of rigid poly(disilaoxaindane) fragments (Figure 7.33b). Curves 1, 1' and 2, 2' in Figure 7.33b display phase transitions at the first and the second scanning at heating, and curves 3 and 3' display the phase transition for scanning at cooling. This clearly displays the effects on glass transition and melting temperatures typical of flexible dimethylsiloxane fragments and polycyclic disilaoxaindane blocks.

For such block-copolymers, formation of the two-phase system indicates their microdomain structure. Note also that thermomechanical studies of structures XIV-XVI copolymers display no melting transition typical of poly(dimethylsiloxane) blocks. Therefore, at high content of rigid polycyclic blocks the copolymer represents a continuous phase with rigid structure occupying the whole volume and discrete aggregates of dimethylsiloxane fragments, distributed in it. Formation of the microdomain structure is observed at definite *m* and *n* values only.

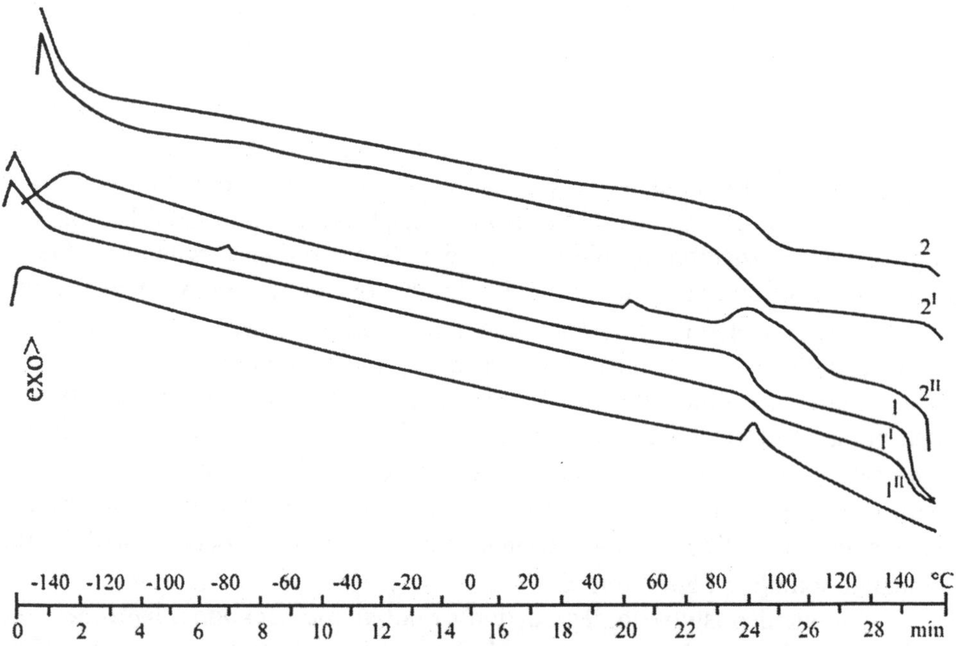

Figure 7.33a. DSC thermograms for structure XVII block-copolymers: 1, 1' – the first scanning at heating and 2, 2' - the second scanning at heating scans for block-copolymers 3 and 8, respectively; 1" and 2" – scanning at cooling (Table 7.19)

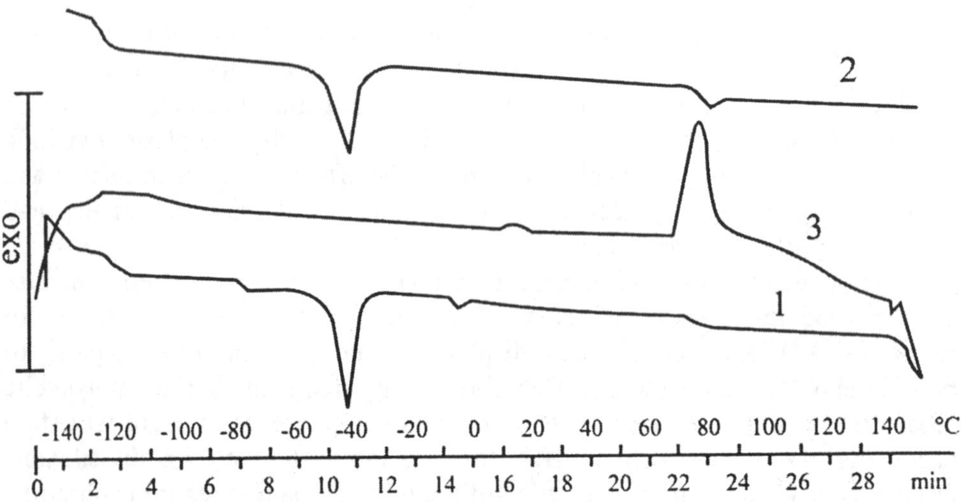

Figure 7.33b. DSC thermograms for structure XVII block-copolymers: 1 and 2 - the first and second scanning at heating, respectively; 3 – scanning at cooling for block-copolymer 10 (Table 7.19)

Thermogravimetric analysis of some block-copolymers was performed on electron thermal weights in argon with simultaneous sampling and analysis of gaseous degradation products. It is known from the literature that introduction of groups different from ≡Si-O bonds into PDMS disturbs degradation of macromolecule by the depolymerization mechanism due to impossibility of complex formation [74, 75]. Moreover, it is known that the type of degradation in ladder polymers (depolymerization or decay by ≡Si-C bond) depends on the origin of substituting groups at silicon atom [76 – 78].

Figure 7.34 indicates that for the majority of studied block-copolymers with organocyclotetrasiloxane fragments in the chain of structure XIV in the temperature range of 350 - 400°C, mass losses are below 5%. Since 450°C, thermal degradation rate of studied block-copolymers is increased and reaches the maximum at 570°C. Mass change curves become supersaturated in the temperature range of 580 - 600°C.

Simultaneously with degradation of polymeric backbone, hydrogen and methane are liberated during thermal degradation of studied polymers [79], accompanied by radical cross-linking by methyl and phenyl groups. For all block-copolymers, hydrogen liberation is initiated at 500°C and reaches the first maximum at 750°C, and the second basic maximum occurs at 800°C. From

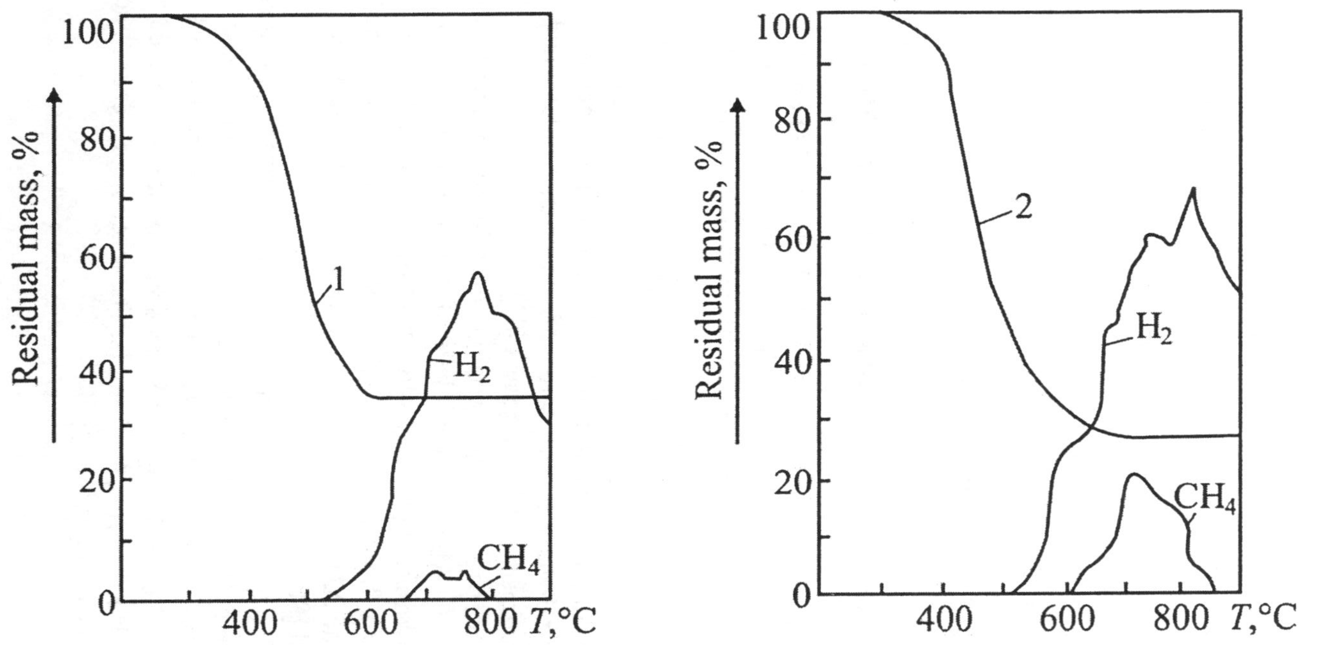

Figure 7.34. Thermogravimetric curves for block-copolymers: 1 – block-copolymer 5; 2 – block-copolymer 8 (Table 7.16, in argon)

block-copolymers methane is liberated at 600°C. Therefore, its quantity is increased with the length of linear fragments. The increase of dimethylsiloxane units' concentration also causes the increase of gaseous degradation products' liberation, primarily, the quantity of liberated methane is increased (Figure 7.33). As a consequence, thermal stability of block-copolymers is decreased. This is also proved by thermal oxidative stability of block-copolymers of structures XVI and XVII.

Synthesized block-copolymers are of interest due to containing reactive cyclosiloxane fragments, which polymerization in the presence of an anionic catalyst may induce disclosure of cyclosiloxane rings with liberation of gaseous products. Thereby the cross-linking frequency may be regulated.

7.1.3. Organosilicon cyclolinear block-copolymers of the $(AB)_mC_n$ type

There are short data on organosilicon block-copolymers of cyclolinear structure of the $(AB)_mC_n$ type in the literature [80 – 89]. It is known that introduction of various cyclic groups to both linear dimethylsiloxane backbone and side chains increases thermal oxidative stability compared with linear unblocked PDMS. This is explained by high stability of phenyl cyclic groups to oxidation and their inhibiting effect on methyl group oxidation [74, 90]. Moreover, introduction of non-Me_2Si-O- groups into linear backbone of PDMS disturbs the spiral-shaped structure of macromolecules and changes physicochemical properties of copolymers [91].

For the purpose of synthesizing cyclolinear orlanosilicon block-copolymers of the $(AB)_mC_n$ type, heterofunctional condensation of rigid cyclosiloxane dihydroxy-containing oligomers of the $(AB)_m$ type, differing by physical parameters from flexible PDMS blocks, synthesized in Sections 2.5 (structures XIII and XIV), 3.2 (structure XI) and Chapter 5 (structures XIV and XV), were studied [80 – 89].

HFC of dihydroxy-containing $(AB)_m$ type oligomers with α,ω–dichlorodimethylsiloxanes proceeded at 1:1 ratio of initial compounds in 50 - 60% anhydrous toluene in the presence of pyridine and temperature range of 20 - 25°C. At the final stage, the reaction mixture was heated up to the boiling point of toluene during 4 – 5 hours. The reaction proceeds in accordance with the following scheme [80 - 89]:

$$xHO-(AB)_{m}-H + xHO-(SiMe_2O)_{n-1}-Cl \xrightarrow[-2kPy\cdot HCl]{2kPy}$$

$$\longrightarrow HO\!\left[(AB)_{m}-(SiMe_2O)_{n}\right]_{x}\!H$$

where:

$(AB)_m =$ 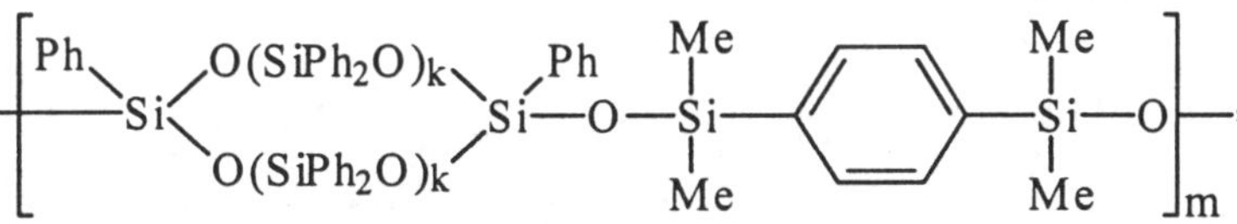 (7.16)

$k = 1$ (XVIII), 2 (XIX); $m = 4 - 16$; $n = 2 - 66$;

$(AB)_m =$ 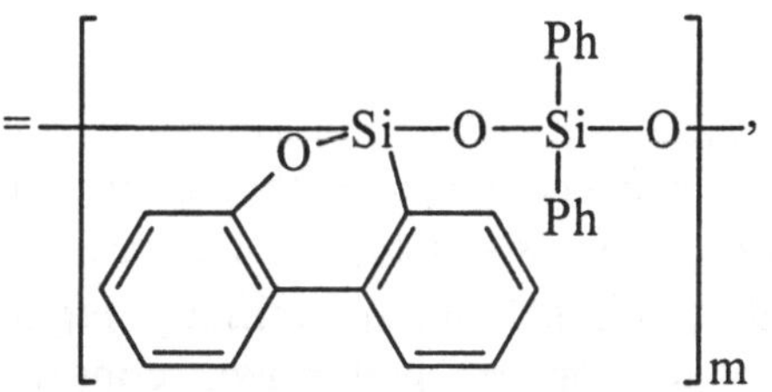 (7.17)

$m = 10, 14$ (XX); $n = 2 - 39$;

$(AB)_m =$ 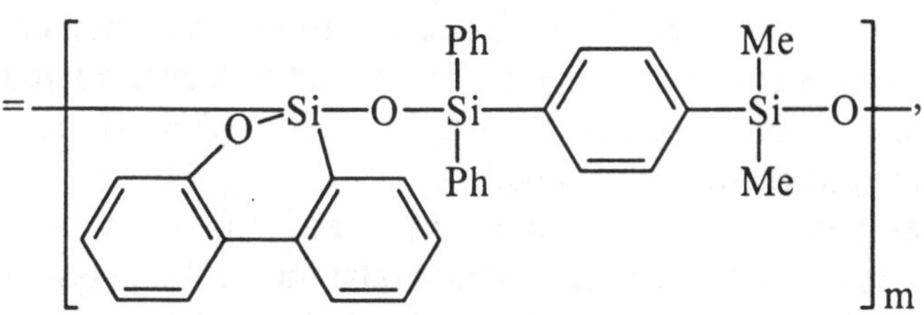 (7.18)

$m = 6 - 12$ (XXI); $n = 2 - 39$;

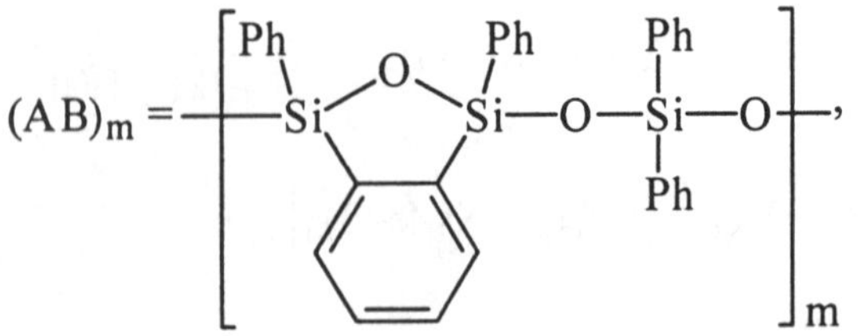

(7.19)

$m = 6 – 9$ (XXII); $n = 2 - 39$;

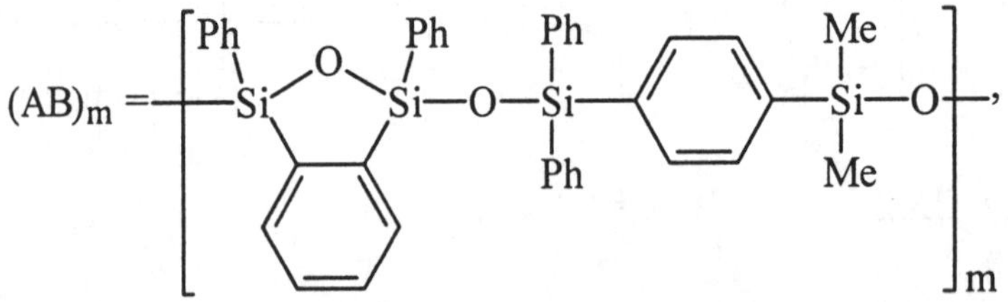

(7.20)

$m = 18$ (XXIII); $n = 2 – 39$.

After re-precipitation from toluene with regard to the length of linear dimethylsiloxane backbone, synthesized copolymers represent solids or liquids with η_{spec} = 0.07 - 2.95, well soluble in ordinary organic solvents. Some physical and chemical parameters of synthesized copolymers are shown in Tables 7.20 - 7.25.

Partial re-precipitation of some copolymers of structures XVIII - XXIII by methanol from toluene gave different fractions. The ultimate analysis of these fractions displays their low difference from one another and initial copolymers, which is low-probable for HFC homofunctional polycondensation, and copolymers with regular disposition of $(AB)_m$-type blocks in dimethylsiloxane backbone are obtained.

The gel permeation chromatographic analysis of some synthesized block-copolymers has indicated their monomodal molecular mass distribution in the range of $\bar{M}_{\omega} = 1.27\times10^5 – 1.98\times10^5$ and polydispersity variation $D = 1.41 – 1.53$.

Table 7.20

Some physical and chemical parameters of structure XVIII block-copolymers [80, 83]:

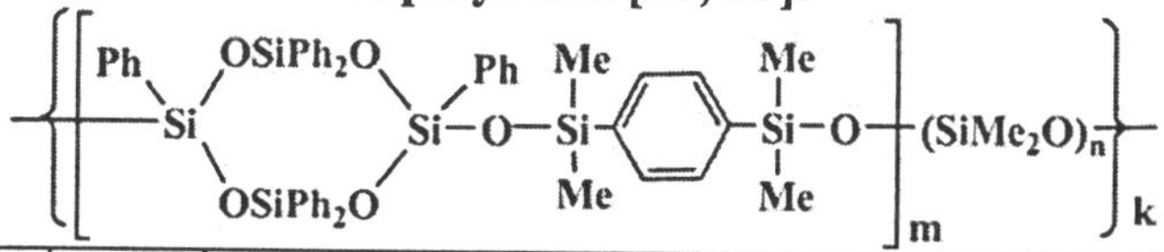

No.	*m*	*n*	Yield, %	η_{spec}	d_1, Å d_1^1, Å	T_g, °C	5% mass losses
1	6	2	78	0.11	10.05	+68	470
2	6	4	79	0.13	-	+57	450
3	6	12	84	0.17	-	+8	430
4	6	32	88	0.23	9.48 7.50	-123 +35	-
5	6	66	90	0.30	9.41 7.54	-123 +25	-
6	16	32	88	0.24	9.83 7.69	-123 +47	460
7	16	38	89	0.26	9.83 7.56	-123 +40	-
8	16	66	90	0.32	9.83 7.56	-123 +36	380

Table 7.21

Some physical and chemical parameters of structure XIX block-copolymers [81, 82]:

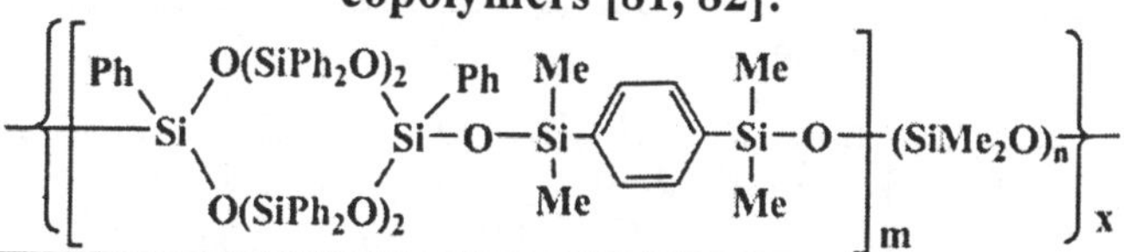

No.	*m*	*n*	Yield, %	η_{spec}	d_1, Å d_1^1, Å	T_g, °C	5% mass losses
1	4	2	79	0.13	10.02	+54	410
2	4	4	90	0.18	-	+40	-
3	4	12	95	0.22	9.75	+10	-
4	4	32	95	0.27	9.65 7.52	-123 +31	400
5	4	66	96	0.29	9.60 7.41	-123 +27	-

Table 7.22

Some physical and chemical parameters of structure XX block-copolymers [84, 85]:

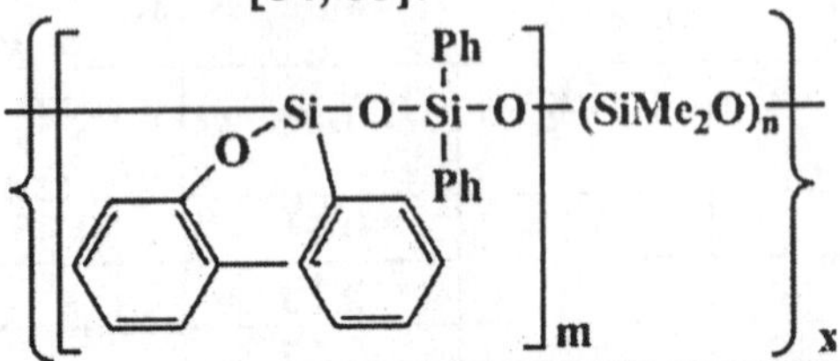

No.	m	n	Yield, %	$[\eta]$, dl/g	d_1, Å d_1^1, Å	T_g, °C, T_{cryst}, °C	T_{melt}, °C T_g, °C	5% mass losses	$\bar{M}_\omega \times 10^{-5}$ (D)
1	10	2	89	0.08	10.35	-	+61÷+65	-	
2	10	10	91	0.12	10.40	-	+42÷+48	360	
3	10	21	93	0.21	10.15 7.70	-115 -60÷-90	-40÷-60 +41÷48	385	
4	10	27	92	0.26	10.10 7.70	-123 -60÷-86	-35÷-57 +42÷+44	-	1.27 (1.41)
5	10	39	94	0.31	9.83 7.53	-123 -	-34÷-54 +39÷+45	-	-
6	14	2	92	0.11	10.10	-	+64÷+69	-	-
7	14	10	91	0.14	10.10	-	+45÷48	395	-
8	14	21	94	0.28	9.83 7.50	-123 -	-40÷-46 +42÷+53	375	-
9	14	27	93	0.30	10.10 7.78	-123 -60÷-90	-39÷-50 +40÷+47	-	1.98 (1.53)
10	14	39	94	0.34	-	-123 -60÷-90	-36÷-52 +35÷+42	-	-

Table 7.23

Some physical and chemical parameters of structure XXI block-copolymers [85, 86]:

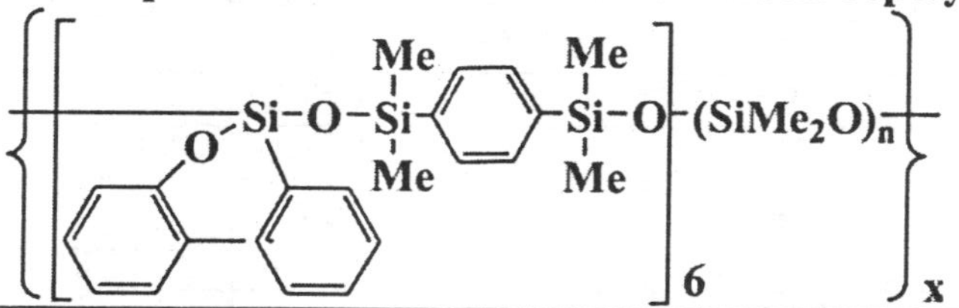

No.	n	Yield, %	η_{spec}	d_1, Å d_1^1, Å	T_g, °C, T_{cryst}, °C	T_{melt}, °C T_g, °C	5% mass losses	Residual mass, %	$\bar{M}_\omega \times 10^{-5}$
1	2	91	0.72	10.45 7.53	-	+44÷+51	420	44	-
2	6	86	0.98	10.04			-	-	4.67
3	21	96	1.57	-	-120 -	-30÷-60 +35÷+50	400	30	5.06
4	39	94	2.95	10.10 7.53	-123 -	-38÷-50 +25÷+50	-	-	-

Table 7.24

Some physical and chemical parameters of structure XXII block-copolymers [88, 89]:

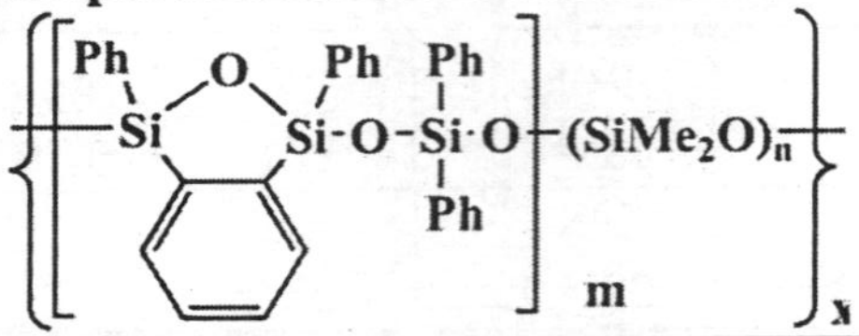

No.	*m*	*n*	Yield, %	[*η*], dl/g	d_1, Å d_1^1, Å	T_g, °C, T_{cryst}, °C	T_{melt}, °C T_g, °C	10% mass losses	Residual mass, %
1	6	2	87	0.11	10.40	-	+110	450	38
2	6	12	92	0.15	-		+81	430	35
3	6	21	94	0.18	10.10 7.53	-123 -85÷-110	-30÷-57 +50÷+60	-	-
4	6	39	95	0.27	9.07 7.7	-123 -	-30÷-50 +58÷+63	400	9
5	9	2	88	0.13	10.65	-	+114	470	41
6	9	12	93	0.17	10.45	-	+85	-	-
7	9	21	94	0.22	10.10 7.65	-123 -95÷-105	-30÷-60 +67÷+73	450	35
8	9	39	96	0.37	9.56 7.87	-123 -80÷-100	-38÷-60 +45÷+54	420	18

For block-copolymer 3, $\bar{M}_\omega = 5.59\times10^4$, $\bar{M}_n = 3.38\times10^4$, $\bar{M}_z = 9.04\times10^4$ ($D = 1.65$).

Table 7.25

Some physical and chemical parameters of structure XXIII block-copolymers [89]:

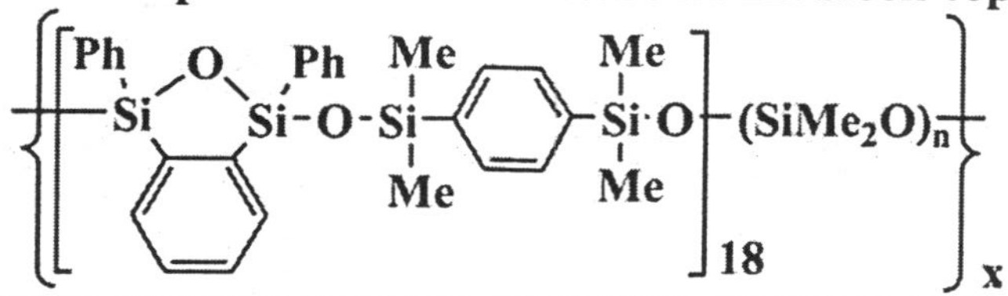

No.	n	Yield, %	η_{spec}	d_1, Å d_1^1, Å	T_g, °C, T_{cryst}, °C	T_{melt}, °C T_g, °C	10% mass losses	Residual mass, %
1	2	88	0.40	10.72	-	+115÷+120	500	38
2	12	92	0.52	-	-	+130÷+134	450	36
3	21	94	0.62	10.40 8.04	-123 -85÷-102	-40÷-56 +77÷+80	-	35
4	39	96	0.83	9.55 7.53	-123 -90÷-100	-40÷-52 +60÷+70	460	32

Note: For block-copolymer 2, $\bar{M}_\omega = 6.67\times10^4$, $\bar{M}_n = 4.35\times10^4$, $\bar{M}_z = 8.75\times10^4$ ($D = 1.53$).

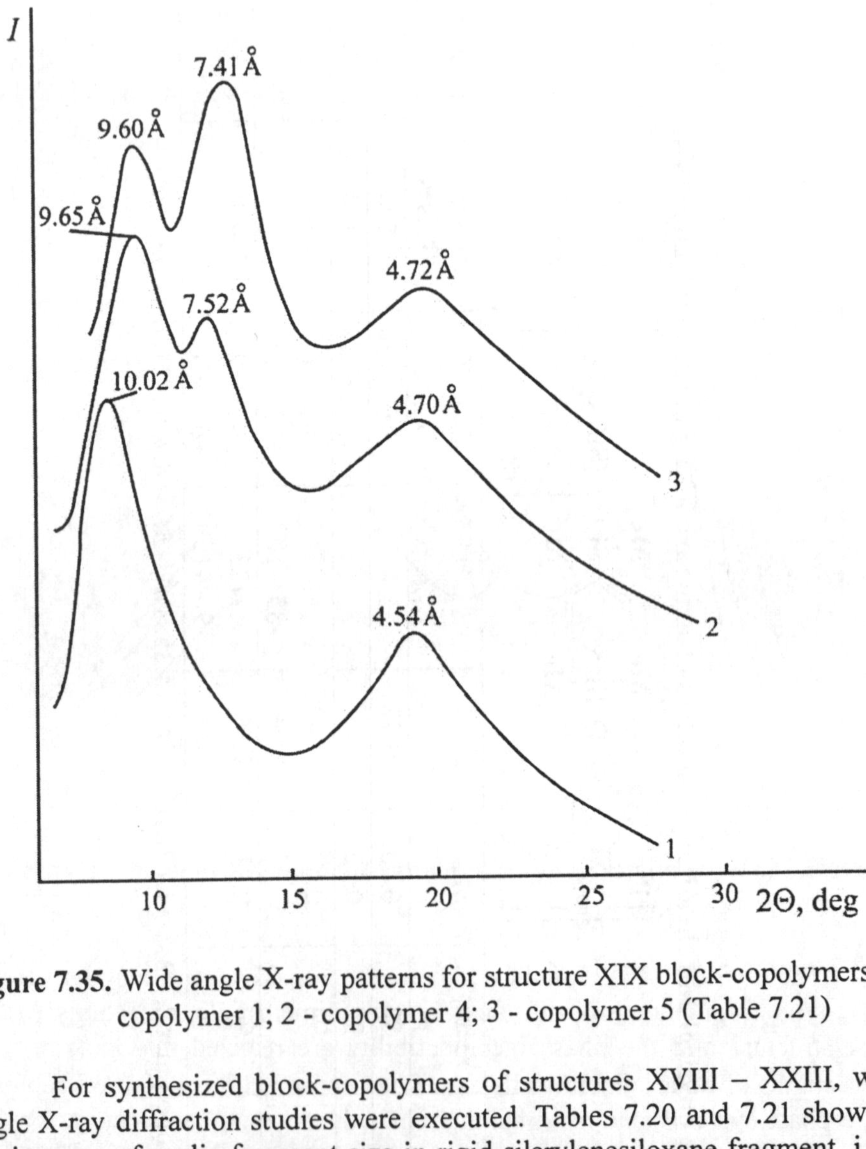

Figure 7.35. Wide angle X-ray patterns for structure XIX block-copolymers: 1 - copolymer 1; 2 - copolymer 4; 3 - copolymer 5 (Table 7.21)

For synthesized block-copolymers of structures XVIII – XXIII, wide-angle X-ray diffraction studies were executed. Tables 7.20 and 7.21 show that the increase of cyclic fragment size in rigid silarylenesiloxane fragment, i.e. at transition from cyclotetrasiloxane fragment (structure XVIII) to cyclohexasiloxane fragment (structure XIX), does not cause increase of the interchain distance, d_1. Similar phenomenon is observed at transition from block-copolymers of structures XX and XXII to the ones of structures XXI and XXIII.

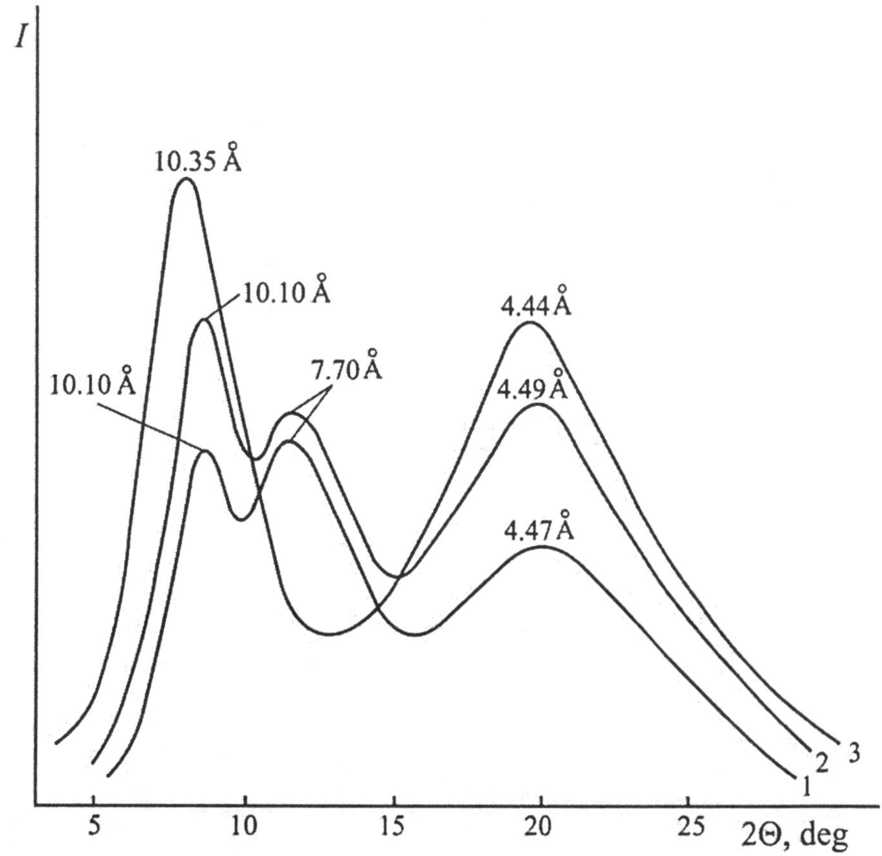

Figure 7.36. Wide angle X-ray patterns for structure XX block-copolymers: 1 - copolymer 4; 2 – copolymer 3; 3 - copolymer 1 (Table 7.22)

Tables 7.20 – 7.25 indicate that with regard to the lengths of rigid $(AB)_m$-type fragments (*m*) and flexible poly(dimethylsiloxane) units (*n*) the block structure and the phase incompatibility are realized in copolymers. At short lengths of linear dimethylsiloxane fragment, $n \leq 12$, copolymers represent single-phase systems, and maximal interchain distances, d_1, are occur at $n = 2$.

Maximums in the ranges 2θ = 8.0 – 8.5° and 2θ = 19 – 20.25° are observed on diffraction patterns. The main maximum at 2θ = 8.0 – 8.5° corresponds to the interchain distance equal d_1 = 10.05 – 10.72 Å. Type of the diffraction patterns proves that at $n = 2$ synthesized copolymers represent single-phase systems. At long lengths of dimethylsiloxane fragments, $n \geq 21$, segregation processes proceed separately for rigid and flexible PDMS blocks, accompanied by the phase separation. Therefore, the third diffraction maximum at $2\theta = 11.0 - 11.5°$ corresponded to the interchain distance of linear PDMS. Figures 7.35 and 7.36 show wide-angle X-ray diffraction patterns for block-copolymers. Occurrence of the third maximum at d_l = 7.41 - 8.04 Å on diffraction patterns proves that the phase separation happens at these lengths of dimethylsiloxane fragments.

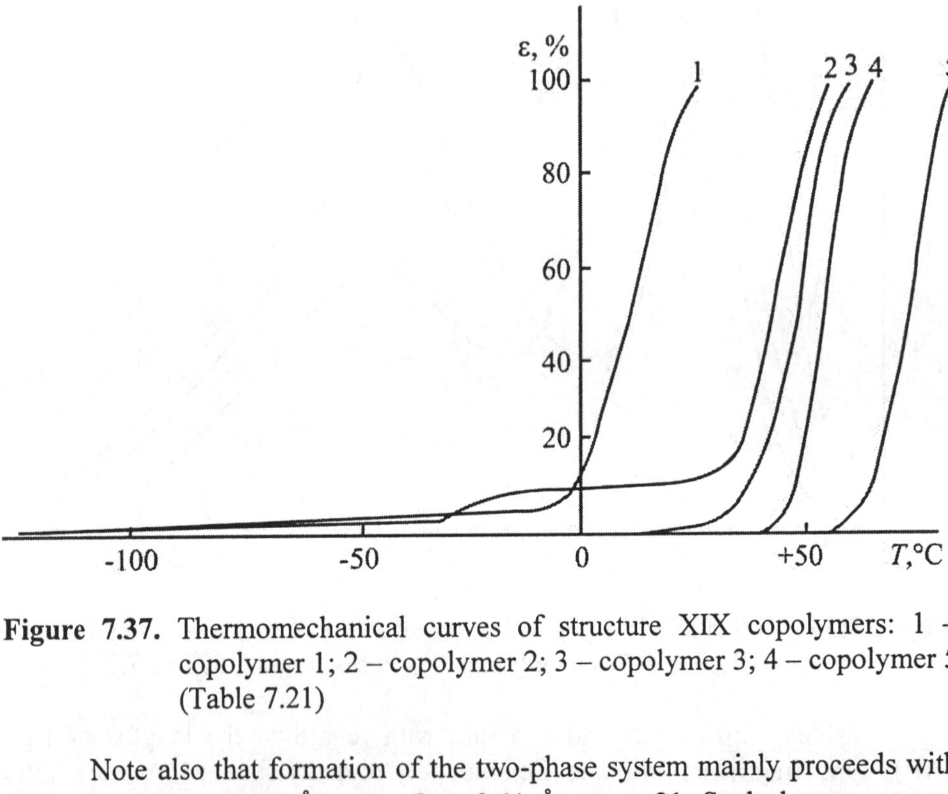

Figure 7.37. Thermomechanical curves of structure XIX copolymers: 1 – copolymer 1; 2 – copolymer 2; 3 – copolymer 3; 4 – copolymer 5 (Table 7.21)

Note also that formation of the two-phase system mainly proceeds with d_l decrease from 10.72 Å at $n = 2$ to 9.41 Å at $n \geq 21$. Such decrease seems reasonable, because for the single-phase system d_l is the mean statistical value characterizing the interchain distance between molecules, close packing of which is, apparently, hindered due to chemical structure heterogeneity. While the interchain distance in PDMS equals ~7.3 Å, i.e. is much lower than for the single-phase structure, one may suggest that d_l decrease at transition to the two-phase state is caused by realization of rigid $(AB)_m$-shaped fragments packing into a separate phase, i.e. in the case of the microdomain structure formation. At constant *n* value, rigid $(AB)_m$-shaped block length increase induces a slight increase of the first diffraction maximum, d_1.

Figure 7.37 shows results of thermomechanical studies of structures XVIII and XIX block-copolymers. Thermomechanical studies prove data of X-ray diffraction investigation. At the length of dimethylsiloxane block $n < 21$ one glass transition point, increased with decrease of flexible PDMS fragment length, *n*, and increase of rigid $(AB)_m$-shaped fragment, occurs. At long lengths of dimethylsiloxane fragments, $n \geq 21$, i.e. in the case of the phase separation, two glass transition temperatures occur, one of which corresponds to T_g of PDMS (-123°C), and the second one relates to the phase formed by rigid silarylenecyclosiloxane blocks and falls within the range of 25 - 50°C (depending on the length of blocks used).

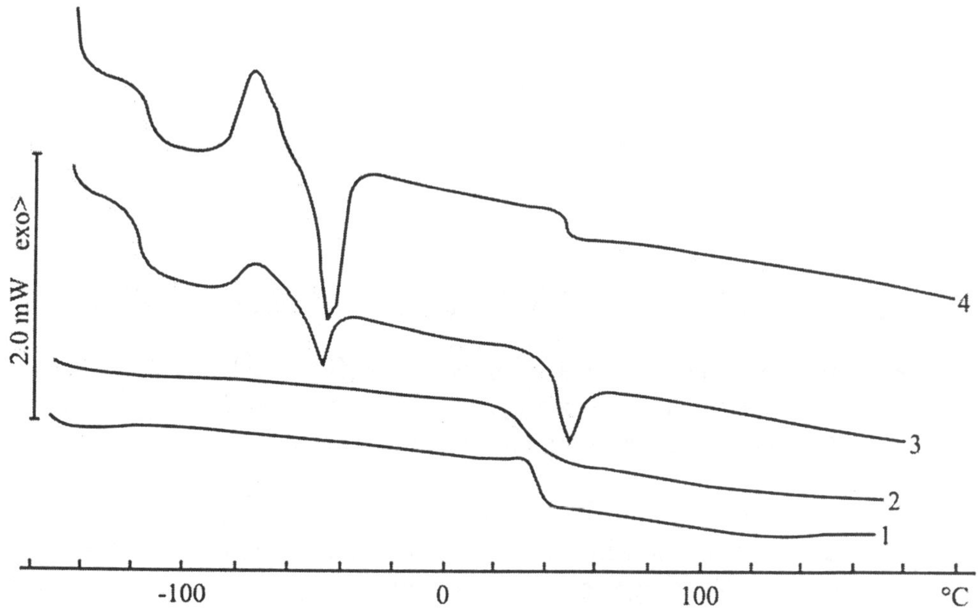

Figure 7.38. DSC thermograms of structure XX block-copolymers: 1 - copolymer 7; 2 - copolymer 2; 3 - copolymer 4; 4 – copolymer 9 (Table 7.22, heating rate 5 deg/min)

Synthesized block-copolymers of structures XX - XXIII were studied by DSC method. At short lengths of flexible dimethylsiloxane unit ($n < 21$) copolymers represent single-phase systems with one transition temperature corresponded to T_g. As the length increases ($n \geq 21$), DSC curves in Figures 7.38 and 7.39 display two glass transition temperature in the range of T_g = -110° ÷ -123°C typical of linear PDMS chain and T_g = +25° ÷ +80°C typical of rigid $(AB)_m$-type fragments.

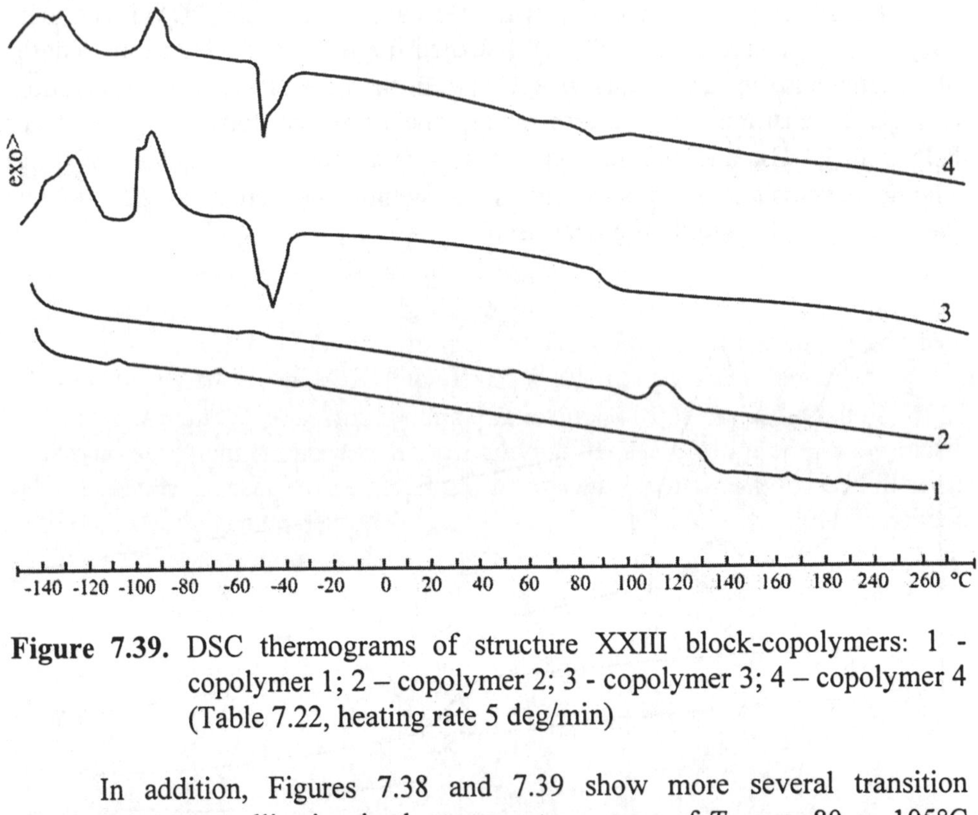

Figure 7.39. DSC thermograms of structure XXIII block-copolymers: 1 - copolymer 1; 2 – copolymer 2; 3 - copolymer 3; 4 – copolymer 4 (Table 7.22, heating rate 5 deg/min)

In addition, Figures 7.38 and 7.39 show more several transition temperatures: crystallization in the temperature range of T_{cryst} = -80 ÷ -105°C and melting in the range of T_{melt} = -30 ÷ -60°C, typical of linear poly(dimethylsiloxane) blocks. As the flexible chain length increases, variation of the melting temperature of linear PDMS blocks is negligible. Note also that DSC curves display no temperature transition typical of melting, T_{melt}, of rigid $(AB)_m$ blocks, as it was observed for dimethylsiloxane-silarylenesiloxane block-copolymers [92].

Thus DSC analysis indicates microdomain structure of copolymers possessing the two-phase structure. At high concentration of rigid blocks, these copolymers represent a continuous phase with discrete PDMS aggregates, distributed in it. Rigid and flexible domains exist at definite lengths of rigid ($m \geq 4$) and flexible ($n \geq 21$) dimethylsiloxane fragment lengths.

Thermogravimetric studies of synthesized block-copolymers indicate their higher thermal oxidative stability, compared with linear PDMS. For these block-copolymers, 10% mass losses occur in 360 - 450°C range. The main degradation process proceeds in the range of 450-540°C, and above 600°C mass losses are negligible. Mass losses of block-copolymers increase with the length of linear PDMS chain. A slight increase of thermal oxidative stability of block-copolymers with the length of rigid fragments is observed. Thus introduction of rigid $(AB)_m$ fragments to the dimethylsiloxane backbone increases thermal oxidative stability of copolymers. This may be explained by high resistance of phenyl groups to oxidation and their inhibiting effect on methyl groups' oxidation [74, 90]. Moreover, introduction of rigid fragments into PDMS backbone disturbs degradation of macromolecules by depolymerization mechanism due to impossibility of complex formation [74, 75].

As copolymers of structures XVIII and XIX are heated up to 110 – 120°C in the presence of 0.1% quaternary ammonium salts, cyclic fragments are disclosed. At the initial stages of polymerization viscosity is increased, and then cross-linked polymers are produced without liberation of gaseous products. The network frequency of synthesized cross-linked copolymers may be varied by the change of linear fragment length, which makes these copolymers more applicable to specialized machinery.

References

1. Tverdokhlebova I.I. and Larina T.A., *Uspekhi Khimii*, 1975, vol. **46**(8), p. 367. (Rus)
2. Tsvetkova V.N., Andrianov K.A., Okhrimenko T.M., Shtennikoba I.N., Fomin G.A., Vitovskaya M.G., Pokhomov V.I., Yarosh A.A., and Andreeva D.N., *Vysokomol. Soedin.*, 1970, vol. **12A**(8), p. 1892. (Rus)
3. Tverdokhlebova I.I., Mamaeva I.I., and Pavlova S.A., *Vysokomol. Soedin.*, 1984, vol. **26A**(9), p. 1971. (Rus)
4. Tverdokhlebova I.I., Larina T.A., Rabkina A.Yu., Mamaeva I.I., Zavin B.G., and Pavlova S.A., *Vysokomol. Soedin.*, 1981, vol. **23B**(4), p. 279. (Rus)
5. Andrianov K.A., Pavlova S.A., Tverdokhlebova I.I., Emelianov V.N., Larina T.A., and Rabkina A.Yu., *Vysokomol. Soedin.*, 1972, vol. **14A**(10), p. 2246. (Rus)
6. Andrianov K.A., Tverdokhlebova I.I., Pavlova S.A., Mamaeva I.I., Rabkina A.Yu., and Larina T.A., *Vysokomol. Soedin.*, 1977, vol. **19B**(6), p. 406. (Rus)
7. Jamakawa H. and Fujiti M., *Macromolecules*, 1974, vol. **7**, p. 128.
8. Kovat J., Fortelny J., and Bohdanecky M., *Makromol. Chem.*, 1977, Bd. **178**(8), S. 2375.
9. Andrianov K.A., Tverdokhlebova I.I., Pavlova S.A., Makarova N.N., Mamaeva I.I., Chekalov A.K., and Menshov V.A., *Vysokomol. Soedin.*, 1978, vol. **20A**(2), p. 377. (Rus)
10. Makarova N.N., *Candidate's Dissertation on Chemistry*, 1971, Moscow, INEOS AS USSR. (Rus)
11. Papkov V.S., *Doctor's Dissertation on Chemistry*, 1979, Moscow, INEOS AS USSR. (Rus)
12. Mukbaniani O.V., *Candidate's Dissertation on Chemistry*, I. Javakhishvili Tbilisi State University, 1977, Tbilisi, Georgia. (Rus)
13. Khananashvili L.M. and Mukbaniani O.V., *Macromolec. Chem. Suppl.*, 1984, No. 6, p. 77.
14. Andrianov K.A., Tikhonov V.S., and Klementiev I.Yu., *Vysokomol. Soedin.*, 1974, vol. **16B**(1), p. 67. (Rus)
15. Volchek B.Z. and Purkina A.B., *Vysokomol. Soedin.*, 1976, vol. **18A**(6), p. 1203. (Rus)
16. Andrianov K.A., Nogaideli A.I., Tsvankin D.Ya., Makarova N.N., and Mukbaniani O.V., *Doklady AN SSSR*, 1976, vol. **229**(6), p. 1353. (Rus)

17. Andrianov K.A., Slonimski G.L., Zhdanov A.A., Levin V.Yu., Tsvankin D.Ya., Kvachev Yu.P., Obolenkova E.S., Makarova N.N., Belavtseva E.M., and Mukbaniani O.V., *Vysokomol. Soedin.*, 1977, vol. **19A**(7), p. 1507. (Rus)
17. *Patent No. 1,433,367*, 1966 (France).
18. Meladze S.M., *Candidate's Dissertation on Chemistry*, I. Javakhishvili Tbilisi State University, 1980, Tbilisi, Georgia. (Rus)
19. Mukbaniani O.V., Karchkhadze M.G., and Khananashvili L.M., *XIth Intern. Symposium on Organosilicon Chemistry*, Montpellier, France, 1996, PB-63.
20. Meladze S.M., Mukbaniani O.V., Makarova N.N., and Khananashvili L.M., *Zh. Obshch. Khim.*, 1980, vol. **50**(11), p. 2493. (Rus)
21. Mukbaniani O.V., *Doctor's Dissertation on Chemistry*, I. Javakhishvili Tbilisi State University, 1993, Tbilisi, Georgia.
23. Khananashvili L.M. and Mukbaniani O.V., *Abstr. Commun. Conf. Regulation of Polymeric Materials Stability*, 1992, Moscow, October 12-15, p. 43.
24. Khananashvili L.M. and Mukbaniani O.V., *Intern. J. Polym. Mater.*, 1994, vol. **27**, p. 31.
25. Andrianov K.A., Pavlova S.S., Zhuravleva I.V., Tolchinski Yu.P., Makarova N.N., and Mukbaniani O.V., *Vysokomol. Soedin.*, 1977, vol. **19A**(6), p. 1387. (Rus)
26. Sobolevski M.V., Skorokhodov I.I., Ditsent V.E., Sobolevskaya A.V., and Efimov V.M., *Vysokomol. Soedin.*, 1969, vol. **11A**(5), p. 1109. (Rus)
27. Andrianov K.A., Papkov V.S., Slonimski G.L., Zhdanov A.A., and Yakushina E.S., *Vysokomol. Soedin.*, 1969, vol. **11A**, p. 2030. (Rus)
28. Hunter M.J., Hyde J.F., Warrick E.L., and Fletcher H.J., *J. Amer. Chem. Soc.*, 1946, vol. **68**, p. 667.
29. Patnode W. and Wilcock D.F., *J. Amer. Chem. Soc.*, 1946, vol. **68**, p. 358.
30. Andrianov K.A., Makarova N.N., and Mukbaniani O.V., *Abst. Commun. 5th Intern. Symp. Organosilicon Chemistry*, 1978, Karsruhe, FRG, p. 66.
31. Andrianov K.A., Nogaideli A.I., Slonimski G.L., Levin V.Yu., Kvachev Yu.P., Makarova N.N., and Mukbaniani O.V., *Vysokomol. Soedin.*, 1976, vol. **13B**(5), p. 359. (Rus)
32. Andrianov K.A., Slonimski G.L., Tsvankin D.Ya., Papkov V.S., Levin V.Yu., Il'ina M.N., and Makarova N.N., *Vysokomol. Soedin.*, 1974, vol. **16B**(3), p. 208. (Rus)

33. Petrova I.M., Makarova N.N., and Andrianov K.A., *Thes. Rep. V All-Union Conference on Organosilicon Compounds Chemistry and Application*, 1980, January 29 – February 1, Tbilisi, Georgia. (Rus)
34. Zhdanov A.A., Zavin A.B., Rabkina A.Yu., and Levin V.Yu., *Vysokomol. Soedin.*, 1985, vol. **28**(2), p. 328. (Rus)
35. Tsvetkov V.N., Frisman E.V., and Boitsova N.N., *Vysokomol. Soedin.*, 1960, No. 2, p. 1001. (Rus)
36. Tsvetkov V.N., *Vysokomol. Soedin.*, 1974, vol. **16A**(7), p. 944. (Rus)
37. Andrianov K.A., Zavin B.G., and Rabkina A.Yu., *Doklady AN SSSR*, 1978, vol. **242**(4), p. 842. (Rus)
38. Tverdokhlebova I.A. and Larina T.A., *Uspekhi Khimii*, 1975, vol. **44**(2), p. 367. (Rus)
39. *A.C. No. 1,137,099*, 1985 (USSR); *Bull. Izobr.*, 1985, No. 4, p. 77. (Rus)
40. Andrianov K.A. and Zhdanov A.A., *Izv. AN SSSR, Ser. Khim.*, 1954, No. 6, p. 1033. (Rus)
41. Andrianov K.A., Pavlova S.A., Tverdokhlebova I.I., Pertsova N.V., and Temnikovski V.A., *Vysokomol. Soedin.*, 1972, vol. **14A**(8), p. 1816. (Rus)
42. Tsvetkov V.N., *Vysokomol. Soedin.*, 1983, vol. **25A**(8), p. 1571. (Rus)
43. Hearsi J.E., *J. Chem. Phys.*, 1964, vol. **40**(5), p. 1506.
44. Bushin S.V., Tsvetkov V.N., Lavrenko E.B., and Emelianov V.N., *Vysokomol. Soedin.*, 1981, vol. **23A**(11), p. 2494. (Rus)
45. Bogdanetsky M., *Macromolecules*, 1983, vol. **16**(9), p. 1483.
46. Pankratova L.N., Sokolov V.N., Mazurina N.V., Obolonkova E.S., Rabkina A.Yu., Kuteinikova L.I., and Zavin B.G., *Vysokomol. Soedin.*, 1997, vol. **39**(10), p. 1711. (Rus)
47. Bernstein V.A., Levin V.Yu., Egorova V.N., Zhdanov A.A., Slonimski G.L., Rabkina A.Yu., Zavin B.G., and Gritsenko O.T., *Vysokomol. Soedin.*, 1987, vol. **29A**(11), p. 2360. (Rus)
48. Obolonkova E.S., Rabkina A.Yu., Zavin V.G., Gritsenko O.T., Slonimski G.L., Zhdanov A.A., and Levin V.Yu., *Vysokomol. Soedin.*, 1986, vol. **28B**(6), p. 460. (Rus)
49. Hsieh B.R., Antoniadis H., Bland D.C., and Feld W.A., *Adv. Mater.*, 1995, vol. **7**(1), p. 36.
50. Monomatsu M., Nie H.Y., Mizutami W., and Tokumoto H., *Japan J. Appl. Phys.*, 1994, vol. **33**(10), p. 3775.
51. *Patent No. 1,423,143*, 1966 (USA).
52. *Patent No. 3,292,737*, 1967 (USA).
53. Dorofeenko L.P., Novinskaya L.A., Dolgoplosk S.B., Kolokoltseva I.G., Mileshkevich V.P., Lobkov V.D., Godovski Yu.K., Metkin I.A.,

Konyukhova E.V., and Volegova I.A., *Thes. Rep. VI All-Union Conference on Organosilicon Compounds Chemistry and Application*, 1986, Riga, p. 90. (Rus)

54. Smirnova T.K., Dolgoplosk S.B., and Mileshkevich V.P., *Thes. Rep. V All-Union Conference on Organosilicon Compounds Chemistry and Application*, 1980, January 29 – February 1, Tbilisi, GSSR. (Rus)
55. Erenburg E.G., Terentieva L.M., Dolgoplosk S.B., Sovchenko V.M., Genkin A.N., and Petrova N.A., *Thes. Rep. VII All-Union Conference on Organosilicon Compounds Chemistry and Application*, Part II, 1980, November 20 – 23, Tbilisi, GSSR, p. 314. (Rus)
56. Shcherbakova T.K., *Candidate's Dissertation on Chemistry*, 1982, S.V. Lebedev All-Union Research Institute of Synthetic Rubber, Leningrad, USSR. (Rus)
57. Andrianov K.A., Zavin B.G., Zhdanov A.A., Evdokimov A.M., Biryukova T.V., and Lavrukhin B.D., *Vysokomol. Soedin.*,1972, vol. **14B**, p. 337. (Rus)
58. Mukbaniani O.V., Scherf U., Karchkhadze M.G., and Tatrishvili T.N., *Int. J. Polym. Mater.*, 2001, vol. **48**(3), p. 311.
59. Andrianov K.A., Tikhonov V.S., and Astapov B.A., *Vysokomol. Soedin.*, 1970, vol. **12B**(8), p. 577. (Rus)
60. Mukbaniani O.V., Khananashvili L.M., Inaridze I.A., *Int. J. Polym. Mater.*, 1994, vol. **24**, p. 211.
61. Mukbaniani O.V., Achelashvili V.A., Karchkhadze M.G., Tkeshelashvili R.Sh., Levin V.Yu., and Khananashvili L.M., *Int. J. Polym. Mater.*, 1992, vol. **18**, p. 129.
62. Andrianov K.A., Nogaideli A.I., Slonimski G.L., Levin V.Yu., Kvachev Yu.P., Makarova N.N., and Mukbaniani O.V., *Vysokomol. Soedin.*, 1976, vol. **13B**(5), p. 359. (Rus)
63. Meladze S.M., Khananashvili L.M., and Mukbaniani O.V., *Vysokomol. Soedin.*, 1984, vol. **26B**(4), p. 250. (Rus)
64. Mukbaniani O.V., Achelashvili V.A., Meladze S.M., and Khananashvili L.M., *Soobshch. AN GSSR*, 1994, vol. **149**(1), p. 89. (Rus)
65. Khananashvili L.M. and Mukbaniani O.V., *Int. J. Polym Mater.*, 1994, vol. **27**, p. 31.
66. *A.C. No. 757,555*, 1980 (USSR); *Bull. Izobr.*, vol. **31**, 1980. (Rus)
67. Mukbaniani O.V., Koyava N.A., Meladze S.M., and Khananashvili L.M., *Abstr. Commun. 6th Intern. Symp. Organosilicon Chemistry*, 1981, Budapest, p. 62.

68. Mukbaniani O.V., Inaridze I.A., Achelashvili V.A., and Khananashvili L.M., *Soobshch. AN GSSR*, 1994, vol. **149**(2), o. 246. (Rus)
69. Mukbaniani O.V., Achelashvili V.A., and Meladze S.M., *VII Int. Symp. Organosilicon Chem.*, 1984, Kyoto, p. 216.
70. Mukbaniani O.V. and Inaridze I.A., *Thes. Rep. VII All-Union Conference on Production Technology Chemistry and Application of Organosilicon Compounds*, 1990, Tbilisi, Part 1, p. 222. (Rus)
71. Mukbaniani O.V., Karchkhadze M.G., and Matsaberidze M.G., *Russian Polymer News*, 2002, vol. **7**(1), p. 7.
72. Mukbaniani O.V., Matsaberidze M.G., Karchkhadze M.G., Khananashvili L.M., and Nasuashvili T., *Proceedings of the Georgian Academy of Sciences*, 2000, vol. **26**(1-2), p. 72.
73. Andrianov K.A., Kononov L.M., and Tsvankin D.Ya., *Vysokomol. Soedin.*, 1968, vol. **10B**, p. 320. (Rus)
74. Thomas T.H. and Kendric T.C., *J. Polym. Sci.*, 1969, vol. **2A**(7), p. 537.
75. Alexander G. and Garzo G., *Chromatography*, 1974, vol. **7**(5), p. 225.
76. Tolchinski Yu.I., *Candidate's Dissertation on Chemistry*, Institute of Organoelement Compounds, AS USSR, 1980, Moscow. (Rus)
77. Tessler M.M., *Amer. Chem. Soc., Polym. Preper.*, 1967, vol. **8**, p. 152.
78. Papkov V.S., Il'ina M.N., Zhdanov A.A., Slonimski G.L., and Andrianov K.A., *Thes. Rep. IV International Symposium on Organosilicon Compounds*, Moscow, 1975, Part II, p. 36. (Rus)
79. Wu T.C. and Hirt C.A., *J. Organomet. Chem.*, 1968, vol. **11**, p. 17.
80. Achelashvili V.A., Mukbaniani O.V., Levin V.Yu., and Khananashvili L.M., *Vysokomol. Soedin.*, 1991, vol. **33A**(2), p. 275. (Rus)
81. Mukbaniani O.V., Karchkhadze M.G., Matsaberidze M.G., Achelashvili V.A., Khananashvili L.M., and Kvelashvili N., *Int. J. Polym. Mater.*, 1998, vol. **41**, p. 103.
82. Mukbaniani O.V., Matsaberidze M.G., and Karchkhadze M.G., *Thes. Rep. All-Union Conference "Condensation Polymers: Synthesis, Structure, Properties"* (to the 90th anniversary of Academician V.V. Korshak), 1999, January 12 – 14, pp. 1 - 16. (Rus)
83. Achelashvili V.A., *Candidate's Dissertation on Chemistry*, I. Javakhishvili Tbilisi State University, 1989, Tbilisi, Georgia. (Rus)
84. Khananashvili L.M., Mukbaniani O.V., Samsonia A.Sh., and Karchkhadze M.K., *XVI Mendeleev's Congress on General and Applied Chemistry*, Saint-Petersburg, May 25 - 26, p. 320. (Rus)
85. Mukbaniani O.V., Samsonia A.Sh., Karchkhadze M.G., and Khananashvili L.M., *J. Appl. Polym. Sci.*, 2002, vol. **84**, p. 9.

86. Mukbaniani O.V., Samsonia A.Sh., and Karchkhadze M.G., *Thes. Rep. All-Union Conference "Condensation Polymers: Synthesis, Structure, Properties"* (to the 90th anniversary of Academician V.V. Korshak), 1999, January 12 – 14, pp. 1 - 15. (Rus)
87. Samsonia A.Sh., Karchkhadze M.G., Mukbaniani O.V., and Khananashvili L.M., *Georgian Egineering News*, 2000, No. 2, p. 144.
88. Mukbaniani O.V., Matsaberidze M.G., and Karchkhazde M.G., *Thes. Rep. All-Union Conference "Condensation Polymers: Synthesis, Structure, Properties"* (to the 90th anniversary of Academician V.V. Korshak), 1999, January 12 – 14, pp. 1 - 16. (Rus)
89. Matsaberidze M.G., *Candidate's Dissertation on Chemistry*, I. Javakhishvili Tbilisi State University, 2001, Tbilisi, Georgia. (Rus)
90. Andrianov K.A., Papkov V.S., Slonimski G.L., Zhdanov A.A., and Yakushina S.E., *Vysokomol. Soedin.*, 1969, vol. **11A**, p. 2030. (Rus)
91. Mileshkevich V.P., *Kauchuk i Rezina*, 1978, No. 6, p. 4. (Rus)
92. Karpova T.N., Sidorovich A.B., Dolgoplosk S.B., Sviridova N.G., and Kuvshinski E.V., *Vysokomol. Soedin.*, 1973, vol. **15A**(4), p. 859. (Rus)